T0222918

Lecture Notes in Artificial Intelligence   10758

Subseries of Lecture Notes in Computer Science

More information about this series at http://www.springer.com/series/1244

Van-Nam Huynh · Masahiro Inuiguchi
Dang Hung Tran · Thierry Denoeux (Eds.)

Integrated Uncertainty in Knowledge Modelling and Decision Making

6th International Symposium, IUKM 2018
Hanoi, Vietnam, March 15–17, 2018
Proceedings

 Springer

Editors
Van-Nam Huynh (iD)
Japan Advanced Institute of Science and
  Technology
Nomi
Japan

Masahiro Inuiguchi
Osaka University
Osaka
Japan

Dang Hung Tran
Hanoi National University of Education
Hanoi
Vietnam

Thierry Denoeux
Université de Technologie de Compiègne
Compiègne
France

ISSN 0302-9743   ISSN 1611-3349 (electronic)
Lecture Notes in Artificial Intelligence
ISBN 978-3-319-75428-4   ISBN 978-3-319-75429-1 (eBook)
https://doi.org/10.1007/978-3-319-75429-1

Library of Congress Control Number: 2018931886

LNCS Sublibrary: SL7 – Artificial Intelligence

Printed on acid-free paper

This Springer imprint is published by the registered company Springer International Publishing AG
part of Springer Nature
The registered company address is: Gewerbestrasse 11, 6330 Cham, Switzerland

Preface

This volume contains the papers that were presented at the 6th International Symposium on Integrated Uncertainty in Knowledge Modelling and Decision Making (IUKM 2018) held in Hanoi, Vietnam, March 15–17, 2018.

The IUKM symposia aim to provide a forum for exchanges of research results and ideas as well as application experiences among researchers and practitioners involved with all aspects of uncertainty modelling and management. Previous editions of the conference were held in Ishikawa, Japan (IUM 2010), Hangzhou, China (IUKM 2011), Beijing, China (IUKM 2013), Nha Trang, Vietnam (IUKM 2015), and Da Nang, Vietnam (IUKM 2016) and their proceedings were published by Springer in AISC 68, LNAI 7027, LNAI 8032, LNAI 9376, and LNAI 9978, respectively.

IUKM 2018 was jointly organized by Hanoi National University of Education (HNUE), Japan Advanced Institute of Science and Technology (JAIST), and Belief Functions and Applications Society (BFAS).

This year the conference received 76 submissions from 11 different countries. Each submission was peer reviewed by at least two members of the Program Committee. After a thorough review process, 39 papers were accepted for presentation at IUKM 2018 and publication in the proceedings.

The IUKM 2018 symposium was partially supported by The National Foundation for Science and Technology Development of Vietnam (NAFOSTED), The U.S. Office of Naval Research Global (ONR Global), and Hanoi National University of Education (HNUE). We are very thankful to the local organizing team from HNUE for their hard work, efficient services, and wonderful local arrangements.

We would like to express our appreciation to the members of the Program Committee for their support and cooperation in this publication. We are also thankful to Alfred Hofmann, Anna Kramer, and their colleagues at Springer for providing a meticulous service for the timely production of this volume. Last, but certainly not the least, our special thanks go to all the authors who submitted papers and all the attendees for their contributions and fruitful discussions that made this conference a great success.

March 2018

Van-Nam Huynh
Masahiro Inuiguchi
Dang Hung Tran
Thierry Denoeux

Organization

General Co-chairs

Dang Hung Tran   Hanoi National University of Education, Vietnam
Thierry Denoeux   University of Technology of Compiègne, France

Advisory Board

Michio Sugeno   European Center for Soft Computing, Spain
Hung T. Nguyen   New Mexico State University, USA; Chiang Mai University, Thailand
Akira Namatame   AOARD/AFRL and National Defense Academy of Japan
Sadaaki Miyamoto   University of Tsukuba, Japan
Van Minh Nguyen   Hanoi National University of Education, Vietnam

Program Co-chairs

Van-Nam Huynh   Japan Advanced Institute of Science and Technology, Japan
Masahiro Inuiguchi   University of Osaka, Japan

Local Arrangements Co-chairs

Minh Lam Duong   Hanoi National University of Education, Vietnam
Thuy Nga Nguyen   Hanoi National University of Education, Vietnam
Thanh Trung Dang   Hanoi National University of Education, Vietnam

Publication and Financial Co-chairs

The Dung Luong   Academy of Cryptography Techniques, Vietnam
Van-Hai Pham   Pacific Ocean University, Nha Trang, Vietnam

Program Committee

Yaxin Bi   University of Ulster, UK
Jurek Blaszczynski   Poznan University of Technology, Poland
Matteo Brunelli   University of Trento, Italy
Tru Cao   Ho Chi Minh City University of Technology, Vietnam
Fabio Cuzzolin   Oxford Brookes University, UK
Quang A. Dang   Vietnam Academy of Science and Technology
Tien-Tuan Dao   University of Technology of Compiègne, France
Yong Deng   Xian Jiaotong University, China
Thierry Denoeux   University of Technology of Compiègne, France

Martin Stepnicka	University of Ostrava, Czech Republic
Noboru Takagi	Toyama Prefectural University, Japan
Kazuhiro Takeuchi	Osaka Electro-Communication University, Japan
Xijin Tang	CAS Academy of Mathematics and Systems Science, China
Yongchuan Tang	Zhejiang University, China
Khoat Than	Hanoi University of Science and Technology, Vietnam
Phantipa Thipwiwatpotjana	Chulalongkorn University, Thailand
Araki Tomoyuki	Hiroshima Institute of Technology, Japan
Vicenc Torra	University of Skovde, Sweden
Dang Hung Tran	Hanoi National University of Education, Vietnam
Seiki Ubukata	Osaka University, Japan
Bay Vo	HUTECH, Vietnam
Guoyin Wang	Chongqing University of Posts and Telecommunications, China
Thanuka Wickramarathne	University of Massachusetts, USA
Zeshui Xu	Sichuan University, China
Koichi Yamada	Nagaoka University of Technology, Japan
Hong-Bin Yan	East China University of Science and Technology, China
Yuji Yoshida	Kitakyushu University, Japan
Chunlai Zhou	Renmin University of China

Sponsoring Organizations

The National Foundation for Science
and Technology Development
of Vietnam (NAFOSTED)

Office of Naval Research Global (ONRG)

Hanoi National University of Education
(HNUE)

Invited Speakers

Using Statistical Machine Learning Techniques to Analyze Satellite Optical Signature

Phan Dao

U.S. Air Force Research Laboratory, USA

Short Biography: Dr. Phan Dao is a researcher in satellite and space object electro-optical signature analysis at the Air Force Research Laboratory, Space Vehicles Directorate. His research includes aspects of space object characterization and space debris environment. His current focus is on the development of analytical tools with increasing reliance on Statistical Machine Learning techniques. He has been with the AFRL for 30 years mainly in Research and Development. Prior to the activities in space technology, he was involved in the development and application of remote sensors and lidar for atmospheric, chemical and particle diagnostics. Prior to AFRL, he was with KMS Fusion, Ann Arbor, MI, developing novel laser concepts. He received a Ph.D. in Physics from the University of Colorado in Boulder, Colorado, in 1985.

Summary: In observation-based science, the objectives of an analysis are often to associate measurables with an unknown state (of a satellite) or to infer latent variables from measurements. Because in satellite analyses, ground truth labels are often unavailable, data-driven techniques to calibrate confidence levels, e.g. cross-validation, are not suitable. On the other hand, analysis results are required to have calibrated and traceable confidence levels. We found some success in using (Bayesian) Statistical Machine Learning techniques such as Gaussian Process, Hidden Markov and other Bayesian techniques in object classification, signature prediction and detection of change points. The techniques generally produce calibrated probabilities. While the signature from a remote GEO object is inherently a time series, spectral and concurrent measurements provide the opportunity and challenge of information fusing which can also be handled with simplicity in ML. We will show some successful applications of ML in photometric analyses of satellites.

Toward Nonlinear Statistics Based on Choquet Calculus

Michio Sugeno

Tokyo Institute of Technology, Japan

Short Biography: After graduating from the Department of Physics, the University of Tokyo, Michio Sugeno worked at a company for three years. Then, he served the Tokyo Institute of Technology as Research Associate, Associate Professor and Professor from 1965 to 2000. After retiring from the Tokyo Institute of Technology, he worked as Laboratory Head at the Brain Science Institute, RIKEN from 2000 to 2005 and, then as Distinguished Visiting Professor at Doshisha University from 2005 to 2010. Finally, he worked as Emeritus Researcher at the European Centre for Soft Computing in Spain from 2010 to 2015. He is Emeritus Professor at the Tokyo Institute of Technology. He was President of the Japan Society for Fuzzy Theory and Systems from 1991 to 1993, and also President of the International Fuzzy Systems Association from 1997 to 1999. He is the first recipient of the IEEE Pioneer Award in Fuzzy Systems with Zadeh in 2000. He also received the 2010 IEEE Frank Rosenblatt Award and Kampé de Feriét Award in 2012.

Summary: In this study we show a first step toward nonlinear statistics by applying Choquet calculus to probability theory. Throughout the study we take a constructive approach. For nonlinear statistics, we first consider a distorted probability space on the nonnegative real line. A distorted probability measure is derived from a conventional probability measure by a monotone transformation with a generator (usually called a distortion function), where we deal with two classes of parametric generators. Next, we explore some properties of Choquet integrals of non-negative continuous functions with respect to distorted probabilities. Then, we calculate basic statistics such as the distorted mean and variance of a random variable for uniform, exponential and Gamma distributions.

In addition, we consider Choquet integrals of real-valued functions to deal with a distorted probability space on the real line. We also calculate basic statistics for uniform and normal distributions.

Computational Decision Making and Data Science – Challenges and Opportunities

Pedja Neskovic

Office of Naval Research, USA

Short Biography: Dr. Pedja Neskovic is a program officer at ONR where he oversees the mathematical data science, and computational methods for decision making programs. He is also an adjunct associate professor of brain science at Brown University and a visiting professor at Johns Hopkins University. Dr. Neskovic received his BSc in theoretical physics from Belgrade University, and his PhD in physics from Brown University. He was a postdoctoral fellow at the Institute for Brain Science at Brown University. Within the scope of his programs, he is addressing various basic research problems. These include methods for the analysis of big data and small data, analysis of complex networks such as social and brain networks, reproducibility in science, and causal inference. He is also interested in developing methods for large-scale distributive decision making that utilize novel crowdsourcing and collaborative techniques.

Summary: We live in an exciting time: in the midst of the new technological revolution. Unlike the previous industrial revolution that extended mostly our physical powers, this revolution has the potential to extend our cognitive powers. One of the basic ingredients that enabled this revolution is the emergence of the so-called "big data": the datasets that are not only large in size but also much more complex and granular than the "old datasets". However, with big data come big problems: many of the standard tools that we used for data analysis are no longer appropriate when applied to big data. In this talk, I will discuss some of the challenges in developing new tools for the analysis of big data and also some of their limitations. In particular, I will address the relations between the predictive and causal modeling, and the reproducibility crisis in science. I will also discuss some of the new opportunities in the area of large scale distributed decision making including the new developments in crowdsourcing, computational social choice and personalized learning.

From Big Data and AI to FinTech

Ye Qiang

Harbin Institute of Technology, China

Short Biography: Dr. Qiang Ye is the Dean and Professor of Information Systems in the School of Management at Harbin Institute of Technology. He had worked in Mccombs School of Business Administration at the University of Texas Austin, Randy School of Management in the University of California San Diego and School of Hotel & Tourism Management at the Hong Kong Polytechnic University as Post Doctoral Fellow, Research Fellow or Visiting Professor.

Dr. Ye is Senior Editor of Journal of Electronic Commerce Research, Area Editor of Electronic Commerce Research and Applications and guest Associate Editor of MIS Quarterly. His research areas of interest include Big Data and Business Analytics, e-Commerce, e-Tourism, and FinTech, et al. He had published about thirty papers in journals including Production Operations Management, Tourism Management and Decision Support Systems, et al. Dr. Ye received the "National Science Fund of China for Distinguished Young Scholars" in 2012 and entitled "Cheung Kong Scholar Professor" by the Ministry of Education of China in 2016.

Summary: Technologies in Business school have evolved from the days of MIS, DSS to Big Data and Business Analytics. The current Artificial Intelligence booming is largely impacting different industries and Business Schools, especially in the finance and accounting areas. As an example, the new buzzword FinTech is catching more and more attentions from both finance industry and business schools. This talk will discuss how the technologies of Internet, Big Data, and Artificial Intelligence influence industries and business school education. It will also explore the challenges and opportunities in research and applications.

Similarly-Based Machine Learning Approaches for Predicting Drug-Target Interactions

Hiroshi Mamitsuka

Kyoto University, Japan

Short Biography: Hiroshi Mamitsuka is a Professor of Bioinformatics Center, Institute for Chemical Research, Kyoto University, being jointly appointed as a Professor of School of Pharmaceutical Sciences of the same university. Also he is a FiDiPro (Finland Distinguished Professor Program) Professor of Department of Computer Science, Aalto University, Finland. His current research interest includes a variety of aspects of machine learning and their diverse applications, mainly cellular- or molecular-level biology, chemistry and medical sciences. He has published more than 100 scientific papers, including those appeared in top-tier conferences or journals in machine learning and bioinformatics, such as ICML, KDD, ISMB, Bioinformatics, etc. Also he has served program committee member of numerous conferences and associate editor of several well-known journals of the related fields. Prior to joining Kyoto University, he worked in industry for more than ten years, on data analytics in business sectors, such as customer/revenue churn, web-access pattern, campaign management, collaborative filtering, recommendation engine, etc.

Summary: Computationally predicting drug-target interactions is useful to discover potential new drugs. Currently, promising machine learning approaches for this issue use not only known drug-target interactions but also drug and target similarities. This idea can be well-accepted pharmacologically, since the two types of similarities correspond to two recently advocated concepts, so-called, the chemical space and the genomic space. I will start this talk by describing detailed background on the problem of predicting drug-target interactions, particularly why similarity-based approaches have been paid attention to now. I will then move on to the existing approaches and their bottlenecks and further present recent factor model-based approaches, which allow low-rank approximation of given matrices, by which the issues of the past methods can be properly considered. Also I note that the problem setting of similarity-based predicting of drug-target interactions is very general, in the sense of binary relations between two sets of events, in which events have similarities each other. This general setting can be found in many applications, such as recommender systems.

Contents

Machine Learning Applications

Statistical Methods

Econometric Applications

Uncertainty Management
and Decision Support

A Granularity Approach to Vague Quantification

Christian G. Fermüller[✉]

Vienna University of Technology, Vienna, Austria
chrisf@logic.at

Abstract. Motivated by the observation that vagueness often involves a shift of underlying granularity levels, we introduce models of quantifiers like 'about one half', 'roughly 10%', etc., that refer to granules of proportionality levels. Corresponding (partial) truth functions are extracted from rough set based models, which may then be systematically mapped into corresponding fuzzy quantifier models.

1   Introduction

Modeling vague quantifier expressions, like few, many, about half or even the classical existential and universal quantifiers when applied to vague predicates is a well known challenge. Such models have applications in data querying, automated reasoning, summarisation, and natural language processing. The standard approach, at least in engineering contexts, is based on fuzzy set theory and uses *fuzzy quantifiers*. There is a considerable amount of literature on fuzzy quantification, summarized, e.g., in the recent survey article [1]; in particular Glöckner's monograph [7] on this topic deserves to be singled out. While this work certainly documents impressive progress, there remain several challenges that call for specific extensions and alternative models of vague quantifiers.

The purpose of this paper is to outline an approach to vague quantification that is not primarily based on fuzzy sets, but rather on granularity as modeled by rough sets. A main motivation for our model is its compatibility with mainstream research in formal semantics of natural language, as pursued in linguistics and analytic philosophy (see, e.g., [8,13,14]). Those approaches stress the digital nature of language and insist on a binary speaker-hearer interface at the outermost level: any declarative sentence is either uttered or not uttered by a speaker and is either accepted or rejected (or, possibly, left unclassified) by the hearer.[1] Rough sets match this feature by focusing on an upper as well as a lower approximation of a given concept, which – as we shall explain – also

---

C. G. Fermüller — This work is supported by Austrian Science Fund (FWF) grant I1827-N25.

[1] Of course, linguists and philosophers recognize the importance of emphasis, hedging, uncertainty, etc., as triggered by vagueness. However these phenomena are modeled beneath or in addition to the outer level of communication via declarative statements.

© Springer International Publishing AG, part of Springer Nature 2018
V.-N. Huynh et al. (Eds.): IUKM 2018, LNAI 10758, pp. 3–14, 2018.
https://doi.org/10.1007/978-3-319-75429-1_1

maps into corresponding criteria for acceptance or rejection of vaguely quantified statements. A probably even more important aspect of our model regards shifting levels of *granularity* arising from vagueness. While it is well understood that vague notions, including vague quantifier expressions, can and should be modeled at different levels of granularity (e.g., by more or less coarse partitions of scales into, e.g., small/medium/large versus very tiny/tiny/moderately small, etc.), fuzzy set theory tends towards ad hoc decisions in this respect. As pointed out by Vetterlein [16], fuzzy sets can be seen as (problematically) mapping arbitrary levels of granularity invariably into the extremely fine grained level of degrees represented by real numbers in the unit interval. The machinery of rough sets directly addresses the challenge of shifting adequate levels of granularity by letting corresponding equivalence relations (or, more generally, similarity relations) arise from the given data, instead of simply imposing it as artifact of the model.

We emphasize that our granularity/rough set based model of vague quantification is by no means incompatible with fuzzy quantification. Rather than interpreting the rough set approach as a strict alternative to the fuzzy set based theory of quantification, we suggest that it provides a useful *augmentation*, able to bring into better view certain features of vague quantification. We argue that, in particular, shifting levels of granularity and a binary interface (of acceptance/rejection of a statement) is modeled more directly by rough sets. As has been pointed out in the literature (see, e.g., [19]) rough sets can be systematically be related to fuzzy sets. This opens up the possibility to translate our rough set based models of quantifiers into fuzzy quantifiers, in addition to mapping rough sets, understood as models of vague predicates, into fuzzy predicates (fuzzy sets).

The paper is organized as follows. In the next section we review some basic notions regarding fuzzy quantification and rough sets. Section 3 introduces, in a step-wise manner, a granularity and hence rough set based quantifier model, where the scope predicates are assumed to be classical (bivalent). Section 4 reviews a rough set approach to vague predicates. In Sect. 5 we indicate how to combine the models of the previous two sections, before mapping our rough set models, that feature truth values gaps, into fuzzy quantifier models in Sect. 6. Section 7 finally provides a short summary and outlook to future work.

2 Basic Notions: Fuzzy Quantification and Rough Sets

The standard approach to vague quantification is based on fuzzy set theory (see, e.g., [1,7,21]). In that approach a unary predicate is interpreted by a *fuzzy set* $\widetilde{B}$, i.e., by a function $\widetilde{B} : D \to [0,1]$, assigning a *membership degree* to every element in the domain D. Crisp (i.e., classical) sets emerge as a special case, where the membership degree is either 0 or 1. A unary *fuzzy quantifier* Q is interpreted by a function $\|Q\| : \widetilde{\mathcal{P}}(D) \to [0,1]$, where $\widetilde{\mathcal{P}}(D)$ denotes the set of all fuzzy sets $\widetilde{B}$ over the domain D. $\|Q\|(\widetilde{B})$ is called the *truth degree* of the formula $Qx\, S_B(x)$, where the *scope predicate* S_B is interpreted by $\widetilde{B}$. Q is called *precise* if the corresponding truth degree is either 0 or 1 (if defined at all). Natural language quantifiers are frequently *binary*; i.e., they have the logical form $Qx(R(x), S(x))$, where the *range predicate* R restricts the set of domain elements to which the scope predicate S is to be applied. Although the concepts developed in the rest of the paper can be generalized to binary quantification, we will focus on unary quantification for sake of conciseness and clarity. The intended applications justify the assumption that the given domain is always *finite*. Moreover, to simplify notation, we identify domain elements with corresponding constant symbols.

Liu and Kerre [10] provide the following useful classification of fuzzy quantification that carries over to the more general setting of vague quantification.

Type I: the quantifier is precise and the scope predicate is crisp;
Type II: the quantifier is precise, but the scope predicate is vague;
Type III: the quantifier is vague, but the scope predicate is crisp;
Type IV: the quantifier as well as the scope predicate are vague.

In contrast to fuzzy sets, *rough sets* provide a model of uncertainty that usually focuses not on vagueness, but on types of uncertainty arising from incomplete data about unknown sets that may well have sharp boundaries (see, e.g., [12] for an overview). We will briefly point out in Sect. 4 that nevertheless also vague predicates can be modeled as rough sets. To this aim let us recall a few basic concepts and notions.

Rough sets usually arise from incomplete data about properties of objects in a given domain D (see, e.g., [12]). These data induce an equivalence relation $\sim \in D \times D$ by classifying objects as equivalent iff the available data ascribe the same properties to them. Let $[x]_E = \{y : y \in D, x \sim y\}$, i.e., the equivalence class containing object x. Then, for any $X \subset D$ we define a pair of *lower* and *upper approximations* with respect to E:

$$\underline{apr}_E(X) = \bigcup\{[x]_E : x \in D, [x]_E \subseteq X\} = \{x \in D : [x]_E \subseteq X\}$$
$$\overline{apr}_E(X) = \bigcup\{[x]_E : x \in D, [x]_E \cap X \neq \emptyset\} = \{x \in D : [x]_E \cap X \neq \emptyset\}$$

The equivalence relation may be understood as a specific *granulation*, partitioning the domain into $[x]_E$ *granules* (equivalence classes). As pointed out, e.g., in [20], proper granulation refers not to isolated partitions, but rather to a whole *hierarchy* of more or less fine grained partitions. We apply this approach below to the specific case of possible proportionality values.

Rough set theory provides a range of variations of those basic concepts, arising from relaxing the conditions on $\sim \in D \times D$ to, e.g., only asking for reflexivity and symmetry. We will briefly allude to such a generalization in Sect. 4, but otherwise stick with the traditional setup sketched above.

3   Vague Proportional Quantifiers and Granularity

Monadic proportional quantifiers determine the truth value of a quantified sentence by the proportion of domain elements that satisfy the scope predicate.

we write $\|G\|_\mathcal{I}$ for the truth value (0 or 1) of the closed formula G under $\mathcal{I}$. If a formula F contains at most one free variable x then $\|F(x)\|_\mathcal{I}$ denotes the *predicate*, i.e., the function of type $D_\mathcal{I} \to \{0,1\}$, induced by $\|F(d)\|_\mathcal{I}$.[2] We define

$$\mathrm{Prop}_\mathcal{I}(F) = \frac{|\{d \in D_\mathcal{I} : \|F(d)\|_\mathcal{I} = 1\}|}{|D_\mathcal{I}|}.$$

The meaning of a *crisp (classical) quantifier* Q is specified by a function $\|Q\|$ that maps every predicate arising from an interpretation $\mathcal{I}$ into $\{0,1\}$.[3] In other words: the truth value $\|QxF(x)\|_\mathcal{I}$ of the sentence $Qx\,F(x)$ under an interpretation $\mathcal{I}$ is obtained as $\|Q\|(\|F(x)\|_\mathcal{I})$. The quantifier Q is called *proportional* if its meaning only depends on the value $\mathrm{Prop}_\mathcal{I}(F) \in [0,1]$.

The notion of a proportional quantifier is quite broad. In the case of type I quantification (i.e., crisp quantifiers as well as predicates) it includes the classical existential quantifier $\exists$ and $\forall$, since the corresponding (classical) truth functions can be specified as

$$\|\exists\|(\|F(x)\|_\mathcal{I}) = \begin{cases} 1 \text{ if } \mathrm{Prop}_\mathcal{I}(F) > 0 \\ 0 \text{ else}, \end{cases} \qquad \|\forall\|(\|F(x)\|_\mathcal{I}) = \begin{cases} 1 \text{ if } \mathrm{Prop}_\mathcal{I}(F) = 1 \\ 0 \text{ else}, \end{cases}$$

respectively, where $\mathcal{I}$ is an interpretation with domain $D_\mathcal{I}$. Further type I quantifiers corresponding to natural language quantifier expressions like, e.g., (precisely) one half, at most 15%, and at least on third arise analogously.

As is well known (see, e.g., [7]) the above definitions straightforwardly generalize to *semi-fuzzy* (type III) quantification by expanding the range of truth values from $\{0,1\}$ to $[0,1]$. While such semi-fuzzy proportionality quantifiers have been studied extensively in the literature, already since Zadeh's landmark paper [21], the challenge to derive corresponding truth functions from first principles of reasoning with imprecise and/or uncertain notions remains active. (We refer to [4] for an approach to this problem that is based on Giles's a dialogue and betting game for Łukasiewicz logic [6].) Here we want to propose a model of type III quantification that is motivated by the observation that the vagueness of classifying notions — typically predicates, but including also quantifier expressions — arises from the possibility or even the need to shift between different levels of granularity (cf. [8,16]).

We start by the observation that the choice of a particular natural language quantifier expression often (implicitly) indicates a certain level of granularity, which in turn entails a corresponding tolerance interval that should be respected in sentence evaluation. For example, the series of proportional quantifier expressions very roughly a third, about a third, 33%, 33.2%, 33.24%, 33.24178% signals a shift from very coarse to very fine grained granularity in judging given values of proportionality as adequate for accepting a corresponding sentence as true. Presumably one would accept a sentence of the form $Qx\,F(x)$, where Q is any of the above expressions, as true, if the proportion of domain elements that have

---

[2] Recall from Sect. 2 that we identify domain elements with constant symbols.

[3] Not every such function is admitted as meaning of a *logical* quantifier. We focus on proportional quantifiers here, which are definitely to be classified as logical.

the property expressed by $F(x)$ is exactly 0.3324178. However, if the proportion is, say, 0.359, then arguably instantiating Q with either the first or the second of the quantifiers in the series still suggests acceptance of the sentence, while the semantics of the remaining quantifiers should lead to rejection. Note that each quantifier induces a partition of the scale $[0,1]$ of possible proportions into *granules of proportionality*. This gets clear if we consider such quantifier expressions not in isolation, but whole sets of "matching" quantifiers that jointly cover the full range $[0,1]$ at a given level of granularity. For sake of concreteness we introduce the following four series of quantifier expressions. (P_i^k denotes the ith quantifier of granularity level k; $\mathcal{P}^k$ specifies the corresponding partition of possible proportions into granules at level k.)

P_i^1 ($i = 0, \ldots, 4$): almost none, few, about half, many, almost all
$$\mathcal{P}^1 = \{[0, \tfrac{1}{8})\} \cup \{[\tfrac{i}{4} - \tfrac{1}{8}, \tfrac{i}{4} + \tfrac{1}{8}) : 1 \le i \le 3\} \cup \{[\tfrac{7}{8}, 1]\}$$
$$= \{[0, 0.125), [0.125, 0.375), [0.375, 0.625), [0.625, 0.875), [0.875, 1]\}$$
P_i^2 ($i = 0, \ldots, 10$): ($\approx$)0%, 10%, 20%, $\ldots$, 90%, ($\approx$)100%[4]
$$\mathcal{P}^2 = \{[0, \tfrac{1}{20})\} \cup \{[\tfrac{i}{10} - \tfrac{1}{20}, \tfrac{i}{10} + \tfrac{1}{20}) : 1 \le i \le 9\} \cup \{[\tfrac{19}{20}, 1]\}$$
P_i^3 ($i = 0, \ldots, 100$): ($\approx$)0%, 1%, 2%, $\ldots$, 99%, ($\approx$)100%
$$\mathcal{P}^3 = \{[0, \tfrac{1}{200})\} \cup \{[\tfrac{i}{100} - \tfrac{1}{200}, \tfrac{i}{100} + \tfrac{1}{200}) : 1 \le i \le 99\} \cup \{[\tfrac{199}{200}, 1]\}$$
P_i^4 ($i = 0, \ldots, 1000$): ($\approx$)0%, 0.1%, 0.2%, $\ldots$, 99.9%, ($\approx$)100%
$$\mathcal{P}^4 = \{[0, \tfrac{1}{2000})\} \cup \{[\tfrac{i}{1000} - \tfrac{1}{2000}, \tfrac{i}{1000} + \tfrac{1}{2000}) : 1 \le i \le 999\} \cup \{[\tfrac{1999}{2000}, 1]\}$$

One may of course augment these four levels of granularity by further ones following this pattern. In any case, the presented partitions into granules of possible proportions are not arbitrary. In particular the principle of rounding any given proportion in $[0,1]$ to the nearest percentage value available in a given series is deeply anchored in our natural language semantics as argued, e.g., in [8]. Clearly, partitions $\mathcal{P}^1, \ldots \mathcal{P}^4$ directly correspond to equivalence relations $E^1, \ldots, E^4$ over the domain $[0,1]$ and thus give rise to talk about 'rough sets of (possible) proportions' connected to quantifier expressions at each granularity level. If we set

$$\|\mathsf{P}_i^k\|(\|F(x)\|_{\mathcal{I}}) = \begin{cases} 1 & \text{if } \mathrm{Prop}_{\mathcal{I}}(F) \in [\mathcal{P}^k]_i \\ 0 & \text{else,} \end{cases}$$

where $[\mathcal{P}^k]_i$ denotes the granule (interval) arising for the ith quantifier in the partition $\mathcal{P}^k$, then the vagueness of the quantifiers is eliminated by systematic precisification. This lack of vagueness related uncertainty is highlighted in the rough set view: e.g., the quantifier $\mathsf{P}_4^2 = 40\%$ is associated with the set of proportions $\in [0.35, 0.45)$. Since this set is an equivalence class with respect to E_2 it coincides with its upper as well as its lower set.

As already indicated, vagueness may be seen as arising from projecting coarser levels of granularity into finer ones. E.g., Thomas Vetterlein writes:

> [...] vagueness reflects the fact that objects can be classified [...] at different levels of granularity. [...] The challenge is to deal with the two or more levels of granularity in a combined formalism ([16], p. 85).

---

[4] Modifiers like approximately ($\approx$) are usually left implicit when communicating by asserting sentences like 10% of the population lives in poverty.

We take up this challenge by employing the machinery of rough set approximation for information granulation. In particular, recall from [20] that, if the equivalence relation E' is obtained from the equivalence relation E by further partitioning the granules (equivalence classes) induced by E ($E' \subseteq E$), then we have

$$\underline{apr}_E(X) \subseteq \underline{apr}_{E'}(X) \subseteq X \subseteq \overline{apr}_{E'}(X) \subseteq \overline{apr}_E(X) \qquad (1)$$

In our case the 'objects to be classified with respect to varying levels of granularity' are possible proportions of domain elements satisfying a given scope predicate. However the successively more fine-grained levels $1, 2, 3, 4$ do not arise from further partitioning coarser levels: there remain possible overlaps between equivalence classes at different levels. For example, our second level of granularity of proportions ($\mathcal{P}^2/E^2$) is not simply a further refinement of the first (coarsest) level ($\mathcal{P}^1/E^1$) since, e.g., 0.124 and 0.126 belong to different equivalence classes with respect to E_1, but not with respect to E_2. But we are only interested in the special case of (1), where X itself is a granule of the more fine grained relation E'. In this case we obtain $X = \underline{apr}_{E'}(X) = \overline{apr}_{E'}(X)$ and $\underline{apr}_E(X) = \emptyset$. In fact the whole relation (1) still applies for $E = E_k, E' = E_{k'}$, where $1 \leq k < k' \leq 4$. Now $\overline{apr}_E(X)$ consists of either a single granule or of exactly two granules with respect to E, depending on whether the (finer) granule X lies completely within a granule of the coarser level or in an overlapping region.

The following *gap-based quantifier model* arises directly from the above considerations. "Gap-based" means that the truth value for borderline cases remains *undefined*. Note that this is in accordance with a number of philosophical and linguistic approaches to vagueness, in particular also supervaluationism, cf. [5]. An additional advantage of our model is that it directly relates to the mentioned principle that vagueness arises from a shift of granularity levels. With respect to two granularity levels $k < k'$ we define

$$\|\mathsf{P}_i^{k/k'}\|(\|F(x)\|_{\mathcal{I}}) = \begin{cases} 1 & \text{if } \mathrm{Prop}_{\mathcal{I}}(F) \in [\mathcal{P}^{k'}]_i \\ 0 & \text{if } \mathrm{Prop}_{\mathcal{I}}(F) \notin \overline{apr}_{E_k}([\mathcal{P}^{k'}]_i) \\ undef. & \text{else} \end{cases}$$

where $E_{k'}$ is the equivalence relation corresponding to granularity level k'. Note that the index i refers to that more fine grained partition of proportionality values.

Example 1. Let us consider the quantifier $\mathsf{P}_6^{1/2}$ as a formal counterpart of the (vague) quantifier expression about half. According to our model, the sentence About half [of the domain elements] have property F is accepted as *definitely true* if at least 45%, but less than 55% domain elements satisfy F. If that percentage is below 37.5% or above (or equal) 62.5%, then the sentence is rejected as *definitely false*, since the upper approximation of $[45, 55]$ with respect to E_1 is the interval $[\frac{1}{2} - \frac{1}{8}, \frac{1}{2} + \frac{1}{8})$. For all other proportions of elements satisfying F, the sentence is classified as *borderline* between true and false; i.e., no (classical) truth value is assigned according to our gap model. $\mathsf{P}_6^{1/2}$ can be seen as a 'gappy' composition of P_3^1 and P_6^2 as illustrated in the following pictures.

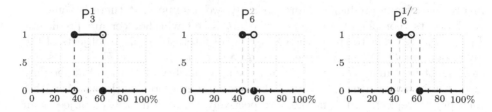

Example 2. For a more fine grained model (compared to that of Example 1) of the quantifier expression 30%, we may consider $P_{30}^{2/3}$, where 30% is interpreted such that 30% [of the domain elements] have property F is evaluated as *definitely false* if the percentage of domain elements satisfying F is smaller than 25% or at least 35%, and is evaluated as *definitely true* if that percentage is at least 29.5%, but less than 30.5%. In all other cases (borderline cases) the truth value of the sentence remains undefined.

4   Vague Predicates Modeled as Rough Sets

Zdzislaw Pawlak, the creator of the concept of rough sets, has demonstrated in [11] how vague models can be represented as rough sets based on tolerance (similarity) relations. We follow Pawlak's account here and just augment it by indicating a possible voting scenario for arriving at tolerance relations that are transitive and thus in fact equivalence (indiscernibility) relations.

Recall the following concepts from [11]. A binary reflexive and symmetric relation $I \subseteq D \times D$ is called a *tolerance relation* on the domain D (also referred to as 'universe'); $x, y \in D$ are called I-*similar* if xIy. Let $I(x) = \{y : yIx\}$ and consider the following definitions of the I-*lower* and I-*upper approximation* of $X \subset D$, respectively:

$$I_*(X) = \{x \in D : I(x) \subseteq X\}$$
$$I^*(X) = \{x \in D : I(x) \cap X \neq \emptyset\}$$

Note that $I_*(X)$ and $I^*(X)$ coincide with $\underline{apr}_I(X)$ and $\overline{apr}_I(X)$, respectively, if I is an equivalence relation. We also define the I-*boundary region* of X by $BN_I(X) = I^*(X) - I_*(X)$. Pawlak [11] defines a range of further notions for classifying vague concepts. For our purposes it suffices to point out that, according to this model, a vague predicate F is identified with an unknown (and possibly even unknowable) crisp set X that is accessible to the agent in question only via $I_*(X)$ and $I^*(X)$.[5]

One might ask how an appropriate similarity relation can be determined in practice. We briefly outline a possible method that is related to Lawry's voting semantics for fuzzy sets [9], but frees the voters from the problematic requirement to either vote for definitive acceptance or else to vote for the rejection of a claim

---

[5] There are obvious connections to epistemic theories of vagueness [15,18] here. Moreover, a similar setup is also discussed in [17].

involving a vague predicate. More precisely, we assume that there is some set A of agents (competent speakers), who are asked whether certain predicates $F_1, \ldots, F_k$ (relevant to the vague concept X of interest) apply to objects of the given domain D. For every object $x \in D$ and every predicate F_i ($1 \leq i \leq k$), each agent in A may vote for or against the applicability of F_i to x, but she may also abstain from voting. Identifying these possibly reactions of the agents with 1 (accept), 0 (reject), and ? (neither), respectively, we obtain for each object x an *answer profile* Σ_x as a function of type $A \to \{1, 0, ?\}^k$. From a rough set point of view, the aggregated voting data induce an equivalence relation: for all $x, y \in D$ we define $x \sim_A y$ iff $\Sigma_x = \Sigma_y$. Note that $\sim_A$ is not only reflexive and symmetric, but also transitive, i.e., an equivalence relation. Similarity relations of this type are called *indistinguishability relations* by Pawlak [11] in this context.

Independently of the manner in which a similarity or indistinguishability relation is extracted from given data, the characterization of vague predicates by lower and upper approximations clearly supports a gap-based account of vagueness. Let us write $\overline{\mathcal{I}}$ for an interpretation that assigns a pair $(I_*(X), I^*(X))$ to $F(x)$ expressing a vague predicate corresponding to the incompletely known set X, circumscribed via the indistinguishability relation I, as indicated above. Then we may straightforwardly translate this approximation into a truth function as follows:

$$\|F(x)\|_{\overline{\mathcal{I}}} = \begin{cases} 1 & \text{if } x \in I_*(X) \\ 0 & \text{if } x \notin I^*(X) \\ undef. & x \in BN_I(X). \end{cases}$$

In Sect. 6 we will indicate how such a gap-based model can be extended to a fuzzy logic model in a natural manner that respects certain principles about degree based reasoning with vague concepts.

5   Combining it All: Type IV Quantification

Before we connect our rough set and granularity based model to (semi-)fuzzy quantification in Sect. 6, we briefly outline how the precise as well as the gap-based meaning functions for vague proportionality quantifiers can be generalized to a scenario with vague scope predicates. In other words, after having discussed type III (and, as a limit case, type I) quantification, we now want to consider type IV (and type II) quantification.

In principle, it should be rather obvious by now how a rough set based model of type IV quantification arises from the concepts discussed so far. Whereas the type III model introduced in Sect. 3 relies on definite proportionality values for the underlying (crisp) predicate, we now generalize to vague scope predicates, which are modeled according to the rough set approach reviewed in Sect. 4. This means that the corresponding truth function for type IV quantifiers can only refer to *upper and lower bounds* for the relevant proportionality values. Instead of just mapping predicate symbols to subsets of the domain $D_{\overline{\mathcal{I}}}$, an interpretation $\overline{\mathcal{I}}$ now assigns a pair $(I_*(X_F), I^*(X_F))$ to any predicate symbol

F where I is an appropriate indistinguishability relation. Instead of $\mathrm{Prop}_{\mathcal{I}}(F)$ we now correspondingly obtain the following *proportionality bounds*:

$$\overline{\mathrm{Prop}_{\overline{\mathcal{I}}}}(F) = \frac{|\{d \in D_{\overline{\mathcal{I}}} : d \in I^*(X_F)\}|}{|D_{\overline{\mathcal{I}}}|} \qquad \underline{\mathrm{Prop}_{\overline{\mathcal{I}}}}(F) = \frac{|\{d \in D_{\overline{\mathcal{I}}} : d \in I_*(X_F)\}|}{|D_{\overline{\mathcal{I}}}|}$$

whereas for the type I and III case we determined the truth values (if any) of quantified statements according to the granules (equivalence classes) which contains the respective proportionality value, we now check which *intervals* of potential proportionality values are fully covered by which granules.

Taking up the proportionality quantifiers of Sect. 3, defined with reference to the partitions $\mathcal{P}^1, \dots \mathcal{P}^4$ covering $[0,1]$ on different granularity levels, we obtain

$$\|\breve{\mathsf{P}}_i^k\|(\|F(x)\|_{\overline{\mathcal{I}}}) = \begin{cases} 1 & \text{if } [\underline{\mathrm{Prop}_{\overline{\mathcal{I}}}}(F), \overline{\mathrm{Prop}_{\overline{\mathcal{I}}}}(F)] \subseteq [\mathcal{P}^k]_i \\ 0 & \text{if } [\underline{\mathrm{Prop}_{\overline{\mathcal{I}}}}(F), \overline{\mathrm{Prop}_{\overline{\mathcal{I}}}}(F)] \subseteq [\mathcal{P}^k]_j, \text{ for some } j \neq i \\ undef. & \text{else,} \end{cases}$$

where $[\mathcal{P}^k]_i$ again denotes the granule arising for the ith quantifier in the partition $\mathcal{P}^k$. Note that the vagueness of scope predicates leads to undefined truth values, where corresponding type III quantifiers remained classical. Of course, the truth functions get even more gappy if we apply the shifting granularity mechanism introduced in Sect. 3 to above quantifier model as follows.

$$\|\breve{\mathsf{P}}_i^{k/k'}\|(\|F(x)\|_{\overline{\mathcal{I}}}) = \begin{cases} 1 & \text{if } [\underline{\mathrm{Prop}_{\overline{\mathcal{I}}}}(F), \overline{\mathrm{Prop}_{\overline{\mathcal{I}}}}(F)] \subseteq [\mathcal{P}^{k'}]_i \\ 0 & \text{if } [\underline{\mathrm{Prop}_{\overline{\mathcal{I}}}}(F), \overline{\mathrm{Prop}_{\overline{\mathcal{I}}}}(F)] \subseteq [\mathcal{P}^k]_j, \text{ for some } j \neq i \\ undef. & \text{else.} \end{cases}$$

6   Generalizing to Fuzzy Quantification

In Sect. 3 we provided a rough set based model of proportionality quantifiers in two stages. First, a family of type I quantifiers P_i^k is obtained from equivalence classes (granules) of possible proportions of domain elements satisfying the (crisp) scope predicate of a given quantified sentence. In the next stage we generalized P_i^k to $\mathsf{P}_i^{k/k'}$, following the idea that a *vague* quantifier expression arises if we relate a coarser level of granularity k to a more fine grained on k'. The vagueness of $\mathsf{P}_i^{k/k'}$ is signified by a borderline region of possible proportions, for which it is neither adequate to judge the corresponding sentence as definitely true (1) nor as definitely false (0). In absence of additional assumptions this induces the gap-based meaning function $\|\mathsf{P}_i^{k/k'}\|$ of Sect. 3.

From the viewpoint of fuzzy logic it is natural to 'bridge the gap' in $\|\mathsf{P}_i^{k/k'}\|$ using intermediate truth values, i.e. values $\in (0,1)$. A naive way of doing so consists in simply replacing '*undef.*' in the definition of $\|\mathsf{P}_i^{k/k'}\|$ by 0.5. The resulting truth functions for $\|P_i^{k/k'}\|$ indeed define semi-fuzzy quantifiers in the sense of, e.g., [7]. However, they are unsatisfying for at least two (related) reasons:

(a) The discontinuity of the function is hardly adequate in presence of the whole real unit interval $[0, 1]$ as possible choices for degrees of truth.
(b) The truth function does not respect the following degree-theoretic principle: The closer we get to definite truth (1) by varying the relevant parameter in a sentence S involving a graded concept, the higher the degree of truth of S.

For a careful and detailed defense of the underlying principles about vagueness we refer to Smith [14], in particular his arguments for the "closeness principle". Here, we just point out that these principles entail the following natural assumption: About half [of the domain elements] have property F is strictly truer in interpretation $\mathcal{I}'$ than in interpretation $\mathcal{I}$ if $\mathrm{Prop}_{\mathcal{I}}(F) < \mathrm{Prop}_{\mathcal{I}'}(F)$ (denoting the corresponding proportions of elements satisfying F in $\mathcal{I}$ and $\mathcal{I}'$, respectively) and both proportions are above the threshold for definitely rejecting, but below the threshold for definitely accepting the sentence.

Fortunately it is straightforward to define truth functions for semi-fuzzy quantifiers $\widetilde{P}_i^{k/k'}$ by linearly interpolating between the definite (0/1) regions. In our concrete scenario of the 4 granularity levels introduced in Sect. 3 we obtain the following definition for $1 \leq k < k' \leq 4$. Adapting the definition to that missing case where $k' = 4$ is a straightforward.) We use p as abbreviation for $\mathrm{Prop}_{\mathcal{I}}(F)$ and assume that the granules $[\mathcal{P}^{k'}]_i$ and $[\mathcal{P}^k]_j$ are the intervals $[l(k', i), u(k', i))$ and $[l(k, j), u(k, j))$, respectively.

$$\|\widetilde{P}_i^{k/k'}\|(\|F(x)\|_{\mathcal{I}}) = \begin{cases} 1 & \text{if } l(k', i) \leq p < u(k', i) \\ 0 & \text{if } p < l(k, j) \text{ or } p \geq u(k, j) \\ \frac{p - l(k,j)}{l(k',i) - l(k,j)} & \text{if } l(k, j) \leq p < l(k', i) \\ \frac{u(k,j) - p}{u(k,j) - u(k',i)} & \text{if } u(k', i) \leq p < u(k, j) \end{cases}$$

where j is the index of the granule of the coarser partition at level k that coincides with the upper approximation of $[\mathcal{P}^{k'}]_i = [l(k', i), u(k', i))$ at that level.

Example 3. Continuing Example 3, we obtain the semi-fuzzy counterpart $\widetilde{P}_6^{1/2}$ of the gap-based (rough set) quantifier $P_6^{1/2}$ ("about half"). Just like for $P_6^{1/2}$, About half [of the domain elements] have property F is accepted as *definitely true* if at least 45%, but less than 55% domain elements satisfy F. Likewise, the sentence is rejected as *definitely false* below 37.5% or above (or equal) 62.5%. In the remaining cases, however, the corresponding degree of truth is obtained by linear interpolation, as illustrated in the following picture:

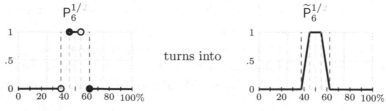

Concerning representations of vague predicates as rough sets (Sect. 4) we may rely on well known concepts for defining corresponding degrees of membership. Once more following Pawlak [11], we define

$$\mu_X^I(x) = \frac{|X \cap I(x)|}{|I(x)|},$$

with respect to a similarity relation I over the domain D. (Recall that $I(x) = \{y : y \in D, yIx\}$). Note that we have

$$I_*(X) = \{x \in D : \mu_X^I(x) = 1\}$$
$$I^*(X) = \{x \in D : \mu_X^I(x) > 0\}$$
$$BN_I(X) = \{x \in D : 0 < \mu_X^I(x) < 1\},$$

which clarifies that the gap-based truth function defined at the end of Sect. 4 directly turns into a corresponding fuzzy model by replacing "*undef.*" by $\mu_X^I(x)$.

Following Glöckner's methodology for systematically lifting semi-fuzzy quantifiers to fuzzy quantifiers via an (axiomatically specified) *quantifier fuzzification mechanism*, we may finally derive fully fuzzy models of quantification with vague proportionality quantifiers and vague predicates (type IV quantification). However, let us point out that those fuzzy models, even where they are adequate for monadic quantifiers, get problematic if they are too naively generalized to binary quantifiers, since (standard) fuzzy quantifiers do not properly take into account semantic dependencies between range and scope predicates (see [2]).

7 Conclusion

It has been pointed out (e.g. in [16] from a mathematical point of view, and in [8] from a linguistic point of view) that an essential feature of vagueness consists in a potential shift of underlying granularity levels. Motivated by this observation we proposed a rough set based approach vague quantification. Corresponding models were devised in a step-wise manner that covers all four types of quantification discussed by Liu and Kerre [10]. Not surprisingly, our models can be turned into fuzzy quantifier models as well in a systematic fashion.

We emphasize that, in contrast to the usual fuzzy logic approach, in our case the truth functions for the family $\widetilde{P}_i^{k/k'}$ of semi-fuzzy quantifiers are not simply defined in an ad-hoc manner, but rather arise from considerations about natural levels of granularity involved in references to proportions of domain elements satisfying scope predicates. However, systematic guidelines for determining linguistically adequate models remain to be explored in further research.

The introduced concepts are intended to stimulate further research on the important topic of computational models of vague quantification. In particular, some natural extensions have not been discussed here for lack of space: generalization to binary quantification, involvement of more than two granularity levels, and the use of general similarity relations. Moreover we plan to further substantiate the claim that our approach supports systematic links to epistemic and to supervaluationist theories of vagueness, not just to degree theoretic accounts.

Last, but not least, it remains to be explored how the suggested quantifier models can be employed in concrete applications. In this vain we plan to apply our approach to data summarization and querying, in a setting recently described for related sampling based quantifier models in [3].

References

1. Delgado, M., Ruiz, M.D., Sánchez, D., Vila, M.A.: Fuzzy quantification: a state of the art. Fuzzy Sets Syst. **242**, 1–30 (2014)
2. Fermüller, C.G.: Combining fuzziness and context sensitivity in game based models of vague quantification. In: Huynh, V.-N., Inuiguchi, M., Denoeux, T. (eds.) IUKM 2015. LNCS (LNAI), vol. 9376, pp. 19–31. Springer, Cham (2015). https://doi.org/10.1007/978-3-319-25135-6_4
3. Fermüller, C.G., Hofer, M., Ortiz, M.: Querying with vague quantifiers using probabilistic semantics. In: Christiansen, H., Jaudoin, H., Chountas, P., Andreasen, T., Legind Larsen, H. (eds.) FQAS 2017. LNCS (LNAI), vol. 10333, pp. 15–27. Springer, Cham (2017). https://doi.org/10.1007/978-3-319-59692-1_2
4. Fermüller, C.G., Roschger, C.: Randomized game semantics for semi-fuzzy quantifiers. Logic J. IGPL **223**(3), 413–439 (2014)
5. Fine, K.: Vagueness, truth and logic. Synthese **30**(3), 265–300 (1975)
6. Giles, R.: A non-classical logic for physics. Stud. Logica **33**(4), 397–415 (1974)
7. Glöckner, I.: Fuzzy Quantifiers: A computational Theory. Studies in Fuzziness and Soft Computing, vol. 193. Springer, Heidelberg (2006). https://doi.org/10.1007/3-540-32503-4
8. Krifka, M.: Approximate interpretation of number words: A case for strategic communication. In: Hinrichs, E., Nerbonne, J. (eds.) Theory and Evidence in Semantics, pp. 109–132. CSLI Publications, Stanford (2009)
9. Lawry, J.: A voting mechanism for fuzzy logic. Int. J. Approximate Reasoning **19**(3–4), 315–333 (1998)
10. Liu, Y., Kerre, E.E.: An overview of fuzzy quantifiers (I) Interpretations. Fuzzy Sets Syst. **95**(1), 1–21 (1998)
11. Pawlak, Z.: Vagueness and uncertainty: a rough set perspective. Comput. Intell. **11**(2), 227–232 (1995)
12. Pawlak, Z., Grzymala-Busse, J., Slowinski, R., Ziarko, W.: Rough sets. Commun. ACM **38**(11), 88–95 (1995)
13. Shapiro, S.: Vagueness in Context. Oxford University Press, Oxford (2006)
14. Smith, N.J.J.: Vagueness and Degrees of Truth. Oxford University Press, Oxford (2008)
15. Sorensen, R.: Vagueness and Contradiction. Clarendon Press, Oxford (2001)
16. Vetterlein, T.: Vagueness: a mathematicians perspective. In: Cintula, P., et al. (eds.) Understanding Vagueness - Logical, Philosophical and Linguistic Perspectives, pp. 67–86. College Publications (2011)
17. Vetterlein, T.: Logic of prototypes and counterexamples: possibilities and limits (2015). https://www.atlantis-press.com/proceedings/ifsa-eusflat-15/
18. Williamson, T.: Vagueness. Routledge, London (2002)
19. Yao, Y.Y.: A comparative study of fuzzy sets and rough sets. Inf. Sci. **109**(1–4), 227–242 (1998)
20. Yao, Y.Y.: Information granulation and rough set approximation. Int. J. Intell. Syst. **16**(1), 87–104 (2001)
21. Zadeh, L.A.: A computational approach to fuzzy quantifiers in natural languages. Comput. Math. Appl. **9**(1), 149–184 (1983)

Linguistic Summarization Based on the Inherent Semantics of Linguistic Words

Thi Lan Pham[1(✉)], Cam Ha Ho[1], and Cat Ho Nguyen[2,3]

[1] The Faculty of Information Technology, HNUE, Hanoi, Vietnam
{ptlan,hahc}@hnue.edu.vn
[2] Institute of Fundamental Research and Application, Duy Tan University,
Danang, Vietnam
ncatho@gmail.com
[3] Institute of Information Technology, VAST, Hanoi, Vietnam

Abstract. Linguistic summarization provides an explicit, concise and easily understandable description expressed in natural language of huge amount of data for human. The result of extracting linguistic summaries is natural language sentences using given sets of words for each numeric attribute. In general, the word-set is always assumed to be of only 7 ± 2 words represented by fuzzy sets. The interpretability of linguistic summarization depends mainly on this fuzzy set representation of the words. In this paper, we consider the given word-set for each attribute as a linguistic frame of cognition LFoC of this attribute whose size depends only on application requirement. We propose a linguistic summary method based on multi-granularity representations of this LFoCs that can preserve order-based semantics relation and generality-specificity relation of words. Theoretically, the number of words in LFoC are not limited. A simulation study using dataset Iris shows that the proposed method can extract sentences using words of length 3 characterizing dataset Iris that other existing ones cannot do.

Keywords: Linguistic summaries · Linguistic frame
Hedge algebra · Fuzzy multi-granularity

1 Introduction

The development of the social-economic system of every nation causes huge increases of databases in their volumes and complexity that make their human users very hard to understand and to draw useful latent information or knowledge. Therefore, extracting potentially useful knowledge in raw databases is intensively studied. Linguistic summarization is a useful methodology to serve for this aim. It may provide explicit, concise and easily understandable descriptions to human [2]. The first idea for linguistic summarization of data was proposed by Yager in 1982 [16]. It has been developed and applied in many fields such as analyzing time series [1, 14, 15], summarization of databases in terms of linguistic sentences [4, 6, 9].

© Springer International Publishing AG, part of Springer Nature 2018
V.-N. Huynh et al. (Eds.): IUKM 2018, LNAI 10758, pp. 15–26, 2018.
https://doi.org/10.1007/978-3-319-75429-1_2

Almost all of studies on extracting linguistic summarization have used fuzzy sets to represent semantics of linguistic words. The size of sets of linguistic words of numeric attributes is limited to have only at most 7 ± 2 words and their fuzzy set representation forms a strong partition of their respective reference domains. A great limitation of recent linguistic summarization approaches is that the word-sets of attributes must be pre-specified and they together with their fuzzy set representations cannot be extensible and scalable. This seems to be not agreement with the reality that human beings as well as experts have right to use any words in their current natural languages to capture a reality semantics in terms of their words and their semantics is in general defined in the whole languages. Therefore, the sizes of the word-sets of attributes are limited by 7 ± 2 is too restrictive and hence it might lead to loses of talent linguistic information involved in databases.

According to our view, as human beings are well familiar with using their languages to recognize the reality, the word-set viewed as linguistic frame of cognitive (LFoC) of each attribute does play a fundamental role when one wish to develop a computational method to simulate the way a human user try to make a summary of a dataset in terms of linguistic sentences. Therefore, how to develop a method to produce fuzzy set representation of the semantic structure of LFoCs of attributes so that can soundly express essential inherent semantics of the LFoCs is crucial and fundamental.

Hedge algebras (HAs) form an algebraic approach to the inherent word-semantics of any linguistic variable, including the fuzzy set based semantics [11]. The HA based approach has been applied to solve successfully many problems (fuzzy control [13], classification [10], regression [11], ...) which justify the essential role of the theory of HAs to make a formal linkage of the desired computational representation of a LFoC with the semantics of the words defined in the context of the entire LFoC.

In the study we propose a linguistic summary method with the following features:

- Ability to exploit the order-based qualitative semantics and the generality-specificity relation (GSp-relation) of the words of the given LFoCs.
- Using the HA-formalism to produce the fuzzy set based semantics of words, including the words of the quantification, from their own qualitative semantics.
- Using extendable fuzzy multi-granularity representation of the attribute LFoCs, i.e. adding new words more specific than the existing words of the LFoC cannot change the fuzzy set semantics of its existing words.

It is shown by experiments made on the dataset Iris that the proposal is able to extract sentences describing specific features of Iris using words of length 3, i.e. they are of high specificity that the existing methods cannot extract, in our opinion.

2   Related Works

2.1   Linguistic Summaries of Data

In this section, we give a short summary of linguistic summarization of a database using linguistic quantified sentences proposed by Yager [16], Yager and Kacprzyk [6] and intensively studied by Kacprzyk's research group, see [4,5,7], for instance. Linguistic quantified sentences are in the form of the so-called protoforms:

$$Q \; y \; are \; S \tag{1}$$

$$Q \; F \; y \; are \; S \tag{2}$$

where: Q is a linguistic quantifier such as *most, many, a half*, F is a qualifier such as *"young employee"*, *"female patients"*, S is a summarizer such as *"very young"* of the variable AGE, *"little high"* of SALARY ... For example, *"most of employees have medium quality"* is in the form of (1) and *"a half of young employees have medium quality"* is in the form of (2). The semantics of the words, e.g. *most, young* and *medium*, in Q, F, S are represented by their respective fuzzy sets.

Each linguistic quantified sentence *Pro* has a validity measure, denoted by $T(Pro)$ taking values in $[0,1]$. Therefore, T may be meant in a more general sense as a quality or goodness of *Pro* which is calculated by Zadehs calculus as follows:

$$T(Q \; y \; are \; S) = \mu_Q \left[\frac{1}{n} \sum_{i=1}^{n} \mu_S(y_i) \right] \tag{3}$$

$$T(Q \; F \; y \; are \; S) = \mu_Q \left[\frac{\sum_{i=1}^{n} (\mu_S(y_i) \wedge \mu_F(y_i))}{\sum_{i=1}^{n} \mu_F(y_i)} \right] \tag{4}$$

where: $\{y_1, y_2, \ldots, y_n\}$ is a data set of objects $y_i's$.

2.2   Enlarged Hedge Algebra

In every word-domain of a linguistic variable $\mathcal{X}$, $Dom(\mathcal{X})$, there is an order-based semantics relation induced by the inherent semantics of words, e.g. for the variable AGE, it is naturally seen that *young* $\leq$ *old*, *very young* $\leq$ *young* and, in contrast, *little young* $\geq$ *young*, *old* $\leq$ *very old*, but however *very old* $\geq$ *old* and *little old* $\leq$ *old*. So, $Dom(\mathcal{X})$ can be viewed as an algebraic order-based structure denoted by $\mathcal{AX}_{en} = (X, G, C, H, \leq)$, where $G = \{c^-, c^+\}$, $c^- \leq c^+$ the set of atomic words considered as the generators of $\mathcal{AX}_{en}$; C is the set of constants $C = \{\mathbf{0}, W, \mathbf{1}\}$ which are, respectively, the least, neutral and greatest words of $Dom(\mathcal{X})$, and are fixed-points of every hedge of H; $H = H^- \cup H^+ \cup \{h_0\}$ is the set of the hedges of $\mathcal{X}$ regarded as unary operations, where H^- and H^+ are respectively the set of the negative hedges, e.g. $H^- = \{Little\}$, and the set of the positive hedges, e.g. $H^+ = \{Very\}$, and h_0 is an artificial hedge to generate the semantics-core of words being mapped to their fuzzy set cores; and $\leq$ is the order-based semantics relation of $\mathcal{X}$. $Dom(\mathcal{X})$ is the set of the strings in the form

$h_n \ldots h_1 c$, $c \in G$, and we have $X = Dom(\mathfrak{X}) = H(G \cup C) = \{\sigma c : c \in G \cup C, \sigma \in H^*\}$. By the specific role of hedges, the structure $\mathcal{AX}_{en}$ is called *enlarged hedge algebra*.

For every word $x \in X$, the set $H(x) = \{\sigma x : \sigma \in H^*\}$ is called the fuzziness model of x. Denote by $|x|$ the length of the string $h_n \ldots h_1 c$, for an integer $k > 0$, put $X_k = \{x \in X : |x| = k\}$ and $X_{(k)} = \{x \in X : |x| \leq k\}$. It can be verified that $\sum_{x \in X_k} H(x) = X$. In any $\mathcal{AX}_{en}$, we have:

- For every hedge $h \in H$ and every word $x \in Dom(\mathfrak{X})$, either $x \leq hx$ or $x \geq hx$.
- X_k and $\{H(x) : x \in X_k\}$ are order isomorphic, i.e. for $\forall x, y \in X_k, x \leq y \Rightarrow H(x) \leq H(y)$.

To deal numerically with $\mathcal{AX}_{en}$, three quantification concepts of $\mathcal{AX}_{en}$ are defined closely to each other as follows:

- *Interval-semantics*: Denote by $\mathbb{P}([0,1])$ the set of all sub-intervals of $[0,1]$, which is the normalization of the universe of the variable $\mathfrak{X}$. By definition, interval-semantics of the words in $Dom(\mathfrak{X})$ is defined by a mapping $\mathfrak{v} : X \to \mathbb{P}([0,1])$ satisfying (i) it is an order-isomorphism, i.e. $x \leq y \Rightarrow \mathfrak{v}(x) \leq \mathfrak{v}(y)$, and (ii) the image $\mathfrak{v}(X)$ is dense in $[0,1]$. Then, $\mathfrak{v}$ is called Interval-Semantically Quantifying Mapping (ISQM) of $\mathcal{AX}_{en}$ and the $\mathfrak{v}$-values of $\mathfrak{v}$ are called the interval-semantics of the word-argument.
- *Fuzziness intervals*: Given an ISQM $\mathfrak{v}$, by (i) and (ii) above, $\mathfrak{v}(H(x))$ is dense in an interval of $\mathbb{P}([0,1])$ which is called the *fuzziness interval* of the word x and denoted by $\mathcal{I}(x)$. A piece of the structure of the fuzziness intervals of X is exposed in Fig. 1.
- *Fuzziness measure* of words. The structure described in Fig. 1 suggests to introduce the concept of *fuzziness measure* of the words in X which is defined for every $x \in X$ by $fm(x) = |\mathcal{I}(x)|$ - the length of $\mathcal{I}(x)$. The fuzziness measure fm of $\mathfrak{X}$ satisfies the following properties:

  (fm1) $fm(\mathbf{0}) + fm(c^-) + fm(W) + fm(c^+) + fm(\mathbf{1}) = 1$
  (fm2) $fm(h_i x) = \mu(h_i).fm(x)$ for all x and i
  (fm3) $\sum_{h \in H^-} \mu(h) + \mu(h_0) + \sum_{h \in H^+} \mu(h) = 1$

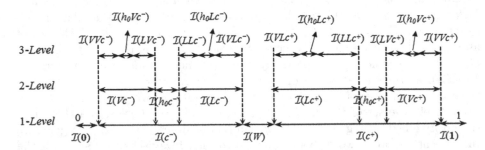

Fig. 1. The fuzziness interval of words in $X_{(3)}$ of the $\mathcal{AX}_{en}$ with $H = \{L, h_0, V\}$

It must be underlined that the quantification described above can be axiomatically developed to become a *sound* theory of the hedge algebra quantification for applications, in which the interval-semantics $\mathfrak{v}(x)$ of any word $x \in X$ can easily be computed based on the structure given in Fig. 1, for given fuzziness parameters comprising $fm(0), fm(c^-), fm(W), fm(c^+), fm(1), \mu(h_i)(h_i \in H)$.

For instance, compute $\mathfrak{v}(Lc^-)$, where Lc^- is in a given $X_{(2)} = \{0, Vc^-, c^-, Lc^-, W, Lc^+, c^+, Vc^+, 1\}$, with the fuzziness parameters given by $fm(0) = 0.01$, $fm(1) = 0.01$, $fm(c^-) = fm(c^+) = 0.44$, $\mu(L) = \mu(V) = 0.4$. Then, we have $fm(W) = 1 - 2 \times 0.01 - 2 \times 0.44 = 0.1$, $\mu(h_0) = 1 - 0.4 - 0.4 = 0.2$. It implies from *(fm2)* that $fm(Vc^-) = 0.4 \times 0.44 = 0.176$, $fm(h_0 c^-) = 0.2 \times 0.44 = 0.088$, $fm(Lc^-) = 0.4 \times 0.44 = 0.176$. Thus, $\mathfrak{v}(0) = [0; 0.01)$, $\mathfrak{v}(Vc^-) = [0.01; 0.177)$, $\mathfrak{v}(h_0 c^-) = [0.177; 0.265)$, $\mathfrak{v}(Lc^-) = [0.265; 0.441)$.

3 Linguistic Summarization Based on the Word-Semantics

3.1 Linguistic Frames of Cognition Based on HA Approach

The study of the interpretability of fuzzy systems was interested since 1990s and attracted large attention of fuzzy community since then based on the standpoint of user centric applications in which it is requires the user understandability of computational representation of fuzzy systems. The interpretability of linguistic summarization is also a significant problem and usually questioned [6]. As the interpretability of fuzzy systems and linguistic summary in the fuzzy set framework, in which word-domains of variables have been not formalized, is examined without any formal basis connected with the inherent semantics of linguistic words, it can be only considered based on the researcher's intuition and, hence, many approaches based on distinct standpoints and many criteria/constraints have been proposed, as discussed in [11].

However, according to our best knowledge, a very interesting and sound viewpoint is proposed by Mencar in [12] in which the interpretability of fuzzy systems is considered to measure how similar between the computational representation of their fuzzy rule bases in terms of fuzzy set expressions that are understandable and handleable by computers and their linguistic descriptions in terms of word expressions that are understandable by human users. This standpoint suggests us to examine a relationship between the fuzzy set representation of the quantified sentences of linguistic summary, which computational methods can handle, and the inherent semantics of linguistic summaries which the human user can capture when he reads them. To examine this relationship, it is necessary to deal immediately with the words and their own qualitative semantics of variables/attributes and, thus, words should be able to be handled as mathematical objects. This is why we propose a linguistic summary approach based on hedge algebras which model word-domains [8,9].

The main idea underlying the HA-approach is that the words in question of a variable should be considered as in a linguistic frame cognition (LFoC) [11] − a linguistic counterpart of a frame of cognition of fuzzy sets designed for the variable, that has its own semantic structure. They are domain-user words in his natural language and therefore the inherent word semantics must be determined *in the context of all words available in his domain language*. This implies that the semantic structures of such LFoCs are very essential. An LFoC of a variable V, denoted by LF_V, understood as a user terminological word-set using the set H of hedges of V, is a collection of words satisfying the following conditions, noting that each word x is written as a string $h_n \ldots h_1 c$, $h_i \in H$, $c \in \{c^+, c^-\}$ is an atomic word:

(i) $\{\mathbf{0}, c^-, W, c^+, \mathbf{1}\} \subseteq LF_V$
(ii) $hx \in LF_V \Rightarrow (\forall h' \in H)(h'x \in LF_V)$
(iii) $x \in LF_V \& x = hx' \& h \in H \Rightarrow x' \in LF_V$

It is observed that the more specificity of a word, the more length of its string representation. Since, as aforementioned, the semantics of words is context-dependent, it is argued in [11] that the semantics of a word of a length $l > 0$ must be defined in the context of all words of length not greater than l. The above conditions guarantee that every LFoC must be of the form of $\{x \in Dom(V) : length(x) \leq l\}$, for some $l > 0$, the set of all words of V of *specificity* degree not greater than l.

In the HA-approach, many properties of the semantics structure of an LFoC can be discovered in terms of certain relations between its words, e.g. the inherent order relation and the generality-specificity relation. Therefore, in the next section, it is necessary to answer the question: is there a fuzzy representation of LFoCs that is able to maintain the relations in terms of relations between the designed fuzzy sets.

3.2   Multi-level Fuzzy Representation of LFoCs

In this section, we present a procedure to construct a multi-level fuzzy representation of a given LFoC which is produced by a method immediately dealing with words and their inherent semantics and, then, analyze some advantages of this semantic representation. Since the theory of hedge algebras provides a formalism to immediately handle the words of a variable, to solve the question mentioned above, it is only necessary to show that one may define relations between the fuzzy sets in the multi-level fuzzy representation of a given LFoC that can represent properly the semantics relations discovered among the words of the LFoC. This can be realized based on the fact that, in the HA-approach, the fuzzy set base semantics of the words of LFoCs must be produced from the inherent semantics of the words of their respective variables, while maintaining the word semantics. It can be done as follows: (i) Provide *appropriate* numeric values of the fuzziness parameters of their variables; (ii) Use these values to compute the *fuzziness intervals* and the *interval-semantics* of the words of the LFoCs;

(iii) Construct a structure of fuzzy sets using the computational semantics of the words computed in (ii) so that the discovered semantic properties of the words can be represented the structure.

In this study, we apply fuzzy sets whose cores are intervals instead of single points, e.g. trapezoids. Assume that V is a linguistic variable, $X_1 = \{0, c^-, W, c^+, 1\}$ - the set of the words of length 1 consisting of constants and the atomic words. Let us assume that the LFoC of linguistic variable V is always of the form $X_{(k)}$.

For given values of the fuzziness parameters $fm(0)$, $fm(c^-)$, $fm(W)$, $fm(c^+)$, $fm(1)$, $\mu(h)(h \in H)$, a multi-granularity representation of LF_V can be constructed by the following procedure.

*Procedure **MGR-LFoC** for constructing a multi-granularity representation of a linguistic frame of cognition.*

- At level 1, compute the fuzziness intervals of the semantics core of the words of X_1 which are denoted by $\mathcal{I}(0), \mathcal{I}(h_0 c^-), \mathcal{I}(W), \mathcal{I}(h_0 c^+), \mathcal{I}(1)$. These intervals are used to interpret as the cores of the trapezoidal fuzzy sets computationally representing the semantics of the words in X_1. They are constructed to form a strong partition of the reference domain of the variable V.
- At level l, $2 \leq l \leq k$, suggested by the existence of the constants 0, W and 1 in the level $l = 1$, we may adopt the assumption that there are artificial constants 0_l, W_l and 1_l at the level l, and we obtain the set $X_l^+ = X_l \cup \{0_l, W_l, 1_l\}$. Similarly as for $l = 1$, compute the fuzziness intervals of the semantics core of the words of X_l^+ and consider them as the core of the trapezoidal fuzzy sets constructed for the respective words in X_l^+ with a note that the fuzziness intervals of the constants of the level l equal the fuzziness intervals of the respective constants of the level 1 [10,11]. Assuming these trapezoids at level l also form a strong partition of the universe of V, they are uniquely defined by their cores, i.e. the fuzziness intervals of the words of X_l^+, see Fig. 2.

Now, similarly as examined in [11] in which the words W_l, $l = 2$ to k, are not taken into account, we will show that the multi-granularity representation of a given LFoC can maintain essential semantic relations of its words: the order-based semantics relation and the generality-specificity relation (GS-relation).

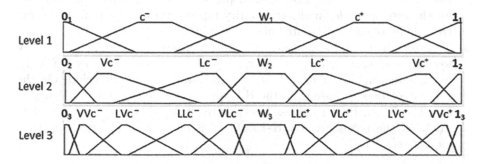

Fig. 2. Multi-granularity representation of LF_V of $\mathcal{AX}_{en}$ with $H = \{V, h_0, L\}$

First, we need define an order relation on the trapezoid fuzzy sets, each of which is characterized by a triple $(a, \mathbf{b}, d)$, where $[a, d]$ is the large base and $\mathbf{b}$ is the small base. The order relation $\leq$ and the GS-relation, denoted by GS, are defined as follows.

Definition 1. *For any two trapezoid fuzzy sets* $(a, \mathbf{b}, c)$ *and* $(a', \mathbf{b}', c')$:

(a) $(a, (\mathbf{b}), c) \leq (a', (\mathbf{b}'), c')$ *if:* (i) *Either at least two parameters necessarily including the core of the first fuzzy set are respectively smaller than the corresponding ones of the second one;* (ii) *Or, for any two fuzzy sets whose both cores are equal, the remaining parameter of the first fuzzy set is smaller than the respective one of the second.*

(b) $GS((a, \mathbf{b}, c), (a', \mathbf{b}', c'))$ *iff* $[a, c] \supset [a', c']$, *i.e. the support of the second trapezoid fuzzy set is included in the first one.*

For illustration (refer to Fig. 2), denoting the trapezoid assigned to the word x as $Trp(x)$, we have:

(i) $Trp(\mathbf{0}_1) \leq Trp(VVc^-)$, as $Core(\mathbf{0}_1) = \mathcal{I}(0) \leq Core(VVc^-) = \mathcal{I}(h_0VVc^-)$ and the first parameter of $Trp(\mathbf{0}_1)$ (i.e. 0) is smaller than the one of $Trp(VVc^-)$ (i.e. $fm(0)$)

(ii) $Trp(\mathbf{0}_3) \leq Trp((0)_2)$, as their cores are the same (i.e. $\mathcal{I}(0)$) and the third parameter of $Trp(\mathbf{0}_3)$ (i.e. $fm(0) + fm(VVVc^-)$) is smaller than the one of $Trp(\mathbf{0}_2)$ (i.e. $fm(0) + fm(VVc^-)$)

(iii) Especially, only the trapezoids $Trp(W_1)$, $Trp(W_2)$ and $Trp(W_3)$ are incomparable as they cannot satisfy (a) of Definition 1.

Theorem 1. *The multi-granularity representation of LFoCs constructed by procedure* **MGR-LFoC** *maintains their order-based semantics relation and the GS-relation.*

Proof

(a) From Fig. 2, it is easy to verify that the assignment x to $Trp(x)$ is an order isomorphism, i.e. for x and y satisfying $x \leq y$, by the above definition, we have $Trp(x) \leq Trp(y)$. The detailed proof can be recovered from the one for the case of *similar* multi-granularity representations given in [11], using triangle fuzzy sets represented also by triples of the form (a, b, c).

(b) To prove the second statement, it is only necessary to verify that for any words x and y satisfying $GS(Trp(x), Trp(y))$, simply denoted by $GS(x, y)$, we have the support of $Trp(x)$, denoted by $supp(Trp(x))$, is included in the support of $Trp(y)$. Indeed, in the HA-approach, $GS(x, y)$ implies that $y = h \ldots h'x$. Then, their fuzziness intervals must satisfy the inclusion $\mathcal{I}(y) \subseteq \mathcal{I}(x)$. By the procedure **MGR-LFoC**, it is obvious that $supp(Trp(y)) \subseteq supp(Trp(x))$.

4   Experiment

This experiment was performed on dataset Iris downloaded from https://archive.
ics.uci.edu/ml/datasets/iris. The patterns of dataset comprise 50 patterns of
Iris-Setosa, 50 patterns of Iris-Versicolor and 50 patterns of Iris-Virginica. To
demonstrate benefits and advantages of the multi-granularity representation of
LFoCs that can be automatically constructed as presented in Sect. 3.2, we perform the proposed linguistic summary method running on the dataset Iris with
the following protoform:

$$Q \text{ of } petal\ width \text{ of an Iris group's name are } S \qquad (5)$$

In this experiment, Q's are taken in the set {*none, very few, few, a half,
many, most, almost*} whose trapezoid semantics are represented in Fig. 3, the
same as in [15] except *none* and *very few* which are not under consideration
there. The quantifier *none* is interpreted as the antonym of *almost*. *Very few* is
assigned to a trapezoid corresponding to a triple $(0.01, [0.06, 0.16], 0.25)$.

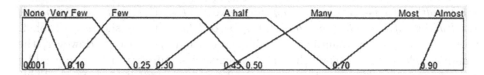

Fig. 3. Fuzzy sets represent for the set of quantifiers

The words in the summarizer S are taken in $LF_{petal\ width} = X_{(3)}$ of an $\mathcal{AX}_{en}$
with $H = \{Little, h_0, Very\}$, $c^- = short$ and $c^+ = long$. The fuzziness parameter
values are selected to be satisfied by the constraints $0.005 \leq fm(0), fm(1) \leq
0.02$; $0.05 \leq fm(medium) \leq 0.2$; $0.35 \leq fm(short), fm(long) \leq 0.5$; $0.1 \leq
\mu(h_0) \leq 0.3$; $0.25 \leq \mu(Very), \mu(Little) \leq 0.45$, which are intuitively and seman-
tically acceptable. Manually, the fuzziness parameter values $fm(0) = 0.01$,
$fm(short) = 0.44$, $fm(medium) = 0.1$, $fm(long) = 0.44$, $fm(1) = 0.01$,
$\mu(Little) = 0.4$, $\mu(h_0) = 0.2$, $\mu(Very) = 0.4$. Its multi-granularity representation of the LFoC of V is just the one given in Fig. 2.

To improve the quality of linguistic summaries we propose another measure,
which can distinguish two summarizers having the same T-value and quantifier.
It is suggested by the observation: Applying T to evaluate a sentence of the
protoform $Q\ y\ are\ S$ with attribute V, the sum $\frac{1}{n}\sum_{i=1}^{n} \mu_S(y_i)$ in (3), called the
support degree of the dataset for S and denoted by $Sup(S)$, which represents
the *average* of the membership degrees of the values $V(y_i)$'s belonging to the
fuzzy set representing the semantics of the summarizer S of V, cannot be able to
identify the distribution of $V(y_i)$ in the support of the fuzzy set representing the
semantics of S. This is because there might exist another summarizer S' that
has the same measure T and quantifier Q as S, i.e. we have $\frac{1}{n}\sum_{i=1}^{n} \mu_{S'}(y_i) =$

$\frac{1}{n}\sum_{i=1}^{n}\mu_S(y_i)$, but S' might be better than S in summarizing. E.g. it appears when the variance of the values $\mu_{S'}(y_i)$'s is greater than the one of $\mu_S(y_i)$'s. Therefore, in this experiment, we propose a measure, called a square support and denoted by $SqSup$, which is calculated by Eq. (6). The inequality $SqSup(S') > SqSup(S)$ means that S' is supported by the dataset more strongly than S.

$$SqSup(S) = \frac{\sum_{i=1}^{n}\mu_S^2(y_i)}{n} \tag{6}$$

Table 1 shows all quantified propositions, except the ones with the quantifier *none*, extracted from the set of values of attribute *petal width* of group Iris-setosa. At word level, if we only use the support Sup, the word *Absolutely short* is the highest. When we use a new support measure $SqSup$, we have $SqSup(absolutely\ short) < SqSup(very\ very\ short)$. This inequality implies that the data set supports the summarizer *very very short* more strongly than *absolutely short*, as shown in Table 1. Therefore, very very short, a word at the most specific level 3, is the best word to describe a specific characteristic of the data set.

At sentence level, lets consider two quantified propositions: (*Pro1*) **Many** of *petal widths of Iris-setosa are* **very very short** ($T = 1$) and (*Pro2*) **Many** of *petal widths of Iris-setosa are* **absolutely short** ($T = 1$). As shown by Table 1, they have the same quantifier and the same validity measure. However, the summary *Pro1* is more exact than *Pro2* because its summarizer is supported by $SqSup$ more strongly than the one of *Pro2* and its specificity is highest. Thus, *Pro1* represents a specific characteristic of the data set in question.

Similarly, Table 2 shows all propositions, except those with the quantifier none, of the data set of values of attribute *petal width* of group Iris-virginica. At word level, the word long has the highest value of $SqSup$ and Sup. The value of $SqSup$ of the word x satisfying $GS(long, x)$ is less than $SqSup(long)$. This reflects exactly the scattered distribution of the set of values for attribute *petal width* of the group Iris-virginica in the right half-side of the domain of *petal width*. In this case, the sentence **Many** of petal width of Iris-virginica are **long** ($T = 1$) is considered as the best linguistic description because of its highest supported summarizer.

Table 1. Propositions are extracted for group Iris-setosa and their quality measures using a multi-granularity representation

Q	S	$SqSup(S)$	$Sup(S)$	T
Many	Absolutely	0.561	0.712	1
Few	Short	0.138	0.288	1
A half	Very short	0.401	0.554	0.976
Many	Very very short	0.672	0.681	1
Few	Little very short	0.131	0.191	0.906

Table 2. Propositions are extracted for group Iris-setosa and their quality measures using a multi-granularity representation

Q	S	$SqSup(S)$	$Sup(S)$	T
Few	Absolutely long	0.163	0.237	1
Many	Long	0.562	0.671	1
Very few	Medium	0.057	0.093	1
Few	Little long	0.238	0.364	1
A half	Very long	0.371	0.502	1
Very few	Very little long	0.049	0.057	0.932
Few	Little little long	0.261	0.341	1
Few	Little very long	0.227	0.345	1
Very few	Very very long	0.132	0.178	0.801

Notice that while the data set of values of attribute *petal width* of the group Iris-setosa needs to be summarized by a word at the *most specific* level, the data set of group Iris-virginica is summarized by a word at the *most general* level. This shows the fact that one cannot pre-specify specific summarizers for a given dataset, instead on can pre-specify a LFoC with a specificity degree l of the attribute in question to mine latent linguistic knowledge in terms of linguistic summaries.

5   Conclusion

In this paper, we propose a novel linguistic summary method that can deal with the inherent semantics of words and can immediately handle words and their computational semantics based on the HA-formalism. Thus, in this method, the semantics of the words of LFoCs and their proper fuzzy representations is crucial to ensure the effectiveness of the method. It is shown that LFoCs involve two essential semantic relations of their own words, the order-based semantics relation and the GS-relation and that the proposed multi-granularity representations in the study can properly maintain these two semantic relations on LFoCs.

To justify the effectiveness of the proposed method, it is performed on the data set Iris to illustrate that the HA-approach and the proposed multi-granularity representations can successfully make use of the advantages and benefits of the semantics structure of LFoCs of attributes of the dataset in question to extract latent knowledge underlying the dataset in terms of qualified propositions that characterizes its specific features. For instance, this shows that words of length 3, i.e. it is of a high specificity that requires two consecutive hedges applied to an atom word, should be used to describe a characteristic of the dataset Iris and, therefore, it seems that the existing linguistic summary methods could not reveal it.

Acknowledgments. This research was supported by Vietnam National Foundation for Science and Technology Development (NAFOSTED) under Grant No 102.01-2017.06; The Vietnam Academy of Science and Technology (VAST) under Grant No VAST01.07/17-18 and the Hanoi National University of Education under No SPHN17-04.

References

1. Almeida, R.J., Lesot, M.J., Bouchon-Meunier, B., Kaymak, U.: Linguistic summaries of categorical time series for septic shock patient data. In: IEEE International Conference on Fuzzy Systems, pp. 1–8 (2013)
2. Boran, F.E., Akey, D., Yager, R.R.: An overview of methods for linguistic summarization with fuzzy sets. Expert Syst. Appl. **61**, 356–377 (2016)
3. Donis-Daz, C.A., Muro, A.G., Bello-Prez, R., Morales, E.V.: A hybrid model of genetic algorithm with local search to discover linguistic data summaries from creep data. Expert Syst. Appl. **41**, 2035–2042 (2014)
4. Kacprzyk, J., Yager, R.R.: Linguistic summaries of data using fuzzy logic. Int. J. Gen Syst **30**, 133–154 (2001)
5. Kacprzyk, J., Zadrozny, S.: Linguistic database summaries and their protoforms: toward natural language based knowledge discovery tools. Inf. Sci. **173**, 281–304 (2005)
6. Kacprzyk, J., Zadrozny, S.: Comprehensiveness and interpretability of linguistic data summaries: a natural language focused perspective. In: IEEE Symposium on Computational Intelligence for Human-like Intelligence (CIHLI), pp. 33–40 (2013)
7. Kacprzyk, J., Zadrozny, S.: Linguistic summarization of the contents of Web server logs via the Ordered Weighted Averaging (OWA) operators. Fuzzy Sets Syst. **285**, 182–198 (2016)
8. Nguyen, C.H., Wechler, W.: Hedge algebras: an algebraic approach to strutures of sets of linguistic domains of linguistic truth variables. Fuzzy Sets Syst. **35**, 281–293 (1990)
9. Nguyen, C.H., Wechler, W.: Extended algebra and their application to fuzzy logic. Fuzzy Sets Syst. **52**, 259–281 (1992)
10. Nguyen, C.H., Tran, T.S., Pham, D.P.: Modeling of a semantics core of linguistic terms based on an extension of hedge algebra semantics and its application. Knowl.-Based Syst. **67**, 244–262 (2014)
11. Nguyen, C.H., Hoang, V.T., Nguyen, V.L.: A discussion on interpretability of linguistic rule based systems and its application to solve regression problems. Knowl.-Based Syst. **88**, 107–133 (2015)
12. Mencar, C., Castiello, C., Cannone, R., Fanelli, A.M.: Interpretability assessment of fuzzy knowledge bases: a cointension based approach. Int. J. Approx. Reason. **52**, 501–518 (2011)
13. Vukadinovi, D., Bai, M., Nguyen, C.H., Vu, N.L., Nguyen, T.D.: Hedge-algebra-based voltage controller for a self-excited induction generator. Control Eng. Pract. **30**, 78–90 (2014)
14. Wilbik, A.: Linguistic summaries of the time series using fuzzy sets and their protoform for performance analysis of mutual funds. Ph.D. Dissertation (2010)
15. Wilbik, A., Keller, J., Bezdek, J.: Linguistic prototypes for data from eldercare residents. IEEE Trans. Fuzzy Syst. **22**(1), 110–123 (2014)
16. Yager, R.R.: A new approach to the summarization of data. Inf. Sci. **28**, 69–86 (1982)

Do It Today or Do It Tomorrow: Empirical Non-exponential Discounting Explained by Symmetry Ideas

Francisco Zapata[1], Olga Kosheleva[1], Vladik Kreinovich[1(✉)],
and Thongchai Dumrongpokaphan[2]

[1] University of Texas at El Paso, El Paso, TX 79968, USA
fazg74@gmail.com, {olgak,vladik}@utep.edu
[2] Department of Mathematics, College of Science, Chiang Mai University,
Chiang Mai, Thailand
tcd43@hotmail.com

Abstract. At first glance, it seems to make sense to conclude that when a 1 dollar reward tomorrow is equivalent to a $D < 1$ dollar reward today, the day-after-tomorrow's 1 dollar reward would be equivalent to $D \cdot D = D^2$ dollars today, and, in general, a reward after time t is equivalent to $D(t) = D^t$ dollars today. This *exponential discounting* function $D(t)$ was indeed proposed by the economists, but it does not reflect the actual human behavior. Indeed, according to this formula, the effect of distant future events is negligible, and thus, it would be reasonable for a person to take on huge loans or get engaged in unhealthy behavior even when the long-term consequences will be disastrous. In real life, few people behave like that, since the actual empirical discounting function is different: it is hyperbolic $D(t) = 1/(1 + k \cdot t)$. In this paper, we use symmetry ideas to explain this empirical phenomenon.

1 Discounting: Theoretical Foundations, Empirical Data, and Related Challenge

What is discounting. Future awards are less valuable than the same size awards given now. This phenomenon is known as *discounting*; see, e.g., [1,3,5–9,11] for details.

Procrastination is an inevitable consequence of discounting. Suppose that we have a task which is due by a certain deadline. This can be a task of submitting a grant proposal, or of submitting a paper to a conference.

In this case, the reward is the same no matter when we finish this task – as long as we finish it before the deadline. Similarly, the overall negative effect caused by the need to do some boring stuff is the same no matter when we do it. But, due to discounting, if we perform this task later, today's negative effect is smaller than if we perform this task today. The further in the future is this negative effect, the smaller is its influence on our today's happiness. Thus, a natural way to maximize today's happiness is to postpone this task as much as possible – which is exactly what people do; see, e.g., [2,7].

© Springer International Publishing AG, part of Springer Nature 2018
V.-N. Huynh et al. (Eds.): IUKM 2018, LNAI 10758, pp. 27–38, 2018.
https://doi.org/10.1007/978-3-319-75429-1_3

A simple theoretical model of discounting. How can we describe discounting in numerical terms? At first glance, providing numerical description for discounting is a straightforward idea.

Indeed, let us assume that 1 dollar tomorrow is equivalent to $D < 1$ dollars today. This is true for every day: 1 dollar at the day $t + 1$ is equivalent to D dollars in day t.

This means, in particular, that 1 dollar in day $t_0 + 2$ is equivalent to D dollars at time $t_0 + 1$. Since 1 dollar on day $t_0 + 1$ is equivalent to D dollars at the initial moment of time t_0, D dollars on day $t_0 + 1$ are equivalent to $D \cdot D$ dollars on day t_0. Thus, we can conclude that 1 dollar at day $t_0 + 2$ is equivalent to D^2 dollars at moment t_0.

Similarly, 1 dollar at moment $t_0 + 3$ is equivalent to D dollars at moment $t_0 + 2$ and thus, to $D \cdot D^2 = D^3$ dollars at moment t_0. In general, by induction over t, we can show that 1 dollar at moment $t_0 + t$ is equivalent to $D(t) \stackrel{\text{def}}{=} D^t$ dollars at the current moment t_0.

We can rewrite the above expression $D(t) = D^t$ as

$$D(t) = \exp(-a \cdot t), \tag{1}$$

where $a \stackrel{\text{def}}{=} -\ln(D)$. Because of this form, this discounting is known as *exponential*.

Practical problem with exponential discounting. At first glance, exponential discounting is a very reasonable idea. However, it has a problem: exponential functions decrease very fast, and for large t, the value $\exp(-a \cdot t)$ becomes indistinguishable from 0.

In practical terms, this means that a person looks for an immediate reward even if there is a significant negative downside in the distant future.

Such behavior indeed happens: a young man takes many loans without taking into account that in the future, he will have to pay; a young person ruins his health by using drugs, not taking into account that in the future, this may lead to a premature death. A person commits a crime without taking into consideration that eventually, he will be caught and punished.

Such behavior does happen, but such behavior is abnormal. Most people do not take an unrealistic amount of loans, most people do not ruin their health during their youth, most people do not commit crimes. This means that for most people, discounting decreases much slower than the exponential function.

So how to describe discounting: empirical data. Empirical data shows that discounting indeed decreases much slower than predicted by the exponential function: namely, 1 dollar at moment $t_0 + t$ is equivalent to

$$D(t) = \frac{1}{1 + k \cdot t} \tag{2}$$

dollars at moment t_0. This formula is known as *hyperbolic discounting*; see, e.g., [1,3,5–9,11].

Problem: how can we explain the empirical data. In principle, there exist many functions that decrease slower than the exponential function $\exp(-a \cdot t)$. So why, out of all these functions, we observe the hyperbolic one?

What we do in this paper. In this paper, we use symmetries to provide a theoretical explanation for the empirical discounting formula. To be more precise, our theoretical explanation leads to a family of functions of which hyperbolic discounting is one of the possibilities.

2   Analysis of the Problem

The idea of a re-scaling. Let $D(t)$ denote the discounting of a reward which is t moments into the future, i.e., the amount of money such that getting $D(t)$ dollars now is equivalent to getting 1 dollar after time t.

By definition, $D(0)$ means getting 1 dollar with no delay, so $D(0) = 1$. It is also reasonable to require that as the time period time t increases, the value of the reward goes to 0, so that $\lim_{t \to +\infty} D(t) = 0$.

It is also reasonable to require that a small change in t should lead to small changes in $D(t)$, i.e., that the function $D(t)$ be differentiable (smooth).

The further into the future we get the reward, the less valuable this reward is now, so the function $D(t)$ is decreasing as the time t increases. Thus, if we further delay all the rewards by some time s, then each value $D(t)$ will be replaced by a smaller value $D(t+s)$. We can describe this replacement as $D(t+s) = F_s(D(t))$, where the function $F_s(x)$ re-scales the original discount value $D(t)$ into the new discount value $D(t + s)$.

For the exponential discounting (1), the re-scaling $F_s(x)$ is linear: $D(t+s) = C \cdot D(t)$, where $C \stackrel{\text{def}}{=} \exp(-a \cdot s)$, so we have $F_s(x) = C \cdot x$. For the hyperbolic discounting (2), the corresponding re-scaling $F_s(x)$ is not linear.

Which re-scaling should we select?

Which re-scalings are reasonable: formulating this question in precise mathematical terms. We want to select some *reasonable* re-scalings. What does "reasonable" mean? Of course, linear re-scalings should be reasonable.

Also, intuitively, if a re-scaling is reasonable, then its inverse should also be reasonable. Similarly, if two re-scalings are reasonable, then applying them one after another should also lead to a reasonable re-scaling. In other words, a composition of two re-scalings should also be reasonable. In mathematical terms, we can conclude that the class of all reasonable re-scalings should be closed under inverse transformation and composition of two mappings. This means that with respect to the composition operation, such re-scalings must form a *group*.

We want to be able to determine the transformation from this group based on finitely many experiments. In each experiment, we gain a finite number of values, so after a finite number of experiments, we can only determine a finite number of parameters. Thus, we should be able to select an element of the desired transformation group based on the values of finitely many parameters.

In mathematical terms, this means that the corresponding transformation group should be *finite-dimensional*.

Summarizing: we want all the transformations $F_s(x)$ to belong to a finite-dimensional transformation group of functions of one variable that contains all linear transformations.

Which re-scalings are reasonable: answer to the question. It is known (see, e.g., [4,10,12]) that the only finite-dimensional transformation groups of functions of one variable that contain all linear transformations are the group of all linear transformations and the group of all fractional-linear transformations

$$\frac{a + b \cdot x}{1 + c \cdot x}.$$

Thus, our informal requirement that each re-scaling is reasonable implies that each re-scaling should be fractionally linear:

$$F_s(x) = \frac{a(s) + b(s) \cdot x}{1 + c(s) \cdot x}.$$

So, we arrive at the following requirement.

3   Definition and the Main Result

Definition 1. *We say that a smooth decreasing function $D(t)$ for which $D(0) = 1$ and $\lim_{t\to\infty} D(t) = 0$ is a* reasonable discounting function *if for every s, there exist values $a(s)$, $b(s)$, and $c(s)$ for which*

$$D(t + s) = \frac{a(s) + b(s) \cdot D(t)}{1 + c(s) \cdot D(t)}. \tag{3}$$

Proposition 1. *A function $D(t)$ is a reasonable discounting function if and only if it has one of the following forms: $D(t) = \exp(-a \cdot t)$, $D(t) = \dfrac{1}{1 + k \cdot t}$, $D(t) = \dfrac{1 + a}{1 + a \cdot \exp(k \cdot t)}$, or $D(t) = \dfrac{a}{(a + 1) \cdot \exp(k \cdot t) - 1}$, for some $a > 0$ and $k > 0$.*

Comment. The first discounting function corresponds to exponential discounting, the second to the hyperbolic discounting, the other two functions correspond to the more general case.

Both exponential and hyperbolic discounting can be viewed as the limit case of the more general formulas. Indeed, in the limit $a \to \infty$, both general expressions tend to the formula $D(t) = \exp(-k \cdot t)$ corresponding to the exponential discounting.

On the other hand, if we tend k to 0, we get $\exp(k \cdot t) \approx 1 + k \cdot t$, so for $a(k) = \alpha \cdot k$, the second general formula takes the form

$$D(t) = \frac{\alpha \cdot k}{(1 + \alpha \cdot k) \cdot (1 + k \cdot t) - 1}.$$

The denominator of this expression has the form

$$1 + \alpha \cdot k + k \cdot t + \alpha \cdot k^2 \cdot t - 1 = \alpha \cdot k + (k + \alpha \cdot k^2) \cdot t,$$

so

$$D(t) = \frac{\alpha \cdot k}{\alpha \cdot k + (k + \alpha \cdot k^2) \cdot t}.$$

Dividing both numerator and denominator of this formula by $\alpha \cdot k$, we get the hyperbolic discounting $D(t) = \dfrac{1}{1 + k' \cdot t}$, with $k' = k + \dfrac{1}{\alpha}$, i.e., in the limit $k \to 0$, with $k' = \dfrac{1}{\alpha}$.

Proof

1°. Let us first show that each of the four functions $D(t)$ listed in the formulation of the Proposition is a reasonable discounting function in the sense of Definition 1.

It is easy to see that all four functions are smooth and decreasing, and that for all of them, we have $D(0) = 1$ and $\lim_{t \to \infty} D(t) = 0$. Let us show, one by one, that each of these four functions satisfies the property (3) for appropriate auxiliary functions $a(s)$, $b(s)$, $c(s)$, and $d(s)$.

1.1°. For $D(t) = \exp(-a \cdot t)$, we have $\exp(a - a \cdot (t + s)) = \exp(-a \cdot s) \cdot \exp(-a \cdot t)$, i.e., $D(t + s) = \exp(-a \cdot s) \cdot D(t)$. Thus, the condition (3) is satisfied for $a(s) = 0$, $b(s) = \exp(-a \cdot s)$, and $c(s) = 0$.

1.2°. For the function $D(t) = \dfrac{1}{1 + k \cdot t}$, we have $1 + k \cdot t = \dfrac{1}{D(t)}$ hence

$$1 + k \cdot (t + s) = (1 + k \cdot t) + k \cdot s = \frac{1}{D(t)} + k \cdot s$$

and thus,

$$D(t + s) = \frac{1}{1 + k \cdot (t + s)} = \frac{1}{\dfrac{1}{D(t)} + k \cdot s}.$$

Multiplying both numerator and denominator of the right-hand side by $D(t)$, we conclude that

$$D(t + s) = \frac{D(t)}{1 + k \cdot s \cdot D(t)}.$$

Thus, the condition (3) is satisfied for $a(s) = 0$, $b(s) = 1$, and $c(s) = k \cdot s$.

1.3°. For the function $D(t) = \dfrac{1 + a}{1 + a \cdot \exp(k \cdot t)}$, we have $1 + a \cdot \exp(k \cdot t) = \dfrac{1 + a}{D(t)}$. Thus,

$$a \cdot \exp(k \cdot t) = \frac{1 + a}{D(t)} - 1 = \frac{1 + a - D(t)}{D(t)},$$

hence

$$a \cdot \exp(k \cdot (t + s)) = \exp(k \cdot s) \cdot (a \cdot \exp(k \cdot t)) = \exp(k \cdot s) \cdot \frac{1 + a - D(t)}{D(t)}$$

$$= \frac{\exp(k \cdot s) \cdot (1 + a) - \exp(k \cdot s) \cdot D(t)}{D(t)}.$$

Thus,

$$1 + a \cdot \exp(k \cdot (t + s)) = 1 + \frac{\exp(k \cdot s) \cdot (1 + a) - \exp(k \cdot s) \cdot D(t)}{D(t)}$$

$$= \frac{\exp(k \cdot s) \cdot (1 + a) - (\exp(k \cdot s) - 1) \cdot D(t)}{D(t)}.$$

Therefore,

$$D(t + s) = \frac{1 + a}{1 + a \cdot \exp(k \cdot (t + s))}$$

$$= \frac{(1 + a) \cdot D(t)}{\exp(k \cdot s) \cdot (1 + a) - (\exp(k \cdot s) - 1) \cdot D(t)}.$$

If we divide both the numerator and the denominator of this fraction by

$$(1 + a) \cdot \exp(k \cdot s),$$

we conclude that

$$D(t + s) = \frac{\exp(-k \cdot s) \cdot D(t)}{1 - \frac{1 - \exp(-k \cdot s)}{1 + a} \cdot D(t)}.$$

This is a formula of type (3), with $a(s) = 0$, $b(s) = \exp(-k \cdot s)$, and

$$c(s) = -\frac{1 - \exp(-k \cdot s)}{1 + a}.$$

1.4°. Finally, for the function $D(t) = \dfrac{a}{(a + 1) \cdot \exp(k \cdot t) - 1}$, we have

$$(a + 1) \cdot \exp(k \cdot t) - 1 = \frac{1}{D(t)},$$

hence

$$(a + 1) \cdot \exp(k \cdot t) = \frac{1}{D(t)} + 1 = \frac{a + D(t)}{D(t)}.$$

Thus,

$$(a + 1) \cdot \exp(k \cdot (t + s)) = \exp(k \cdot s) \cdot (a \cdot \exp(k \cdot t)) = \exp(k \cdot s) \cdot \frac{a + D(t)}{D(t)}$$

$$= \frac{\exp(k \cdot s) \cdot a + \exp(k \cdot s) \cdot D(t)}{D(t)}.$$

So, we have

$$(a+1) \cdot \exp(k \cdot (t+s)) + 1 = \frac{\exp(k \cdot s) \cdot a + \exp(k \cdot s) \cdot D(t)}{D(t)} + 1$$

$$= \frac{\exp(k \cdot s) \cdot a + (\exp(k \cdot s) + 1) \cdot D(t)}{D(t)}.$$

Thus,

$$D(t+s) = \frac{a}{(a+1) \cdot \exp(k \cdot (t+s)) + 1}$$

$$= \frac{a \cdot D(t)}{\exp(k \cdot s) \cdot a + (\exp(k \cdot s) + 1) \cdot D(t)}.$$

If we divide both the numerator and the denominator of this fraction by

$$\exp(k \cdot s) \cdot a,$$

we conclude that

$$D(t+s) = \frac{\exp(-k \cdot s) \cdot D(t)}{1 + \frac{1 + \exp(-k \cdot s)}{a} \cdot D(t)}.$$

This is a formula of type (3), with $a(s) = 0$, $b(s) = \exp(-k \cdot s)$, and

$$c(s) = \frac{1 + \exp(-k \cdot s)}{a}.$$

$2°$. So, for all four cases, we have proved that the corresponding functions $D(t)$ are reasonable discounting functions. Let us now show that, vice versa, if $D(t)$ is a reasonable discounting function in the sense of Definition 1, then it has one of the four forms listed in the formulation of Proposition 1.

$2.1°$. Let us assume that $D(t)$ is a reasonable discounting function. Thus, by definition of a reasonable discounting function, there exists functions $a(s)$, $b(s)$, and $c(s)$ for which, for all t and s, we have the property (3). Let us first prove that in this case, we have $a(s) = 0$.

Indeed, in the limit when $t \to \infty$, we have $D(t) \to 0$ and $D(t+s) \to 0$. Tending both sides of the equality (3) to the limit, we conclude that $a(s) = 0$. Thus, the formula (3) can be reformulated in an equivalent simplified form:

$$D(t+s) = \frac{b(s) \cdot D(t)}{1 + c(s) \cdot D(t)}. \tag{4}$$

$2.2°$. Let us now prove that the functions $b(s)$ and $c(s)$ are differentiable.

By definition of a reasonable discounting function, the function $D(t)$ is differentiable. Now, if we multiply both sides of the formula (4) by the denominator of the right-hand side, we get:

$$D(t + s) + c(s) \cdot D(t) \cdot D(t + s) = b(s) \cdot D(t).$$

If we now move all the terms containing $b(s)$ and $D(s)$ to the left-hand side and all the other terms to the right-hand side, we conclude that

$$b(s) \cdot (-D(t)) + c(s) \cdot (D(t) \cdot D(t + s)) = -D(t + s).$$

Thus, for each s, if we take two different values $t = t_1$ and $t = t_2$, we will get a system of two linear equations from which we can determine the two unknowns $b(s)$ and $c(s)$:

$$b(s) \cdot (-D(t_1)) + c(s) \cdot (D(t_1) \cdot D(t_1 + s)) = -D(t_1 + s);$$

$$b(s) \cdot (-D(t_2)) + c(s) \cdot (D(t_2) \cdot D(t_2 + s)) = -D(t_2 + s).$$

The solution of a system of linear equation can be explicitly described by the Cramer's rule. According to this rule, we have a differentiable formula that describes the solution to a system in terms of the coefficients at the unknowns and of the right-hand sides. In our case, both coefficients and right-hand sides are differentiable functions of s – since the discounting function $D(t)$ is differentiable. Thus, we conclude that the functions $b(s)$ and $c(s)$ are also differentiable.

2.3°. Let us now use the fact all the functions $D(t)$, $b(s)$, and $c(s)$ are differentiable to transform a difficult-to-solve functional equation (4) into a easier-to-solve differential equation. For this purpose, let us differentiate both sides of the formula (4) with respect to s and take $s = 0$.

After differentiation, we get the following differential equation:

$$D'(t + s) = \frac{b'(s) \cdot D(t)}{1 + c(s) \cdot D(t)} - \frac{b(s) \cdot D(t) \cdot c'(s) \cdot D(t)}{(1 + c(s) \cdot D(t))^2},$$

where b', c', and D' denote the derivatives of the corresponding functions. Let us now take $s = 0$. In this case, $D(t + s) = D(t)$, so we have $b(0) = 1$ and $c(0) = 1$. Thus, we get:

$$D'(t) = b'(0) \cdot D(t) - c'(0) \cdot (D(t))^2,$$

i.e.,

$$\frac{dD}{dt} = B \cdot D - C \cdot D^2, \tag{5}$$

where we denoted $B \overset{\text{def}}{=} b'(0)$ and $C \overset{\text{def}}{=} c'(0)$.

2.4°. Let us now analyze the differential equation (5). The Eq. (5) has two parameters B and C. They cannot be both equal to 0, since then (5) would imply that

$D(t)$ is a constant, not depending on time t at all – but we know that the function $D(t)$ is decreasing. Thus, one of these two coefficients has to be different from 0. We therefore have three possible cases:

- the case when $B \neq 0$ and $C = 0$,
- the case when $B = 0$ and $C \neq 0$, and
- the case when $B \neq 0$ and $C \neq 0$.

Let us consider these three cases one by one.

2.5°. Let us first consider the case when $B \neq 0$ and $C = 0$. In this case, the Eq. (5) has the form $\dfrac{dD}{dt} = B \cdot D$. Moving all the terms containing D to the left-hand side and all the other terms to the right-hand side, we conclude that $\dfrac{dD}{D} = B \cdot dt$. Integrating both sides, we get $\ln(D) = B \cdot t + C_0$ for some constant C_0. Thus, for $D(t) = \exp(\ln(D(t)))$, we get the formula

$$D(t) = \exp(C_0) \cdot \exp(B \cdot t).$$

From the condition $D(0) = 1$, we conclude that $\exp(C_0) = 1$ and thus, $D(t) = \exp(B \cdot t)$. From the requirement that the function $D(t)$ be decreasing, we conclude that $B < 0$, i.e., that $B = -a$ for some $a > 0$, and $D(t) = \exp(-a \cdot t)$.

2.6°. Let us now consider the case when $B = 0$ and $C \neq 0$. In this case, the Eq. (5) has the form $\dfrac{dD}{dt} = -C \cdot D^2$. Moving all the terms containing D to the left-hand side and all the other terms to the right-hand side, we conclude that $\dfrac{dD}{D^2} = -C \cdot dt$. Integrating both sides, we get $-\dfrac{1}{D} = -C \cdot t + C_0$ for some constant C_0. Thus, $D(t) = \dfrac{1}{-C_0 + C \cdot t}$. From the condition $D(0) = 1$, we conclude that $C_0 = -1$ and thus, $D(t) = \dfrac{1}{1 + C \cdot t}$. From the requirement that the function $D(t)$ be decreasing, we conclude that $C > 0$. This is the hyperbolic expression for $k = C$.

2.7°. Let us now consider the generic case when $B \neq 0$ and $C \neq 0$. In this case, the Eq. (5) has the form $\dfrac{dD}{dt} = B \cdot D - C \cdot D^2$, i.e., equivalently, that $\dfrac{dD}{dt} = -C \cdot (D^2 - r \cdot D)$, where we denoted $r \overset{\text{def}}{=} \dfrac{B}{C}$. Moving all the terms containing D to the left-hand side and all the other terms to the right-hand side, we conclude that

$$\frac{dD}{D^2 - r \cdot D} = -C \cdot dt. \tag{6}$$

Here, $D^2 - r \cdot D = D \cdot (D - r)$. We can therefore represent the fraction $\dfrac{1}{D^2 - r \cdot D}$ as a linear combination of the fractions $\dfrac{1}{D}$ and $\dfrac{1}{D - r}$:

$$\frac{1}{D^2 - r \cdot D} = \frac{C_1}{D} + \frac{C_2}{D - r}. \tag{7}$$

Multiplying both sides of the equality (7) by the denominator of the left-hand side, we conclude that $1 = C_1 \cdot (D-r) + C_2 \cdot D$, i.e., that $1 = -C_1 \cdot r + (C_1 + C_2) \cdot D$. This equality must hold for all D, so for $D = 0$, we get $1 = -C_1 \cdot r$ and $C_1 = -\dfrac{1}{r}$ and for $D = 1$, we get $C_2 = -C_1$ and thus, $C_2 = \dfrac{1}{r}$. Thus, the formula (6) takes the form

$$\frac{1}{r} \cdot \left(\frac{dD}{D-r} - \frac{dD}{D} \right) = -C \cdot dt.$$

Multiplying both sides by $r = \dfrac{B}{C}$ and taking into account that $r \cdot C = \dfrac{B}{C} \cdot C = B$, we conclude that

$$\frac{dD}{D-r} - \frac{dD}{D} = -B \cdot dt.$$

Integrating both sides, we get

$$\ln(D - r) - \ln(D) = -B \cdot t + C_0 \tag{8}$$

for some constant C_0. Here,

$$\ln(D - r) - \ln(D) = \ln\left(\frac{D-r}{D} \right) = \ln\left(1 - \frac{r}{D} \right).$$

Thus, the formula (8) takes the form

$$\ln\left(1 - \frac{r}{D} \right) = -B \cdot t + C_0.$$

Applying $\exp(x)$ to both sides, we conclude that

$$1 - \frac{r}{D} = C' \cdot \exp(-B \cdot t),$$

where $C' \stackrel{\text{def}}{=} \exp(C_0)$. Thus,

$$\frac{r}{D} = 1 - C' \cdot \exp(-B \cdot t),$$

and

$$D(t) = \frac{r}{1 - C' \cdot \exp(-B \cdot t)}.$$

For $t = 0$, we get $1 = D(0) = \dfrac{r}{1 - C'}$, hence $r = 1 - C'$, and we have

$$D(t) = \frac{1 - C'}{1 - C' \cdot \exp(-B \cdot t)}. \tag{9}$$

Let us now analyze what happens in the limit when $t \to \infty$, The corresponding asymptotics depends on where $B > 0$ or $B < 0$. If $B > 0$, we get $\exp(-B \cdot t) \to 0$, hence the expression (9) tends to $1 - C'$. The fact that this limit is 0 implies that $1 - C' = 0$, but in this case, $D(t)$ would be identically 0, and this is not the case (e.g., $D(0) = 1 \neq 0$). Thus, $B < 0$, i.e., $B = -k$ for some $k > 0$, and hence,

$$D(t) = \frac{1 - C'}{1 - C' \cdot \exp(k \cdot t)}.$$

Here, we cannot have $0 < C' < 1$, because otherwise, we will have the value $t = \frac{1}{k} \cdot \ln\left(\frac{1}{C'}\right)$ for which the denominator is 0 and for which, therefore, $D(t)$ is not defined. We cannot have $C' = 0$ – then $D(t)$ would not depend on time t at all; similarly, we cannot have $C' = 1$. So, must have either $C' < 0$ or $C' > 1$. When $C' < 0$, i.e., when $C' = -a$ for some $a > 0$, we get the expression

$$D(t) = \frac{1 + a}{1 + a \cdot \exp(k \cdot t)}.$$

When $C' > 1$, i.e., $C' = 1 + a$ for some $a > 0$, we get

$$D(t) = \frac{a}{(a + 1) \cdot \exp(k \cdot t) - 1}.$$

The proposition is proven.

4   Conclusions

For any person, an award that will only be delivered in the future is less valuable that the same amount delivered right away. To take this difference into account, economists use a special *discounting function* $D(t)$, so that an amount A delivered t moments in the future is equivalent to the discounted amount $D(t) \cdot A$ delivered today.

From the theoretical viewpoint, it seems that a natural discounting function is exponential $D(t) = D^t$: it corresponds, e.g., to the idea that a person can deposit the current amount $D(t) \cdot A$ in the bank, an after time t, get the value $(1 + q)^t \cdot D(t) \cdot A$, where q is the bank's interest rate. It is reasonable to equate this amount to the original amount A, getting $D(t) = D^t$ for $D = \frac{1}{1 + q}$.

Empirical research of human behavior shows, however, that in practice, when people make decisions, they use a different discounting function $D(t) = \frac{1}{1 + k \cdot t}$. To the best of our knowledge, so far, there has not been any quantitative theoretical explanation of this empirical phenomenon.

In this paper, we provide such an explanation. In this explanation, we use natural symmetry ideas.

Acknowledgments. We acknowledge the support of Chiang Mai University, Thailand. This work was also supported in part by the US National Science Foundation grant HRD-1242122.

The authors are thankful to the anonymous referees for valuable suggestions.

References

1. Critchfield, T.S., Kollins, S.H.: Temporal discounting: basic research and the analysis of socially important behavior. J. Appl. Behav. Anal. **34**, 101–122 (2001)
2. Durakiewicz, C.: A universal law of procrastination. Phys. Today **69**(2), 11–12 (2016)
3. Frederick, S., Loewenstein, G., O'Donoghue, T.: Time discounting: a critical review. J. Econ. Lit. **40**, 351–401 (2002)
4. Guillemin, V.M., Sternberg, S.: An algebraic model of transitive differential geometry. Bull. Am. Math. Soc. **70**(1), 16–47 (1964)
5. King, G.R., Logue, A.W., Gleiser, D.: Probability and delay in reinforcement: an examination of Mazur's equivalence rule. Behav. Processes **27**, 125–138 (1992)
6. Kirby, K.N.: Bidding on the future: evidence against normative discounting of delayed rewards. J. Exp. Psychol. (Gen.) **126**, 54–70 (1997)
7. Konig, C.J., Kleinmann, M.: Deadline rush: a time management phenomenon and its mathematical description. J. Psychol. **139**(1), 33–45 (2005)
8. Mazur, J.E.: An adjustment procedure for studying delayed reinforcement. In: Commons, M.L., Mazur, J.E., Nevin, J.A., Rachlin, H. (eds.) Quantitative Analyses of Behavior: The Effect of Delay and Intervening Events, vol. 5. Erlbaum, Hillsdale (1987)
9. Mazur, J.E.: Choice, delay, probability, and conditional reinforcement. Animal Learn. Behav. **25**, 131–147 (1997)
10. Nguyen, H.T., Kreinovich, V.: Applications of Continuous Mathematics to Computer Science. Kluwer, Dordrecht (1997)
11. Rachlin, H., Raineri, A., Cross, D.: Subjective probability and delay. J. Exp. Anal. Behav. **55**, 233–244 (1991)
12. Singer, I.M., Sternberg, S.: Infinite groupsof Lie and Cartan, Part I. Journal d'Analyse Mathematique **15**, 1–113 (1965)

Estimation of Interval Probability Distribution in Categorical Data Using Maximum Entropy

Bingyi Kang[1] and Yong Deng[1,2(✉)]

[1] School of Computer and Information Science, Southwest University,
Chongqing 400715, China
ydeng@swu.edu.cn, prof.deng@hotmail.com
[2] Institute of Fundamental and Frontier Science,
University of Electronic Science and Technology of China,
Chengdu 610054, China

Abstract. Estimating probability distribution is a primary requirement in the application of statistic based. We proposed a framework of estimating interval probability distribution in the category data with unknown information using the maximum entropy in the framework of Dempster-Shafer theory. The proposed method can approximately generate the optimal interval probability distribution considering the unknown information. Two applications are used to illustrate the procedure of the proposed framework.

1 Introduction

Estimating probability distribution is a primary requirement in the statistical applications, such as statistical process control [1], medical diagnosis [2]. The classical approach to dealing with the estimation of probability (or proportion) distribution is the maximum likelihood method (MLE) [3,4]. Another process to handle this problem is the maximum entropy method (MEM), which has obtained plenty of attention recent decades of years [5,6]. MEM is regarded as a more fundamental mechanism to estimate the probability distribution [7]. In paper [1], the authors investigated what are the probability distributions like using the MEM with the different constraint of the variables. Of which the normal Gaussian distribution is obtained using the mechanics of MEM when the mean and variance of the variables are given, i.e., $f\left(x|\mu,\sigma^2\right) = \frac{1}{\sqrt{2\pi}\sigma}e^{-\frac{(x-\mu)^2}{2\sigma^2}}$ is derived from $\max - \int f\left(x\right)\log f\left(x\right)dx$ such that $\omega_1 = E\left(X\right), \omega_2 = E\left[\left(X - \omega_1\right)^2\right]$. Other types of distributions are also derived from MEM under different constraints in [1], such as Laplace distribution, exponential distribution, Gamma distribution, Beta distribution, geometric distribution, and Pareto distribution. It is noted that the MEM has got lots of achievements in the statistical process control [8], especially a method of approximately estimating the optimum probability distribution using the maximum Shannon entropy considering the confidence interval (CI) is proposed [9] to deal with the process control of category data.

© Springer International Publishing AG, part of Springer Nature 2018
V.-N. Huynh et al. (Eds.): IUKM 2018, LNAI 10758, pp. 39–47, 2018.
https://doi.org/10.1007/978-3-319-75429-1_4

From the literature review, we find that limited work has been done to deal with the estimation of the probability distribution of unknown information. Unknown information is a pervasive phenomenon [10], especially in some situations of limited knowledge and unavailable detecting equipment. For example, a doctor wants to estimate the occurrence rate of a disease. Due to somewhat special reason, he doesn't know whether or not the candidate is infected. In this article, we use the Dempster-Shafer framework to represent the unknown information. We propose an optimal method of estimating the interval probability distribution considering the unknown information in the framework of Dempster-Shafer theory using the maximum entropy. The main steps are concluded as follows: we at first use the framework of Dempster-Shafer theory to represent the unknown (unknown) information. Secondly, we obtain the confidence interval (CI) using the MLE. Thirdly, we estimate the optimum interval probability distribution using the maximum entropy mechanism.

The paper is organized as follows. The preliminaries introduce the framework of Dempster-Shafer evidence theory, the uncertainty measure (entropy), and interval probability distribution in Sect. 2. Section 3 proposes the approximate method of estimating interval probability distribution based on the maximum entropy subjecting to some entailment constraints. Section 4 presents two simple applications of the proposed method. Finally, this paper is concluded in Sect. 5.

2   Preliminaries

In this section, some preliminaries are briefly introduced.

2.1   Framework of Dempster-Shafer Evidence Theory

Dempster-Shafer theory (short for D-S theory) is presented by Dempster and Shafer [11,12]. Some basic concepts in D-S theory are introduced.

Let X be a set of mutually exclusive and collectively exhaustive events, indicated by

$$\Theta = \{\theta_1, \theta_2, \cdots, \theta_i, \cdots, \theta_{|X|}\} \tag{1}$$

where set Θ is called a frame of discernment (FOD). The power set of X is indicated by 2^X, namely

$$2^\Theta = \{\emptyset, \{\theta_1\}, \cdots, \{\theta_{|\Theta|}\}, \{\theta_1, \theta_2\}, \cdots, \{\theta_1, \theta_2, \cdots, \theta_i\}, \cdots, \Theta\} \tag{2}$$

For a frame of discernment $\Theta = \{\theta_1, \theta_2, \cdots, \theta_{|\Theta|}\}$, a mass function is a mapping m from 2^Θ to $[0,1]$, formally defined by:

$$m: \quad 2^\Theta \to [0,1] \tag{3}$$

which satisfies the following condition:

$$m(\emptyset) = 0 \quad and \quad \sum_{A \in 2^X} m(A) = 1 \tag{4}$$

Given a belief structure m on Θ, the belief function and plausibility function which are in one-to-one correspondence with m can be defined respectively as:

$$Bel\,(A) = \sum_{B \subseteq A} m\,(B) \tag{5}$$

$$Pl\,(A) = \sum_{B \cap A \neq \emptyset} m\,(B) = 1 - \sum_{B \cap A = \emptyset} m\,(B) \tag{6}$$

In the next subsection, we introduce an uncertainty (entropy) measure, SU, which has some good qualities to handle the uncertainty information, such as (1) probability consistency, i.e., when m reduces to a Bayesian belief structure, SU is identical to Shannon entropy. (2) The range of measure SU is between 0 and $|\Theta|$, $|\Theta|$ is the cardinality of FOD. (3) If a BPA m is a categorical BPA focusing on A, then the uncertainty of m is $|A|$. The properties of SU is consistent with the axiomatic requirement of range to measure uncertainty in [14].

2.2 Uncertainty Measure (Entropy) in Evidence Theory

Let m be a belief structure on the discernment frame $\Theta = \{\theta_1, \theta_2, \cdots, \theta_n\}$. The interval probabilities-based uncertainty measure of m, which is proposed by Wang and Song [13], can be expressed as below:

$$SU\,(m) = \sum_{i=1}^{n} \left[-\frac{Bel\,(\theta_i) + Pl\,(\theta_i)}{2} \log_2 \frac{Bel\,(\theta_i) + Pl\,(\theta_i)}{2} \right. $$
$$\left. + \frac{Pl\,(\theta_i) - Bel\,(\theta_i)}{2} \right] \tag{7}$$

where $Bel\,(\theta_i)$ is the belief of single element θ_i, and $Pl\,(\theta_i)$ is the plausibility of single element θ_i.

2.3 Addition Operation of Interval Numbers

Let $A = [a_1, a_2]$, $B = [b_1, b_2]$ two interval numbers, the addition of A and B is denoted as

$$A + B = [a_1 + b_1, a_2 + b_2] \tag{8}$$

2.4 Interval Probability Distribution

Let $\Theta = \{\theta_1, \theta_1, \cdots, \theta_n\}$ be the frame of discernment, $[a_i, b_i]$ $(i = 1, 2, \cdots, n)$ be n intervals with $0 \leq a_i \leq b_i \leq 1$. $P\,(\theta_i) = [a_i, b_i]$ constitute an interval probability distribution on Θ such that [13]:

$$(1) \sum_{i=1}^{n} a_i \leq 1 \text{ and } \sum_{i=1}^{n} b_i \geq 1 \tag{9}$$

$$(2) \sum_{i=1}^{n} b_i - (b_k - a_k) \geq 1 \text{ and } \sum_{i=1}^{n} a_i + (b_k - a_k) \leq 1 \; \forall k \in \{1, \cdots, n\} \tag{10}$$

$$(3)\ P\,(H) = 0, \forall H \notin \Theta \tag{11}$$

In the next section, we propose the framework to estimate the interval probability distribution with unknown information. This method is different from the method of Pignistic probability transformation, which assumes the belief is promotional to the cardinality of the proposition. However, the assumption always includes controversy. Interval distribution estimation is a possible way to handle the uncertain information considering the influence of sample size.

3 Proposed Framework of the Interval Distribution Estimation Problem

First, we present the confidence interval for the proposition considering the influence of sample size.

3.1 Generate the Confidence Interval for the Proposition

We firstly determine the CI of an estimate of $\mathbf{p_1}$ using i.i.d. sample $\{x_1, \ldots, x_2\}$ from distribution $\mathbf{p_1}$, which is based on the method of the maximum likelihood estimation (MLE) introduced in [9]. Then, we use the proposed method in the next subsection constrained with the CI to formulate the problem of estimating the distribution. Define $\hat{\mathbf{p}}_{\mathbf{mle}}$ as the MLE estimator of $\mathbf{p_1}$ from sample $\{x_1, \ldots, x_2\}$ as follow,

$$\hat{\mathbf{p}}_{\mathbf{mle, i}} = \frac{\sum_{k=1}^{N} \mathrm{I}\left(x_k = i\right)}{N} \tag{12}$$

where I is the indicator function, which is equal to one when the condition within the parentheses is satisfied and is zero otherwise, N is the sample size.

In this article, we applied the CI on $\hat{\mathbf{p}}$ defined in [9] as $\hat{l}_i\left(\kappa\right) \leq \hat{p}_i \leq \hat{u}_i\left(\kappa\right)$, where

$$\hat{l}_i\left(\kappa\right) = \max\left\{0, \ \hat{p}_{mle, i} - \kappa\sqrt{\frac{\hat{p}_{mle, i}\left(1 - \hat{p}_{mle, i}\right)}{N}}\right\} \tag{13}$$

$$\hat{u}_i\left(\kappa\right) = \min\left\{1, \ \hat{p}_{mle, i} + \kappa\sqrt{\frac{\hat{p}_{mle, i}\left(1 - \hat{p}_{mle, i}\right)}{N}}\right\} \tag{14}$$

where $\sqrt{\frac{\hat{p}_{mle, i}\left(1 - \hat{p}_{mle, i}\right)}{N}}$ is the standard derivation or standard error (SE); κ is a tuning parameter, which relates to the confidence level. If we want to obtain a larger CI, a larger value of κ is used and vice versa. A larger κ implicates that the CI we obtained has a higher confidence level and vice versa (the parameter κ is setting to one in this article).

Next, we propose the framework of generating interval probability distribution with unknown information.

3.2   Proposed Framework of Generating Interval Probability Distribution with Unknown Information

The estimation of interval probability distribution is to maximize the entropy using Eq. (7) subject to the following six constraints: Condition 1 and condition 2 are generated from the CI using Eqs. (8), (13) and (14). Condition 3 means that the lower probability $Bel(\theta_i)$ and upper probability $Pl(\theta_i)$ of the interval probability must be ranged in [0, 1] for θ_i $(i = 1, \cdots, n)$ are singleton elements. Condition 4, condition 5 and condition 6 stand for the generated interval probability distribution must satisfy the basic constraints of the interval probability distribution from Eqs. (9) to (11).

$$Max\ SU\left(m\right) = Max\left\{\sum_{i=1}^{n}\left[-\frac{Bel(\theta_i)+Pl(\theta_i)}{2}\log_2\frac{Bel(\theta_i)+Pl(\theta_i)}{2} + \frac{Pl(\theta_i)-Bel(\theta_i)}{2}\right]\right\}$$

$subject\ to$

$$\begin{cases} 1.\ Bel\left(\theta_i\right) \in \left[\hat{l}_{\theta_i}, \hat{u}_{\theta_i}\right] \\[2mm] 2.\ Pl\left(\theta_i\right) \in \left[\sum_{\theta_i\subseteq A}\hat{l}_A, \sum_{\theta_i\subseteq A}\hat{u}_A\right] \\[2mm] 3.\ 0 \leq Bel\left(\theta_i\right) \leq Pl\left(\theta_i\right) \leq 1 \\[2mm] 4.\ \sum_{i=1}^{n}Bel\left(\theta_i\right) \leq 1; \sum_{i=1}^{n}Pl\left(\theta_i\right) \geq 1 \\[2mm] 5.\ \sum_{i=1}^{n}Bel\left(\theta_i\right) + \left[Pl\left(\theta_k\right) - Bel\left(\theta_k\right)\right] \leq 1, \forall k = 1, \cdots, n \\[2mm] 6.\ \sum_{i=1}^{n}Pl\left(\theta_i\right) - \left[Pl\left(\theta_k\right) - Bel\left(\theta_k\right)\right] \geq 1, \forall k = 1, \cdots, n \end{cases}$$

$$(15)$$

The possible optimal solution is $P^*(\theta_i) = [Bel^*(\theta_i), Pl^*(\theta_i)]$, the analytic solution $P^*(\theta_i)$ is hard to obtain. In this article, we use the special optimization tool (genetic algorithm) to solve this optimization problem.

If the sample size N is infinite (large enough), the influence of interval confidence will be faded, i.e., $\hat{l}_i(\kappa) = \hat{u}_i(\kappa) = \hat{p}_{mle,i}$. The condition 1 and condition 2 in Eq. (15) degenerate into $Bel(\theta_i) = \hat{l}_{\theta_i} = \hat{u}_{\theta_i}$ and $Pl(\theta_i) = \sum_{\theta_i\subseteq A}\hat{l}_A = \sum_{\theta_i\subseteq A}\hat{u}_A$. Therefore, here, $Bel(\theta_i)$ is the belief function of θ_i, and $Pl(\theta_i)$ is the plausibility function of θ_i. According to the Theorem 1 in [13], i.e., for a belief structure m on the discernment frame $\Theta = \{\theta_1, \theta_2, ..., \theta_n\}$, all belief intervals $[Bel(\theta_i), Pl(\theta))]$ $(i = 1, 2, ..., n)$ constitute an interval probability distribution on Θ. Hence, the possible optimal solution is $P^*(\theta_i) = [Bel(\theta_i), Pl(\theta_i)]$ if the sample size N is infinite (large enough), where $Bel(\theta_i) = \hat{l}_{\theta_i} = \hat{u}_{\theta_i}$ and $Pl(\theta_i) = \sum_{\theta_i\subseteq A}\hat{l}_A = \sum_{\theta_i\subseteq A}\hat{u}_A$.

In the next section, we use two applications to illustrate the proposed method.

4   Applications of Estimating Interval Probability Distribution

4.1   Case of Estimating Interval Probability Distribution of Disease Diagnosis with Unknown Information

Assume a manager wants to estimate the occurrence rate of some disease in some regions. Suppose n = 20 people are selected at random to check whether they suffer a certain disease. Let Y be the event that a randomly selected participant is surely affected by the disease; Let N be the event that a randomly selected participant is not affected by the disease; Let $\Theta = \{Y, N\}$ be the event that the doctor does not know whether the selected participant suffers the disease. The data are collected given in Table 1. According to Eq. (12), the estimated proportion $\hat{p}_{mle}$ of the diseased people, not diseased people, and unknown is given in Table 1. According to Eqs. (13) and (14), the estimated CI ($[\hat{l}(\kappa), \hat{u}(\kappa)]$) of the diseased people, not diseased people, and unknown is respectively given in Table 1, where κ is setting to 1 in this case.

Table 1. Count of selected people in suffering (Y), not suffering (N) the disease, unknown (Θ), and it's confidence interval

Item	Number	$\hat{p}_{mle}$	SE	$\hat{l}(\kappa)$	$\hat{u}(\kappa)$
Y	4	0.2	0.089	0.111	0.289
N	14	0.7	0.102	0.598	0.802
Θ	2	0.1	0.067	0.033	0.167

Then we apply the framework of maximum entropy with six constraints formulated in Eq. (15) to obtain the optimal interval probability distribution (Interval P). The results are shown in Table 2 through the optimization tool of genetic algorithm.

We investigate the influence of the size of the samples on the proposed method. We assume the proportion $[p(Y)\, p(N)\, p(\theta)] = [0.2, 0.7, 0.1]$ is unchanged and the size of the samples is changing from 10 to 100 with the increment 10. The trend of the approximately estimated interval probability distributions is shown in Fig. 1. From Fig. 1, we can see that the probability intervals (yellow regions) are trending to convergence with the increase of the sample size.

Table 2. Estimated interval probability distribution

Item	$\hat{l}(\kappa)$	$\hat{u}(\kappa)$	$Bel \in$	$Pl \in$	Interval P
Y	0.111	0.289	[0.111, 0.289]	[0.144, 0.456]	[0.110, 0.3]
N	0.598	0.802	[0.598, 0.802]	[0.631, 0.969]	[0.699, 0.891]

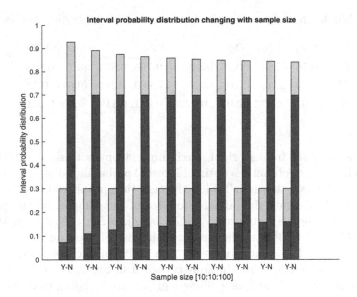

Fig. 1. Interval probability distribution changing with sample size from 10 to 100 (Color figure online)

4.2   Case of Estimating Interval Probability Distribution of Goods Quality Detection with Unknown Information

Assume a manage want to estimate the quality of the produced goods. Suppose n = 100 goods are randomly selected. The quality of the goods is categorized into three classes: Good (G), Medium (M), and Bad (B). For somewhat special reasons, the quality of some goods is hard to make a distinction (the categorization is a intersection between the neighbor classes). The data are collected given in Table 3. According to Eq. (12), the estimated proportion $\hat{p}_{mle}$ of the good (G), medium (M), bad (B), unknown good or medium (G, M), unknown medium or bad (M, B) is given in Table 3. According to Eqs. (13) and (14), the estimated CI ($[\hat{l}(\kappa), \hat{u}(\kappa)]$) is respectively given in Table 3.

Table 3. Count of selected goods in good (G), medium (M), bad (B), unknown good or medium (G, M), unknown medium or bad (M, B) and the generated CI

Item	Number	$\hat{p}_{mle}$	SE	$\hat{l}(\kappa)$	$\hat{u}(\kappa)$
G	80	0.8	0.04	0.76	0.84
M	10	0.1	0.03	0.07	0.13
B	5	0.05	0.022	0.028	0.072
(G, M)	2	0.02	0.014	0.006	0.034
(M, B)	3	0.03	0.017	0.013	0.047

Table 4. Approximately estimated interval probability distribution

Item	$\hat{l}(\kappa)$	$\hat{u}(\kappa)$	$Bel \in$	$Pl \in$	Interval P
G	0.76	0.84	[0.76, 0.84]	[0.766, 0.874]	[0.76, 0.874]
M	0.07	0.13	[0.07, 0.13]	[0.089, 0.211]	[0.071, 0.19]
B	0.028	0.072	[0.028, 0.072]	[0.041, 0.119]	[0.05, 0.051]

Then we apply the framework of maximum entropy with six constraints formulated in Eq. (15) to obtain the optimal interval probability distribution (Interval P). The results are shown in Table 4 through the optimization tool of genetic algorithm. We investigate the influence of size of the samples on the proposed method. We assume the proportion $[p(G), p(M), p(B), p(G, M), p(M, B)] = [0.8, 0.1, 0.05, 0.02\ 0.03]$ is unchanged and the size of the samples is changing from 10 to 100 with the increment 10. The trend of the approximately estimated interval probability distributions is shown in Fig. 2. From Fig. 2, we can see that the probability intervals (yellow regions) are trending to converge with the increasing of the sample size.

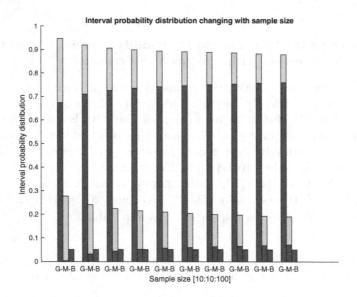

Fig. 2. Interval probability distribution changing with sample size from 10 to 100 (Color figure online)

5   Conclusion

Estimating probability distribution is a primary requirement in the applications of statistic based. Commonly, the estimation of the probability distribution is derived from the accurate statistic results based on the randomly selected samples. However, we sometimes can not make a distinction between the categories of some samples, which is a normal phenomenon e.g. in the disease diagnosis for the limited knowledge of the experts or unavailable detecting equipment. We proposed a framework of estimating interval probability distribution in the category data with unknown information using the maximum entropy in the framework of Dempster-Shafer theory. The proposed method can approximately generate the optimum interval probability distribution considering the unknown information. Two applications are used to illustrate the procedure of the proposed framework.

Acknowledgment. The work is partially supported by National Natural Science Foundation of China (Grant Nos. 61573290, 61503237), and China Scholarship Council.

References

1. Alwan, L.C., Ebrahimi, N., Soofi, E.S.: Information theoretic framework for process control. Eur. J. Oper. Res. **111**, 526–542 (1998)
2. Kononenko, I.: Inductive and Bayesian learning in medical diagnosis. Appl. Artif. Intell. **7**, 317–337 (1993)
3. Jiao, J., Venkat, K., Han, Y., Weissman, T.: Maximum likelihood estimation of functionals of discrete distributions. IEEE Trans. Inf. Theory. **63**, 6774–6798 (2017)
4. Kaplan, E.L., Meier, P.: Nonparametric estimation from incomplete observations. J. Am. Stat. Assoc. **53**, 457–481 (1958)
5. Levy, W.B., Delic, H.: Maximum entropy aggregation of individual opinions. IEEE Trans. Syst. Man Cybern. **24**, 606–613 (1994)
6. Myung, I.J., Ramamoorti, S., Bailey, A.D.: Maximum entropy aggregation of expert predictions. Manage. Sci. **42**, 1420–1436 (1996)
7. Jaynes, E.T.: Information theory and statistical mechanics. Phys. Rev. **106**, 620–630 (1957)
8. Das, D., Zhou, S.: Statistical process monitoring based on maximum entropy density approximation and level set principle. IIE Trans. **47**, 215–229 (2015)
9. Das, D., Zhou, S.: Detecting entropy increase in categorical data using maximum entropy distribution approximations. IISE Trans. **49**, 827–837 (2017)
10. Li, C.C.: The incomplete binomial distribution. In: Kojima, K. (ed.) Mathematical Topics in Population Genetics, vol. 1, pp. 337–366. Springer, Heidelberg (1970). https://doi.org/10.1007/978-3-642-46244-3_11
11. Dempster, A.P.: Upper and lower probabilities induced by a multivalued mapping. Ann. Math. Statist. **38**, 325–339 (1967)
12. Shafer, G.: A Mathematical Theory of Evidence, vol. 42. Princeton University Press, Princeton (1976)
13. Wang, X., Song, Y.: Uncertainty measure in evidence theory with its applications. In: Applied Intelligence, pp. 1–17 (2017)
14. Deng, X., Deng, Y.: On the axiomatic requirement of range to measure uncertainty. Phy. A **406**, 163–168 (2014)

Measuring Efficiency Intervals in Multiple Groups with Privacy Concerns

Tomoe Entani(✉) (iD)

University of Hyogo, Kobe, Japan
entani@ai.u-hyogo.ac.jp

Abstract. The organization is willing to take data from the other organizations for the better analysis, when the self-accumulated data only gives the limited findings for instance in case of focusing on a specific small group. The multi-accumulated data can be available for the analysis, if the different several organizations accumulated similar data independently. However, the organizations need to care for privacy since the original data could include a kind of personal information. Under such a privacy concern, this paper proposes the method for efficiency evaluation with multi-accumulated data based on Data Envelopment Analysis (DEA), which is a non-parametric technique to measure efficiency of a decision making unit (DMU) relatively in a group. The efficiency interval of a DMU in a group by referencing DMUs in another group as well as DMUs in its own group is obtained, even if the group cannot access to the original data of another group. Instead, the group takes the information of the efficient frontier of another group denoted as the weight set, from which the group cannot guess the original data of another group. As a result, three kinds of efficiency intervals for a DMU are obtained: the efficiency in its own group, that in another group, and that in the integrated group. Comparing them can give us a rich and useful information on the DMU from wide viewpoint.

Keywords: Data Envelopment Analysis · Efficiency interval
Efficient frontier · Multiple groups · Privacy concerns

1 Introduction

Various methods, techniques and tools have been developed in data collection and analysis and this trend continues further in highly information-oriented society. Simultaneously, our concern on personal identification becomes higher. It is not an issue to concern on data privacy as far as an organization analyzes data which is properly self-accumulated. When it comes to analyze multi-accumulated data, which different several organizations accumulate, we need to be careful for privacy.

It often happens that the data size becomes too small for the detailed analysis focusing on a specific group. When the data processing with own data only gives

© Springer International Publishing AG, part of Springer Nature 2018
V.-N. Huynh et al. (Eds.): IUKM 2018, LNAI 10758, pp. 48–59, 2018.
https://doi.org/10.1007/978-3-319-75429-1_5

the limited findings, the organization is willing to take data from the other organizations for the better analysis. Some studies to evaluate the units in case of two or more groups have been done from the view point of group characteristic [1,5] and also to distinguish each characteristic from group characteristic [2,3]. They assumed that one can access data of all the groups without caring for the inter-group data usage. However, it is not straightforward for both organizations to give and take the original data in some real situations. The original data would include a kind of personal information, which can be used in each organization but often cannot be open to other organizations. This study undertakes such a problem as, for instance, the evaluation of the financial advisors (FA) in two banks. The information on personnel assessment is an example of data which is used only in the organization and cannot be easily accessed by the other organizations. The banks introduce the same e-learning system and a FA takes several courses on financial products and communication skills. The assessment of a FA is based on the following criteria: the amount of sales, the number of sales, and the number of new customers. In other words, this evaluation system has two kinds of inputs on courses and three kinds of outputs on sales, which are considered as cost and performance, respectively, and efficiency is denoted as cost per performance. Each bank accumulates data on two inputs and three outputs of its FAs so as to evaluate them relatively by comparing them each other. The findings are useful and reliable if there are plenty of FAs under the similar situation in each bank. Otherwise, the bank needs more data for fair evaluations. However, it is not easy for the bank to take data from another bank, since a bank does not often want to open its data including sales information to the competitor. The data on sales is the other example which can be used only in the organization. This study proposes one of the approaches to use multi-accumulated data with privacy concerns from the point of view of efficiency evaluation.

In efficiency literature, an efficient frontier is important and core. It represents a production function or a transformation function, and the efficiency of a unit is measured by the function. In parametric methods, a production function to represent the frontier of the production possibility set is derivative. It is straightforward and unambiguous. While, in non-parametric methods, it is no need to impose a specific function for frontier theoretically. It is an advantage not to require specification of function, although a change in a variable may cause the shift of frontier. In case of more than two organizations, each could have its inherent frontier by parametric or non-parametric method and evaluate the units in own organization based on it. Therefore, the evaluations of the units are peculiar to the organization. This study contributes to encourage an organization to take data from the other organizations. As a result, a unit in an organization has three kinds of efficiency: the efficiency in its own organization, the efficiency if it were in the other organization, and the efficiency if two organizations were integrated. The last two kinds of efficiency are easily obtained when an organization can access to data of the other organizations or a third party accumulate data of all organizations once, analyze it, and return the results to each organization. Otherwise, some considerations are needed to keep secret of the original individual data each other. One of the solutions is to replace the original data into a different form of data. From this viewpoint, the efficient frontier for each organization depending on its data is preferable to the specific global efficient frontier for all units in both organizations. Thus, this study focuses on non-parametric methods, especially Data Envelopment Analysis (DEA).

DEA is one of non-parametric techniques and evaluates the relative efficiency of Decision Making Units (DMUs) [4]. It focuses on the group of DMUs and constructs an efficient frontier for them. The efficient frontier consists of some DMUs in the group, and they are efficient in other words. Generally, the efficiency of each DMU in the group is obtained by comparing to the nearest efficient frontier, so that it is projected to the point on the frontier. Therefore, DEA gives optimistic evaluations to DMUs in a group. On the contrary, pessimistic evaluation is obtained by projecting a DMU to the point on the farthest efficient frontier. The efficiency interval is defined as an interval from the pessimistic evaluation to the optimistic evaluation in Interval DEA [6]. It represents the possible efficiency values of a DMU as an interval. The formulations to derive an efficiency interval are revisited from the viewpoint of a weight set.

This paper is organized as follows. DEA and Interval DEA is briefly explained in the next section. In Sect. 3, an efficient frontier, which consists of the efficient DMUs, is denoted as a set of the input and output weight vectors. The weight set is in replace of the original data in giving and taking information among the organizations. Then, three kinds of efficiency in own organization, another organization, and integrated organization are shown in Sect. 4. A unit may find something new, which is different from what was found in its organization alone, and it could be helpful to understand its evaluation and proper position among the similar units. In Sect. 5, the proposed method is illustrated through a numerical example and the results are compared to those when data are fully open. The last section concludes this paper.

2 Data Envelopment Analysis: DEA

In DEA, there are n DMUs, each of which produces n_o outputs from n_i inputs in a similar manner, for $i = 1, \ldots, n$. The given input and output vectors of DMU_i are denoted as $\boldsymbol{x}_i$ and $\boldsymbol{y}_i$, respectively, where $\boldsymbol{x}_i = (x_{i1}, \ldots, x_{in_i})^t$ and $\boldsymbol{y}_i = (y_{i1}, \ldots, y_{in_o})^t$. The efficiency value of DMU_j in the group is obtained as follows.

$$\theta_j^* = \max_{v_j, u_j} \frac{\boldsymbol{u}_j^t \boldsymbol{y}_j}{\boldsymbol{v}_j^t \boldsymbol{x}_j},$$

$$\text{s.t.} \ \frac{\boldsymbol{u}_j^t \boldsymbol{y}_i}{\boldsymbol{v}_j^t \boldsymbol{x}_i} \leq 1, \forall i \in N \tag{1}$$

$$\boldsymbol{u}_j, \boldsymbol{v}_j \geq \boldsymbol{0},$$

where $N = \{1, \ldots, n\}$ and the variables are the input and output weight vectors, $\boldsymbol{v}_j = (v_{j1}, \ldots, v_{jn_i})^t$ and $\boldsymbol{u}_j = (u_{j1}, \ldots, u_{jn_o})^t$, respectively. The efficiency of DMU_i is denoted as a ratio of weighted sum of outputs to that of inputs: $u_j^t y_i / v_j^t x_i$. The input and output weight vectors are unique to DMU_j. They are obtained from the optimistic viewpoint for DMU_j by maximizing its efficiency under the condition that those of all DMUs are less than 1. This problem is equivalent to the following LP problem.

$$\theta_j^* = \max_{v_j, u_j} u_j^t y_j,$$
$$\text{s.t. } v_j^t x_j = 1,$$
$$u_j^t y_i - v_j^t x_i \leq 0, \forall i \in N \tag{2}$$
$$u_j, v_j \geq 0,$$

where the denominator of the objective function of (1) is assumed to be 1. Then, the efficiency values of all DMUs, $\theta_j^*, \forall j \in N$, are obtained by solving n LP problems of (2).

We can find the best relative efficiency value of a DMU as the optimistic efficiency by (2), although we cannot find the worst one yet. Then, the other efficiency values could be obtained from the different viewpoints and such possible efficiency values could give us more detailed observation on a DMU. In Interval DEA [6], the efficiency value is redefined as follows.

$$\theta_j^* = \max_{v_j, u_j} \frac{\frac{u_j^t y_j}{v_j^t x_j}}{\max_{i \in N} \frac{u_j^t y_i}{v_j^t x_i}}, \tag{3}$$
$$\text{s.t. } u_j, v_j \geq 0,$$

where the relative efficiency of DMU_j comparing to all the other DMUs is maximized. Thus, it derives the optimistic evaluation for DMU_j in the same concept as that of (1). It is transformed into LP problem of (2), since it becomes equal to (1) by constraining the denominator is less than 1.

On the contrary, the pessimistic evaluations in Interval DEA is defined as follows.

$$\theta_{j*} = \min_{v_j, u_j} \frac{\frac{u_j^t y_j}{v_j^t x_j}}{\max_{i \in N} \frac{u_j^t y_i}{v_j^t x_i}}, \tag{4}$$
$$\text{s.t. } u_j, v_j \geq 0,$$

where the relative efficiency is minimized. The objective function value is obtained as

$$\theta_{j*} = \min_i \theta_j^i, \tag{5}$$

where

$$\theta_j^i = \min_{v_j, u_j} \frac{u_j^t y_j}{v_j^t x_j},$$
$$\text{s.t. } \frac{u_j^t y_i}{v_j^t x_i} = 1, \tag{6}$$
$$u_j, v_j \geq 0.$$

It is transformed into an LP problem in the similar way as transforming (1) to (2). The pessimistic efficiency θ_{j*} is obtained by solving at most n LP problems of (6), $\forall i \in N$. Thus, the efficiency interval of DMU_j is obtained by solving at most $(n + 1)$ LP problems: (2) for its upper bound and (6) for its lower bound. Such an efficiency interval is reasonable and acceptable for DMU_j, since it includes all the possible efficiency values comparing to all DMUs in the group.

Furthermore, DMU_j could be interested in referencing DMUs in other groups if there are two or more similar groups. The group cannot always access to the input and output vectors of DMUs in the other groups. Therefore, it needs what is in replace of the original input and output vectors and does not allow to estimate the original data. For this purpose, this study uses the input and output weight vectors.

3 Set of Weight Vectors on Efficient Frontier

We induce a set of the efficient DMUs, whose upper bounds of the efficiency intervals are 1, as follows.

$$K = \{k | \theta_k = 1 \text{ by } (2)\}. \tag{7}$$

The efficient frontier consists of DMUs in K and a part of their linear combinations. It is considered that any input and output vectors inside the efficient frontier is possible, so that the efficient frontier frames the production possibility set for the group of DMUs. The efficiency values of other DMUs are obtained by projecting to the efficient frontier.

Figure 1 shows two kinds of outputs of 6 DMUs, whose single inputs are equal. Let focus on 4 DMUs, 1 to 4, illustrated as □. They are in the same group, Group A, and the other 2 DMUs, 5 and 6 as ○, are considered later. The efficient frontier for this group consists of them: $K = \{2, 4\}$, as dotted line, since the upper bounds of efficiency intervals of DMU_2 and DMU_4 by (2) are 1. The upper bounds of efficiency intervals of DMUs inside the efficient frontier are less than 1.

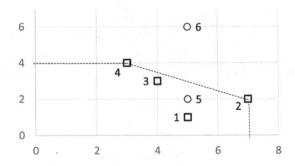

Fig. 1. Two outputs of 4 DMUs in Group A and two added DMUs

We focus on the efficient DMUs in K, since the efficient frontier consists of them and frames the group of DMUs. Therefore, the efficient frontier is strongly dependent on the input and output vectors of DMUs in the group. In other words, it can be a representative of their original input and output vectors. One of the approaches to denote the efficient frontier is to derive the interval marginal weights for referencing efficient DMUs [7]. They are denoted as interval in order to reflect all the possibilities. The interval marginal weights with the efficient DMUs in a group of DMUs are given to the other groups, if there is no restrictions to access to the original input and output vectors among the groups. In this study, instead of the marginal weights for the DMUs, we derive a set of the input and output weight vectors with which a DMU is efficient, in order to concern privacy of the original data. In other words, the relation of each input or output to efficiency value is considered, so that each DMU needs not to be disclosed. One of the input and output weight vectors for $DMU_k, k \in K$ is 1, is obtained as follows.

$$\begin{aligned}
\max_{v_k^t, u_k^t} \; & u_{kq}, \\
\text{s.t. } & \boldsymbol{u}_k^t \boldsymbol{y}_k = 1, \\
& \boldsymbol{v}_k^t \boldsymbol{x}_k = 1, \\
& \boldsymbol{u}_k^t \boldsymbol{y}_i - \boldsymbol{v}_k^t \boldsymbol{x}_i \le 0, \forall i \in N \\
& \boldsymbol{u}_k, \boldsymbol{v}_k \ge \boldsymbol{0},
\end{aligned} \tag{8}$$

where the weight of output q, u_{kq}, is focused on and maximized. Denote the optimal solutions as input and output weight vector $\overline{\boldsymbol{v}}_k^q$ and $\overline{\boldsymbol{u}}_k^q$.

Similarly, we minimize the weight of output q and denote the optimal solutions as $\underline{\boldsymbol{v}}_k^q$ and $\underline{\boldsymbol{u}}_k^q$. Moreover, the weights of the other outputs and those of the inputs are considered. Then, there are $2(n_i + n_o)$ kinds of input and output weight vectors for $DMU_k, k \in K$. In total, there are $2|K|(n_i + n_o)$, where $|K|$ is the cardinality of K, input and output weight vectors, and some of them are the same. By reducing redundancy, the set of the input and output weight vectors to denote the efficient frontier is denoted as follows.

$$W = \{(\boldsymbol{u}, \boldsymbol{v}) | (\boldsymbol{u}, \boldsymbol{v}) \in (\overline{\boldsymbol{u}}_k^r, \overline{\boldsymbol{v}}_k^r) \cup (\underline{\boldsymbol{u}}_k^r, \underline{\boldsymbol{v}}_k^r), \forall k \in K, \forall r \in N_i \cup N_o \}, \tag{9}$$

where $N_i = \{1, \ldots, n_i\}$, and $N_o = \{1, \ldots, n_o\}$. Each input and output weight vector in W determines a part the efficient frontier for the group of DMUs. The original input and output vectors of DMUs in the group are unpredictable from W.

Denote the cardinality of W as $|W|$. There are $|W|$ kinds of efficiency values for a DMU, since a ratio of the weighted outputs to that of input is obtained with each element of W. The efficiency interval of $DMU_{k'}$, whose input and output vectors are $\boldsymbol{x}_{k'}$ and $\boldsymbol{y}_{k'}$, respectively, is obtained as $[\underline{\theta}_{k'}, \overline{\theta}_{k'}]$, where

$$\underline{\theta}_{k'} = \min_{(u,v) \in W} \frac{\boldsymbol{u}^t \boldsymbol{y}_{k'}}{\boldsymbol{v}^t \boldsymbol{x}_{k'}}, \; \overline{\theta}_{k'} = \max_{(u,v) \in W} \frac{\boldsymbol{u}^t \boldsymbol{y}_{k'}}{\boldsymbol{v}^t \boldsymbol{x}_{k'}}. \tag{10}$$

When $DMU_{k'}$ is the group of DMUs: $k' \in N$, the upper bound of efficiency interval equals to the optimal objective function value of (2): $\overline{\theta}_{k'} = \theta_{k'}^*$. The lower bound equals to that of (5) with (6): $\underline{\theta}_{k'} = \theta_{k'*}$, since the efficiency value of a DMU, DMU_i or DMU_k, is constrained to be 1 in (6) or (8), respectively.

It is noted that the upper bound of efficiency interval $\overline{\theta}_{k'}$ is not always less than 1 when $DMU_{k'}$ is not in the group: $k' \notin N$.

As for 4 DMUs in Group A in Fig. 1, the set of output weight vectors is obtained by (9) as $W = \{(0.143, 0), (0.091, 0.182), (0, 0.25)\}$ corresponding to the efficient DMUs in $K = \{2, 4\}$. For instance, efficiency values of DMU_3 are 0.572, 0.91, and 0.75 with each output weight vector in W, respectively, so that its efficiency interval is obtained as $[\underline{\theta}_3, \overline{\theta}_3] = [0.572, 0.91]$ by (10). Its upper bound corresponds to $(0.091, 0.182)$, with which the upper bounds of efficiency intervals of both DMU_2 and DMU_4 are 1. It mentions that DMU_3 is projected to their linear combination from the optimistic viewpoint. While, its lower bound corresponds to $(0.143, 0)$, with which the upper bound of DMU_2 is 1. It is more pessimistic for DMU_3 to be compared to DMU_2 than DMU_4. It is apparent from Fig. 1, where DMU_3 is away from DMU_2 comparing to DMU_4.

4   Three Kinds of Efficiency Intervals Depending on Groups

Let assume two groups G1 and G2, which are consisting of n^1 and n^2 DMUs, respectively: $DMU_{k^1}, \forall k^1 \in \{1, \ldots, n^1\}$ and $DMU_{k^2}, \forall k^2 \in \{n^1+1, \ldots, n^1+n^2\}$. First, G1 evaluates a unit, DMU_{k^1}, with the self-accumulated input and output vectors and obtains its efficiency interval $[\underline{\theta}_{k^1}^1, \overline{\theta}_{k^1}^1]$ by (10) with weight set $W = W^{k1}$. This is the first kind of efficiency interval, which is based on its own group. It represents the possible efficiency values of DMU_{k^1} relatively to n^1 DMUs in G1 and the other group G2 is ignored.

Next, G1 tries to evaluate DMU_{k^1} referencing n^2 DMUs in G2, although G2 does not want to give G1 its DMUs' input and output vectors with privacy concerns. Instead, G2 may give G1 its set of the input and output weight vectors by (9), W^2, since it does not allow G1 to estimate each DMU's input and output vectors. It is useful enough for G1 to reference G2 in the evaluations of its DMUs. The weight set determines the efficient frontier for G2, so that it can be used as a representative of G2. Then, the second kind of efficiency interval of DMU_{k^1} is obtained as $[\underline{\theta}_{k^1}^2, \overline{\theta}_{k^1}^2]$ by (10) with $W = W^2$. It is based on G2. We compare two kinds of efficiency intervals. In case of $\overline{\theta}_{k^1}^2 < \overline{\theta}_{k^1}^1$, there is a DMU in G2 which has similar trend to DMU_{k^1} and superior to DMU_{k^1}, so that DMU_{k^1} is unique only in G1. In case of $\underline{\theta}_{k^1}^2 < \underline{\theta}_{k^1}^1$, DMU_{k^1} is more similar to DMUs in G2 than to those in G1. While, in case of $\underline{\theta}_{k^1}^2 > \underline{\theta}_{k^1}^1$, there is an efficient DMU in G2 which is different from DMU_{k^1}. In this way, the efficiency interval in another group could give a new evaluation to a DMU.

Last, assume the integrated group G consisting of G1 and G2 from a wide viewpoint. The third kind of efficiency interval of DMU_{k^1} is obtained by referencing $(n_1 + n_2)$ DMUs as $[\underline{\theta}_{k^1}, \overline{\theta}_{k^1}]$, where

$$\underline{\theta}_{k^1} = \min\{\underline{\theta}_{k^1}^1, \underline{\theta}_{k^1}^2\}, \quad \overline{\theta}_{k^1} = \min\{\overline{\theta}_{k^1}^1, \overline{\theta}_{k^1}^2\}.$$

The both bounds are the minimums of those in G1 and G2, because the efficiency interval becomes smaller when a DMU is added into the group.

$$[\underline{\theta}_{k'}^{+}, \overline{\theta}_{k'}^{+}] \leq [\underline{\theta}_{k'}, \overline{\theta}_{k'}] \leftrightarrow \overline{\theta}_{k'}^{+} \leq \overline{\theta}_{k'}, \ \underline{\theta}_{k}^{+} \leq \underline{\theta}_{k'}, \tag{11}$$

where $[\underline{\theta}_{k'}, \overline{\theta}_{k'}]$ is with $N = \{1, \ldots, n\}$ and $[\underline{\theta}_{k'}^{+}, \overline{\theta}_{k'}^{+}]$ is with $N^{+} = \{1, \ldots, n, (n+1)\}$. This is apparent by the denominators of the objective functions of (3) and (4). The denominator with the added DMU is equal or larger than that without it. It is also explained with Fig. 1 by assuming 4 DMUs: $N = \{1, \cdots, 4\}$. The dotted line illustrates the efficient frontier for them. It is stable, when a DMU inside the dotted line such as DMU_5 is added. Its efficiency interval obtained by projecting to the dotted line is [0.5, 0.819], whose bounds equal to the optimal objective function values of (2) and (5) with $N = \{1, \cdots, 4, 5\}$. While, when the added DMU is outside of the dotted line such as DMU_6, its efficiency interval by projecting to the dotted line is [0.714, 1.547], whose upper bound is more than 1. It implies that DMU_6 is surely efficient relatively to 4 DMUs in $N = \{1, \cdots, 4\}$. When we evaluate DMU_4 by adding DMU_6: $N^{+} = \{1, \cdots, 4, 6\}$ again, it is apparent that DMU_4 is not on efficient frontier anymore. Its efficiency interval [0.429, 1] with N is renewed into efficiency interval [0.429, 0.667] with N^{+}, so that its upper bound is reduced from 1 to 0.667 because of the more efficient DMU_6.

In this way, three kinds of efficiency intervals for a DMU are obtained. They are based on its own group, another group, and the integrated group. Moreover, the both bound of the efficiency interval in the integrated group equal to the minimums of those in own group and in another group.

5   Numerical Example

There are two groups of DMUs, Group A and Group B, with one input and two outputs shown in Table 1. The bottom row, DMU_9, shows the average of Group B. They are illustrated in Fig. 2, where $\square$, $\bullet$, and $\blacktriangle$ correspond to Group A, Group B, and the average of Group B. It is noted that DMU_4 and DMU_5 are the same. The goal is to evaluate 4 DMUs in Group A referencing 4 DMUs in Group B as well, under the condition that Group A cannot access to the input and output vectors of Group B.

At first, Group A finds the efficient DMUs by (7) as $K^{A} = \{2, 4\}$. The efficient frontier for Group A consists of them and is illustrated as dotted line in Fig. 2. After that, it derives the set of the input output weight vectors for the efficient frontier by (9) as $W^{A} = \{(0.143, 0), (0.091, 0.182), (0, 0.25)\}$, and obtains the efficiency intervals with the input and output weight vector in W^{A} as follows.

$$\begin{aligned} &[\underline{\theta}_{1}^{A}, \overline{\theta}_{1}^{A}] = [0.25, 0.714], \ [\underline{\theta}_{2}^{A}, \overline{\theta}_{2}^{A}] = [0.5, 1], \\ &[\underline{\theta}_{3}^{A}, \overline{\theta}_{3}^{A}] = [0.571, 0.909], \ [\underline{\theta}_{4}^{A}, \overline{\theta}_{4}^{A}] = [0.429, 1]. \end{aligned} \tag{12}$$

Next, Group A wants to reference 4 DMUs in Group B, although Group B does not give their original input and output with privacy concerns. Instead, Group A asks Group B for the information on its efficient frontier, which is a set of the input

Table 1. Input and output data and efficiency interval

Group	DMU	Input	Output		Efficiency interval in Group AB
{A, B}	i	x_{i1}	y_{i1}	y_{i2}	
A	1	1	5	1	[0.167, 0.714]
A	2	1	7	2	[0.333, 1]
A	3	1	4	3	[0.5, 0.7]
A	4	1	3	4	[0.429, 0.75]
B	5	1	3	4	[0.429, 0.75]
B	6	1	6	4	[0.667, 1]
B	7	1	5	5	[0.714, 1]
B	8	1	2	6	[0.286, 1]
A$^+$	9	1	4	4.75	-

and output weight vectors. Group A or the others cannot estimate each input and output vector from the weight set. For this request, Group B finds the efficient DMUs by (7) as $K^B = \{6, 7, 8\}$, and derives the set input output weight vectors by (9) as $W^B = \{(0.167, 0), (0.1, 0.1), (0, 0.167), (0.05, 0.15)\}$. When Group A takes W^B from Group B, it obtains the efficiency intervals by (10) with $W = W^B$ as follows.

$$[\underline{\theta}_1^B, \overline{\theta}_1^B] = [0.167, 0.833], \quad [\underline{\theta}_2^B, \overline{\theta}_2^B] = [0.333, 1.667],$$
$$[\underline{\theta}_3^B, \overline{\theta}_3^B] = [0.5, 0.7], \qquad [\underline{\theta}_4^B, \overline{\theta}_4^B] = [0.5, 0.75]. \tag{13}$$

The efficiency intervals in (13) are compared to those in (12). As for one of the efficient DMUs, DMU_4, its upper bound and lower bound of efficiency interval are decreased and increased, respectively, by replacing Group A into Group B. The upper bound of efficiency interval of DMU_4 evaluated in Group B is less than that in Group A: $\overline{\theta}_4^A > \overline{\theta}_4^B$. It mentions that DMU_4 could be evaluated better in its group, Group A, than the other group, Group B. In other words, there were a DMU in Group B which could be better than DMU_4. Moreover, its lower bound in Group B is more than that in Group A: $\underline{\theta}_4^A < \underline{\theta}_4^B$, so that DMU_4 would be more similar to DMUs in Group B than those in Group A. These facts are synthesized to Fig. 2, where the point on the linear combination of DMU_7 and DMU_8 dominates DMU_4, and DMU_4's □ looks closer to •s than to the other □s. While, another efficient DMU in Group A, DMU_2, is dissimilar to DMUs in Group B, so that its lower bound of efficiency interval in Group A is more than that in Group B: $\underline{\theta}_2^A > \underline{\theta}_2^B$. We find in Fig. 2 that DMU_8 in Group B is far away from DMU_2. Its upper bound $\overline{\theta}_2^B$ is more than 1, so that if DMU_2 were in Group B, it would have been on the efficient frontier and been unique. As for DMU_3, its upper bound and lower bound of efficiency interval are decreased by replacing Group A into Group B. It mentions that there were a DMU in Group B which could be better than DMU_3 but DMU_3 was not very similar to DMUs in Group B. The optimistic efficiency which is the upper bound of the efficiency interval, In this way, the efficiency interval could give more detailed observation than the optimistic efficiency value which is its upper bound and often focused on.

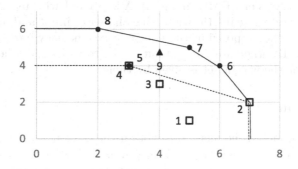

Fig. 2. Two outputs of DMUs in Group A and Group B

We cannot estimate the original input and output vectors of DMUs in Group B from their average, DMU_9. Therefore, let assume that Group A takes the input and output vector of DMU_9, as a representative of Group B, instead of the efficient frontier information W^B. There are two ways for Group A to use DMU_9. One is to obtain its efficiency interval with W^A as $[\underline{\theta}_9^A, \overline{\theta}_9^A] = [0.571, 1.227]$, whose upper bound is more than 1. This implies that Group B is superior to Group A, although DMU_4 in Group A is equals to DMU_5 in Group B. It could not be possible to compare each DMU in Group A to Group B. The other way is to recalculate the efficiency intervals of 4 DMUs in Group A referencing Group A^+ of $DMU_k, \forall k \in \{1, 2, 3, 4, 9\}$. They are obtained as follows.

$$[\underline{\theta}_1^{A^+}, \overline{\theta}_1^{A^+}] = [0.211, 0.714], \quad [\underline{\theta}_2^{A^+}, \overline{\theta}_2^{A^+}] = [0.421, 1],$$
$$[\underline{\theta}_3^{A^+}, \overline{\theta}_3^{A^+}] = [0.571, 0.792], \quad [\underline{\theta}_4^{A^+}, \overline{\theta}_4^{A^+}] = [0.429, 0.842]. \tag{14}$$

They are smaller than those in (12) because of the added DMU_9 as in (11). It is difficult to distinguish the influence by DMU_9 of Group B, since the efficiency intervals in (14) are also influenced by Group A. We find that DMU_2 is still efficient, but cannot find whether DMU_2 is unique in Group B or not.

Last, the efficiency intervals of 4 DMUs in Group A referencing the integrated group, Group AB of $DMU_k, k \in \{1, \ldots, 8\}$, are obtained as the minimums of those in (12) and (14) as follows.

$$[\underline{\theta}_1^{AB}, \overline{\theta}_1^{AB}] = [0.167, 0.714], \quad [\underline{\theta}_2^{AB}, \overline{\theta}_2^{AB}] = [0.333, 1],$$
$$[\underline{\theta}_3^{AB}, \overline{\theta}_3^{AB}] = [0.5, 0.7], \quad\quad [\underline{\theta}_4^{AB}, \overline{\theta}_4^{AB}] = [0.429, 0.75]. \tag{15}$$

It should be noted that they are obtained without using the original input and output vectors of DMUs in Group B. This is an advantage of the proposed method with privacy concerns. For a comparison, the right column of Table 1 shows the efficiency intervals of 8 DMUs in Group AB by (10) with W from all DMUs, which are equal to those by (2) and (5) with (6). The efficient frontier for Group AB is illustrated as rigid line in Fig. 2. There are 4 efficient DMUs. The efficiency intervals of 4 DMUs in Group A in Table 1 with using the original data of Group B and those in (15) with using the set of input and output weight vectors are equal. The proposed method derives the efficiency interval of a DMU in a group when the original input and output vectors of some other DMUs are not accessible but the weight set is given by them.

6 Conclusion

This paper has proposed the method to obtain the efficiency interval of a DMU by referencing DMUs in another group as well as DMUs in its own group. It is applicable when the original input and output vectors of a group is not easily accessible from the other groups with privacy concerns. Such a condition as handling a kind of personal data often happens in real situations. For the efficiency analysis with this kind of inter-group data usage, the efficient frontier, instead of the original data, has been considered. We denoted the efficient frontier for a group as the set of input and output weight vectors, with which the efficiency value of at least one DMU reaches 1 and those of the other DMUs are less than 1. The weight vector set corresponds to the efficient DMUs so that the first step is to find them. Then, the efficiency interval of a DMU, which equals to that by Interval DEA, is obtained. Its upper and lower are the minimum and maximum of efficiency values with the elements in the weight set. On one hand, the weight set can be a representative of a group, since the efficient frontier is strongly depend on the original data. On the other hand, it does not allow us to estimate the original data of each DMU, so that it is easier for a group to give or take it than the original data. Hence, there are three kinds of efficiency intervals for a DMU based on its own group, another group, and the integrated group. Comparing the efficiency intervals from different viewpoints can give us a rich and useful information on the evaluation of a DMU. The advantages of the proposed method for multi-accumulated data are to refrain from giving and taking the original data and to derive an efficiency evaluation based on another group.

Acknowledgment. This work was partially supported by JSPS KAKENHI Grant Number JP16K01251.

References

1. Agrell, P.J., Steuer, R.E.: ACADEA-decision support system for faculty performance reviews. J. Multi-criteria Decis. Anal. **9**, 191–204 (2000)
2. Aparicio, J., Crespo-Cebada, E., Pedraja-Chaparro, F., Santin, D.: Comparing school ownership performance using a pseudo-panel database: a Malmquist-type index approach. Eur. J. Oper. Res. **256**, 533–542 (2017)

3. Camanho, A.S., Dyson, R.G.: Data envelopment analysis and Malmquist indices for measuring group performance. J. Prod. Anal. **26**, 35–49 (2006)
4. Charnes, A., Cooper, W.W., Rhodes, E.: Measuring the efficiency of decision making units. Eur. J. Oper. Res. **2**(6), 429–444 (1978)
5. Cook, W.D., Ruiz, J., Sirvent, I., Zhu, J.: Within-group common benchmarking using DEA. Eur. J. Oper. Res. **256**, 901–910 (2017)
6. Entani, T., Maeda, Y., Tanaka, H.: Dual models of Interval DEA and its extension to interval data. Eur. J. Oper. Res. **136**(1), 32–45 (2002)
7. Ouellette, P., Vigeant, S.: From partial derivatives of DEA frontiers to marginal products, marginal rates of substitution, and returns to scale. Eur. J. Oper. Res. **253**, 880–887 (2016)

An Evidence Theoretic Approach to Interval Analytic Hierarchy Process

Masahiro Inuiguchi$^{(\boxtimes)}$

Graduate School of Engineering Science, Osaka University,
Toyonaka, Osaka 560-8531, Japan
inuiguti@sys.es.osaka-u.ac.jp
http://www-inulab.sys.es.osaka-u.ac.jp/?page=en-inuiguchi

Abstract. The interval Analytic Hierarchy Process (AHP) has been proposed to express vague evaluations of the decision maker as a normalized interval weight vector. In this paper, we apply the evidence theory to the representation of the vague evaluations as an alternative way. Accordingly, in the proposed approach, a basic probability assignment (BPA) is used for representing vague priority weights instead of a normalized interval weight vector. We formulate the problem estimating a BPA from a given pairwise comparison matrix as a linear programming problem. We investigate the relation between the proposed approach and the interval AHP. We show that the formulated BPA estimation problem is equivalent to the problem of estimating a normalized interval weight vector with an additional constraint.

Keywords: Interval priority weight · Dempster-Shafer theory
Linear programming · Dominance relation · Robust evaluation

1 Introduction

Analytic Hierarchy Process (AHP) [1] is one of the most useful tools to evaluate the alternatives under multiple criteria. It evaluates the priority weights of alternatives and criteria from a pairwise comparison matrix (PCM) in multiple criteria decision problems. The PCM is given by a decision maker and its (i, j) component shows the relative importance of the i-th alternative/criterion to the j-th one. From a given PCM, the priority weights are estimated by the maximum eigenvalue method [1] or by the geometric mean method [2]. However, a given PCM is seldom consistent because of the vagueness of human judgements. The degree of inconsistency is measured by the consistency index and the estimated priority weights are adopted for decision analysis when the consistency index is in the allowable range.

From a different viewpoint that the inconsistency of a PCM comes from the vague human evaluation, the interval AHP [3] was proposed to estimate a normalized interval weight vector from a given PCM and support the decision

This work was partially supported by JSPS KAKENHI Grant Number 17K18952.

V.-N. Huynh et al. (Eds.): IUKM 2018, LNAI 10758, pp. 60–71, 2018.
https://doi.org/10.1007/978-3-319-75429-1_6

making in consideration of vagueness of the evaluation. However, it is shown that the estimated normalized interval weight vector does not reflect the vagueness well (see [4]). Several alternative interval weight estimation methods [5] have been proposed. On the other hand, it is shown that the normalized interval weight vector does not always exist for any given interval weight vector and that the normalized conditions do not specify a unique interval weight vector even when there exists a normalized interval weight vector for the given interval weight vector (see [6]).

In this paper, we apply evidence theory [7], also called Dempster-Shafer theory, to the representation of the vague evaluation of the decision maker. The interval weights can be seen as a model of imprecise probability distribution. Therefore, it is conceivable to apply the evidence theory to estimating imprecise weights from a given PCM because the evidence theory is also a model of imprecise probability. In the evidence theory, imprecise probabilities are expressed by a basic probability assignment (BPA). Accordingly, a BPA is used for representing vague priority weights in the proposed approach.

Although the evidence theory has been applied to multiple criteria decision making problems, the way of the application is different. In the previous applications [8], a BPA is utilized to express the imprecise evaluation of an alternative in view of each specified criterion and combination rules of BPAs are used to obtain the comprehensive evaluation of the alternative. In the proposed approach, we use a BPA for representing priority weights and estimate a BPA over criteria from a given PCM. The proposed approach can be used for the estimation of a BPA over alternatives on each criterion as the conventional AHP does. However, in order to avoid the a labyrinthine argument not related to the main result of this paper, we concentrate on the estimation of a BPA over criteria from a given PCM. The belief and plausibility functions are defined from a BPA. Those functions correspond to the lower and upper bounds of interval weight vectors, respectively. The application of the evidence theory has the advantage that the normalization condition of a BPA is well defined. By the comparison between the interval AHP and the AHP based on the evidential theory, we may have some new perspective on the normalization condition in the interval AHP.

We formulate the estimation problem of a BPA from a given PCM as a linear programming (LP) problem in parallel to the estimation problem of interval weights from a give PCM. Because there are many optimal solutions to the LP problem, we consider the set of optimal solutions, i.e., a set of BPAs. Using the sets of BPAs showing weights of criteria, we describe the way to make a robust estimation of dominance relation between alternatives when their scores in each criterion are given. The dominance relation is estimated by solving several LP problems.

We investigate the relations between the interval AHP approach and the evidence theoretic approach to AHP. We show that the BPA estimation problem is equivalent to the problem of estimating a normalized interval weight vector with an additional constraint. Moreover, we show that this equivalence holds also for the robust estimation of the dominance relation between alternatives.

As the result, it is shown that the evidence theoretic approach to AHP differs from the interval AHP in virtue of an additional condition.

This paper is organized as follows. In next section, we briefly review the interval AHP. In Sect. 3, we formulate a new interval weight estimation method which is more appropriate in the treatment of the normalization condition. In Sect. 4, we formulate an estimation method of a BPA under a crisp PCM. In Sect. 5, we investigate the relations between interval AHP and the proposed evidence theoretic approach.

2   The Interval AHP

2.1   The Conventional Interval Weight Estimation Method

We introduce the interval AHP with PCMs for multiple criteria decision problem with n criteria, $f_1, f_2, \ldots, f_n$. For convenience, we define $N = \{1, 2, \ldots, n\}$, $K = \{1, 2, \ldots, k\}$ and $N \setminus j = N \setminus \{j\} = \{1, 2, \ldots, j-1, j+1, \ldots, n\}$ for $j \in N$.

In AHP, the decision problem is structured hierarchically as criteria, subcriteria and alternatives. At each node except leaf nodes of the hierarchical tree, a weight vector $\boldsymbol{w}$ for criteria or alternatives is obtained from a PCM A. In this paper, we mainly discuss about the estimation of a weight vector $\boldsymbol{w}$ from a given PCM A. The (i, j) component a_{ij} of A shows the relative importance of the i-th criterion/alternative to the j-th one. Theoretically, we assume $a_{ij} = w_i/w_j$, $i, j \in N$. From this meaning, we assume the reciprocity of A, i.e., $a_{ij} = 1/a_{ji}$, $i \neq j$, $i, j \in N$. If a_{ij}, $i, j \in N$ are obtained exactly, the strong transitivity $a_{ij} = a_{il}a_{lj}$, $i, j, l \in N$ is satisfied. However, human evaluation is not very accurate so that strong transitivity is usually unsatisfied. In the conventional AHP [1], a_{ij}, $i, j \in N$, $i \neq j$ are assumed to be approximations of w_i/w_j. Then, $\boldsymbol{w}$ is estimated by minimizing the errors in a_{ij}, $i, j \in N$, $i \neq j$. The eigenvector method [1] and geometric mean method [2] are frequently used for the estimation of $\boldsymbol{w}$ in AHP. The consistency index of the given pairwise comparison matrix A is defined by $C.I. = (\lambda_{\max} - n)/(n - 1)$, where $\lambda_{\max}$ is the maximal eigenvalue of A. If $C.I.$ is not greater than 0.1 or 0.15, it is often considered that the obtained vector $\boldsymbol{w}$ is acceptable.

On the other hand, in the interval AHP, it is assumed that decision maker's evaluation is intrinsically vague so that it cannot be expressed well by a unique priority weight vector $\boldsymbol{w}$. From this point of view, an interval weight vector $\boldsymbol{W} = (W_1, W_2, \ldots, W_n)^{\mathrm{T}}$ instead of a crisp weight vector $\boldsymbol{w}$ is estimated from the given PCM A, where $W_i = [w_i^{\mathrm{L}}, w_i^{\mathrm{R}}]$, $i \in N$ and $w_i^{\mathrm{L}} \leq w_i^{\mathrm{R}}$, $i \in N$. To fit in the given PCM A, the interval priority weight vector $\boldsymbol{W}$ is required to satisfy

$$\frac{w_i^{\mathrm{L}}}{w_j^{\mathrm{R}}} \leq a_{ij} \leq \frac{w_i^{\mathrm{R}}}{w_j^{\mathrm{L}}}, \; i, j \in N, \; i < j. \tag{1}$$

We note that by the reciprocity, $a_{ij} = 1/a_{ji}$, $i, j \in N$, we only consider $i, j \in N$ such that $i < j$. Corresponding to the normalization condition of $\boldsymbol{w}$ in the conventional AHP, the interval priority weight vector $\boldsymbol{W}$ is required to satisfy

$$\sum_{j \in N \setminus i} w_j^{\mathrm{R}} + w_i^{\mathrm{L}} \geq 1, \ i \in N, \tag{2}$$

$$\sum_{j \in N \setminus i} w_j^{\mathrm{L}} + w_i^{\mathrm{R}} \leq 1, \ i \in N. \tag{3}$$

(2) and (3) ensure that, for any $w_i^{\circ} \in W_i$, there exist $w_j \in W_j$, $j \in N \setminus i$ such that $\sum_{j \in N \setminus i} w_j + w_i^{\circ} = 1$ (see [9]). Namely, any values in W_i, $i \in N$ are meaningful (there is no ineffective subarea in $\boldsymbol{W}$).

Then, an interval weight vector $\boldsymbol{W}$ is estimated by solving the following linear programming (LP) problem:

$$\text{minimize} \ \ d(\boldsymbol{W}) = \sum_{i \in N} (w_i^{\mathrm{R}} - w_i^{\mathrm{L}}), \tag{4}$$

$$\text{subject to (1), (2), (3),} \ \epsilon \leq w_i^{\mathrm{L}} \leq w_i^{\mathrm{R}}, \ i \in N, \tag{5}$$

where ϵ is a very small positive number. Let $\boldsymbol{W}^{\mathrm{conv}}$ be an optimal solution to this LP problem.

Once an interval weight vectors are obtained, we can define a dominance relation between alternatives. We assume a simple hierarchical tree having criteria $f_1, f_2, \ldots, f_n$ and alternatives $o_1, o_2, \ldots, o_k$, only. When a crisp weight of a criterion f_i is given as $w_i \geq 0$ ($\sum_{i \in N} w_i = 1$), the comprehensive evaluation $s(o_p)$ of alternative o_p, $p \in K$ is obtained by $s(o_p) = \sum_{i \in N} w_i f_i(o_p)$, where $f_i(o_p)$ represents a score of alternative o_p in view of criterion f_i. Then the dominance relation between alternatives o_p and o_q is obtained by $o_p \succeq o_q$ (o_p is not worse than o_q) if and only if $s(o_p) \geq s(o_q)$.

In the current setting, we have only an interval weight vectors. We consider the worst case so that we obtain a robust dominance relation between alternatives o_p and o_q. Namely, we define $o_p \succeq^{\mathrm{conv}} o_q$ (o_p is certainly not worse than o_q) by

$$o_p \succeq^{\mathrm{conv}} o_q \Leftrightarrow \forall \boldsymbol{w} \in \boldsymbol{W}^{\mathrm{conv}} \text{ such that } \mathbf{e}^{\mathrm{T}} \boldsymbol{w} = 1; \ \sum_{i \in N} w_i (f_i(o_p) - f_i(o_q)) \geq 0,$$

$$\tag{6}$$

where $\mathbf{e}^{\mathrm{T}} = (1, 1, \ldots, 1) \in \mathbf{R}^n$. We can easily confirm that dominance relation $\succeq^{\mathrm{conv}}$ is reflexive and transitive. Thus, $\succeq^{\mathrm{conv}}$ is a preorder.

2.2 The Interval Estimation Method Treated in This Paper

It is shown that the interval weight vector $\boldsymbol{W}^{\mathrm{conv}}$ estimated by solving LP problem (4)–(5) does not reflect the vagueness of the decision maker's evaluation well. The authors [4,5] proposed several approaches improving the quality of the estimation. Moreover, the normalization condition (2) and (3) of interval vectors is controversial (see [6]). For example, consider a normalized interval vector $\boldsymbol{W}$ satisfying

$$\sum_{j \in N \setminus i} w_j^{\mathrm{R}} + w_i^{\mathrm{L}} > 1, \ i \in N \quad \text{or} \quad \sum_{j \in N \setminus i} w_j^{\mathrm{L}} + w_i^{\mathrm{R}} < 1, \ i \in N, \tag{7}$$

where the i-th component W_i of $\boldsymbol{W}$ is represented as $W_i = [w_i^{\mathrm{L}}, w_i^{\mathrm{R}}]$, $i \in N$. In this case, there exist many $\bar{\boldsymbol{W}} = (\bar{W}_1, \bar{W}_2, \ldots, \bar{W}_n)^{\mathrm{T}}$ satisfying the normalization condition and

$$\frac{\bar{w}_i^{\mathrm{L}}}{\bar{w}_j^{\mathrm{R}}} = \frac{w_i^{\mathrm{L}}}{w_j^{\mathrm{R}}}, \ \frac{\bar{w}_i^{\mathrm{R}}}{\bar{w}_j^{\mathrm{L}}} = \frac{w_i^{\mathrm{R}}}{w_j^{\mathrm{L}}}, \ \forall i, j \in N, \ i \neq j, \tag{8}$$

where $\bar{W}_i = [\bar{w}_i^{\mathrm{L}}, \bar{w}_i^{\mathrm{R}}]$, $i \in N$. Namely, even if the ratios between interval weights W_i and W_j for all $i, j \in N$ such that $i \neq j$ are given, the normalized interval weight vector satisfying the given ratios is not unique (see [6]).

From the viewpoint of the non-uniqueness, we consider the interval weight vector $\boldsymbol{W}$ obtained by solving the following optimization problem in this paper:

$$\text{minimize} \ \ \delta(\boldsymbol{W}) = \frac{\displaystyle\sum_{i \in N} \sum_{j=i+1}^{n} (\delta_{ij} + \delta_{ji})}{\displaystyle\sum_{i \in N} (w_i^{\mathrm{L}} + w_i^{\mathrm{R}})}, \tag{9}$$

$$\text{subject to} \ \sqrt{a_{ji}} w_i^{\mathrm{L}} + \delta_{ij} = \sqrt{a_{ij}} w_j^{\mathrm{R}}, \ \delta_{ij} \geq 0, \ i, j \in N, \ i \neq j,$$
$$(2), \ (3), \ \epsilon \leq w_i^{\mathrm{L}} \leq w_i^{\mathrm{R}}, \ i \in N, \tag{10}$$

where we note that the reciprocity $a_{ij} = 1/a_{ji}$ and thus, the first constraint corresponds to (1). Indeed, (1) is equivalent to $\sqrt{a_{ji}} w_i^{\mathrm{L}} \leq \sqrt{a_{ij}} w_j^{\mathrm{R}}$, $i, j \in N$, $i \neq j$. Introducing slack variables $\delta_{ij} \geq 0$, we obtain the first constraint. In this constraint, w_i^{L} and w_i^{R} are treated symmetrically. We still impose the normalization condition expressed by (2) and (3). The numerator of objective function $\delta(\boldsymbol{W})$ represents the sum of deviations between a_{ij} and $w_i^{\mathrm{L}}/w_j^{\mathrm{R}}$ (or equivalently, deviations a_{ij} and $w_i^{\mathrm{R}}/w_j^{\mathrm{L}}$), $i, j \in N$, $i \neq j$, while the denominator is twice the sum of center values $\frac{1}{2}(w_i^{\mathrm{L}} + w_i^{\mathrm{R}})$ of interval weights. Namely, because 2 is a constant, the objective function means the ratio of the sum of the deviations to the sum of center values. Considering the non-uniqueness of the normalized interval weight vector, such a ratio is adopted for the objective function.

Because Problem (9)–(10) is a linear fractional programming (LFP) problem, it can be reduced to an LP problem (see [10]). There may exist many optimal solutions to LFP Problem (9)–(10). Then, consider all optimal solutions to make a robust decision analysis. Let $\mathcal{W}^{\mathrm{int}}$ be the set of optimal solutions of LFP problem (9)–(10).

The dominance relation between alternatives o_p and o_q associated with interval weights estimated by LFP problem (9)–(10) is defined by

$$o_p \succeq^{\mathrm{int}} o_q \Leftrightarrow \forall \boldsymbol{W} \in \mathcal{W}^{\mathrm{int}}, \ \forall \boldsymbol{w} \in \boldsymbol{W} \text{ such that } \mathbf{e}^{\mathrm{T}} \boldsymbol{w} = 1,$$
$$\sum_{i \in N} w_i (f_i(o_p) - f_i(o_q)) \geq 0, \tag{11}$$

Let $\hat{\delta}$ be the optimal values of LFP problem (9)–(10). Then the right-hand side of (11) can be confirmed by solving the following LP problem:

$$\text{minimize } dif(o_p, o_q) = \sum_{i \in N} w_i(f_i(o_p) - f_i(o_q)), \tag{12}$$

$$\text{subject to } \sqrt{a_{ji}}w_i^L + \delta_{ij} = \sqrt{a_{ij}}w_j^R, \; \delta_{ij} \geq 0, \; i,j \in N, \; i \neq j,$$
$$(2) \; (3), \; \epsilon \leq w_i^L \leq w_i^R, \; i \in N, \tag{13}$$
$$\sum_{i \in N} \sum_{j=i+1}^{n} (\delta_{ij} + \delta_{ji}) = \hat{\delta} \sum_{i \in N} (w_i^L + w_i^R).$$

Let $dif^*(p, q)$ be the optimal value of LP problem (13). If $dif^*(p, q) \geq 0$, $o_p \succeq^{\text{int}} o_q$ holds, and otherwise, $o_p \succeq^{\text{int}} o_q$ does not hold. We note that $\succeq^{\text{int}}$ is a preorder.

3 Evidence Theoretic Approach to the Interval AHP

In the evidence theory [7], probability masses are assigned not only to elements, i.e., singletons but also to nonempty subset of the universal set Ω (called frame of discernment). Namely, we consider a basic probability assignment (BPA) $m :$ $2^\Omega \to [0, 1]$ satisfying

$$m(\emptyset) = 0, \quad \sum_{B \subseteq 2^\Omega} m(B) = 1. \tag{14}$$

The mass $m(B)$ is understood as a semi-mobile probability mass (see [11]), which implies that we do not know precise probability mass assignment $m_B(\omega) \geq 0$ to each element $\omega \in B$ but $\sum_{\omega \in B} m_B(\omega) = m(B)$. We assume Ω is a finite set, i.e., $|\Omega| < +\infty$. When $\Omega = \{\omega_1, \omega_2, \ldots, \omega_n\}$, the set of all conceivable probability distributions $\boldsymbol{p} = (p_1, p_2, \ldots p_n)^\mathrm{T}$ under a given BPA m is obtained as

$$\mathcal{P}(m) = \left\{ (p_1, p_2, \ldots p_n)^\mathrm{T} \; \middle| \; p_i = \sum_{B : \omega_i \in B} m_B(\omega_i), \right.$$
$$\left. \sum_{i : \omega_i \in B} m_B(\omega_i) = m(B), \; m_B(\omega_i) \geq 0, \; i \in N, \; B \in 2^\Omega \right\}. \tag{15}$$

Using a BPA m, we calculate the belief $Bel : 2^\Omega \to [0, 1]$ and the plausibility $Pl : 2^\Omega \to [0, 1]$ by

$$Bel(B) = \sum_{C \subseteq B} m(C), \quad Pl(B) = \sum_{C : C \cap B \neq \emptyset} m(C). \tag{16}$$

$Bel(B)$ and $Pl(B)$ can be seen as the lower and upper probability of B, respectively.

We assume that the decision maker's evaluation given by a PCM A can be expressed by a BPA $m : 2^\Omega \to [0, 1]$, where $\Omega = \{\omega_1, \omega_2, \ldots, \omega_n\}$. From a given

PCM A, the BPA m can be estimated by solving the following LFP problem:

$$\text{minimize } \delta'(m) = \frac{\sum_{i \in N} \sum_{j=i+1}^{n} (\delta_{ij} + \delta_{ji})}{\sum_{i \in N} (Bel(\{\omega_i\}) + Pl(\{\omega_i\}))}, \tag{17}$$

$$\text{subject to } \sqrt{a_{ji}} Bel(\{\omega_i\}) + \delta_{ij} = \sqrt{a_{ij}} Pl(\{\omega_j\}),$$
$$\delta_{ij} \geq 0, \ i,j \in N, \ i \neq j, \ m(\emptyset) = 0, \ m(\{\omega_i\}) \geq \epsilon, \ i \in N, \tag{18}$$
$$\sum_{B \subseteq \Omega} m(B) = 1, m(B) \geq 0, \ B \subseteq \Omega,$$

where $Bel(\{\omega_i\}) = m(\{\omega_i\})$, $i \in N$ and $Pl(\{\omega_j\}) = \sum_{B \subseteq \Omega : \omega_j \in B} m(B)$, $j \in N$. LFP problem (17)–(18) corresponds to LFP problem (9)–(10) as $Bel(\{\omega_i\})$ and $Pl(\{\omega_i\})$ correspond to w_i^{L} and w_i^{R}, respectively. Moreover, LFP problem (17)–(18) can be reduced to an LP problem by a well-known reduction method of the linear fractional programming [10].

FLP problem (17)–(18) has a lot of variables $m(B)$, $B \subseteq \Omega$ and δ_{ij}, $i,j \in N$, $i \neq j$ but the constraint corresponding to the normalization condition is naturally introduced. Similar to LFP problem (9)–(10), LFP problem (17)–(18) may have multiple optimal solutions, too. For making a robust decision analysis, we consider all optimal solutions. Let $\mathcal{W}^{\text{evi}}$ be the set of optimal solutions of LFP problem (17)–(18).

Now we consider the dominance relation between alternatives associated with the BPA m estimated by LFP problem (17)–(18). Because we assume a set of optimal BPAs, the calculation become complex. The dominance relation between p-th and q-th alternatives o_p and o_q can be defined by

$$o_p \succ^{\text{evi}} o_q \Leftrightarrow \forall m \in \mathcal{W}^{\text{evi}}, \ \forall \boldsymbol{w} \in \mathcal{P}(m); \ \sum_{i \in N} w_i(f_i(o_p) - f_i(o_q)) \geq 0. \tag{19}$$

Let $\check{\delta}$ be the optimal value of LFP problem (17)–(18). Then, the right-hand side of (19) can be confirmed by solving the following LP problem:

$$\text{minimize } dis(p,q) = \sum_{i \in N} (f_i(o_p) - f_i(o_q)) \sum_{B : \omega_i \in B \subseteq \Omega} m_B(\omega_i), \tag{20}$$

$$\text{subject to } \sqrt{a_{ji}} m(\omega_i) + \delta_{ij} = \sqrt{a_{ij}} \sum_{B : \omega_j \in B \subseteq \Omega} m(B), \ \delta_{ij} \geq 0, \ i,j \in N, \ i \neq j,$$

$$\sum_{i : \omega_i \in B} m_B(\omega_i) = m(B), \ m_B(\omega_i) \geq 0, \ \sum_{B \subseteq \Omega} m(B) = 1,$$
$$m(\{\omega_i\}) \geq \epsilon, \ m(\emptyset) = 0, \ i \in N, \ B \subseteq \Omega, \tag{21}$$
$$\sum_{i \in N} \sum_{j=i+1}^{n} (\delta_{ij} + \delta_{ji}) = \check{\delta}.$$

Let $dis^*(p,q)$ be the optimal value of LP problem (21). If $dis^*(p,q) \geq 0$, $o_p \succeq^{\text{evi}} o_q$ holds, and otherwise, $o_p \succeq^{\text{evi}} o_q$ does not hold. We note that $\succeq^{\text{int}}$ is a preorder.

4   Relations Between Two Approaches

In this section, we investigate the relations between the interval AHP and the evidence-theoretic AHP. In other words, we investigate the relations between optimal interval weight vectors to LFP problem (17)–(18) and optimal BPAs to LFP problem (17)–(18).

For a normalized interval weight vector $W = (W_1, W_2, \ldots, W_n)^{\mathrm{T}}$, we have the following proposition.

Proposition 1. *Let* $W = (W_1, W_2, \ldots, W_n)^{\mathrm{T}}$ *be an interval weight vector satisfying normalization conditions (2) and (3), where* $W_i = [w_i^{\mathrm{L}}, w_i^{\mathrm{R}}]$, $i \in N$. *If* W *satisfies*

$$\sum_{i \in N} (w_i^{\mathrm{L}} + w_i^{\mathrm{R}}) \geq 2, \tag{22}$$

there exists a BPA $m : 2^{\Omega} \to [0, 1]$ *such that*

$$Bel(\{\omega_i\}) = m(\{\omega_i\}) = w_i^{\mathrm{L}}, \ Pl(\{\omega_i\}) = \sum_{B : \omega_i \in B} m(B) = w_i^{\mathrm{R}}, \ i \in N, \tag{23}$$

where $\Omega = \{\omega_1, \omega_2, \ldots \omega_n\}$.

Proof. Let $d_i = w_i^{\mathrm{R}} - w_i^{\mathrm{L}}$, $i \in N$. From (22), we obtain

$$\sum_{i \in N} d_i = \sum_{i \in N} (w_i^{\mathrm{R}} - w_i^{\mathrm{L}}) \geq 2 \left(1 - \sum_{i \in N} w_i^{\mathrm{L}} \right). \tag{$*$}$$

From (3), we obtain

$$d_i = w_i^{\mathrm{R}} - w_i^{\mathrm{L}} \leq 1 - \sum_{i \in N} w_i^{\mathrm{L}}.$$

We consider a construction of a BPA $m : 2^{\Omega} \to [0, 1]$ satisfying (23). We define $m(\emptyset) = 0$ and $m(\{\omega_i\}) = w_i^{\mathrm{L}}$, $i \in N$. By this definition, the first equality of (23) is satisfied. Therefore, we define $m(B)$ for $B \subseteq \Omega$ with $|B| \geq 2$ to satisfy the second equality of (23). We define

$$\kappa = \left\lceil \frac{\displaystyle\sum_{i \in N} d_i}{1 - \displaystyle\sum_{i \in N} w_i^{\mathrm{L}}} \right\rceil,$$

$$\iota(k) = \min \left\{ l \ \middle| \ \sum_{i=1}^{l} d_i > (k-1) \left(1 - \sum_{i \in N} w_i^{\mathrm{L}} \right) \right\}, \ k = 1, 2, \ldots, \kappa,$$

where $\lceil r \rceil$ is the minimum integer not less than or equal to r. Then we define $m(B)$, $B \subseteq \Omega$ such that $|B| \geq 2$ by the following procedure (see Fig. 1): Initialize $i = 1$, $q = \kappa$, $s = 0$, $\eta(k) = \iota(k)$, $k = 1, 2, \ldots, q$ and $u_k = \sum_{t=1}^{\iota(k)} d_t - (k-1)$ $\left(1 - \sum_{t \in N} w_t^{\mathrm{L}} \right)$, $k = 1, 2, \ldots, q$. Initialize $m(B) = 0$ for all $B \subseteq \Omega$ such that $|B| \geq 2$. Repeat procedure (a) to (d) until $s \geq 1 - \sum_{i \in N} w_i^{\mathrm{L}}$;

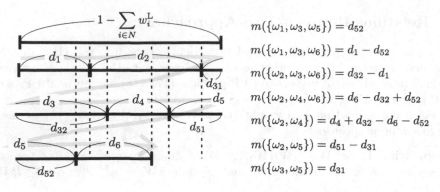

$$m(\{\omega_1, \omega_3, \omega_5\}) = d_{52}$$
$$m(\{\omega_1, \omega_3, \omega_6\}) = d_1 - d_{52}$$
$$m(\{\omega_2, \omega_3, \omega_6\}) = d_{32} - d_1$$
$$m(\{\omega_2, \omega_4, \omega_6\}) = d_6 - d_{32} + d_{52}$$
$$m(\{\omega_2, \omega_4\}) = d_4 + d_{32} - d_6 - d_{52}$$
$$m(\{\omega_2, \omega_5\}) = d_{51} - d_{31}$$
$$m(\{\omega_3, \omega_5\}) = d_{31}$$

Fig. 1. Construction of $m : 2^{\Omega} \to [0, 1]$ satisfying (23)

(a) Calculate $B_i = \{\omega_{\iota(k)} \mid k = 1, 2, \ldots, q\}$. Define $m(B_i) = \min_{k=1,2,\ldots,q} u_k$.
(b) Update $s = s + m(B_i)$, $u_k = u_k - m(B_i)$, $k = 1, 2, \ldots, q$.
(c) For $k = 1$ to q,
   if $u_k = 0$ then update $\eta(k) = \eta(k) + 1$ and $u_k = \min\left(d_{\eta(k)}, 1 - s - \sum_{t \in N} w_t^{\mathrm{L}}\right)$.
(d) $i = i + 1$. If $q = \kappa$ and $\eta(q) > n$ then $q = q - 1$.

$\square$

We have the following proposition for a BPA $m : 2^{\Omega} \to [0, 1]$.

Proposition 2. *Let $m : 2^{\Omega} \to [0, 1]$ be a BPA with $\Omega = \{\omega_1, \omega_2, \ldots, \omega_n\}$. Then, the interval weight vector $\boldsymbol{W} = (W_1, W_2, \ldots, W_n)^{\mathrm{T}}$ defined by $W_i = [w_i^{\mathrm{L}}, w_i^{\mathrm{R}}] = [Bel(\{\omega_i\}), Pl(\{\omega_i\})]$, $i \in N$ satisfies normalization conditions (2) and (3) as well as (22).*

Proof. The satisfaction of normalization conditions (2) and (3) is shown as follows:

$$w_i^{\mathrm{L}} + \sum_{j \in N \setminus i} w_j^{\mathrm{R}} = Bel(\{\omega_i\}) + \sum_{j \in N \setminus i} Pl(\{\omega_j\})$$

$$= m(\{\omega_i\}) + \sum_{j \in N \setminus i} \sum_{B : \omega_j \in B} m(B) \geq \sum_{B \subseteq \Omega, B \neq \emptyset} m(B) = 1, \ i \in N,$$

$$w_i^{\mathrm{R}} + \sum_{j \in N \setminus i} w_j^{\mathrm{L}} = Pl(\{\omega_i\}) + \sum_{j \in N \setminus i} Bel(\{\omega_j\})$$

$$= \sum_{B : \omega_i \in B} m(B) + \sum_{j \in N \setminus i} m(\{\omega_j\}) \leq \sum_{B \subseteq \Omega, B \neq \emptyset} m(B) = 1, \ i \in N.$$

Equation (22) is obtained as follows:

$$\sum_{i \in N} (w_i^{\mathrm{L}} + w_i^{\mathrm{R}}) = \sum_{i \in N} \left(m(\{\omega_i\}) + \sum_{B : \omega_i \in B} m(B) \right) \geq 2 \sum_{B \subseteq \Omega, B \neq \emptyset} m(B) = 2.$$

$\square$

From Propositions 1 and 2, we obtain the following corollary

Corollary 1. *LFP problem (17)–(18) is equivalent to the following LFP problem:*

$$minimize \;\; \delta(\boldsymbol{W}) = \frac{\displaystyle\sum_{i \in N} \sum_{j=i+1}^{n} (\delta_{ij} + \delta_{ji})}{\displaystyle\sum_{i \in N} (w_i^{\mathrm{L}} + w_i^{\mathrm{R}})}, \tag{24}$$

$$subject \; to \; (10), \;\; \sum_{i \in N} (w_i^{\mathrm{L}} + w_i^{\mathrm{R}}) \geq 2, \tag{25}$$

where the equivalence implies that any optimal solution to LFP problem (17)–(18) is constructed from an optimal solution to LFP problem (24)–(25), and vice versa.

Owing to Corollary 1, LFP problem (17)–(18) can be reduced to an LFP problem with much fewer decision variables. Moreover, we can expect that the evaluation of dominance $o_p \succeq^{\mathrm{evi}} o_q$ can be done by simpler LP problem (12)–(13) with an additional constraint $\sum_{i \in N} w_i^{\mathrm{L}} + w_i^{\mathrm{R}} \geq 2$, respectively.

Now, we investigate the equivalence between LP problem (20)–(21) and LP problem (12)–(13) with an additional constraint $\sum_{i \in N} w_i^{\mathrm{L}} + w_i^{\mathrm{R}} \geq 2$.

Proposition 3. *Let $m : 2^{\Omega} \to [0, 1]$ be a BPA with $\Omega = \{\omega_1, \omega_2, \ldots, \omega_n\}$. Create a normalized interval vector $\boldsymbol{W} = (W_1, W_2, \ldots, W_n)^{\mathrm{T}}$ by $W_i = [w_i^{\mathrm{L}}, w_i^{\mathrm{R}}] = [Bel(\{\omega_i\}), Pl(\{\omega_i\})]$, $i \in N$. For a normalized crisp weight vector $\boldsymbol{w} = (w_1, w_2, \ldots, w_n)^{\mathrm{T}}$ such that $w_i \in W_i$ and $\sum_{i \in N} w_i = 1$. The existence of $m_i(B) \geq 0$, $i \in N$, $B \subseteq \Omega$ satisfying the following equation is not guaranteed.*

$$\sum_{\omega_i \in B} m_i(B) = m(B), \; B \subseteq \Omega, \;\; \sum_{B : \omega_i \in B} m_i(B) = w_i, \; i \in N. \tag{26}$$

Proof. We give a counter example. Consider a BPA $m : 2^{\Omega} \to [0,1]$ with $n = 4$: $m(\{\omega_2\}) = 0.05$, $m(\{\omega_3\}) = 0.15$, $m(\{\omega_1, \omega_2\}) = 0.4$, $m(\{\omega_1, \omega_3\}) = 0.2$, $m(\{\omega_1, \omega_4\}) = 0.1$, $m(\{\omega_2, \omega_3\}) = 0.1$ and $m(B) = 0$ for other $B \subseteq \Omega$. Then, we have $W_1 = [0, 0.7]$, $W_2 = [0.05, 0.55]$, $W_3 = [0.15, 0.45]$ and $W_4 = [0, 0.1]$. For a normalized crisp vector $\boldsymbol{w} = (0.05, 0.5, 0.45, 0)^{\mathrm{T}}$ satisfying $w_i \in W_i$, $i \in N$, there is no solution $m_i(B) \geq 0$, $B \subseteq \Omega$ satisfying (26). Indeed, we have $m_i(\{\omega_i\}) = m(\{\omega_i\})$, $i = 2, 3$, and from $w_3^{\mathrm{R}} = 0.45$, $m_3(\{\omega_1, \omega_3\}) = 0.2$ and $m_3(\{\omega_2, \omega_3\}) = 0.1$. $m_3(\{\omega_2, \omega_3\}) = 0.1$ implies $m_2(\{\omega_2, \omega_3\}) = 0$ and eventually, from $w_2^{\mathrm{R}} = 0.55$, it requires $m_2(\{\omega_1, \omega_2\}) = 0.45$. On the other hand, we should have $m_1(\{\omega_1, \omega_2\}) + m_2(\{\omega_1, \omega_2\}) = 0.4$. Because $m_1(\{\omega_1, \omega_2\}) \geq 0$, no solution exists. $\square$

Proposition 3 implies that $\boldsymbol{w}$ is not consistent with BPA m even if it is consistent with the normalized interval vector $\boldsymbol{W}$ defined by $W_i = [w_i^{\mathrm{L}}, w_i^{\mathrm{R}}] = [Bel(\{\omega_i\}), Pl(\{\omega_i\})]$, $i \in N$. From this result, we may think that we do not have the equivalence between LP problem (20)–(21) and LP problem (12)–(13) with an additional constraint $\sum_{i\in N} w_i^{\text{L}} + w_i^{\text{R}} \geq 2$. However, owing to the multiplicity of the optimal solutions to LP problem (20)–(21), we obtain the equivalence as shown in the following proposition.

Proposition 4. *Let $\boldsymbol{W} = (W_1, W_2, \ldots, W_n)^{\text{T}}$ be a normalized interval weight vector. For any normalized weight vector $\boldsymbol{w} = (w_1, w_2, \ldots, w_n)^{\text{T}}$ satisfying $w_i \in W_i$ and $\sum_{i\in N} w_i = 1$, there exists a BPA $m : 2^{\Omega} \to [0,1]$ satisfying (26) and $W_i = [w_i^{\text{L}}, w_i^{\text{R}}] = [Bel(\{\omega_i\}), Pl(\{\omega_i\})]$, $i \in N$.*

Proof. Let $p_i = w_i - w_i^{\text{L}}$ and $q_i = w_i^{\text{R}} - w_i$, $i \in N$. We construct $m : 2^{\Omega} \to [0,1]$ satisfying

$$Bel(\{\omega_i\}) = m(\{\omega_i\}) = Bel(\{\omega_i\}) = w_i^{\text{L}}, \ Pl(\{\omega_i\}) = \sum_{B:i\in B} m(B) = w_i^{\text{R}}, \quad (*)$$

For this purpose, we consider $\bar{q}_j^i$, q_j^i, $i, j \in N$, $j \neq i$ satisfying

$$p_i = \sum_{j\in N\setminus i} \bar{q}_j^i, \ \sum_{i\in N\setminus j} \bar{q}_j^i + \sum_{i\in N\setminus j} q_j^i = q_j, \ \bar{q}_j^i + q_j^i \leq p_i, \ \bar{q}_j^i \geq 0, \ q_j^i \geq 0, \ i,j \in N, \ i \neq j.$$

$$(**)$$

The existence of a solution to $(**)$ is guaranteed by normalization conditions (2), (3), $\sum_{i\in N} w_i = 1$ and $\sum_{i\in N}(w_i^{\text{L}} + w_i^{\text{R}}) \geq 2$. Indeed, we obtain $\sum_{j\in N\setminus i} q_j \geq p_i$ from (3) and $\sum_{i\in N} w_i = 1$; $\sum_{i\in N} q_i \geq \sum_{i\in N} p_i$ from $\sum_{i\in N} w_i = 1$ and $\sum_{i\in N}(w_i^{\text{L}} + w_i^{\text{R}}) \geq 2$; and $q_i \leq \sum_{j\in N\setminus i} p_j$ from (2) and $\sum_{i\in N} w_i = 1$. The existence of $\bar{q}_j^i$, $i,j \in N$, $i \neq j$ is guaranteed by $\sum_{j\in N\setminus i} q_j \geq p_i$ and $\sum_{i\in N} q_i \geq \sum_{i\in N} p_i$. Fixing $\bar{q}_j^i$, $i,j \in N$, $i \neq j$ satisfying $p_i = \sum_{j\in N\setminus i} \bar{q}_j^i$, $i \in N$ and $\sum_{i\in N\setminus j} \bar{q}_j^i \leq q_j$, $j \in N$, the existence of $\bar{q}_j^i$, $i,j \in N$ is guaranteed by $q_i \leq \sum_{j\in N\setminus i} p_j$, $i \in N$.

We construct $m : 2^{\Omega} \to [0,1]$ from $\bar{q}_j^i$, q_j^i, $i,j \in N$, $j \neq i$. Initialize $m(B) = 0$, $\forall B \subseteq \Omega$. Let $r = \left[\left(\sum_{i\in N}\sum_{j\in N\setminus i} q_j^i\right)/\left(\sum_{i\in N} p_i\right)\right]$. Define $r_i^j = q_j^i$, $i,j \in N$, $i \neq j$. For $i = 1$ to n, set $t = 0$ and do (a1) to (a2): (a1) For $j = 1$ to n, set $k = 1$, $u = 1$ and do (b1) to (b2): (b1) Set $l = 1$ and $s = \bar{q}_j - t$. (b2) If $s > 0$, let $B = \{\omega_i, \omega_j\}$ and do (c1) to (c3): (c1) If $k \notin \{i,j\}$ and $r_k^i > 0$, update $B = B \cup \{\omega_k\}$, $s = \min(s, r_k^i)$ and $l = l + 1$. (c2) If $l > r$, do (d1) to (d3): (d1) Update $m(B) = m(B) + s$ and $t = t + s$. (d2) For $u = 1$ to k if $r_u^i > 0$ update $r_u^i = r_u^i - s$. (d3) Reset $k = 1$, $u = 1$ and return to (b1). (c3) Update $k = k + 1$ and go to (c1). (a2) Update $m(\{\omega_i\}) = w_i^{\text{L}}$.

The constructed BPA m satisfies (26) with $m_i(B)$ defined during the construction of m with $p_i = \sum_{j\in N\setminus i} \bar{q}_j^i$. $Bel(\{\omega_i\}) = w_i^{\text{L}}$, $i \in N$ is satisfied by the definition of BPA m and $Pl(\{\omega_i\}) = w_i^{\text{R}}$, $i \in N$ is satisfied from $\sum_{i\in N\setminus j} \bar{q}_j^i + \sum_{i\in N\setminus j} q_j^i = q_j$, $j \in N$. □

Example 1. Let $\boldsymbol{W} = (W_1, W_2, W_3, W_4)^{\text{T}}$ with $W_1 = [0, 0.2]$, $W_2 = [0.05, 0.35]$, $W_3 = [0.15, 0.85]$ and $W_4 = [0, 0.8]$, and $\boldsymbol{w} = (0.2, 0.35, 0.15, 0.3)^{\text{T}}$. Then, we have $p_1 = 0.2$, $p_2 = 0.3$, $p_3 = 0$, $p_4 = 0.3$, $q_1 = 0$, $q_2 = 0$, $q_3 = 0.7$ and $q_4 = 0.5$. We obtain a solution to $(**)$ in the proof of Proposition 4 as $\bar{q}_3^1 = 0.2$, $\bar{q}_4^2 = 0.3$, $\bar{q}_3^4 = 0.3$, $q_4^1 = 0.2$, $q_3^2 = 0.2$ and other variables take zeros. Then, we obtain $m(\{\omega_1, \omega_3, \omega_4\}) = 0.2$, $m(\{\omega_2, \omega_3, \omega_4\}) = 0.2$, $m(\{\omega_2, \omega_4\}) = 0.1$, $m(\{\omega_3, \omega_4\}) = 0.3$, $m(\{\omega_2\}) = 0.05$, $m(\{\omega_3\}) = 0.15$ and $m(B) = 0$ for other subsets B of Ω. This BPA m satisfies (26) and $W_i = [w_i^L, w_i^R] = [Bel(\{\omega_i\}), Pl(\{\omega_i\})]$, $i \in N$.

5 Conclusions

We applied the evidence theory to the representation of vague evaluations of the decision maker in the setting of AHP. We formulated the problem estimating a BPA from a given PCM as an LFP problem in parallel to the problem estimating a normalized interval weight vector. We investigated relations between BPAs and normalized interval weight vectors estimated from those problems. We showed that the problem estimating a BPA is equivalent to the problem estimating a normalized interval weight vector with an additional constraint. This equivalence holds not only for the problems formulated in this paper but also other formulations of BPA estimation problems under a PCM. Investigating other relations between BPAs and normalized interval vectors is a future topic.

References

1. Saaty, T.L.: The Analytic Hierarchy Process. McGraw-Hill, New York (1980)
2. Saaty, T.L., Vargas, C.G.: Comparison of eigenvalue, logarithmic least squares and least squares methods in estimating ratios. Math. Model. **5**, 309–324 (1984)
3. Sugihara, K., Tanaka, H.: Interval evaluations in the analytic hierarchy process by possibilistic analysis. Comput. Intell. **17**, 567–579 (2001)
4. Inuiguchi, M., Innan, S.: Improving interval weight estimations in interval AHP by relaxations. J. Adv. Comput. Intell. Intell. Inf. **21**(7), 1135–1143 (2017)
5. Inuiguchi, M., Innan, S.: Comparison among several parameter-free interval weight estimation methods from a crisp pairwise comparison matrix. In: USB Proceedings of the 14th International Conference on Modeling Decisions for Artificial Intelligence, pp. 61–76 (2017)
6. Inuiguchi, M.: Non-uniqueness of interval weight vector to consistent interval pairwise comparison matrix and logarithmic estimation methods. In: Huynh, V.-N., Inuiguchi, M., Le, B., Le, B.N., Denoeux, T. (eds.) IUKM 2016. LNCS (LNAI), vol. 9978, pp. 39–50. Springer, Cham (2016). https://doi.org/10.1007/978-3-319-49046-5_4
7. Shafer, G.: A Mathematical Theory of Evidence. Princeton University Press, Princeton (1976)
8. Xu, D.-L.: An introduction and survey of the evidential reasoning approach for multiple criteria decision analysis. Ann. Oper. Res. **195**, 163–187 (2012)
9. Entani, T., Inuiguchi, M.: Pairwise comparison based interval analysis for group decision aiding with multiple criteria. Fuzzy Sets Syst. **271**, 79–96 (2015)
10. Charnes, A., Cooper, W.W.: Programming with linear fractional functionals. Nav. Res. Logist. Q. **9**, 181–186 (1962)
11. Ishizuka, M., Fu, K.S., Yao, J.P.: Inference procedures under uncertainty for the problem-reduction Method. Inf. Sci. **28**, 179–206 (1982)

Clustering and Classification

Methods for Clustering Categorical and Mixed Data: An Overview and New Algorithms

Sadaaki Miyamoto[1]([✉]), Van-Nam Huynh[2], and Shuhei Fujiwara[1]

[1] University of Tsukuba, Tsukuba, Japan
miyamoto.sadaaki.fu@u.tsukuba.ac.jp
[2] Japan Advanced Institute of Science and Technology, Nomi, Japan
huynh@jaist.ac.jp

Abstract. Methods of clustering for categorical and mixed data are considered. Dissimilarities for this purpose are reviewed and different classes of algorithms according to different classes of similarities are discussed. Details of several algorithms are then given, which include agglomerative hierarchical clustering, K-means and related methods such as K-medoids and K-modes, and methods of network clustering. The way how the combinations of existing ideas leads to new algorithms is discussed.

Keywords: Data clustering · Categorical data · Mixed data
Network clustering · K-medoids

1 Introduction

Data clustering has now become a standard technique of data mining, and yet it has a number of unique characteristics different from other methods of supervised and unsupervised classification. One of those characteristics is that different types of data are assumed to be given for analysis: not only the Euclidean space but also other spaces and moreover general types of dissimilarities can be used as measures of relatedness between a pair of objects. On the other hand, a standard class of clustering algorithms of *agglomerative hierarchical clustering* has an unique and useful form of output that is called a *dendrogram*. The dendrogram is popular in various fields of applications and its usefulness could not be ignored.

In this paper we give a brief overview of methods of clustering for non-Euclidean models in the sense that given data types of an object for clustering is categorical or mixed; a mixed data type consists of categorical data and numerical data at the same time. First dissimilarities for categorical and mixed data are discussed. Then three classes of methods of clustering are introduced, which are agglomerative hierarchical clustering, non-hierarchical methods for Euclidean data, and non-hierarchical methods for non-Euclidean data. The best known method of K-means clustering is in the second class, related methods of the third class is considered which includes K-median, K-modes, and K-medoids.

V.-N. Huynh et al. (Eds.): IUKM 2018, LNAI 10758, pp. 75–86, 2018.
https://doi.org/10.1007/978-3-319-75429-1_7

Consideration of these methods stimulates the development of new linkage methods in the first class. These considerations then lead us to related algorithms such as a generative model and a method of fuzzy clustering. Moreover network clustering is briefly mentioned. The way how these methods lead to the developemnt of new methods for categorical and mixed data is discussed.

2   Categorical Data and Dissimilarities

Let $\mathcal{X} = \{x_1, x_2, \ldots, x_N\}$ be a finite set of objects for clustering. Assume that $\mathcal{A} = \{A_1, \ldots, A_M\}$ be another set of attributes. For each A_j, an associated set Z_j which contains all values that A_j takes. Thus, every $z \in Z_j$ is a value of attribute A_j. For each x_i and A_j, the value of object x_i concerning attribute A_j is given by $x_i^j \in Z_j$. We write $A_j(x_i) = x_i^j$, as A_j is a mapping $(A_j \colon \mathcal{X} \to Z_j)$. Alternatively, we express $x = (x^1, \ldots, x^M)$ as a kind of vectors, although its components are not necessarily numbers. Concretely, Z_j can be *numerical* when its elements are numerical values: $Z_j \subseteq \mathbf{R}$. Or Z_j can be *symbolical* when its elements are symbols and not numerical. Note also that a symbol represents a category, and hence the words 'categorical' and 'symbolical' are used for the same meaning herein. Let us write a typical case as $Z_j = \{t_1, \ldots, t_q\}$ where the elements t_l are symbols.

Assume that clusters denoted by $G_1, \ldots, G_K$ are disjoint subsets of $\mathcal{X}$ such that the union of clusters covers the whole set:

$$\bigcup_{i=1}^{K} G_i = \mathcal{X}, \quad G_i \cap G_j = \emptyset \ (i \neq j). \tag{1}$$

Moreover the collection of clusters is denoted by $\mathcal{G} = \{G_1, \ldots, G_K\}$.

Similarity or dissimilarity is a key concept in data clustering. We assume a similarity measure $s(x, x')$ or a dissimilarity measure $d(x, x')$ is defined between a pair of objects $x, x' \in \mathcal{X}$. The difference between similarity and dissimilarity is that two objects are similar or near when similarity between them has a high value while they are similar when dissimilarity value is lower. Accordingly,

$$s(x, x) = \max_{x' \in \mathcal{X}} s(x, x'), \quad d(x, x) = \min_{x' \in \mathcal{X}} d(x, x').$$

Symmetric property is also assumed for the both measures:

$$s(x, x') = s(x', x), \quad d(x, x') = d(x', x). \tag{2}$$

Note that the triangular inequality is not assumed: the triangular inequality in general is not especially useful in clustering.

2.1   Measures of Dissimilarity

We mostly use dissimilarity and refer to similarity only when necessary. How to define an appropriate dissimilarity is a first problem to be considered in clustering. Sometimes $d(x, x')$ is directly given without referring to their attributes, as in the case of network clustering [5,20,21]. We, however, assume that dissimilarities are defined by the observation of attribute values x_i^j. Since we have different types of sets Z_j in general, different kinds of dissimilarities should be considered. We therefore assume that $d_j(x,y)$ for $x, y \in Z_j$, and consider how d_j should be defined.

Let us consider the most frequent case of an Euclidean space $\boldsymbol{R}^M$. In this case $x, y \in \boldsymbol{R}$ for all attributes and we set

$$d_j(x,y) = (x - y)^2, \quad for \ all \ 1 \le j \le M,$$

and the dissimilarity is given by

$$d(x, x') = \sum_{j=1}^{M} d_j(x^j, y^j) = \sum_{j=1}^{M} (x^j - y^j)^2, \tag{3}$$

for $x = (x^1, \ldots, x^M)$ and $x' = (y^1, \ldots, y^M)$. We also write $d(x, x') = \|x - x'\|^2$ using the Euclidean norm symbol. Note that the squared Euclidean norm is used instead of the norm itself.

Let us suppose that data are of a mixed type, i.e., some Z_j is numerical while another Z_l is symbolical. A simple definition of a dissimilarity is that

$$d_j(x,y) = \frac{1}{2}|x - y|, \tag{4}$$

i.e., $d_j(x,y)$ is the difference between the two numerical values, while

$$d_l(t_h, t_k) = \begin{cases} 1 & (t_h \neq t_k), \\ 0 & (t_h = t_k). \end{cases} \tag{5}$$

and define

$$d(x, x') = \sum_{j=1}^{M} d_j(x^j, y^j). \tag{6}$$

Let us assume that there is no Z_l of the set of symbols, then all attribute values are numerical but we do not have a squared Euclidean dissimilarity, but instead we have the L_1-norm:

$$d(x, x') = \frac{1}{2} \sum_{j=1}^{M} |x^j - y^j| = \frac{1}{2}\|x - y\|_{L_1}.$$

On the other hand, if all attribute values are symbolic, Eq. (6) consists of (5) alone. There is an interesting relationship between the latter two. Let us convert x into 0/1 numerical values. Actually, only one of $t_1, \ldots, t_M$, say t_1, represents the object and hence we can write $x = \{t_1\}$ or $x = (1, 0, \ldots, 0)$ using 0 and 1. Suppose $x' = \{t_2\}$ or $x' = (0, 1, 0, \ldots, 0)$. Then it is easy to see that

$$d_l(t_1, t_2) = \frac{1}{2}|x - x'|$$

If all attributes are symbolical, we have $d(x, x') = \frac{1}{2}\|x - x'\|_{L_1}$. Thus, the weight $\frac{1}{2}$ in (4) is justified.

2.2   Minimization Problems

For later use, we consider optimization problems: $\min\limits_{x \in R^M} \sum\limits_{y \in \mathcal{X}} d(x, y)$, in case when all values are numerical, and we assume the both cases of an Euclidean space (3) and L_1-space (6).

In the Euclidean space, it is easy to see that the solution is the average: $x = \dfrac{1}{|\mathcal{X}|} \sum\limits_{y \in \mathcal{X}} y$. where $|\mathcal{X}|$ is the number of elements in $\mathcal{X}$.

On the other hand, if L_1-space is used, the solution is the median. Each component of the median is defined independently. Let the first component (corresponding to A_1) is $x_1^1, x_2^1, \ldots, x_M^1$. Sort this set of real numbers into ascending order and the result is $y_1, \leq y_2 \leq \cdots \leq y_M$. Then the median for the first component is $y_{[M/2]+1}$. Other components are calculated in the same manner.

There is still other minimization problems. Suppose all data are symbolic, we consider

$$\min_{x \in \mathcal{Z}} \sum_{x' \in \mathcal{X}} d(x, x'), \tag{7}$$

where $\mathcal{Z} = Z_1 \times \cdots \times Z_M$. Note that $d(x, x')$ is defined by (6) and (5). To solve this problem, let the frequency of occurrences of $y_k \in Z_j$ be f_k on $\mathcal{Z}$. Thus we have a histogram $(f_1/y_1, \ldots, f_L/y_L)$ for X_j. Assume that the maximum of $f_1, \ldots, f_L$ is f_h, then the mode is written as

$$\text{mode}(\mathcal{X}, Z_j) = (\arg\max\{f_1, \ldots, f_L\}, \max\{f_1, \ldots, f_L\}) = (h, f_h), \tag{8}$$
$$\arg\text{mode}(\mathcal{X}, Z_j) = h, \tag{9}$$
$$\text{value}\,\text{mode}(\mathcal{X}, Z_j) = f_h. \tag{10}$$

Then it is easy to see that the solution of (7) is given by

$$(\text{mode}(\mathcal{X}, Z_1), \ldots, \text{mode}(\mathcal{X}, Z_M)).$$

These minimization problems with their solutions are useful in considering K-modes and related clustering problems.

3   Algorithms of Clustering

Two major methods are agglomerative hierarchical clustering and the K-means.

3.1   Agglomerative Hierarchical Clustering

The agglomerative hierarchical algorithm [1,10,18] is one of best known methods of clustering. It uses a measure $d(G_i, G_j)$ of an inter-cluster dissimilarity. The following is a general description of the agglomerative hierarchical algorithm [18]. Note that initial clusters $\mathcal{G}(0) = \{G(0)_1, \ldots, G(0)_{C_0}\}$ are assumed to be given. Typically, $G(0)_i = \{x_i\}$, but we assume other cases later.

AHC (Agglomerative Hierarchical Clustering)

AHC1: Each object forms an initial cluster: $G_i = G(0)_i$, $(i = 1, \cdots, N)$. $C = N$, (C is the number of clusters). For all $G_i, G_j \in \mathcal{G}$, let $d(G_i, G_j) = d(x_i, x_j)$.

AHC2: Find the pair of clusters of minimum dissimilarity:

$$(G_q, G_r) = \arg \min_{G_i, G_j \in \mathcal{G}} d(G_i, G_j) \tag{11}$$

$$m_C = d(G_q, G_r) \tag{12}$$

Add $G' = G_q \cup G_r$ to $\mathcal{G}$ and delete G_q, G_r from $\mathcal{G}$. Let $C = C - 1$. If $C = 1$, output clusters as a *dendrogram* and stop.

AHC3: Update dissimilarity $d(G, G')$ between the merged cluster G' and all other clusters $G \in \mathcal{G}$. Go to **AHC2**.

End of AHC.

Here, $m_N, \ldots, m_2$ are called the levels of merging clusters.

We have several *linkage methods* to update dissimilarity $d(G, G')$ in **AHC3**, from which the single linkage, the average linkage, and the Ward method are mentioned here.

Single linkage: $d(G_i, G_j) = \min\limits_{x \in G_i, y \in G_j} d(x, y)$.

Average linkage: $d(G_i, G_j) = \dfrac{1}{|G_i||G_j|} \sum\limits_{x \in G_i, y \in G_j} d(x, y)$.

Ward method: Assume

$$E(G) = \sum_{x_k \in G} \|x_k - M(G)\|^2.$$

Let

$$d(G_i, G_j) = E(G_i \cup G_j) - E(G_i) - E(G_j).$$

where $M(G)$ is the centroid of G: $M(G) = \sum\limits_{x_k \in G} \dfrac{x_k}{|G|}$, and $\|\cdot\|$ is the Euclidean norm: this method assumes that the objects are points in an Euclidean space.

They moreover use the following formulas of updating in **AHC3** in which $d(G, G')$ is expressed using $d(G, G_q)$, $d(G, G_r)$, and so on.

Updating formula of the single linkage:

$$d(G, G') = \min\{d(G, G_q), d(G, G_r)\}.$$

Updating formula of the Ward method:

$$d(G, G') = \frac{(|G_q| + |G|)d(G_q, G) + (|G_r| + |G|)d(G_r, G) - |G|d(G_q, G_r)}{|G_q| + |G_r| + |G|}.$$

The updating formula of the average linkage is omitted here. See, e.g., [18] for more detail.

The single linkage and the Ward method are two popular algorithms in agglomerative hierarchical clustering. The former is known to have best theoretical properties [18], while the Ward method has been considered to be practically useful by researchers in applications.

3.2   The K-means and Related Methods

We assume that objects $x_1, \ldots, x_N$ are in a space $\mathbf{S}$ whose distance is defined by the dissimilarity $d(x, y)$. Consider the next problem of *alternate minimization.*

K-means prototype algorithm.
Step 0. Give an initial partition $G_1, \ldots, G_K$ of $\{x_1, \ldots, x_N\} \subseteq \mathbf{S}$.
Step 1. Let

$$v_i = \arg \min_{v \in \mathbf{S}} \sum_{x_k \in G_i} d(x_k, v), \quad i = 1, 2, \ldots, K. \tag{13}$$

Step 2. Allocate each x_k $(k = 1, \ldots, N)$ to the cluster of the nearest center v_i:

$$x_k \to G_i \iff v_i = \arg \min_{1 \leq j \leq K} d(x_k, v_j). \tag{14}$$

Step 3. If $(v_1, \ldots, v_K)$ is convergent, stop. Else go to **step 1**.
End K-means prototype.

The above algorithm describes a family of different methods.

The method of K-means [15] is the most popular clustering algorithm. It assumes that the objects $x_1, \ldots, x_N$ are points in an Euclidean space. Hence we assume $\mathbf{S} = \mathbf{R}^p$ with $d(x, y) = \|x - y\|^2$. Accordingly the center of a cluster (13) is the centroid:

$$v_i = M(G_i) = \frac{1}{|G_i|} \sum_{x_k \in G_i} x_k.$$

Thus the K-means prototype algorithm is reduced to the K-means algorithm.

K-median and K-mode algorithms are derived likewise. If L_1-space is used, then v_i is given by the median described above; if the data are categorical and the dissimilarity is given by (6), then the center v_i is given by the mode for cluster G_i:

$$v_i^j = \begin{cases} 1 & (j = \arg \text{mode}(G_i, Z_j)), \\ 0 & (\text{otherwise}). \end{cases} \tag{15}$$

and we have the K-mode algorithm.

Moreover if we have mixed data in which numerical data has L_1-norm, then the resulting algorithm has the mixture of the median and the mode corresponding to the data types.

There is still another method of the K-means family, in which $\mathbf{S}$ is the set of objects itself: $\mathbf{S} = \mathcal{X}$ with the general dissimilarity $d(x, x')$. In such a case the space is a weighted network and accordingly the element v_i corresponds to an object which satisfies

$$v_i = \arg \min_{v \in \mathcal{X}} \sum_{x_k \in G_i} d(x_k, v). \tag{16}$$

The above defined object for G_i is called the medoid [14] for cluster G_i. Thus the algorithm gives the method of K-medoids. It is obvious to see $v_i \in G_i$.

3.3   Network Clustering

The last method of K-medoids is an algorithm of network clustering in the sense that no other space than just the weighted network is given. There are other methods that should be mentioned in addition.

DBSCAN [9] is known to be an efficient algorithm that searches clusters of *core-points* on a weighted graph. This method has been proved to be a variation of the single linkage that connects core-points and node-points [16].

Newman's method [20,21] of hierarchical clustering and its non-hierarchical version [5] use the modularity index in a network; they automatically determine the number of clusters by optimizing the index. It seems that the modularity index works effectively in the both algorithms, but the non-hierarchical algorithm is faster and appropriate for handling large-scale data sets. On the other hand, the hierarchical version can output a dendrogram, but the shape of the dendrogram by this method is very different from those by the traditional linkage method, as we will see later in an example, and Newman's method may not be useful in understanding subcluster structures in a dendrogram.

3.4   Fuzzy Clustering

The method of fuzzy c-means [3,4,8,12,19] has been popular among researchers in at least two senses. First, the method gives fuzzy clusters instead of crisp clusters with much more information on the belongingness of an object to a cluster. Second, the algorithm is known to have high robustness over the K-means algorithm as to the variation of initial values and also noises and outliers. The robustness concerning outliers may still be improved by using fuzzy clustering and noise clustering [6,7].

Moreover the method of fuzzy c-means using an entropy term generalizes the Gaussian mixture model (see, e.g., [19]) and thus shows the expressive power of the fuzzy clustering model. Recently, Honda *et al.* [12] showed the multinomial mixture model for categorical data can be generalized by using a fuzzy co-clustering model.

4   Development of New Algorithms

We consider new algorithms on the basis of the above methods.

4.1   Fuzzy Clustering

Fuzzy clustering for categorical and mixed data can be studied by a similar way as the fuzzy c-means. The objective function is as follows:

$$J(U, V) = \sum_{i=1}^{c} \sum_{k=1}^{N} (u_{ki})^m d(x_k, v_i), \quad (m > 1). \tag{17}$$

where $d(x_k, v_i)$ is given by (6). U has the constraint: $\{u_{ki} \geq 0$ for all k, i, $\sum_{j=1}^{c} u_{kj} = 1$ for all $j\}$, while $V = (v_1, \ldots, v_c)$ does not have a constraint. The alternate minimization $\min_U J(U, V)$ and $\min_V J(U, V)$ while other variable is fixed to the last optimal solution is iteratively applied to $J(U, V)$ until convergence. There is no guarantee that the converged solution is the optimal solution for $J(U, V)$, but the solutions are empirically satisfactory.

For the present case of (17), the optimal solution U is:

$$u_{ki} = \frac{d(x_k, v_i)^{-\frac{1}{m-1}}}{\sum_{j=1}^{c} d(x_k, v_j)^{-\frac{1}{m-1}}}, \tag{18}$$

which is essentially the same as that for the standard Euclidean space, while the optimal solution V is different from the Euclidean case, and hence we should consider the case of L_1-space, that of categorical data, and that of medoids $(\mathbf{S} = \mathcal{X})$.

For $\mathbf{S} = \mathbf{R}^p$ with L_1 norm, we can use a weighted median algorithm [17]. For the case of medoids, the algorithm is essentially the same as the crisp case, i.e., we search the minimum of $\sum_k (u_{ki})^m d(x_k, v)$ with respect to v.

Since both a medoid and center are good representatives of a cluster, we can consider a new algorithm of the two representatives: Let $v_i = (v_i', v_i'')$ and assume that v_i' is a non-medoid center and v_i'' is a medoid, we define a new dissimilarity

$$d'(x_k, v_i) = \alpha d(x_k, v_i') + (1 - \alpha) d(x_k, v_i''), \tag{19}$$

with $0 < \alpha < 1$. If $d(x_k, v_i)$ is the L_1-distance, then v_i' is a weighted median and v_i'' is a medoid for G_i.

Such an algorithm using two representatives for a cluster have been developed for non-symmetric measure of dissimilarity [11,13]. Since we do not consider a non-symmetric measure here, we omit the detail.

4.2   Two-Stage Algorithms

A multi-stage algorithm can be a useful procedure when large-scale data should be handled. Consider a case when a large number of objects are gathered into a medium number of clusters using K-means, and then the centers are made into clusters using the same algorithm. In such a case K-medoids are also appropriate, since an object is made as a representative of a cluster.

Tamura *et al.* [22] proposed a two-stage procedure in which the first stage uses a p-pass K-means (i.e., a K-means procedure where the number of iterations is p; $p = 1$ or 2 is usually used.) in the first stage with the initial selection of centers using K-means++ [2], and the Ward method is used for the second stage. There is no loss of information because K-means and Ward method are based on the same criterion of the squared sum of errors from the center.

Two-stage algorithms of a median-Ward method and a medoid-Ward method can moreover be developed: the median-Ward method uses the one-pass K-median and an agglomerative procedure like the Ward method which uses L_1-norm throughout the procedure. The medoid-Ward method uses K-medoids in the first stage and an agglomerative procedure like the Ward method which uses an arbitrary dissimilarity.

In the latter method we use an objective function

$$J(\mathcal{G}) = \sum_{k=1}^{C} \sum_{x_l \in G_k} d(x_l, v_k),$$

where v_k is the medoid for G_k. Moreover we assume

$$\mathcal{G}[i,j] = \mathcal{G} \cup \{G_i \cup G_j\} - \{G_i\} - \{G_j\}, \quad d(G_i, G_j) = J(\mathcal{G}[i,j]) - J(\mathcal{G}).$$

The dissimilarity $d(G_i, G_j)$ is used in AHC algorithm of the medoid-Ward method. It is immediate to observe $d(G_i, G_j) \geq 0$. Note that if we set $d(x_l, v_k) = \|x_l - v_k\|_{L_1}$, we have the median-Ward method.

The last two of the median-Ward and the medoid-Ward algorithms have the drawback of much calculation in the second stage, but the number of objects are not many in the second stage and hence we can manage the processing time practically, but large-scale problems of millions of objects still have the fundamental problem of the inefficiency of calculations.

An Example: We show an example of network clustering on Twitter data [11]. Figure 1 shows the dendrogram output using Newman's method. The Twitter data of the graph with the number of nodes is 1744 and 108312 edges. The data consists of 5 political parties in Japan. The details are given in [11] and omitted here. The adjacency matrix A is made into the similarity matrix $S = A + A^2/2$. Apart from the large number of nodes, it is hard to observe subcluster structures in this dendrogram.

Figure 2 shows the dendrogram using the average linkage to the same data. The result shows subcluster structures but due to the large number of nodes, to observe the details is difficult.

Figure 3 shows the result using the two-stage procedure of medoid-Ward method. Subclusters are more clearly shown in this figure. The initial objects are summarized into 100 small clusters in the first stage.

Note that five clusters are observed in the latter two figures, while they are not clear in the first figure.

4.3 Use of Core Points

The concept of core points was introduced in DBSCAN [9], which is an important idea for effectively reducing the number of points for clustering.

In order to define a core point, a neighborhood $N(x; \epsilon)$ is defined: $N(x; \epsilon) = \{y \in \mathcal{X} : d(x, y) \leq \epsilon\}$. Let L be a positive integer. If $|N(x; \epsilon)| \geq L$ (the number

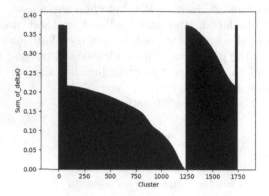

Fig. 1. Dendrogram from party data using newman method

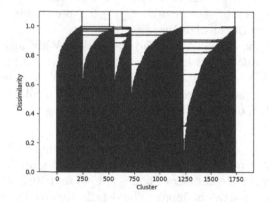

Fig. 2. Dendrogram from party data using an AHC algorithm. Average linkage was used.

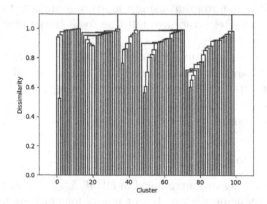

Fig. 3. Dendrogram from party data using a two-stage method

of points in the neighborhood is greater than or equal to L), then x is called a core point. This means that a core point has a enough number of points in its neighborhood. On the other hand, an isolated point is inclined to be a non-core point. The algorithm of DBSCAN starts from a core point and searches another core point in the neighborhood and connects it into the same cluster until no point is connected. The final point connected may not be a core-point. The final point is called a node point.

Let us consider the single linkage clustering in which only core points are clustered. Moreover merging in **AHC** is stopped when the level of merging m_C becomes lower than ϵ, and then the obtained clusters are output. We now have the following proposition.

Proposition 1. *Let the clusters obtained by DBSCAN be $G_1, \ldots, G_K$, and let the clusters obtained by the single linkage (with the stopping parameter ϵ) be $F_1, \ldots, F_K$. Take an arbitrary G_i. Then there is F_j such that $F_j \subseteq G_i$. Moreover if $F_j \neq G_i$, any $x \in G_i - F_j$ is a node point.*

This proposition implies that the result of DBSCAN is similar to clusters obtained from the single linkage for core points. In such a case a non-core point are allocated to a cluster of core points using a simple allocation rule such as the k-nearest neighbor rule.

5   Conclusion

An overview toward new algorithms for clustering categorical and mixed data has been given. Basic methods are reviewed and new methods are shown, which includes a two-stage agglomerative hierarchical algorithm with an example on Twitter and a theoretical results on the relation between DBSCAN and the single linkage.

An important problem of validation of clusters was not discussed, since this problem should be considered in a specific context of a practical application.

To handle a large-scale problem is still difficult in the sense that more efficient algorithms should be developed and also the interpretation problem of clusters should be solved. The latter problem needs knowledge of application domains.

Possible applications of methods herein include not only the categorical and mixed data, but also network clustering such as SNS (Social Networking Services) analysis.

Acknowledgment. This paper is based upon work supported in part by the Air Force Office of Scientific Research/Asian Office of Aerospace Research and Development (AFOSR/AOARD) under award number FA2386-17-1-4046.

References

1. Anderberg, M.R.: Cluster Analysis for Applications. Academic Press, New York (1973)
2. Arthur, D., Vassilvitskii, S.: k-means++: The advantages of careful seeding. In: Proceedings of SODA 2007, pp. 1027–1035 (2007)
3. Bezdek, J.C.: Fuzzy Mathematics in Pattern Classification, Ph.D. Thesis, Cornell University, Ithaca, NY (1973)
4. Bezdek, J.C.: Pattern Recognition with Fuzzy Objective Function Algorithms. Kluwer, Norwell (1981)
5. Blondel, V.D., Guillaume, J.L., Lambiotte, R., Lefebvre, E.: Fast unfolding of communities in large networks. J. Stat. Mech. Theory Exp. P10008 (2008)
6. Davé, R.N.: Characterization and detection of noise in clustering. Pattern Recogn. Lett. **12**, 657–664 (1991)
7. Davé, R.N., Krishnapuram, R.: Robust clustering methods: a unified view. IEEE Trans. Fuzzy Syst. **5**(2), 270–293 (1997)
8. Dunn, J.C.: A fuzzy relative of the ISODATA process and its use in detecting compact well-separated clusters. Cybern. Syst. **3**, 32–57 (1973)
9. Ester, M., Kriegel, H.P., Sander, J., Xu, X.: A density-based algorithm for discovering clusters in large spatial databases with noise. In: KDD 1996, pp. 226–231 (1996)
10. Everitt, B.S.: Cluster Analysis, 3rd edn. Arnold, London (1993)
11. Fujiwara, S.: Hierarchical Clustering for Directed Network Data, Master's thesis. University of Tsukuba, Master's Program in Risk Engineering (2017). (in Japanese)
12. Honda, K., Oshio, S., Notsu, A.: Fuzzy co-clustering induced by multinomial mixture model. J. Adv. Comput. Intell. Intell. Inform. **19**(6), 717–726 (2015)
13. Kaizu, Y., Miyamoto, S., Endo, Y.: Hard fuzzy C-Medoids for asymmetric networks. In: Proceedings of 16th World Congress of the International Fuzzy Systems Association (IFSA 2015), 30 June–July 3, Gijon, Spain, pp. 435–440 (2015)
14. Kaufman, L., Rousseeuw, P.J.: Finding Groups in Data: An Introduction to Cluster Analysis. Wiley, New York (1990)
15. MacQueen, J.B.: Some methods of classification and analysis of multivariate observations. In: Proceedings of the Fifth Berkeley Symposium on Mathematical Statistics and Probability, vol. 1 (University of California Press 1967), pp. 281–297 (1967)
16. Miyahara, S., Miyamoto, S.: A family of algorithms using spectral clustering and DBSCAN. In: Proceedings of 2014 IEEE International Conference on Granular Computing (GrC 2014), Noboribetsu, Hokkaido, Japan, pp. 196–200, 22–24 October 2014
17. Miyamoto, S., Agusta, Y.: An efficient algorithm for ℓ_1 fuzzy c-means and its termination. Control Cybern. **24**(4), 421–436 (1993)
18. Miyamoto, S.: Fuzzy Sets in Information Retrieval and Cluster Analysis. Springer, Heidelberg (1990)
19. Miyamoto, S., Ichihashi, H., Honda, K.: Algorithms for Fuzzy Clustering. Springer, Heidelberg (2008)
20. Newman, M.E.J., Girvan, M.: Finding and evaluating community structure in networks. Phys. Rev. E **69**(2), 026113 (2004)
21. Newman, M.E.J.: Fast algorithm for detecting community structure in networks. Phys. Rev. E **69**, 066133 (2004)
22. Tamura, Y., Miyamoto, S.: A method of two stage clustering using agglomerative hierarchical algorithms with one-Pass k-Means++ or k-Median++. In: Proceedings of 2014 IEEE International Conference on Granular Computing (GrC2014), Noboribetsu, Hokkaido, Japan, pp. 281–285, 22–24 October 2014

An Efficient Clustering Algorithm
Based on Grid Density and its Application
in Human Mobility Analysis

Chonghui Guo[1(✉)], Zhenna Na[1], Leilei Sun[2], and Xiaoguang Chen[1]

[1] Institute of Systems Engineering, Dalian University of Technology,
Dalian 116024, Liaoning, China
dlutguo@dlut.edu.cn, 644083826@mail.dlut.edu.cn
[2] School of Economics and Management, Tsinghua University,
Beijing 100084, China
sunll@sem.tsinghua.edu.cn

Abstract. Density Peaks based Clustering (DPC) is a recently proposed clustering algorithm, which is realized by first selecting some representative objects named density peaks, then assigning each remaining objects to one of the density peaks. Different from classical centroid-based clustering algorithms, DPC can find arbitrary-shaped clusters, and no predefined initial centroid set is required. However, a key disadvantage of the DPC lies in its computational complexity. DPC requires computation of two indicators for each data object. When the number of data increases, the computational complexity of DPC grows dramatically, which limits the application in many real-world problems. For example, when we use the taxi drop-offs to analyze the human mobility, DPC cannot be directly used due to the large number of taxi drop-off records. This paper proposes an efficient DPC algorithm based on grid density. By partitioning the effective data space into a desirable number of grids, two indicators of each grid are computed, as the number of grids is much smaller than that of data objects, a great amount of computational time and memory space can be saved. In experiments, we compare Grid-DPC with K-centers, affinity propagation and DPC on both synthetic and publicly available datasets. Results demonstrate that Grid-DPC can achieve comparable clustering performance with the classical DPC. We also employee Grid-DPC to analyze large-scale taxi records of a city in China and of New York Manhattan area. The discovered human mobility zones have great potential in urban planning and can help taxi drivers make better routing decisions.

Keywords: Clustering · DPC algorithm · Grid-DPC algorithm
Human mobility

1 Introduction

Clustering is the process of dividing a collection of points into several clusters, where the similarity of data points in the same cluster is high, while the similarity of data points in different clusters is low [1]. Now a great number of clustering algorithms have

© Springer International Publishing AG, part of Springer Nature 2018
V.-N. Huynh et al. (Eds.): IUKM 2018, LNAI 10758, pp. 87–100, 2018.
https://doi.org/10.1007/978-3-319-75429-1_8

been proposed, which can be divided into the following categories: the grid-based method, the density-based method, the partitioning method, the hierarchical method and the model-based method.

Clustering based on representative points is one of the most important clustering methods. K-means algorithm is a typical method which is based on representative points [2]. This algorithm uses the distance as a similarity evaluation index. At the first step, the algorithm randomly selects K points as the initial centers of the clusters. Then the algorithm assigns the remaining data points to the nearest clusters according to its distance from the center of each cluster. When all the data points are assigned, an iteration is completed and the new cluster center is calculated. The new cluster centers are obtained by calculating the average of each cluster. K-centers is very similar to K-means method, with the only difference that the way of choosing the cluster centers is different [3]. The K-centers method uses the real data points which are the nearest to the average cluster centers to be the cluster centers. K-centers is a simple and efficient method, but it has the shortcomings that the K value must be specified in advance and the choice of the initial cluster centers can have a great influence on the clustering results.

Affinity Propagation algorithm (AP) is a relatively new algorithm published in Science in 2007 [4]. The AP algorithm has been used in many fields since it was put forward. A significant advantage of this algorithm is that there is no need to specify the number of clusters in advance. Alternatively, it uses self-similarity to determine the number of clusters. Though it has remarkable success in practical applications, it can only discover spherical clusters and it is not suitable for arbitrary-shaped clustering problems. Besides, the computational efficiency of the algorithm is relatively high and it needs to occupy large memory.

In 2014, Alex et al. proposed a clustering method by Fast Search and Find of Density Peaks [5], which is referred to as DPC. The algorithm is based on a simple assumption that the density of the cluster centers is always higher than the density of the surrounding points, and the distance between the cluster centers is relatively large. By calculating the density and the high density distance of each data point, the algorithm can find out the cluster centers. It clusters centers through the decision graph and then the remaining non-center points are assigned to the corresponding clusters. The DPC algorithm does not need to be iteratively calculated and can be used to identify clusters of arbitrary shapes. However, when the algorithm is used on the large-scale datasets, calculating the distance between data points is very time-consuming and because of the limitation of the computer memory, the memory overflow problem occurs when the amount of data comes to a certain extent. So the algorithm greatly limits the use on large-scale datasets.

In order to overcome this difficulty, the Grid-DPC algorithm is proposed to reduce the running time and lower the memory demand. Our new algorithm first divides the data space into grid units, and then all the operations are based on the grid units. Our new method greatly reduces the running time when dealing with large-scale datasets.

Our Grid-DPC algorithm can be used to find the human mobility patterns by clustering the taxi datasets. Human mobility studies have attracted the attention of people in different fields and human mobility patterns can be used to explore the rhythm of a city. The taxi datasets in a city can be used as an important source of data for analyzing the human mobility patterns and it can accurately record the locations where the passengers get on and get off the taxis. Therefore, we can find the hot spots of cities by studying the datasets of the taxis. Also, the laws of human activities can be found by clustering the taxi dataset of a city.

So this paper is organized as follows: Sect. 2 introduces the DPC clustering algorithm in detail. In Sect. 3, our new clustering algorithm Grid-DPC which is inspired by the grid-based clustering method is fully introduced. At the same time, in order to verify the validity of the new algorithm, we make a theoretical complexity analysis. Section 4 validates the effectiveness and the efficiency of the Grid-DPC algorithm by testing it on the synthetic datasets compared with K-centers, AP and DPC clustering algorithms. Then the algorithm is applied on two large-scale taxi datasets in a city in China and in New York Manhattan area to discover the hot spots and the human mobility zones of the city in Sect. 5. Finally, the conclusions are concluded in Sect. 6.

2   DPC Clustering Algorithm

Clustering by fast search and find of density peaks was presented by Alex et al. in 2014. This algorithm is based on a basic assumption that the points have higher density are always surrounded by the points with lower density, and the distance between the cluster centers is relatively large. The DPC algorithm has two important parameters called the local density ρ_i and the high density distance δ_i. For any point i, its local density ρ_i has two methods to calculate which are called the cut-off kernel and the Gaussian kernel. The definition of the cut-off kernel is

$$\rho_i = \sum_j \chi(d_{ij} - d_c) \tag{1}$$

where

$$\chi(x) = \begin{cases} 0 & x > 0 \\ 1 & x \le 0 \end{cases} \tag{2}$$

d_c is a cutoff distance needs to be specified in advance and d_{ij} is the distance between the point i and j. The Gaussian kernel is defined as

$$\rho_i = \sum_j e^{-(\frac{d_{ij}}{d_c})^2} \tag{3}$$

It is easy to find that the cut-off kernel is a discrete value in contrast with the Gaussian kernel with a continuous value. Hence, the probability that different points have the same density in the latter method is smaller. For any point i, its high density distance can be defined as

$$\delta_i = \begin{cases} \max_j d_{ij} \\ \min_{j:\rho_j > \rho_i} (d_{ij}) \end{cases} \tag{4}$$

which means that when point i has the highest local density, δ_i represents the maximum distance between point i and other points, which means that $i = \arg \max_j \rho_j$. Otherwise, δ_i represents the smallest distance between point j and point i if the local density of point j is bigger than point i. Then the cluster centers can be found by drawing the decision graph or by calculating the value of γ_i in formula (5) and sorting it in descending order. Obviously, the bigger the value of γ_i, the more likely it is to become a cluster center for point i. This means that by sorting the value of γ_i in descending order, the cluster centers can be selected directly. Finally, all the remaining data points can be assigned to the nearest cluster centers.

$$\gamma_i = \rho_i \delta_i \tag{5}$$

The DPC algorithm is able to cluster datasets of arbitrary shapes. However, the computational complexity of the algorithm will show a rapid growth with the increase of data number. When dealing with large-scale datasets, due to the limited computing ability, the running time is very long. Also the process of computing will consume much memory space. When the amount of data comes to a certain extent, the memory overflow problem will occur. Therefore, the ability of the algorithm to deal with the large-scale datasets is relatively limited. So the Grid-DPC algorithm is proposed to solve this problem.

3   Grid-DPC Clustering Algorithm

To overcome the difficulties that the DPC algorithm cannot cluster large-scale datasets, this section combines the idea of the grid-based clustering algorithm and proposes the grid-based density peak clustering algorithm (Grid-DPC). At the same time, in order to verify the validity of the new algorithm, we make a detailed theoretical analysis from two aspects: time complexity and space complexity.

3.1   The Main Idea of the Grid-DPC

To improve the computational efficiency of the DPC algorithm, the data points need to be divided into grid units first. Each dimension is divided into m pieces, where m is the parameter that needs to be specified in advance. Here we set D to be the number of grid units, then when dealing with two-dimensional dataset, D equals m^2. After determining the number of grids that need to be divided, the data space is evenly divided into equal size of the grid units and the center point of each grid unit is selected as the representative point of the grid. Then all the operations are carried out on the grid units.

The Grid-DPC algorithm still needs to calculate the local density ρ_i and the high density distance δ_i. Unlike the DPC algorithm, each grid unit of the Grid-DPC algorithm is replaced by a representative point, and the local density of the representative point is not the number of points within a certain radius around the point. Instead, the Gaussian kernel function is used to calculate the local density as follows:

$$\rho_i = \sum_{j=1}^{N} e^{-(\frac{d_{ij}}{d_c})^2} \tag{6}$$

where

$$d_{ij} = \|Grid(i) - X(j)\|^2 \tag{7}$$

$Grid(i)$ is the coordinate of the grid representative point i, $X(j)$ is the coordinate of the data point j in the whole dataset, d_c is the cutoff distance which is specified in advance and N is the number of points in the dataset. Then the high density distance of all the representative points is calculated as follows:

$$\delta_i = \begin{cases} \max_j d_{ij} \\ \min_{j:\rho_j > \rho_i} (d_{ij}) \end{cases} \tag{8}$$

where

$$d_{ij} = \|Grid(i) - Grid(j)\|^2 \tag{9}$$

point i and point j are both the representative points. It can be concluded that when the representative point i has the highest local density, δ_i represents the biggest distance between all the representative points and point i, which means that $i = \arg\max_j \rho_j$.

Otherwise, δ_i represents the smallest distance between point j and point i if the local density of the representative point j is bigger than point i.

After getting two parameters ρ_i and δ_i, γ_i can be calculated by

$$\gamma_i = \rho_i \delta_i \tag{10}$$

where i is the representative point. This means that by multiplying the two parameters and sorting γ_i in descending order, the first K grid representative points are automatically selected as K cluster centers. Also, the decision graph can be drawn to select the K cluster centers.

Finally it comes to the process of the class label transferring. There are two steps to assign the remaining data points. First, the remaining representative points are assigned to the nearest cluster centers. Then the class labels of all the representative data points are transferred to all the data points within the corresponding grids, thus completing the whole clustering process.

3.2 Description of the Grid-DPC

The Grid-DPC algorithm needs to input three parameters in advance, which are the cluster number K, the cutoff distance d_c and the grid number D. The cutoff distance d_c is a robust parameter. The cluster number K and the grid number D need to be determined by the actual situation. So the whole process of the Grid-DPC algorithm can be described as follows:

Algorithm Grid-DPC

Input: Cutoff distance d_c, cluster number K and gird number D;

Output: Clustering result c;
Steps:
1: Divide dataset X into D grid units and select the center point of each grid unit as the representative point;
2: Calculate the local density ρ_i and the high density distance δ_i of each representative grid i;
3: Calculate γ_i by (10) and sort it in descending order to select K representative points as the cluster centers;
4: Assign the remaining grid representative points and all the non-representative points to the cluster centers they belong to;

3.3 Analysis of the Algorithm Complexity

In order to verify the effectiveness of the Grid-DPC algorithm, its space complexity and time complexity are used to compare with the original DPC algorithm. Let N be the total number of data points and D be the total grid number, so it is obvious that $N \gg D$.

The optimization of the Grid-DPC algorithm in space complexity is mainly reflected in the following: the maximum space required by the DPC algorithm lies in the process of constructing the distance matrix and the space complexity in this process is $O(N^2)$, The Grid-DPC algorithm also requires a distance matrix that stores the distance between all the data points and D grid representative points, but the space complexity that the Grid-DPC algorithm needs is only $O(DN)$.

The optimization of the Grid-DPC algorithm in time complexity is mainly analyzed in the following two aspects: (a) When calculating the local density of the algorithm, the Grid-DPC algorithm calculates the local density of all the grid representative points, so the time complexity is $O(DN)$, but the time complexity of the DPC algorithm when calculating the local density is $O(N^2)$. (b) In the process of calculating the high density distance, the time complexity of the Grid-DPC algorithm is $O(DN)$ while the time complexity of the Grid-DPC algorithm is $O(N^2)$.

For the reason that $N \gg D$, the algorithm Grid-DPC is superior to the original DPC algorithm in terms of time complexity and space complexity.

4 Computational Experiments

In this section, we use the synthetic datasets to test the performance of the Grid-DPC algorithm compared with DPC, AP and K-centers. The evaluation criteria is the Accuracy and the NMI which will be introduced in the following. All the experiments are conducted on a PC with Intel Core i7-6700 3.4 GHz processer and 8 G memory. The programming language is MATLAB. Other clustering algorithms like AP, K-centers and DPC are used to give a contrast to the Grid-DPC algorithm. For the reason that the number of clusters in each synthetic dataset is specified, so AP algorithm needs to adjust the parameter P when using it. There is no need to specify any parameter of K-centers in advance, but the algorithm always achieves different clustering results every time using it. So the algorithm repeats 100 times to get the average clustering results.

4.1 Evaluation Criteria

Here, the evaluation of the clustering results uses two criteria, the Accuracy and the normalized mutual information (NMI). The Accuracy is a relatively intuitive evaluation criteria, which is defined as follows:

$$\text{Accuracy} = \frac{\sum_{i=1}^{n} \delta(\hat{c}_i, c_i)}{n} \tag{11}$$

where $\hat{c}_i$ is the real class label of point i, c_i is the clustering label of point i. The function $\delta(x, y)$ is used to compare the consistency of x and y, which means that if $x = y$, then $\delta(x, y) = 1$, otherwise $\delta(x, y) = 0$.

The normalized mutual information is an evaluation criteria based on information theory [6]. It evaluates the consistency of the clustering results by calculating the mutual information between the clustering results and the real class labels. NMI is defined as follows.

$$\text{NMI} = \frac{I(\mathbf{c}, \hat{\mathbf{c}})}{\sqrt{H(\mathbf{c})H(\hat{\mathbf{c}})}} \tag{12}$$

where $I(\mathbf{c}, \hat{\mathbf{c}})$ represents the mutual information between the clustering label $\mathbf{c}$ and the real class label $\hat{\mathbf{c}}$. $H(\cdot)$ represents the information entropy of a single vector.

In the experiment, the running time is also a criteria to evaluate the efficiency of the algorithm. Therefore, the Accuracy, the NMI and the running time test the consistency and efficiency of the algorithms.

4.2 Synthetic Datasets

The synthetic experiment uses datasets like Flame [7], D31 [8], R15 [9], Aggregation [10] and Jain [11] to verify the effectiveness of the Grid-DPC algorithm, which contain 240, 3100, 600, 788 and 373 records. All the datasets are tagged so the number of each cluster is specified in advance. At the same time, all the datasets are multiplied by 10, 20, 30, and 50 with some noise added on them to observe the clustering results when the amount of data increases. The way of adding noise can be seen as follows:

$$X^* = X + 0.1 * E \tag{13}$$

where E is the randomly generated noise matrix, X is the original matrix and X^* is the new matrix with some noise. d_c is chosen as the value that maximizes the clustering accuracy. When the amount of the data is more than 30000, the clustering algorithms like AP and DPC almost reach the upper limit of the memory, so the results of this kind of datasets are not shown. The grid number D used in the second experiment is 400. The clustering results are shown in Table 1.

Table 1. The clustering results of synthetic datasets

Flame	time	acc	Flame*10	time	acc	Flame*20	time	acc	Flame*30	time	acc	Flame*50	time	acc
DPC	0.032	1	DPC	3.6	1	DPC	15.04	1	DPC	34.08	1	DPC	96.063	1
Grid-DPC	0.446	0.9875	Grid-DPC	3.08	0.9958	Grid-DPC	5.489	0.9917	Grid-DPC	8.4437	0.9958	Grid-DPC	13.927	0.9875
K-centers	0.038	0.84	K-centers	3.528	0.8339	K-centers	13.827	0.8315	K-centers	31.619	0.8374	K-centers	90.088	0.8295
AP	0.24	0.8417	AP	26.817	0.8313	AP	592.244	0.8167	AP	—	—	AP	—	—

D31	time	acc	D31*10	time	acc	D31*20	time	acc	D31*30	time	acc	D31*50	time	acc
DPC	6.109	0.9668	DPC	—	—	DPC	—	—	DPC	—	—	DPC	—	—
Grid-DPC	3.6426	0.9174	Grid-DPC	35.9987	0.9049	Grid-DPC	71.4179	0.9178	Grid-DPC	106.593	0.9125	Grid-DPC	177.915	0.9179
K-centers	5.576	0.8252	K-centers	—	—	K-centers	—	—	K-centers	—	—	K-centers	—	—
AP	105.326	0.9761	AP	—	—	AP	—	—	AP	—	—	AP	—	—

R15	time	acc	R15*10	time	acc	R15*20	time	acc	R15*30	time	acc	R15*50	time	acc
DPC	0.391	0.9949	DPC	22.764	0.9967	DPC	94.627	0.9967	DPC	220.076	0.9967	DPC	1199.4	0.9967
Grid-DPC	1.038	0.9975	Grid-DPC	7.088	0.9615	Grid-DPC	13.406	0.9633	Grid-DPC	20.6217	0.9633	Grid-DPC	34.2312	0.9717
K-centers	0.4	0.8826	K-centers	21.728	0.8267	K-centers	85.841	0.8302	K-centers	201.485	0.8253	K-centers	1811.6	0.8236
AP	3.464	0.9967	AP	392.921	0.9967	AP	—	—	AP	—	—	AP	—	—

Aggregation	time	acc	Aggregation*10	time	acc	Aggregation*20	time	acc	Aggregation*30	time	acc	Aggregation*50	time	acc
DPC	0.387	0.9949	DPC	39.995	0.9949	DPC	159.172	0.9949	DPC	385.889	0.9949	DPC	—	—
Grid-DPC	1.0732	0.9975	Grid-DPC	9.276	0.9975	Grid-DPC	17.8545	0.9975	Grid-DPC	27.4327	0.9979	Grid-DPC	43.2887	1
K-centers	0.37	0.8892	K-centers	37.632	0.8837	K-centers	148.657	0.8829	K-centers	358.372	0.8799	K-centers	—	—
AP	1.853	0.9074	AP	—	—	AP	—	—	AP	—	—	AP	—	—

Jain	time	acc	Jain*10	time	acc	Jain*20	time	acc	Jain*30	time	acc	Jain*50	time	acc
DPC	0.09	0.9571	DPC	9.076	0.9571	DPC	36.279	0.9571	DPC	83.653	0.9571	DPC	238.267	0.9571
Grid-DPC	0.5953	0.933	Grid-DPC	4.3852	0.933	Grid-DPC	8.7964	0.933	Grid-DPC	13.0351	0.9383	Grid-DPC	21.2817	0.9383
K-centers	0.084	0.8606	K-centers	8.473	0.8606	K-centers	33.599	0.8606	K-centers	78.348	0.8606	K-centers	225.078	0.8606
AP	0.457	0.8606	AP	345.657	0.8606	AP	—	—	AP	—	—	AP	—	—

The experimental results show that the running time of Grid-DPC is much less than the time of DPC, AP and K-centers with the increase of the data number. For example, in the R1550 dataset, the running time of Grid-DPC is 34.2312 s while DPC and K-centers use more than 1000 s to get the results. However, the accuracy of Grid-DPC is not as good as DPC in some datasets, this is because the grid number 400 is chosen to balance the time and the Accuracy of the Grid-DPC. For this reason, the R1530 and R1550 datasets are used to test the clustering results and the running time when the grid number is different. Table 2 shows the results of the Accuracy and the running time of the R1530 and R1550 datasets. Figure 1 separately describe the Accuracy and the running time of the clustering results. It can be seen that the reason the Accuracy of the Grid-DPC algorithm is not better than DPC is that the grid number is not large enough. When increasing the gird number of Grid-DPC, the Accuracy can be close to the Accuracy of DPC, but this will lead to a rapidly increase of the running time. Also through varying the grid number D from 100 to 1600 in the two datasets, it is obvious that the computational time is increasing rapidly. So it is necessary to consider how to maintain a balance between the accuracy and the running time of the algorithm. This means that in order to ensure the less running time, sometimes giving up a little accuracy is possible and necessary.

Table 2. The Accuracy and the running time of the R1530 and the R1550 datasets

Dataset	Method	Grid number(D)	Accuracy	Time
R1530	Grid-DPC	100	0.8911	**5.0502**
	Grid-DPC	400	0.9633	20.6217
	Grid-DPC	900	0.9911	47.5404
	Grid-DPC	1600	0.9917	85.4904
	DPC	–	**0.9967**	220.076
R1550	Grid-DPC	100	0.9078	**8.4165**
	Grid-DPC	400	0.9717	34.2312
	Grid-DPC	900	0.9842	79.4331
	Grid-DPC	1600	0.9933	139.9579
	DPC	–	**0.9967**	1199.4

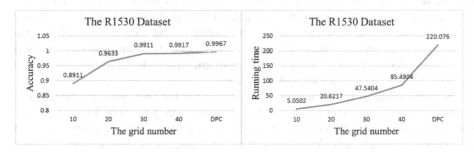

(a) The trend of Accuracy on R1530 dataset   (b) The trend of running time on R1530 dataset

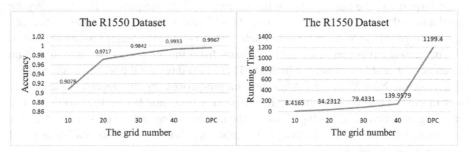

(c) The trend of Accuracy on R1550 dataset   (d) The trend of running time on R1550 dataset

Fig. 1. The Accuracy and running time of R1530 and R1550 datasets

5 Applications on Taxi Datasets

In recent years, the studies of human mobility have been more and more popular. The Grid-DPC algorithm can be used to find the human mobility patterns by clustering the taxi datasets. In this section, two taxi datasets are used to verify the validity of the Grid-DPC algorithm on large-scale datasets and find the hot spots of cities. As a means of transport, taxi occupies a very important role in the daily life. The locations where passengers get on and get off the taxis also contain rich information.

In the first experiment of the taxi dataset, the locations of latitude and longitude of a city in China will be used to verify the effectiveness of the Grid-DPC algorithm. Also the hot spots in the city can be explored by visualizing the clustering results. The first dataset comes from 4285 taxis of the city for one week from January 1th to January 7th in 2014. By pre-processing the dataset, the total number of the effective location points is 1.87 millon. In the experiment, the grid number is set to 100 * 100 according to the actual situation of the city. All the data points are divided into 20 clusters by taking the K value as 20.

The clustering results are visualized and the results are shown in Fig. 2. The points shown in Fig. 2(a) and (b) are the grid centers of each grid which have higher density, and the data points and clustering results are depicted on the actual map. Figure 2(a) illustrates the clustering result with no height, while Fig. 2(b) shows the height of the grid centers. The height of each point is measured by the local density of the grid center, so it represents the heat of a location to some extent. At the same time, the heat map is also depicted to find the hottest areas of the city in Fig. 2(c) and twenty hot areas of the city can be seen through Fig. 2(d).

From the result it can be concluded that most of the top ten regions are the business districts, which is consistent with the will of shopping during the New Year's Day. It is obvious that the heat in the business districts is not the same. So by ranking the heat of the business districts, reliable advice can be provided to businessmen to choose a suitable location. Since the selected date of data is in the period of the New Year's Day, so the airport and the railway station have relatively high frequency. By clustering the taxi dataset, the results can help the taxi drivers understand which places of the city have more people to take the taxis and help taxi drivers make better routing decisions.

The second experiment uses the taxi dataset from the New York Manhattan area on December 2th and December 6th in 2015. These two days are the working day and the off day for the New York area. After the data pre-processing, the record of the yellow taxis in New York on December 2th comes to a total number of 0.75 million. The following statistical result in Fig. 3 shows the number of records in each hour. Also the record on December 6th comes to a total number of 0.8 million and the statistical result is also shown in Fig. 3.

In order to obtain distributions of the taxis on different time, three periods 4–5, 8–9 and 19–20 on December 2th and 0–1, 6–7 and 12–13 on December 6th are selected to be clustered separately to observe the evolution of the distribution. It can be seen from Fig. 3 that the three periods on December 2th correspond to the minimum number of data points, the largest number of data points when getting on work and the largest number of data points when getting off work. Also the three periods on December 6th

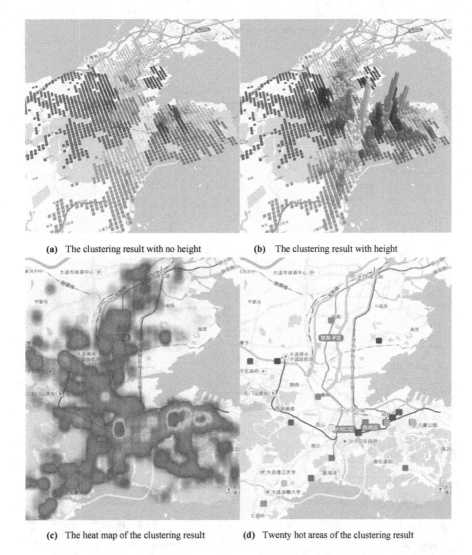

(a) The clustering result with no height   (b) The clustering result with height

(c) The heat map of the clustering result   (d) Twenty hot areas of the clustering result

Fig. 2. Clustering results of the first taxi dataset

represent the minimum number of the data points, the largest number of data points during the day time and the largest number of data points in the evening. In this experiment, the number of clusters K is selected as 10. The clustering results are shown in Fig. 4.

The clustering results in Fig. 4 only show the longitude range from −74.02182 to −73.94789 and the latitude range from 40.700542 to 40.766252. By observing the clustering result, it can be found that during 4:00 to 5:00 of the working day, most of the cluster centers are restaurants and public transport stations, and two parks are also displayed in the cluster results. In the 8:00–9:00 period, nearly half of the cluster

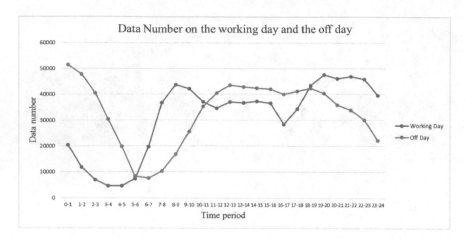

Fig. 3. Data number in each hour on the working day and the off day

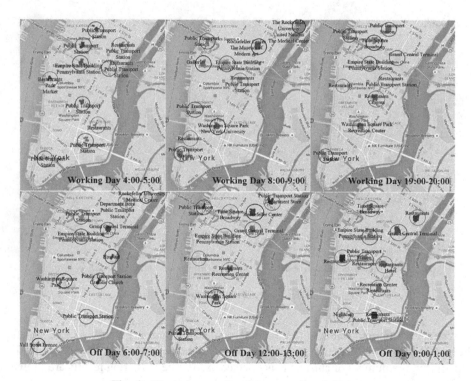

Fig. 4. Clustering results of the second taxi dataset

centers are public transport stations and restaurants, at the same time two universities, the Empire State Building, Rockefeller Center and Washington Square are also displayed in the cluster results. For the time of 19:00 to 20:00 in the working day, most of the cluster centers are still restaurants and public transport stations, while the Grand Central Station and the Broadway emerge as the new cluster centers.

Through the comparison of these three time periods, it can be concluded that public transport stations and restaurants occupied most of the cluster centers, while the Pennsylvania station is an important transportation hub for all of the three periods. In the early peak and late peak period, the ferry terminal and the Washington Square are also the cluster centers. With the emergence of early peak, the heat of the landmark buildings is gradually rising. Also with the emergence of late peak, the recreation centers in the cluster results gradually emerge.

On the off day, the period 6:00–7:00 has the least amount of data number in the day, in which the public transport stations occupy most of the cluster centers. At the same time, the Wall Street Bronze, the Grand Central Terminal, hospitals and the other landmarks are also shown in the cluster results. For the time of 12:00 to 13:00 on the off day, public transport stations, restaurants and shopping centers occupy most of the cluster centers. In the meanwhile, the Broadway, the Rockefeller Center, the Washington Square and the other landmark buildings are also shown in the cluster results. During the 0:00–1:00 period of the off day, restaurants occupy most of the cluster centers, while the nightclubs, the Broadway and two famous train stations are also shown in the clustering results.

By comparing the three periods of the off day, it can be concluded that the two train stations in Manhattan area have always been the cluster centers in all of the three periods, which shows the importance of the two stations on the off day. The cluster centers of 6:00–7:00 period are mainly the public transport stations, while in the 12:00–13:00 and 0:00–1:00 time period, the restaurants and the recreation centers gradually emerge in the cluster results. For the reason that 0:00–1:00 period is the midnight on the off day, restaurants and recreation centers such as nightclubs, the Broadway and cinemas are particularly prominent.

By comparing 4:00–5:00 period of the working day and 6:00–7:00 period of the off day, it can be found that the cluster centers of 4:00–5:00 period are mainly the public transport stations and the restaurants, while in the 6:00–7:00 period of the off day the cluster centers are mainly public transport stations and few restaurants. By comparing the peak hours of 8:00–9:00 period of the working day and 12:00–13:00 period of the off day, it can be seen that there are several recreation centers, department stores and the Broadway as the new cluster centers of the off day. And then by comparing the late peak of the working day and the off day, it can be found that some restaurants are added in the cluster centers on the off day. This is consistent with people's law of activity on the working day and the off day.

The experiments on these two taxi datasets demonstrate the efficiency and the effectiveness of the Grid-DPC algorithm, which shows the superiority of the algorithm in large-scale datasets. Also, the cluster centers reflect the human mobility patterns in the cities and can help taxi drivers make better decisions.

6 Conclusions

An efficient DPC algorithm based on grid density is proposed in this paper. This algorithm divides the whole dataset into grid units and all the operations are conducted based on grid units including the calculation of local density ρ_i and high density distance δ_i. The running time of the Grid-DPC algorithm is much less than the DPC algorithm when dealing with large-scale datasets but the accuracy of Grid-DPC is basically the same as DPC.

The performance of our algorithm is tested on both synthetic and publicly available datasets. The experimental results show that the running time of our Grid-DPC algorithm has a great advantage compared to DPC, AP and K-centers. The experimental results in the taxi datasets depict the hot spots in the cities through clustering the locations of getting on and getting off the taxis. The discovered human mobility zones have great potential in urban planning and facility location for governments, which also can help taxi drivers make better routing decisions.

Acknowledgements. This work was supported in part by the Natural Science Foundation of China [Grant Numbers 71771034].

References

1. Han, J., Kamber, M., Pei, J.: Data Mining: Concepts and Techniques, 3rd edn., pp. 443–496. Morgan Kaufmann Publishers Inc., Burlington (2011)
2. MacQueen, J.B.: Some methods for classification and analysis of multivariate observations. In: Proceedings of the Fifth Berkeley Symposium on Mathematical Statistics and Probability, vol. 1, pp. 281–297. University of California Press (1967)
3. Kaufman, L., Rousseeuw, P.J.: Finding Groups in Data: An Introduction to Cluster Analysis. Wiley, New York (1990)
4. Frey, B.J., Dueck, D.: Clustering by passing messages between data points. Science **315** (5814), 972–976 (2007)
5. Rodriguez, A., Laio, A.: Clustering by fast search and find of density peaks. Science **344** (6191), 1492–1496 (2014)
6. Vinh, N.X., Epps, J., Bailey, J.: Information theoretic measures for clusterings comparison: is a correction for chance necessary? In: Proceedings of the 26th Annual International Conference on Machine Learning, pp. 1073–1080. ACM (2009)
7. Fu, L., Medico, E.: FLAME, a novel fuzzy clustering method for the analysis of DNA microarray data. BMC Bioinform. **8**(1), 3 (2007)
8. Lancichinetti, A., Fortunato, S., Kertész, J.: Detecting the overlapping and hierarchical community structure in complex networks. New J. Phys. **11**(3), 033015 (2009)
9. Veenman, C.J., Reinders, M.J.T., Backer, E.: A maximum variance cluster algorithm. IEEE Trans. Pattern Anal. Mach. Intell. **24**(9), 1273–1280 (2002)
10. Gionis, A., Mannila, H., Tsaparas, P.: Clustering aggregation. ACM Trans. Knowl. Discov. Data (TKDD) **1**(1), 341–352 (2007)
11. Jain, A.K., Law, M.H.C.: Data clustering: a user's dilemma. In: Pal, S.K., Bandyopadhyay, S., Biswas, S. (eds.) PReMI 2005. LNCS, vol. 3776, pp. 1–10. Springer, Heidelberg (2005). https://doi.org/10.1007/11590316_1

A Multi-class Support Vector Machine Based on Geometric Margin Maximization

Yoshifumi Kusunoki$^{(\boxtimes)}$ and Keiji Tatsumi

Graduate School of Engineering, Osaka University,
2-1, Yamadaoka, Suita, Osaka 565-0871, Japan
{kusunoki,tatsumi}@eei.eng.osaka-u.ac.jp

Abstract. Support vector machines (SVMs) are popular supervised learning methods. The original SVM was developed for binary classification. It selects a linear classifier by maximizing the geometric margin between the boundary hyperplane and sample examples. There are several extensions of the SVM for multi-class classification problems. However, they do not maximize geometric margins exactly. Recently, Tatsumi and Tanino have proposed multi-objective multi-class SVM, which simultaneously maximizes the margins for all class pairs. In this paper, we propose another multi-class SVM based on the geometric margin maximization. The SVM is formulated as minimization of the sum of inverse-squared margins for all class pairs. Since this is a nonconvex optimization problem, we propose an approximate solution. By numerical experiments, we show that the propose SVM has better performance in generalization capability than one of the conventional multi-class SVMs.

1 Introduction

Support vector machines (SVMs) are popular supervised learning methods in machine learning. The original SVM [3,9] was proposed for binary classification problems, i.e., tasks to learn a classifier separating two group of examples. The SVM determines a linear discriminant function based on the principle of margin maximization. It means that the linear function is selected such that the minimum distance between sample examples and the hyperplane associated with the linear function is maximized. It is based on geometric motivation that the neighborhood of a labeled example includes those of the same label as that example. In another viewpoint, the margin maximization is regarded as regularization in model selection.

There are various extensions of the binary SVM for multi-class classification problems, i.e., there are more than two groups which should be separated. A simple approach of extension is one-against-all (OAA) [7], which reduces a k-class problem to k binary problems to separate one class from the others, and applies the binary SVM to these problems. Another major approach is all-together (AT), which is formulated as minimization of the sum of regularization terms and errors of a classifier. Since optimization problems of AT-SVM includes more decision variables and constraints, computational costs are higher than OAA-SVM.

© Springer International Publishing AG, part of Springer Nature 2018
V.-N. Huynh et al. (Eds.): IUKM 2018, LNAI 10758, pp. 101–113, 2018.
https://doi.org/10.1007/978-3-319-75429-1_9

On the other hand, Doğan et al. [4] showed that AT-SVMs outperform OAA-SVM in classification accuracy in the case of the linear kernel. (And they are comparable in the case of the RBF kernel.)

The above mentioned multi-class SVMs do not exactly maximize geometric margins. Especially, the existing AT-SVMs are derived from regularized model selection, instead of geometric interpretation. Recently, Tatsumi and Tanino [8] have pointed out geometric margin maximization in multi-class problems, and formulated a multi-objective optimization problem which simultaneously maximizes all of the class pair margins. The model is called multi-objective multi-class SVM (MMSVM).

To obtain Pareto solutions of multi-objective optimization problems, we need some scalarization method. Because of nonconvexity of MMSVM, almost all conventional scalarization methods cannot be computed efficiently. Tatsumi and Tanino [8] used the ε-constraint method, and showed that obtained classifiers have better classification accuracy than those of AT- and OAA-SVMs. However, the method needs high computational effort to find good parameter ε, and cannot maximize the margins of class pairs uniformly like weighted-sum scalarization.

In this paper, we propose a multi-class SVM which is another scalarizing formulation of MMSVM. It minimizes the sum of inverse-squared margins for all class pairs. To overcome nonconvexity of the scalarized MMSVM, we linearize its nonconvex parts and solve the modified convex optimization problem. Consequently, we obtain an approximation solution for the original problem. Moreover, we show an upper bound of the ratio of the objective function value of the approximation solution to the optimal value of the original problem. A special case of the proposed multi-class SVM coincides with the conventional AT-SVM in [2,9,10]. By numerical experiments, we show that the proposed multi-class SVM outperforms the AT-SVM in generalization capability.

We demonstrate that large margin classifiers can be obtained by the proposed SVM. See Fig. 1. The left and right figures show classification boundaries of Wine Data Set obtained by AT-SVM and the proposed SVM, respectively. The tables after the figures show the values of margins for three class pairs $(1, 2)$, $(1, 3)$ and $(2, 3)$. We can see that all of the margins of the classifier by the proposed SVM are larger than those by AT-SVM.

This paper is organized as follows. In Sect. 2, binary and multi-class SVMs are introduced. In Sect. 3, we discuss the proposed multi-class SVM. We formulate the model minimizing the sum of inverse-squared margins, and propose the approximate solution. In Sect. 4, numerical experiments are presented to examine performance of the proposed SVM. Finally, in Sect. 5, concluding remarks are provided.

2   Multi-class SVMs

2.1   Multi-class Classification

In this paper, we deal with classification problems of supervised learning. Let n-dimensional real space $\mathbf{R}^n$ be an input space and $C = \{1, 2, \ldots, c\}$, $c \geq 2$ be a

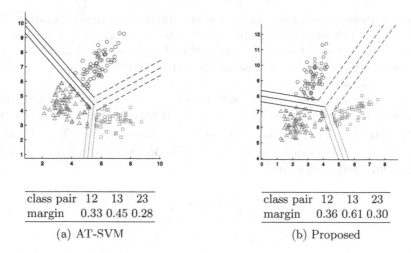

class pair	12	13	23
margin	0.33	0.45	0.28

(a) AT-SVM

class pair	12	13	23
margin	0.36	0.61	0.30

(b) Proposed

Fig. 1. Separating lines obtained by AT-SVM (the left figure) and the proposed SVM (the right figure) for Wine Data Set. There are three classes—class 1: blue circles, class 2: orange triangles, class 3: yellow squares. The solid lines are the separating line with the margins of class pair 12. The broken lines are of class pair 13. The dotted lines are of class pair 23. The data are plotted in the 2-dimensional affine subspace passing through 3 normal vectors of 3 classes. (Color figure online)

class label set. A classification problem is to find a function $D : \mathbf{R}^n \to C$ from m input vectors $x^1, \ldots, x^m \in \mathbf{R}^n$ and class labels $y_1, \ldots, y_m \in C$. Such a function D is called a classifier. $S = ((x^1, y_1), \ldots, (x^m, y_m))$ is called a training set. We aim to find a function having high classification accuracy, i.e., it can correctly assign class labels to (unseen) input vectors. Let $M = \{1, \ldots, m\}$ be the index set of the training set. For $p \in C$, we define $M^p = \{i \in M \mid y_i = p\}$.

When $c = 2$, the problem is called binary classification. On the other hand, when $c \geq 3$, it is called multi-class classification. The SVM proposed in this paper can solve multi-class classification problems.

We consider a linear classifier D given in the following form: for $x \in \mathbf{R}^n$,

$$D(x) = \underset{p \in C}{\operatorname{argmax}} \{f_p(x) = (w^p)^\top x + b_p\}, \tag{1}$$

where $w^1, \ldots, w^c \in \mathbf{R}^n$ and $b_1, \ldots, b_c \in \mathbf{R}$. If there is more than one label p whose value $f_p(x)$ is the maximum, we arbitrarily select one label among them. Each $f_p(x)$ is called a linear discriminant function for the class label p. We propose a method to learn the parameters $(w^1, b_1), \ldots, (w^p, b_p)$ from the training set S, to construct the linear classifier D with high accuracy.

2.2   SVMs

SVMs (Support Vector Machines) are methods for binary classification to learn linear classifiers from examples. First, we mention the SVM for linearly separable binary classification problems. In binary classification, a linear classifier D is reduced to the following form: letting $f(x) = w^\top x + b$, $D(x) = 1$ if $f(x) > 0$; $D(x) = -1$ if $f(x) < 0$. Here, we suppose the class labels are 1 and -1. w and b are parameters of the linear classifier D. Input vectors x with $f(x) = 0$ are arbitrarily classified.

SVM selects a classifier whose boundary hyperplane having the largest margin. A margin of a hyperplane $f(x) = 0$ is the distance between the hyperplane and the nearest input vector in the training S, namely, $\frac{\min_{i \in M} |w^\top x^i + b|}{\|w\|}$, where $\| \cdot \|$ is the Euclidean norm. The largest-margin classifier is obtained by solving the following optimization problem.

$$\begin{aligned} \underset{w,b}{\text{minimize}} \quad & \frac{1}{2} \left(\frac{\min_{i \in M} |w^\top x^i + b|}{\|w\|} \right)^{-2} \\ \text{subject to} \quad & y_i(w^\top x^i + b) > 0, \ i \in M \end{aligned} \qquad (2)$$

Here, we consider the problem minimizing the inverse-squared margin instead of maximizing the margin. The constraint ensures that the selected hyperplane $w^\top x + b = 0$ correctly classifies all training points. The objective function is invariant if (w, b) is multiplied by a positive value. Hence, without loss of generality, we can fix $\min_{i \in M} |w^\top x^i + b| = 1$, and the above optimization problem is equivalent to the following.

$$\begin{aligned} \underset{w,b}{\text{minimize}} \quad & \frac{1}{2} \|w\|^2 \\ \text{subject to} \quad & y_i(w^\top x^i + b) \geq 1, \ i \in M. \end{aligned} \qquad (3)$$

For $i \in M$, let α_i be the optimal dual variable with respect to the constraint $y_i(w^\top x^i + b) \geq 1$. A training input vector x^i is called a support vector if $\alpha_i > 0$[1]. The optimal hyperplane $w^\top x + b = 0$ depends on the set of support vectors only.

The model (3) has no feasible solution if the positive class $(y_i = 1)$ and the negative class $(y_i = -1)$ cannot be separated by any hyperplanes. Additionally, even if two classes are separable, a better hyperplane may be obtained by taking account of input vectors near to the hyperplane. To archive these ideas, we consider errors for training examples, and minimization of the sum of the errors. In this paper, we use the squared hinge loss function to assess the errors.

$$\underset{w,b}{\text{minimize}} \quad \frac{1}{2} \|w\|^2 + \frac{\mu^2}{2} \sum_{i \in M} L(y_i, f(x^i)), \qquad (4)$$

where $L(y, f(x)) = (\max\{0, 1 - y(w^\top x + b)\})^2$. SVMs with tolerance of errors are called soft-margin. (On the other hand, the model (3) is called hard-margin.) In this formulation, the first term $\|w\|^2/2$ of the margin minimization can be regarded as a regularization term to prevent overfitting of classifiers. μ is a

---

[1] Roughly speaking, it is equivalent to $y_i(w^\top x^i + b) = 1$.

hyperparameter to adjusting the effect of the sum of losses. It is equivalent to,

$$\underset{w,b,\xi}{\text{minimize}} \quad \frac{1}{2}\|w\|^2 + \frac{\mu^2}{2} \sum_{i \in M} \xi_i^2 \tag{5}$$

$$\text{subject to} \quad y_i(w^\top x^i + b) + \xi_i \geq 1, \ i \in M,$$

where $\xi = (\xi_1, \ldots, \xi_m)$ is the vector of additional decision variables.

2.3   Multi-class SVMs

We extend the SVM model (3) for multi-class problems. Let $f_p(x) = (w^p)^\top x + b_p$ be a linear discriminant function of class label $p \in C$. Additionally, let $C^{\bar{2}} = \{pq \mid p, q \in C, p < q\}$ be the set of class label pairs. For each class label pair $pq \in C^{\bar{2}}$, the boundary hyperplane separating two sets of p and q is $f_{pq}(x) = (w^p - w^q)^\top x + (b_p - b_q) = 0$. Similarly to (2), the optimization problem to minimize the sum of inverse-squared margins for all $pq \in C^{\bar{2}}$ is formulated as follows.

$$\underset{(w^p, b_p)}{\text{minimize}} \quad \frac{1}{2} \left(\sum_{pq \in C^{\bar{2}}} \frac{\min_{i \in M^{pq}} |(w^p - w^q)^\top x^i + b_p - b_q|}{\|w^p - w^q\|} \right)^{-2} \tag{6}$$

$$\text{subject to} \quad (w^p - w^q)^\top x^i + b_p - b_q > 0, \ i \in M^p, \ pq \in C^{\bar{2}},$$

$$(w^q - w^p)^\top x^i + b_q - b_p > 0, \ i \in M^q, \ pq \in C^{\bar{2}}.$$

By the same reduction as (3), we fix $\min_{i \in M^{pq}} |(w^p - w^q)^\top x^i + b_p - b_q| = 1$. Then, we obtain the following optimization problem.

$$\underset{(w^p, b_p)}{\text{minimize}} \quad \frac{1}{2} \sum_{pq \in C^{\bar{2}}} \|w^p - w^q\|^2 \tag{7}$$

$$\text{subject to} \quad (w^p - w^q)^\top x^i + b_p - b_q \geq 1, \ i \in M^p, \ pq \in C^{\bar{2}},$$

$$(w^q - w^p)^\top x^i + b_q - b_p \geq 1, \ i \in M^q, \ pq \in C^{\bar{2}}.$$

The multi-class SVM to construct a classifier using the linear discriminant functions obtained by problem (7) is called AT-SVM (All-Together SVM). The obtained classifier correctly separates all training input vectors. However, it may not be a margin maximization solution [8].

AT-SVM is also extended to soft-margin cases. There are several soft-margin models considering types of loss functions and functions to aggregate losses of training examples [4]. In this paper, we consider the following soft-margin model using the squared hinge loss function.

$$\underset{(w^p, b_p), \xi}{\text{minimize}} \quad \frac{1}{2} \sum_{pq \in C^{\bar{2}}} \|w^p - w^q\|^2 + \frac{1}{2} \sum_{pq \in C^{\bar{2}}} \left(\sum_{i \in M^p} \xi_{qi}^2 + \sum_{i \in M^q} \xi_{pi}^2 \right)$$

$$\text{subject to} \quad (w^p - w^q)^\top x^i + b_p - b_q + \xi_{qi} \geq 1, \ i \in M^p, \ pq \in C^{\bar{2}},$$

$$(w^q - w^p)^\top x^i + b_q - b_p + \xi_{pi} \geq 1, \ i \in M^q, \ pq \in C^{\bar{2}},$$

(8)

where $\xi = ((\xi_{1i})_{i \in M \setminus M^1}, \ldots, (\xi_{ci})_{i \in M \setminus M^c})$.

3   Multi-class SVM Maximizing Geometric Margins

3.1   Geometric Margin Maximization

In this paper, we propose a new multi-class SVM based on minimization of the sum of inverse-squared margins (6). Let $s_{pq} = (\min_{i \in M^p \cup M^q} | (w^p - w^q)^\top x^i + b_p - b_q |)^2$ for $pq \in C^{\bar{2}}$, and $s = (s_{12}, \ldots, s_{1c}, s_{23}, \ldots, s_{(c-1)c})$. The model (6) is reformulated as follows.

$$\underset{(w^p, b_p), s}{\text{minimize}} \quad \frac{1}{2} \sum_{pq \in C^{\bar{2}}} \frac{\|w^p - w^q\|^2}{s_{pq}}$$

$$\text{subject to} \quad (w^p - w^q)^\top x^i + b_p - b_q \geq \sqrt{s_{pq}} > 0, \ i \in M^p, \ pq \in C^{\bar{2}},$$

$$(w^q - w^p)^\top x^i + b_q - b_p \geq \sqrt{s_{pq}} > 0, \ i \in M^q, \ pq \in C^{\bar{2}}.$$

(9)

Let $((w^p, b_p)_{p \in C}, s)$ be a feasible solution of (9) and $a > 0$. Then, $((aw^p, ab_p)_{p \in C}, a^2 s)$ is also a feasible solution and the objective function is invariant for the multiplication of a. Hence, without loss of generality, we can add constraints $s_{pq} \geq 1$ for $pq \in C^{\bar{2}}$.

$$\underset{(w^p, b_p), s}{\text{minimize}} \quad \frac{1}{2} \sum_{pq \in C^{\bar{2}}} \frac{\|w^p - w^q\|^2}{s_{pq}}$$

$$\text{subject to} \quad (w^p - w^q)^\top x^i + b_p - b_q \geq \sqrt{s_{pq}}, \ i \in M^p, \ pq \in C^{\bar{2}}, \qquad \text{(P1)}$$

$$(w^q - w^p)^\top x^i + b_q - b_p \geq \sqrt{s_{pq}}, \ i \in M^q, \ pq \in C^{\bar{2}},$$

$$s_{pq} \geq 1, \ pq \in C^{\bar{2}}.$$

Let OPT1 be the optimal value of (P1).

3.2   Approximate Solutions

The optimization problem (P1) is nonconvex, because of $\sqrt{s_{pq}}$ in the right hand sides of the first and second constraints. Nonconvexity causes for difficulty in solving optimization problems. Hence, we replace $\sqrt{s_{pq}}$ with an affine function of s_{pq}, and make (P1) convex.

First, we put additional constraints $s_{pq} \leq \rho^2$ for $pq \in C^{\bar{2}}$, where ρ is a hyperparameter.

$$
\begin{aligned}
\underset{(w^p, b_p), s}{\text{minimize}} \quad & \frac{1}{2} \sum_{pq \in C^{\bar{2}}} \frac{\|w^p - w^q\|^2}{s_{pq}} \\
\text{subject to} \quad & (w^p - w^q)^\top x^i + b_p - b_q \geq \sqrt{s_{pq}}, \ i \in M^p, \ pq \in C^{\bar{2}}, \\
& (w^q - w^p)^\top x^i + b_q - b_p \geq \sqrt{s_{pq}}, \ i \in M^q, \ pq \in C^{\bar{2}}, \\
& 1 \leq s_{pq} \leq \rho^2, \ pq \in C^{\bar{2}}.
\end{aligned}
\tag{P2}
$$

Let OPT2(ρ) be the optimal value of (P2) with ρ. We have OPT2(ρ) $\geq$ OPT1. Let $((\bar{w}^p, \bar{b}_p)_{p \in C}, \bar{s})$ be an optimal solution of (P1) and $\bar{s}_{\min} = \min_{pq \in C^{\bar{2}}} \bar{s}_{pq}$. Then, $((\bar{w}^p / \sqrt{\bar{s}_{\min}}, \bar{b}_p / \sqrt{\bar{s}_{\min}})_{p \in C}, \bar{s}/\bar{s}_{\min})$ is also an optimal solution of (P1). If $\max_{pq \in C^{\bar{2}}} \bar{s}_{pq} / \bar{s}_{\min} < \rho^2$ then $((\bar{w}^p / \sqrt{\bar{s}_{\min}}, \bar{b}_p / \sqrt{\bar{s}_{\min}})_{p \in C}, \bar{s}/\bar{s}_{\min})$ is feasible for (P2). Therefore, it is optimal for (P2).

We replace $\sqrt{s_{pq}}$ in (P2) with $\frac{s_{pq} + \rho}{1 + \rho}$, and obtain the following optimization problem.

$$
\begin{aligned}
\underset{(w^p, b_p), s}{\text{minimize}} \quad & \frac{1}{2} \sum_{pq \in C^{\bar{2}}} \frac{\|w^p - w^q\|^2}{s_{pq}} \\
\text{subject to} \quad & (w^p - w^q)^\top x^i + b_p - b_q \geq \frac{s_{pq} + \rho}{1 + \rho}, \ i \in M^p, pq \in C^{\bar{2}}, \\
& (w^q - w^p)^\top x^i + b_q - b_p \geq \frac{s_{pq} + \rho}{1 + \rho}, \ i \in M^q, pq \in C^{\bar{2}}, \\
& 1 \leq s_{pq} \leq \rho^2, \ pq \in C^{\bar{2}}.
\end{aligned}
\tag{P3}
$$

This is a second-order cone programming, which is a kind of convex optimization problems and can be easily solved by several software packages. Let OPT3(ρ) be the optimal value of (P3) with ρ.

Figure 2 shows the relation between $\sqrt{s}$ and $(s + \rho)/(1 + \rho)$. In the section $1 \leq s \leq \rho^2$, it holds that $\sqrt{s} \geq (s + \rho)/(1 + \rho)$. Hence, we have OPT3($\rho$) $\leq$ OPT2(ρ). It leads that we obtain a lower bound of the optimal value of (P2) by solving the convex optimization problem (P3).

Let $((\bar{w}^p, \bar{b}_p)_{p \in C}, \bar{s})$ be an optimal solution of (P3) with respect to ρ. For each $pq \in C^{\bar{2}}$, we define $s'_{pq} = \min_{i \in M^p \cup M^q} \left((w^p - w^q)^\top x^i + b_p - b_q \right)^2$. Remarking $1 \leq s'_{pq} \leq \rho^2$ for $pq \in C^{\bar{2}}$, solution $((\bar{w}^p, \bar{b}_p)_{p \in C}, s')$ is feasible for (P2) with respect to ρ. We evaluate optimality of the solution $((\bar{w}^p, \bar{b}_p)_{p \in C}, s')$. For each $pq \in C^{\bar{2}}$, the following inequality holds.

$$
\begin{aligned}
\frac{\|\bar{w}^p - \bar{w}^q\|^2}{s'_{pq}} &= \frac{\|\bar{w}^p - \bar{w}^q\|^2}{\min_{i \in M^{pq}} ((\bar{w}^p - \bar{w}^q)^\top x^i + \bar{b}_p - \bar{b}_q)^2} \\
&= \frac{\|\bar{w}^p - \bar{w}^q\|^2}{\bar{s}_{pq}} \frac{\bar{s}_{pq}}{\min_{i \in M^{pq}} ((\bar{w}^p - \bar{w}^q)^\top x^i + \bar{b}_p - \bar{b}_q)^2}
\end{aligned}
$$

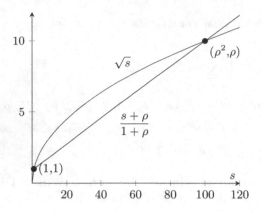

Fig. 2. Approximating $\sqrt{s}$ by $(s + \rho)/(1 + \rho)$ (when $\rho = 10$).

$$\leq \bar{s}_{pq} \left(\frac{1 + \rho}{\bar{s}_{pq} + \rho} \right)^2 \frac{\|\bar{w}^p - \bar{w}^q\|^2}{\bar{s}_{pq}}$$

$$\leq \max_{1 \leq s \leq \rho^2} s \left(\frac{1 + \rho}{s + \rho} \right)^2 \frac{\|\bar{w}^p - \bar{w}^q\|^2}{\bar{s}_{pq}} = \frac{(1 + \rho)^2}{4\rho} \frac{\|\bar{w}^p - \bar{w}^q\|^2}{\bar{s}_{pq}}.$$

Therefore,

$$\text{OPT2}(\rho) \leq \sum_{pq \in C^{\bar{2}}} \frac{\|\bar{w}^p - \bar{w}^q\|^2}{s'_{pq}} \leq \frac{(1 + \rho)^2}{4\rho} \sum_{pq \in C^{\bar{2}}} \frac{\|\bar{w}^p - \bar{w}^q\|^2}{\bar{s}_{pq}} = \frac{(1 + \rho)^2}{4\rho} \text{OPT3}(\rho).$$

We define $\theta(\rho) = \frac{(1+\rho)^2}{4\rho}$. Consequently, the optimal value of (P2) is at most the optimal value of (P3) multiplied by $\theta(\rho)$.

Summarizing the above discussion, we have the following theorem.

Theorem 1. *We have* $1 \leq \frac{OPT2(\rho)}{OPT3(\rho)} \leq \theta(\rho)$. *Moreover, suppose that there exists an optimal solution* $((w^p, b_p)_{p \in C}, s)$ *of* (P1) *such that* $\frac{\max_{pq \in C^{\bar{2}}} s_{pq}}{\min_{pq \in C^{\bar{2}}} s_{pq}} \leq \rho^2$. *Then, we have* $1 \leq \frac{OPT1}{OPT3(\rho)} \leq \theta(\rho)$.

This theorem implies that we obtain an approximation solution for (P1) and (P2) with ρ by solving convex problem (P3) with ρ, and the ratio of approximation is at most $\theta(\rho)$.

The upper bound function θ monotonically increases with respect to ρ. The relationship of ρ and θ is shown in Fig. 3. $\theta(\rho)$ is approximated by $\rho/4 + 1/2$, i.e., the upper bound of the ratio of approximation deteriorates linearly with respect to ρ. On the other hand, the range of s_{pq} in (P2) and (P3) increases quadratically.

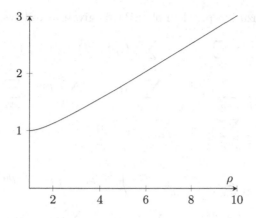

Fig. 3. Function $\theta(\rho)$, $\rho \in [1, 10]$.

When $\rho = 1$, (P2) and (P3) are reduced to (7). In other words, the proposed SVM is an extension of AT-SVM. Additionally, in the binary case: $C^2 = \{12\}$, the assumption of Theorem 1 holds for any ρ.

Corollary 1. *For binary classification problems, we have OPT1 = OPT2(1) = OPT3(1).*

3.3 Soft Margins

In this section, we consider a soft-margin formulation in the proposed multi-class SVM. Similarly to (8), we introduce slack variables ξ_{ip} for $p \in C$ and $i \in M \setminus M^p$. The soft-margin model for (P3) is defined as follows.

$$\underset{(w^p, b^p), \xi, s}{\text{minimize}} \quad \frac{1}{2} \sum_{pq \in C^2} \frac{\|w^p - w^q\|^2}{s_{pq}} + \frac{\mu^2}{2} \sum_{pq \in C^2} \frac{\sum_{i \in M^p} \xi_{qi}^2 + \sum_{i \in M^q} \xi_{pi}^2}{s_{pq}}$$

$$\text{subject to} \quad (w^p - w^q)^\top x^i + b^p - b^q + \xi_{qi} \geq \frac{s_{pq} + \rho}{1 + \rho}, \ pq \in C^2, \ i \in M^p, \quad \text{(SP3)}$$

$$(w^q - w^p)^\top x^i + b^q - b^p + \xi_{pi} \geq \frac{s_{pq} + \rho}{1 + \rho}, \ pq \in C^2, \ i \in M^q,$$

$$1 \leq s_{pq} \leq \rho^2, \ pq \in C^2.$$

In the same manner, we can define the soft-margin models for (P1) and (P2). Theorem 1 also holds in the soft-margin case without any modification.

The dual optimization problem of (SP3) is given as follows.

$$
\begin{aligned}
\underset{\alpha,(\beta^{pq}),\gamma,\delta}{\text{minimize}} \quad & -\sum_{i\in M}\sum_{p\in C\setminus\{y_i\}} \alpha_{pi} + \sum_{pq\in C^{\bar{2}}} \left((\rho^2-1)\gamma_{pq} + \delta_{pq} \right)
\end{aligned}
$$

$$
\text{subject to} \quad -\sum_{i\in M^p}\sum_{q\neq p}\alpha_{qi}x^i + \sum_{i\in M\setminus M^p}\alpha_{pi}x^i - \sum_{q>p}\beta^{pq} + \sum_{q<p}\beta^{qp} = 0, \ p\in C,
$$

$$
-\sum_{i\in M^p}\sum_{q\neq p}\alpha_{qi} + \sum_{i\in M\setminus M^p}\alpha_{pi} = 0, \ p\in C,
$$

$$
\frac{1}{1+\rho}\left(\sum_{i\in M^p}\alpha_{qi} + \sum_{i\in M^q}\alpha_{pi} \right) + \gamma_{pq} - \delta_{pq} \geq 0, \ pq\in C^{\bar{2}},
$$

$$
2\delta_{pq} \geq \|\beta^{pq}\|^2 + \frac{1}{\mu^2}\sum_{i\in M^p}\alpha_{qi}^2 + \frac{1}{\mu^2}\sum_{i\in M^q}\alpha_{pi}^2, \ pq\in C^{\bar{2}},
$$

$$
\alpha_{pi} \geq 0, \ p\in C\setminus\{y_i\}, \ i\in M; \quad \gamma_{pq} \geq 0, \ pq\in C^{\bar{2}},
$$

$$
\text{(SD3)}
$$

where $\alpha = ((\alpha_{1i})_{i\in M\setminus M^1}, \ldots, (\alpha_{ci})_{i\in M\setminus M^c})$, $\beta^{pq} \in \mathbf{R}^n$ for $pq \in C^{\bar{2}}$, $\gamma = (\gamma_{12}, \ldots, \gamma_{1c}, \gamma_{23}, \ldots, \gamma_{(c-1)c})$ and $\delta = (\delta_{12}, \ldots, \delta_{1c}, \delta_{23}, \ldots, \delta_{(c-1)c})$. In some software packages, the dual problem (SD3) can be solved more efficiently than the primal problem (SP3), since the dual problem has the smaller size of constraints[2], which significantly affects the speed of interior point methods. The dual problem is not a standard second-order cone programming, since α is included in the intersection of quadratic cones and cones of nonnegative regions. However, it is effectively handled in interior point methods.

3.4 The Proposed Method

We describe a training procedure using our SVM. It includes two phase. In the first phase, given the hyperparameters $\rho \geq 1$ and $\mu > 0$, we solve the optimization problem (SP3), and obtain $(\bar{w}^p, \bar{b}_p)_{p\in C}$ and $\bar{\xi}$. Calculate $\hat{s}_{pq}$ for $pq \in C^{\bar{2}}$:

$$
\hat{s}_{pq} = \min\{ \min_{i\in M^p} \left((\bar{w}^p - \bar{w}^q)^\top x^i + \bar{b}_p - \bar{b}_q + \bar{\xi}_{iq} \right)^2,
$$

$$
\min_{i\in M^q} \left((\bar{w}^q - \bar{w}^p)^\top x^i + \bar{b}_q - \bar{b}_p + \bar{\xi}_{ip} \right)^2 \}.
$$

In the second phase, we solve the soft-margin version of (P1) with $s_{pq} = \hat{s}_{pq}$ for $pq \in C^{\bar{2}}$, and obtain $(\hat{w}^p, \hat{b}_p)_{p\in C}$ and $\hat{\xi}$.

---

[2] To convert (SP3) to the primal form of second-order cone programming in [1], we need additional constraints $w^{pq} = w^p - w^q$ for $pq \in C^{\bar{2}}$.

As mentioned in Introduction, Fig. 1 demonstrates that the proposed method archives larger margins than AT-SVM ($\rho = 1$).

In the proposed SVM, regularization terms $\|w^p - w^q\|^2$ are divided by s_{pq}. s_{pq} is the minimum value of squared differences of discriminant functions $\left(f_p(x^i) - f_q(x^i)\right)^2$ for $i \in M^p \cup M^q$. If s_{pq} is large, we can say that the distance between two classes of p and q is large. In that case, the value $1/s_{pq}$ is small. Hence, the regularization by $\|w^p - w^q\|^2$ gives little effect when the distance of two classes of p and q is large. Since the regularization term $\|w^p - w^q\|^2$ is scaled by s_{pq}, we call the proposed SVM AT-SVM-SR (AT-SVM using Scaled Regularization terms).

4 Numerical Experiments

To examine generalization capability of the proposed SVM, we performed numerical experiments using 13 benchmark data sets in UCI Machine Learning Repository [5]. We compared classifiers obtained by AT-SVM-SR with $\rho = 100$ and AT-SVM (i.e. AT-SVM-SR with $\rho = 1$). To solve optimization problem (SD3), we used software package MOSEK [6]. Accuracy of classifiers was measured by 10-fold cross-validation with balancing class distribution.

We adapted the SVMs to nonlinear classification by kernel methods. The RBF kernel $k(x, y) = \exp\left(-\frac{\|x-y\|^2}{2\sigma^2}\right)$ was used in the experiments, where $x, y \in \mathbf{R}^n$ are input vectors and σ is a parameter to control distances of feature vectors of examples. Furthermore, the feature vectors were projected to 200-dimensional real space by the kernel principal component analysis.

The parameter σ of the RBF kernel was varied in $\{1, 2, 5, 10, 20, 50, \ldots, 1 \times 10^4, 2 \times 10^4, 5 \times 10^4\}$. The hyperparameter μ of the SVMs was varied in $\{1, 10, \ldots, 1 \times 10^4\}$.

For each benchmark data set, we performed two experiments. In one experiments, we did scaling the set of values of each variable so that the mean is 0 and the standard deviation is 1. In the other experiments, we did not that scaling.

Table 1 shows classification errors of classifiers measured in the numerical experiments. The first column shows the names of data sets with the numbers of sample examples and class labels. The next two columns show the results without scaling data sets. The last two columns show those with scaling. For each of non-scaling and scaling sections, we show the results of AT-SVM and AT-SVM-SR with $\rho = 100$. Each entry of the table shows the best (smallest) error in all of combinations of hyperparameters σ and μ. The selected values of σ and μ following the best error. The numbers in bold type mean the best results (the smallest errors) for each dataset. In 7 data sets, whose names are shown in bold, AT-SVM-SR archived better classifiers than AT-SVM. On the other hand, in 2 data sets, whose names are shown in italic, AT-SVM archived better classifiers. We can say that the generalization capability of AT-SVM-SR is better than AT-SVM in general.

Table 1. The classification errors of AT-SVM and AT-SVM-SR with $\rho = 100$. m and c in the first column indicate the numbers of objects and classes, respectively. The entry corresponding to each data set (non-scaling or scaling) and each method shows the classification error obtained by the cross validation. The selected parameters σ and μ for each result are shown after the error.

Data set	Non-scaling		Scaling	
(m, c)	AT-SVM	AT-SVM-SR	AT-SVM	AT-SVM-SR
Balance-scale	**0.0**	**0.0**	**0.0**	**0.0**
$(625, 3)$	$(1 \times 10^1, 1 \times 10^4)$	$(1 \times 10^1, 1 \times 10^4)$	$(1 \times 10^2, 1 \times 10^4)$	$(5 \times 10^1, 1 \times 10^4)$
Car	0.6	**0.3**	0.8	0.8
$(1728, 4)$	$(5 \times 10^0, 1 \times 10^4)$	$(1 \times 10^0, 1 \times 10^4)$	$(2 \times 10^0, 1 \times 10^4)$	$(2 \times 10^0, 1 \times 10^4)$
CNAE-9	5.5	5.6	4.9	**4.8**
$(1080, 9)$	$(5 \times 10^3, 1 \times 10^4)$	$(5 \times 10^3, 1 \times 10^4)$	$(5 \times 10^4, 1 \times 10^4)$	$(5 \times 10^4, 1 \times 10^4)$
Dermatology	**2.2**	**2.2**	**2.2**	2.5
$(366, 6)$	$(5 \times 10^4, 1 \times 10^4)$	$(5 \times 10^4, 1 \times 10^4)$	$(5 \times 10^4, 1 \times 10^4)$	$(5 \times 10^4, 1 \times 10^4)$
DNA	4.3	**4.2**	4.3	4.3
$(3186, 3)$	$(5 \times 10^1, 1 \times 10^1)$	$(5 \times 10^4, 1 \times 10^4)$	$(2 \times 10^4, 1 \times 10^3)$	$(2 \times 10^4, 1 \times 10^3)$
Iris	**2.0**	2.7	2.7	2.7
$(150, 3)$	$(5 \times 10^3, 1 \times 10^4)$	$(2 \times 10^4, 1 \times 10^4)$	$(2 \times 10^3, 1 \times 10^4)$	$(2 \times 10^3, 1 \times 10^4)$
Movement	**9.4**	10.6	10.3	10.8
$(360, 15)$	$(1 \times 10^0, 1 \times 10^1)$	$(1 \times 10^0, 1 \times 10^1)$	$(5 \times 10^0, 1 \times 10^1)$	$(5 \times 10^0, 1 \times 10^1)$
Optdigits	**1.0**	**1.0**	1.4	1.5
$(5620, 10)$	$(2 \times 10^1, 1 \times 10^1)$	$(5 \times 10^1, 1 \times 10^1)$	$(1 \times 10^1, 1 \times 10^1)$	$(1 \times 10^1, 1 \times 10^1)$
Page-blocks	3.7	3.5	3.0	**2.9**
$(5473, 5)$	$(2 \times 10^4, 1 \times 10^4)$	$(2 \times 10^4, 1 \times 10^4)$	$(2 \times 10^1, 1 \times 10^3)$	$(2 \times 10^1, 1 \times 10^3)$
Segment	3.1	**2.8**	3.2	3.2
$(2310, 7)$	$(2 \times 10^3, 1 \times 10^4)$	$(5 \times 10^1, 1 \times 10^1)$	$(1 \times 10^1, 1 \times 10^2)$	$(2 \times 10^2, 1 \times 10^4)$
Semeion	4.7	**4.2**	4.5	4.6
$(1593, 10)$	$(5 \times 10^0, 1 \times 10^1)$	$(5 \times 10^0, 1 \times 10^1)$	$(1 \times 10^1, 1 \times 10^1)$	$(1 \times 10^1, 1 \times 10^1)$
Vowel	0.6	0.7	0.9	**0.4**
$(990, 11)$	$(1 \times 10^0, 1 \times 10^1)$	$(1 \times 10^0, 1 \times 10^1)$	$(2 \times 10^0, 1 \times 10^4)$	$(2 \times 10^0, 1 \times 10^1)$
Wine	2.8	2.8	**1.1**	**1.1**
$(178, 3)$	$(2 \times 10^4, 1 \times 10^4)$	$(5 \times 10^3, 1 \times 10^4)$	$(5 \times 10^4, 1 \times 10^4)$	$(2 \times 10^0, 1 \times 10^4)$

5   Concluding Remarks

In this paper, we have proposed AT-SVM-SR, which is a new multi-class SVM derived from geometric margin maximization. In AT-SVM-SR, linear classifiers are provided by approximate solutions for the optimization problem of minimization of the sum of inverse-squared margins. Using Wine Data Set, we have demonstrated that the proposed AT-SVM-SR can obtain a classifier with larger margins comparing AT-SVM. The numerical experiments have shown that generalization capability of AT-SVM-SR outperforms that of AT-SVM in several data sets. One of the future work is detailed investigation on characteristics of AT-SVM-SR, e.g. the relationship between ρ and classification accuracy.

References

1. Andersen, E., Roos, C., Terlaky, T.: On implementing a primal-dual interior-point method for conic quadratic optimization. Math. Program. **95**(2), 249–277 (2003)
2. Bredensteiner, E.J., Bennett, K.P.: Multicategory classification by support vector machines. Comput. Optim. Appl. **12**(1), 53–79 (1999)
3. Cortes, C., Vapnik, V.N.: Support-vector networks. Mach. Learn. **20**(3), 273–297 (1995)
4. Doğan, Ü., Glasmachers, T., Igel, C.: A unified view on multi-class support vector classification. J. Mach. Learn. Res. **17**(45), 1–32 (2016)
5. Lichman, M.: UCI machine learning repository (2013). http://archive.ics.uci.edu/ml
6. MOSEK ApS: The MOSEK optimization toolbox for MATLAB manual. Version 7.1 (Revision 28) (2015). http://docs.mosek.com/7.1/toolbox/index.html
7. Rifkin, R., Klautau, A.: In defense of one-vs-all classification. J. Mach. Learn. Res. **5**, 101–141 (2004)
8. Tatsumi, K., Tanino, T.: Support vector machines maximizing geometric margins for multi-class classification. TOP **22**(3), 815–840 (2014)
9. Vapnik, V.N.: Statistical Learning Theory. A Wiley-Interscience Publication, New York (1998)
10. Weston, J., Watkins, C.: Support vector machines for multi-class pattern recognition. In: ESANN, pp. 219–224 (1999)

A New Context-Based Similarity Measure for Categorical Data Using Information Theory

Thanh-Phu Nguyen[1]([✉]), Mina Ryoke[2], and Van-Nam Huynh[1]

[1] School of Knowledge Science, Japan Advanced Institute of Science and Technology,
1-1 Asahidai, Nomi, Ishikawa 923-1292, Japan
{ntphu,huynh}@jaist.ac.jp
[2] Graduate School of Business Sciences, University of Tsukuba, Tokyo, Japan
ryoke@mbaib.gsbs.tsukuba.ac.jp

Abstract. Similarity is a common notion in many fields including machine learning and data mining. For numerical data, similarity measures are relatively straightforward due to their designed metrics in numerical space. However, with categorical data, measures to quantify their resemblance are still not well understood. In this research, we propose a new similarity measure based on information theoretic approach that could be able to integrate context information into the quantification of similarity between categorical data. The evaluation experiment conducted on classification task shows that the effectiveness of our proposed measure is competitive with other current state-of-the-art similarity measures.

Keywords: Similarity measure · Categorical data
Classification task · Information theory · k Nearest neighbors

1 Introduction

The idea of similarity can be found in a vast range of scientific researches. Especially, similarity measures play an important role in many machine learning and data mining techniques. Furthermore, methods for quantifying the resemblance between objects depend largely on the types of processed data. Two major types of data which can normally be observed in common studies are numerical (quantitative) and categorical (qualitative) data. For numerical data, the similarity measures are rather easy to understand due to the inherent order of the data. Specifically, Euclidean distance is usually used to compute the similarity (dissimilarity) between numerical objects as described in Fig. 1. Other common used measures are Manhattan or more general Minkowski distances.

However, for categorical data, there is no order between categorical values. Therefore, similarity measures for numerical data cannot be applied directly to this kind of data. Currently, label encoding is usually used as a common practice to simply converts each categorical value to a number. By handling categorical data in this way, it would be possible to apply geometric based measures such

© Springer International Publishing AG, part of Springer Nature 2018
V.-N. Huynh et al. (Eds.): IUKM 2018, LNAI 10758, pp. 114–125, 2018.
https://doi.org/10.1007/978-3-319-75429-1_10

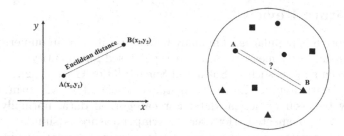

Fig. 1. Quantify similarities for numerical and categorical data

as Euclidean distance to quantify their similarity. Nevertheless, this approach does not have a firm rationale behind when adding ordinal information into categorical data regardless of their nature. Another approach which is also usually recommended to tackle the problem is doing dummy (indicator) coding for categorical values before applying numerical similarity measures [1]. However, it was pointed out that this method is not effective and may cause the loss of several important featured information that belongs to categorical data type [2].

On the other hand, many efforts have been put into quantifying the similarity between categorical data without sacrificing their characteristics by exploiting other facts such as their frequency or relationship information. Particularly, in numerical taxonomy, the agreement in possessing an uncommon value for an attribute is a more convincing evidence of affinity than the agreement in having a common value [2]. Moreover, two attribute values turn to be more similar when they appear with similar distributions of values in a set of correlated context attributes [3]. Currently, those two main ideas have become key intuitions for most of the existing similarity measures that are designed for categorical data.

In this paper, we first conduct a comprehensive survey that focuses on classifying similarity measures for categorical data based on their characteristics and approaches. Then we propose a new similarity measure for categorical data that is able to integrate relationship information between categorical attributes based on information theoretic approach. In particular, by extending the measure of Nguyen and Huynh [4], the relationship between attributes is included by estimating the information amount of the appearance of co-occurrence pairs between highly correlated attributes. Finally, we conduct a comparative experiment on popular used similarity measures for categorical data to evaluate their efficiency under the scheme of classification task using kNN technique.

The rest of this paper is organized as follows. In Sect. 2, a literature review is conducted on the development of similarity/dissimilarity measures for categorical data that focuses mainly on different aspects to classify available measures. New trends on the improvement of current similarity measures are also mentioned in this part. After that, details of our new proposed similarity measure are described in Sect. 3. In Sect. 4, information about experiments for evaluating our new proposed measure is included with the detailed methodology and results. Finally, in Sect. 5, we make some remarks and conclusions on this research and mention about our future work.

2   Literature Review

Similarity notion is popular used in many fields. Specifically, in numerical taxonomy, several similarity measures have been proposed for quantifying the resemblance between species such as Sokal and Sneath [5] or Goodall [2]. In machine learning and data mining, Euclidean distance is usually used to measure the dissimilarity between numerical data. For categorical data, normally Overlap measure [6] is implemented. However, Overlap measure is simple and cannot include the information of frequency of categories as well as relationships between attributes.

2.1   Similarity Measures Classification

Depend on characteristics or approaches of similarity/dissimilarity measures, they are classified into different groups. Below are three common ways for classifying similarity measures of categorical data.

1. According to Boriah et al. [7], similarity measures for categorical data can be categorized by the way they fill in the similarity matrix (Table 1).

Table 1. Similarity matrix for a single categorical attribute

	a	b	c	d
a	$S(a,a)$	$S(a,b)$	$S(a,c)$	$S(a,d)$
b		$S(b,b)$	$S(b,c)$	$S(b,d)$
c			$S(c,c)$	$S(c,d)$
d				$S(d,d)$

Consequently, similarity measures are classified into three groups. The first group is those that fill the diagonal entries only including Overlap, Goodall [2] and Gambaryan [8]. The second group is filling off-diagonal entries measures (Eskin [9], IOF [7], OF [7] and Burnaby [10]). And the last group contains measures that fill both diagonal and off-diagonal entries (Lin [11], Lin1 [7], Smirnov [12] and Anderberg [13]).

2. In Ring et al. [3], distance (dissimilarity) measures are classified into four categories: (I) without considering context attributes, (II) considering all context attributes, (III) considering a subset of context attributes and (IV) based on entire objects instead of individual attributes. Same idea with Ring et al., however Alamuri et al. [14] is simpler, measures are classified into only two groups: context-free and context-sensitive. Figure 2 shows the classification for unsupervised similarity measures for categorical data according to Alamuri et al.

3. Similarity measures could also be classified based on their approaches such as frequency (OF, IOF and Eskin), probabilistic (Goodall, Smirnov and Anderberg) or information theoretic (Lin, Lin1, Nguyen and Huynh [4], Burnaby and Gambaryan) approaches [7].

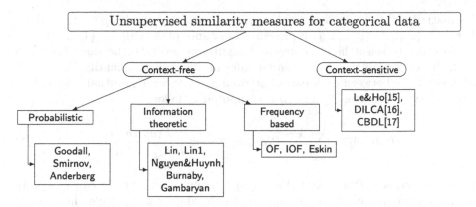

Fig. 2. Taxonomy of unsupervised similarity measures for categorical data

2.2 Hybrid Approach

Recently, several efforts have been made to include not only the frequency information of categories but also the co-occurrence information between categorical attributes into the similarity measures such as Morlini and Zani [18] and Jia et al. [19]. Morlini and Zani took the information theoretic approach, but they considered all possible co-occurrences which lead to a high computational complexity. Besides, Jia et al. used the probabilistic approach to solve the problem.

3 Proposed Similarity Measure

Based on the work of Nguyen and Huynh [4], we propose a new similarity measure which can integrate the information of relations between categorical attributes into the process of quantifying the similarity between categorical values.

Specifically, we extend Nguyen and Huynh measure by considering not only the information of single values but also their relations with other attributes through co-occurrences. The proposed method can reduce the computational complexity by considering only a group of attributes that have a high degree of correlation with the computed attribute. In the case that all the attributes are totally independent with each other, the proposed measure will be reduced to Nguyen and Huynh measure that only considers the information amount in the appearance of attribute values.

Generally, the proposed similarity measure can be split into two steps: the first step is selecting highly correlated attributes, the second step is quantifying the similarity between data objects. Details of those two steps are described in the following parts.

3.1 Correlated Attributes Selection

Consider a dataset X containing n instances and described by d attributes. Let $A = \{A_1, ..., A_k, ..., A_d\}$ be the set of d attributes with $k \in \{1, ..., d\}$.

Each attribute A_k has a domain of $\{a_{k1}, ..., a_{kp}, ..., a_{k|A_k|}\}$ with $p \in \{1, ..., |A_k|\}$. Let $x_i = \{x_{i1}, ..., x_{ik}, ..., x_{id}\}$ represent an instance of X with $i \in \{1, ..., n\}$.

In order to select highly correlated attributes, we take the same approach with Jia et al. [19] by using mutual information [20] to quantify the degree of dependence between each pair of attributes. Specifically, mutual information between two attributes A_k and A_l is computed as follows.

$$I(A_k, A_l) = \sum_{p=1}^{|A_k|} \sum_{q=1}^{|A_l|} P(a_{kp}, a_{lq}) * \log \frac{P(a_{kp}, a_{lq})}{P(a_{kp}) * P(a_{lq})} \tag{1}$$

where the terms of $P(a_{kp})$ and $P(a_{lq})$ respectively stand for the frequency probabilities of attribute values a_{kp} and a_{lq} in the data set X; while the term of $P(a_{kp}, a_{lq})$ is the joint probability of those two attribute values. More specifically, we have:

$$P(a_{kp}) = P(A_k = a_{kp}|X) = \frac{count(A_k = a_{kp})}{|X|} \tag{2}$$

$$P(a_{lq}) = P(A_l = a_{lq}|X) = \frac{count(A_l = a_{lq})}{|X|} \tag{3}$$

$$P(a_{kp}, a_{lq}) = P(A_k = a_{kp} \wedge A_l = a_{lq}|X) = \frac{count(A_k = a_{kp} \wedge A_l = a_{lq})}{|X|} \tag{4}$$

However, mutual information value is not normalized and it increases with the number of available values of each attribute. Therefore, a normalized version of mutual information named interdependence redundancy measure proposed by Au et al. [21] is utilized. The formulation of interdependence redundancy measure is as below.

$$R(A_k, A_l) = \frac{I(A_k, A_l)}{H(A_k, A_l)} \tag{5}$$

where $H(A_k, A_l)$ is the joint entropy between two attributes A_k and A_l which is calculated by the following formula.

$$H(A_k, A_l) = -\sum_{p=1}^{|A_k|} \sum_{q=1}^{|A_l|} P(a_{kp}, a_{lq}) * \log P(a_{kp}, a_{lq}) \tag{6}$$

According to Au et al. [21], the interdependence redundancy measure reflects the degree of deviation from independence between attributes. Particularly, two attributes A_k and A_l are strictly dependent if $R(A_k, A_l) = 1$ and statistically independent when $R(A_k, A_l) = 0$. In other cases, A_k and A_l are partially dependent. If two attributes are dependent on each other, they are more correlated with each other when compared to two independent attributes. The interdependence redundancy measure is therefore able to evaluate the interdependence or correlation of attributes. A large value of R implies a high degree of dependence between attributes. With the interdependence redundancy measure, a matrix R

of interdependence redundancy values between all pairs of attributes could be formed. Then for each attribute A_k, a relation set S_k which contains attributes that are highly correlated with A_k can be generated by the following expression:

$$S_k = \{A_l | R(A_k, A_l) > \beta, 1 \leq l \leq d\} \tag{7}$$

where β is a threshold value set by users, d is the number of attributes of the data set X.

3.2 Quantify the Similarity Between Data Objects

Given a data set X that contains n instances and is described by d attributes. The similarity between two instances x_i and x_j in the dataset X could be quantified by the following general formulation:

$$S(x_i, x_j) = \sum_{k=1}^{d} \frac{1}{d} * sim_{A_k}(x_{ik}, x_{jk}) \tag{8}$$

while the similarity between two values of attribute A_k is computed by:

$$sim_{A_k}(x_{ik}, x_{jk}) = \sum_{A_l \in S_k} \sum_{a \in dom(A_l)} \frac{1}{|S_k|} \frac{1}{|A_l|} \frac{2 \log P(\{x_{ik}, x_{jk}\}|a)}{\log P(x_{ik}|a) + \log P(x_{jk}|a)} \tag{9}$$

where

$$P(\{x_{ik}, x_{jk}\}|a) = \begin{cases} P(x_{ik}|a) + P(x_{jk}|a) & \text{if } x_{ik} \neq x_{jk} \\ P(x_{ik}|a) = P(x_{jk}|a) & \text{if } x_{ik} = x_{jk} \end{cases} \tag{10}$$

3.3 Characteristics

It is easy to see that the new proposed similarity measure defined in Eq. (8) satisfies the following conditions:

1. $S(x_i, x_j) \geq 0$ for each x_i, x_j with $i, j \in \{1, 2, ..., n\}$
2. $S(x_i, x_i) = 1$ with $i \in \{1, 2, ..., n\}$
3. $S(x_i, x_j) = S(x_j, x_i)$ for each x_i, x_j with $i, j \in \{1, 2, ..., n\}$

3.4 Algorithm and Computational Complexity

Given a data set X that contains n instances and is described by d attributes. Based on the two steps defined in previous parts, the algorithm for the proposed measure can be formulated as in the Algorithm 1.

About the computational complexity of the proposed measure, the steps of computing the interdependence redundancy matrix and calculating the similarity between two instances have the same complexity of $O(nd^2)$. In total, the proposed measure has the complexity of $O(nd^2)$. In the case that attributes of data set X are totally independent with each other, the complexity of the algo-rithm is equal to the complexity of Nguyen and Huynh measure [4] which is $O(nd)$. It is evident that our proposed measure has a reasonable computational complexity comparing to other common used similarity measures and is much less than the measure of Morlini and Zani [18] which is $O(nd^d)$.

Algorithm 1. Quantifying the similarity of categorical data

Input: Data set $X = \{x_1, x_2, ..., x_n\}$.
Output: Similarity between two data objects x_i, x_j with $i, j \in \{1, 2, ..., n\}$.
Step 1: Select highly correlated attributes
1: For each pair of attributes (A_k, A_l) with $k, l \in \{1, 2, ..., d\}$, calculate $R(A_k, A_l)$ according to Eq. (5).
2: Construct interdependence redundancy matrix.
3: Generate relation set S_k for each attribute A_k following Exp. (7).
Step 2: Quantify the similarity between data objects
4: Choose two data objects x_i, x_j from X.
5: Let $S(x_i, x_j) = 0$.
6: for $k = 1$ to d do
7:     if $S_k = \emptyset$ then
8:         $sim_{A_k}(x_{ik}, x_{jk}) = \frac{2 \log P(\{x_{ik}, x_{jk}\})}{\log P(x_{ik}) + \log P(x_{jk})}$
9:     else
10:         $sim_{A_k}(x_{ik}, x_{jk}) = 0$
11:         for each $A_l \in S_k$ do
12:             for each $a \in dom(A_l)$ do
13:                 $sim_{A_k}(x_{ik}, x_{jk}) = sim_{A_k}(x_{ik}, x_{jk}) + \frac{1}{|S_k|} \frac{1}{|A_l|} \frac{2 \log P(\{x_{ik}, x_{jk}\}|a)}{\log P(x_{ik}|a) + \log P(x_{jk}|a)}$
14:             end for
15:         end for
16:     $S(x_i, x_j) = S(x_i, x_j) + sim_{A_k}(x_{ik}, x_{jk})$
17: end for
18: return $S(x_i, x_j)/d$

---

4   Evaluation

4.1   kNN Classification

To evaluate the efficiency of our newly proposed measure, kNN classification task on public data sets is utilized as a basement for evaluation purpose. The accuracy of the experiment is considered as a standard of performance efficiency.

Specifically, kNN is a simple but robust technique that is usually used in clas-sification task. In the kNN technique, the class of a new data object is predicted based on the majority of classes of its k nearest neighbors (Fig. 3).

In our experiment, for simplifying the problem, the number of considered nearest neighbors is set to 7. And each measure is evaluated with 10 times 10 fold cross-validation per data set.

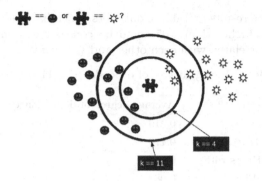

Fig. 3. kNN classification problem [22]

4.2 Data Sets

We select 11 data sets from UCI machine learning repository [23] for testing. The chosen data sets contain not only categorical attributes but also real and integer attributes (Table 2). For numerical values, a discretization tool of Weka [24] is utilized to discretize numerical values into equal intervals that in turn could be treated as categorical values.

Table 2. Details of 11 data sets from UCI

Data set	Inst.	Attr.	Classes	Data types	Missing values
Audiology-std	226	69	24	Categorical	Yes
Balance-scale	625	4	3	Categorical	No
Dermatology	366	33	6	Categorical, integer	Yes
Hayes-roth	160	5	3	Categorical	No
Iris	150	4	3	Real	No
Post-operative	90	8	3	Categorical, integer	Yes
Soybean-sml	47	35	4	Categorical	No
Teaching-asst	151	5	3	Categorical, integer	No
Tictactoe	958	9	2	Categorical	No
Wine	178	13	3	Integer, real	No
Zoo	101	17	7	Categorical, integer	No

4.3 Data Dependence Analysis

In order to prove the effectiveness of our proposed measure when considering the relationship information between categorical attributes into the process of quantifying the similarity, we compute the average degree of dependence of selected data sets and combine those information when analyzing with the final classification results. The average degree of dependence of 11 data sets are calculated by averaging interdependence redundancy values of all distinctive pairs of attributes. From the results in Table 3, data sets which have the value of average dependence degree higher than 0.05 could be considered as having attributes which are highly correlated with each other and vice versa.

Table 3. Average degree of dependence of 11 data sets

Data set	Average degree of dependence
Iris	0.219
Teaching-asst	0.177
Hayes-roth	0.113
Wine	0.089
Zoo	0.080
Dermatology	0.052
Soybean-sml	0.041
Post-operative	0.014
Audiology-std	0.009
Tictactoe	0.006
Balance-scale	0.000

4.4 Hyper-parameters Tunning

With our new proposed measure, the β parameter is tunned by using grid search and cross validation techniques. The chosen β value is the one which yields the highest accuracy. Specifically, value of β is varied in the range of $[0, 1]$ with the step of 0.1. The results of some noticeable cases are described in Fig. 4. From the results, it can be seen that with some highly correlated data sets such as iris or hayes-roth, the accuracy increases when β decreases. It means that more co-occurrence information is included when considering the similarity between categorical objects. By contrast, for data sets having low correlated degree between attributes (post-operative), the accuracy decreases when β value decreases. Moreover, for totally independent data sets like balance-scale, it is almost no change when β varies.

4.5 Experimental Results

The experiment is run with 10 times 10 fold cross-validation strategy for each measure per data set. That means we have 100 results for each pair of similarity measure and data set. The final results are calculated by averaging the values of 100 results. Final experimental results are displayed in Table 4 with the highest accuracy results bold typed.

From the experimental results, there is no best similarity measure for all categorical data sets. However, it can be seen that the similarity measures which consider context information tend to perform better with highly correlated data sets than the ones which do not consider context information.

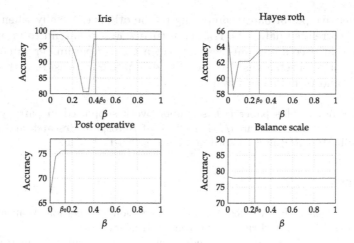

Fig. 4. Hyper-parameter tunning for threshold value β

The efficiency of our new proposed measure is competitive with other common used measures. It also has relatively good results on both high and low correlated data sets with the largest average accuracy in total of 11 data sets.

Table 4. Experimental results of 11 data sets

Data set	Proposed	Jia et al.	DILCA	Nguyen and Huynh	OF	Eskin	Overlap
Iris	**95.133**	90.533	93.600	94.133	92.733	92.933	93.000
Teaching-asst	52.125	46.563	**53.875**	49.563	39.813	49.125	49.000
Hayes-roth	**72.643**	68.071	65.500	69.143	29.357	33.571	60.500
Wine	94.389	90.333	**94.556**	93.833	92.611	92.555	93.722
Zoo	91.818	87.545	92.363	**93.727**	85.091	92.182	91.091
Dermatology	**98.622**	96.568	98.162	98.135	58.676	95.270	96.595
Soybean-sml	**100.000**	**100.000**	**100.000**	**100.000**	99.600	99.800	**100.000**
Post-ope	70.000	64.778	68.667	67.222	**71.778**	69.111	70.333
Audiology-std	65.435	58.261	59.783	**68.261**	54.348	61.000	64.000
Tictactoe	99.917	**99.969**	87.458	99.917	76.406	94.719	94.958
Balance-scale	79.349	79.317	77.063	78.524	78.889	**79.413**	79.032
Average	**83.585**	80.176	81.002	82.951	70.846	78.153	81.112

5   Remarks and Conclusions

In this paper, we have proposed a new unsupervised similarity measure for categorical data based on the information theoretic approach. The new proposed measure could be able to integrate the information of relations between attributes through co-occurrence content. In order to evaluate our measure, an experiment has been conducted to compare its effectiveness with other categorical similarity measures. The results have shown that the new proposed measure has a competitive performance comparing to the others especially when handles with highly correlated data sets. For the future work, we intend to improve our measure so that it could automatically select a good β parameter value according to different data sets. It is also useful to evaluate our measure with different applications and techniques as well.

Acknowledgment. This paper is based upon work supported in part by the Air Force Office of Scientific Research/Asian Office of Aerospace Research and Development (AFOSR/AOARD) under award number FA2386-17-1-4046.

References

1. Cohen, J., Cohen, P.: Applied Multiple Regression/Correlation Analysis for the Behavioral Sciences. L. Erlbaum Associates, Hillsdale (1983)
2. Goodall, D.W.: A new similarity index based on probability. Biometrics **22**(4), 882–907 (1966)
3. Ring, M., Otto, F., Becker, M., Niebler, T., Landes, D., Hotho, A.: ConDist: a context-driven categorical distance measure. In: Appice, A., Rodrigues, P.P., Santos Costa, V., Soares, C., Gama, J., Jorge, A. (eds.) ECML PKDD 2015. LNCS (LNAI), vol. 9284, pp. 251–266. Springer, Cham (2015). https://doi.org/10.1007/978-3-319-23528-8_16
4. Nguyen, T.-H.T., Huynh, V.-N.: A k-means-like algorithm for clustering categorical data using an information theoretic-based dissimilarity measure. In: Gyssens, M., Simari, G. (eds.) FoIKS 2016. LNCS, vol. 9616, pp. 115–130. Springer, Cham (2016). https://doi.org/10.1007/978-3-319-30024-5_7
5. Sokal, R.R., Sneath, P.H.A.: Principles of Numerical Taxonomy. W. H. Freeman, San Francisco (1961)
6. Stanfill, C., Waltz, D.: Toward memory-based reasoning. Commun. ACM **29**(12), 1213–1228 (1986)
7. Boriah, S., Chandola, V., Kumar, V.: Similarity measures for categorical data: a comparative evaluation. In: Proceedings of the 2008 SIAM International Conference on Data Mining, pp. 243–254 (2008)
8. Gambaryan, P.: A mathematical model of taxonomy. SSR **17**(12), 47–53 (1964)
9. Eskin, E., Arnold, A., Prerau, M., Portnoy, L., Stolfo, S.: A geometric framework for unsupervised anomaly detection: detecting intrusions in unlabeled data. In: Applications of Data Mining in Computer Security. Kluwer (2002)
10. Burnaby, T.P.: On a method for character weighting a similarity coefficient, employing the concept of information. J. Int. Assoc. Math. Geol. **2**(1), 25–38 (1970)
11. Lin, D.: An information-theoretic definition of similarity. In: Proceedings of the 15th International Conference on Machine Learning, pp. 296–304. Morgan Kaufmann (1998)
12. Smirnov, E.S.: On exact methods in systematics. Syst. Zool. **17**(1), 1–13 (1968)
13. Anderberg, M.R.: Cluster Analysis for Applications. Academic Press, New York (1973)
14. Alamuri, M., Surampudi, B.R., Negi, A.: A survey of distance/similarity measures for categorical data. In: 2014 International Joint Conference on Neural Networks (IJCNN), pp. 1907–1914, July 2014
15. Le, S.Q., Ho, T.B.: An association-based dissimilarity measure for categorical data. Pattern Recogn. Lett. **26**(16), 2549–2557 (2005)

16. Ienco, D., Pensa, R.G., Meo, R.: From context to distance: learning dissimilarity for categorical data clustering. ACM Trans. Knowl. Discov. Data **6**(1), 1:1–1:25 (2012)
17. Khorshidpour, Z., Hashemi, S., Hamzeh, A.: Distance learning for categorical attribute based on context information. In: 2010 2nd International Conference on Software Technology and Engineering, vol. 2, pp. V2-296–V2-300, October 2010
18. Morlini, I., Zani, S.: A new class of weighted similarity indices using polytomous variables. J. Classif. **29**(2), 199–226 (2012)
19. Jia, H., Cheung, Y., Liu, J.: A new distance metric for unsupervised learning of categorical data. IEEE Trans. Neural Netw. Learn. Syst. **27**(5), 1065–1079 (2016)
20. MacKay, D.J.C.: Information Theory, Inference & Learning Algorithms. Cambridge University Press, New York (2002)
21. Au, W.H., Chan, K.C.C., Wong, A.K.C., Wang, Y.: Attribute clustering for grouping, selection, and classification of gene expression data. IEEE/ACM Trans. Comput. Biol. Bioinform. **2**(2), 83–101 (2005)
22. Machine Learning with Python: k-Nearest Neighbor Classifier. http://www.python-course.eu/k_nearest_neighbor_classifier.php
23. Lichman, M.: UCI machine learning repository (2013). http://archive.ics.uci.edu/ml
24. Hall, M.A., Holmes, G.: Benchmarking attribute selection techniques for discrete class data mining. IEEE Trans. Knowl. Data Eng. **15**(6), 1437–1447 (2003)

Vector Representation of Abstract Program Tree for Assessing Algorithm Variety for the Same Purpose

Yoshiki Mashima$^{(\boxtimes)}$ and Kazuhiro Takeuchi

Graduate School of Engineering, Osaka Electro-Communication University,
Neyagawa, Osaka, Japan
mi17a004@oecu.jp

Abstract. There are various ways to realize programs to meet the same requirements. Therefore, for evaluation of a program code, it is necessary not only to satisfy the requirement for it but also to evaluate how it achieves the purpose. We think that the knowledge of such alternative ways to achieve the same purpose is important for program education and software engineering. In this paper, we propose a method to analyze how the program achieves requirements. We propose particularly a vector representation that appropriately indicates the structure of an abstract syntax tree. We confirmed that our proposal not only analyzes fundamental programs as appears in textbooks effectively but also classifies algorithms that are adopted in various programs submitted to a programming contest. Based on this confirmation, we further investigated the relationship between the class name of the Java language and the program structure. As a result, it was shown that classes with similar program structure are named by certain similar linguistic expressions. Therefore, we conclude that our proposed method is a useful basis for representing the diversity of programs on vector space.

Keywords: Vector representation · Program mapping
Program structure tree

1 Introduction

Programs are created to achieve their purposes. One of the viewpoints for evaluating whether a program achieves the purpose or not is to confirm it by prepare test cases. That is, we can examine the program by checking the pair of input and output with the prepared correct cases one by one. However, there is another viewpoint how the program achieves its purpose. The latter viewpoint is useful for assessing the quality of software and for improving better methods of teaching programs. Since various kinds of programs satisfy the same requirements, due to the nature of the programming, we think that knowledge of alternative methods to achieve the same requirement is important for students and teachers.

On the other hand, purposes are defined for every part of programs. We can commonly see the purposes as program comments with natural language. Expert programmers can recognize the function of the program by reading the descriptions of natural language comment and program codes. In general, program codes are described in programming languages, which is an artificial language, and has a structure as represented by an abstract syntax tree.

In this paper, we analyze examples employed different algorithm despite their purposes are same. In particular, we propose a vector representation that appropriately indicates the structure of an abstract syntax tree in order to define the similarity between them.

2 Related Works

In software development, the work of changing the source code accompanying the version upgrade of the program is not only costly but also a human error which overlooks the changed part easily occurs. Therefore, many studies are conducted to inform the programmer in advance of the program correction point, and to automatically detect and correct bugs [1–3].

Murakami et al. [4] is conducting the detection of the method of the program to be modified in the future. Concretely, using multiple regression analysis, they detect changes in methods. However, the input value of multiple regression analysis is a numerical regression. Therefore, in order to analyze the program by multiple regression analysis, it is necessary to convert the program into a numerical representation. They use the number of nonterminal symbols used in the abstract syntax tree to convert the program into a vector representation that is a numeric representation. And they detect changes in the method.

Mou et al. [5] are trying to analyze programs using Deep Learning. That is, by inputting programs converted to numerical regression, they classify and investigate programs using Deep Learning. Concretely, like Murakami et al., they convert programs into vector expressions based on abstract syntax trees, and let Deep Learning learn programs. As a result, they acquire the distributed representation learning the program structure. Using this acquired distributed representation, they investigated relations among nodes of abstract syntax trees and classify programs.

Gvero and Kuncak [6] is developing a system that automatically generates function calls executable by programs by inputting natural language. That is, even without knowing the functions that the developer wants to use, this system can output executable programs by input in natural language. This system is called "anyCode" and is implemented by correspondence between natural language and programs. For correspondence between natural language and programs, they use JDT-AST-Parser [7] which generates abstract syntax tree and Github Java Corpus [8] which is resource of java program collected by Allamanis et al.

3   Our Proposal

3.1   Point of Interest in Program Description

We think that there are at least two important points in the description of the program. The first point is what kind of data structure is prepared to achieve the description purpose of the program. Because we think that the required data structure is determined to some extent according to the means to achieve the purpose. For example, we consider a swap program that exchanges multiple numeric data. We think that the data structure of the swap program is generally an int type array and int type variables. The purpose of preparing an int type array is because it is necessary to store numerical data. And the purpose of preparing int type variables is because it is necessary to use variables for subscripts for referring to arrays and to use as evacuation. Therefore, we think that what kind of data structure is prepared is important.

The second point is how to use the prepared data structure. In general, the program makes the computer achieve the purpose by using the data structure prepared. Therefore, we think that the difference in how to achieve the purpose is expressed in how to use the data structure. For example, we consider bubble sort algorithm and quick sort algorithm of a method of sorting numerical data. The bubble sort algorithm is an algorithm that rearranges numerical values by sequentially scanning the stored data structures from the beginning to the end and comparing adjacent numerical values. This algorithm requires an int type variable for subscript reference for array scanning. On the other hand, the quick sorting algorithm is an algorithm that recursively divides an array into a plurality of arrays and rearranges the divided arrays. In this algorithm as well, an int type variables are used as well. However, there are variables that are different from the bubble sort algorithm and usage, such as variables for axes dividing the array. We think that the difference in use purpose creates differences in how variables are using. Therefore, we think that the difference in use method of the prepared data structure is important.

Based on those two points, we suggest a vector representation of the program.

3.2   Forming Vector from Abstract Syntax Tree

The vector representation proposed in this paper is based on AST (abstract syntax tree) of programs. In the process of the vectorization, we convert AST of a program code into bipartite graph representation.

A bipartite graph is a graph structure having two sets of node V_1 and V_2, with valid edges drawn from V_1 to V_2. The proposed bipartite graph is a graph representation expressing how the defined data structure is used. Specifically, V_1 is set as the type of the defined data structure, and V_2 is set as the control structure and method type. V_2 expresses how the defined data structure is used. We call V_2 a function pattern. Also, in order to show the importance of each valid edge, we assign weight to the edge. The edge weight is defined as the number of occurrences by a combination of the defined data structure and the function pattern. The reason for attaching the importance is that the appearance frequency of the used method of the defined data structure is considered to represent the feature in the program.

Figure 1 shows the flow of vectorization of programs via bipartite graphs by using a part of BubbleSort program and a part of QuickSort program written in Java language. In this figure, the type of the data structure and the type of the function pattern are extracted from each program. Then, the appearance frequency of the appearing combination pattern is counted. A bipartite graph is building based on these types and patterns. Valid edges are drawn from the node set of the defined data structure to the node set of the function pattern. Based on the built bipartite graph, vector components are determined, and edge weights are used for vector elements.

If there is an edge that appears in a certain program but does not appear in a different program, the vector element is set to 0. Thus, when there is a common edge in each program, the vector component is a component contributing to the similarity between programs. Conversely, when there are the specific edges to each program, those become the vector component expressing the feature of each program.

4   Experiments

As an experiment of our proposal, we conduct classification with the principal component analysis (PCA) to the following three types of programs set.

- Textbook programs: programs described in textbooks for program beginners to begin with
- Task-driven programs: programs that solves the same problem
- General programs: programs which is considered to be commonly used

We used the Java language in this experiment and used JDT-AST-Parser which is provided as an Eclipse plugin for building abstract syntax tree. Also, We experimented by narrowing down the control structure to iterative syntax (for) and conditional branching syntax (if). In other words, for statements, while statements, and do-while statements are integrated into the iterative syntax (for), and if statements and switch statements are integrated into the conditional branch syntax (if).

4.1   Clustering Experiment on Textbook Programs

The purpose of this experiment is to evaluate the proposed method by clustering using programs with the same function name. In other words, we evaluate the proposed vector expression by visually confirming the relationship by the difference in algorithms. Therefore, we did the experiment which clustering is performed by using textbook programs focused on algorithms in a sort field. In this experiment, we used sort programs described in four textbooks for beginner programmers [9–12].

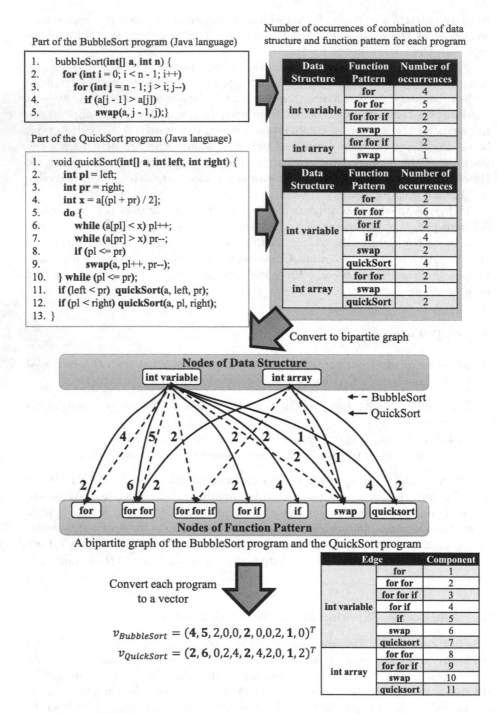

Fig. 1. Vectorization of program by proposed method via bipartite graph

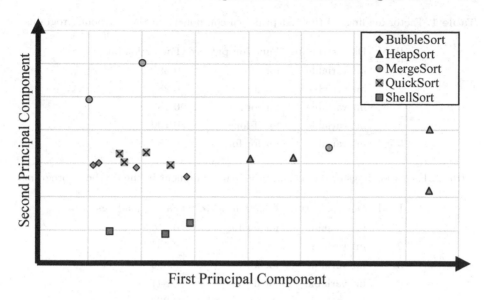

Fig. 2. Result of PCA of textbook programs

Firstly, we build a bipartite graph of the whole program using the proposed bipartite graph. Secondly, we make the vector for each program and make a matrix created by linking the vectors. The rows represent each program and the columns represent the components of each vector. By using this matrix, we confirm the similarity of each textbook programs based on PCA.

The experiment's result of clustering is shown in Fig. 2. This figure shows the result of two-dimensional plots by PCA. Each plotted point shows different textbook programs and expresses each sorting algorithm as an attribute of the plot. From this figure, we can confirm that the ShellSort programs and QuickSort programs are plotted in close positions. Also, in the QuickSort programs and the HeapSort programs, we can confirm plotted in different positions. Therefore, we can see that the proposed method expresses that the structural representation is different. However, in the HeapSort programs and the MergeSort programs, we can confirm that some programs exist plots which are not plotted in close positions.

Then, we investigate principal components in order to examine what kind of patterns are affected in the entire textbook programs. Tables 1, 2 summarize the top five factor loadings by the absolute value of each vector component in the PCA for each axis. From these tables, the first principal component shows that the component which int type variable is used in the iterative syntax is the most important factor. In other words, we can consider that the axis of the first principal component is the number of times that the int type variable is used in the iterative syntax. In the second principal component, we can confirm that the

Table 1. Factor loadings of the first principal component in the textbook programs

Rank	Data structure	Function pattern	Factor loadings
1	int variable	for	0.899
2	int variable	for if	0.283
3	int variable	for for	0.187
4	int variable	for if for	0.100
5	int variable	for for for	0.087

Table 2. Factor loadings of the second principal component in the textbook programs

Rank	Data structure	Function pattern	Factor loadings
1	int variable	for for for	0.518
2	int variable	if for	0.450
3	int variable	for for	0.395
4	int variable	for if	0.341
5	int variable	if	0.269

component which the int type variable is used in the multiple control structures is an important factor. In other words, we can consider that the axis of the second principal component is the number of times that the int type variable is used in the multiple control structures.

4.2 Clustering Experiment on Task-Driven Programs

In this experiment, the proposed method is evaluated by using the program in the case of achieving the same task. In other words, we confirm whether the proposed method can express the difference of the program implementation method. The content of the assignment is that when five numerical data are given, those are rearranged in ascending order. The total number of program data was 140 samples [13]. The experiment procedure was same as for the textbook programs.

Figure 3 shows the result of two-dimensional plots by PCA. Each point shows submitted programs. An attribute of the plots is whether the program has a structure as shown in the Fig. 4 or a program using a prepared library in advance. From this figure, we can confirm that it is classified into programs having a bubble sort structure shown in the figure and programs simply described using prepared library functions.

We perform a clustering with principal components in order to investigate what kind of patterns are affected in those programs. Tables 3, 4 summarize the top five factor loading From these tables, the same components as those of the first principal component in the textbook programs are listed at the top. We can confirm that the second principal component is the component which int type data structure is used by multiple control structures to the higher level. That is,

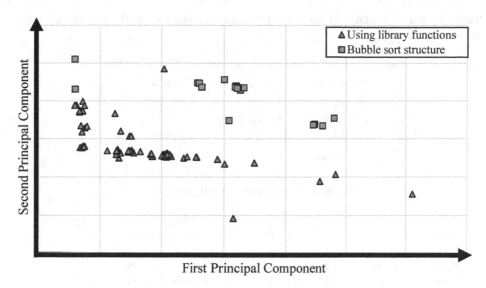

Fig. 3. Result of PCA of task-driven programs

```
Set numbers in the array x
for each of array index A {
    for each of array index B {
        Set variable to x[A]
        Set x[A] to x[B]
        Set x[B] to variable }}
```

Fig. 4. Bubble sort structure

Table 3. Factor loadings of the first principal component of the task-driven programs

Rank	Data structure	Function pattern	Factor loadings
1	int variable	for	0.914
2	int variable	for for if	0.258
3	int array	for	0.191
4	int array	for for if	0.167
5	int variable	for if	0.122

the axis of the second principal component is considered to be the number of times that the int type data structure is used by multiple control structures.

4.3 Clustering Experiment on General Programs

In the third experiment, the proposed method is evaluated by using program groups of the same class name. That is, for each identifier used in the program,

Table 4. Factor loadings of the second principal component of the task-driven programs

Rank	Data structure	Function pattern	Factor loadings
1	int variable	for for	0.636
2	int variable	for for if	0.561
3	int array	for for if	0.397
4	int variable	for if	0.240
5	int variable	for	0.158

Table 5. Data set of general programs

Class name	Max of LOC.	Min LOC.	Median LOC.	Mean LOC.	Number of programs
Server	2758.0	2	55	146.1	498
Client	4822.0	2	67	181.4	580
Parser	20702	2	87	467.0	465
Token	1221.0	2	41	86.04	443

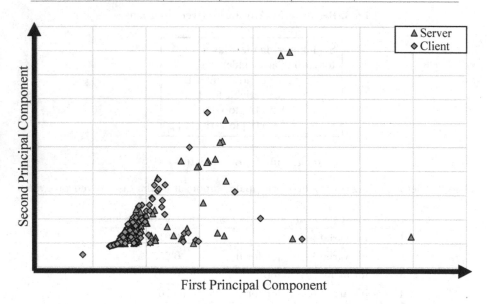

Fig. 5. Result of PCA of Server class and Client class

we investigate whether the program is classified by the proposed method. We used a part of programs collected by Allamanis and Sutton [8] and we used class names "Server", "Client", "Parser", and "Token". The details of the programs are shown in a Table 5. The experiment procedure is the same as for the textbook programs.

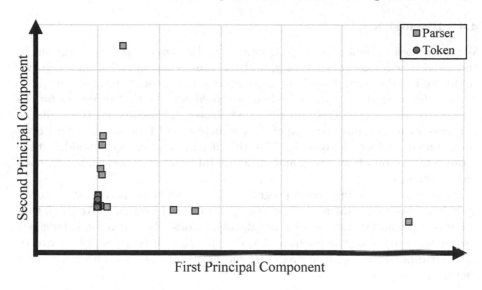

Fig. 6. Result of PCA of Parser class and Token class

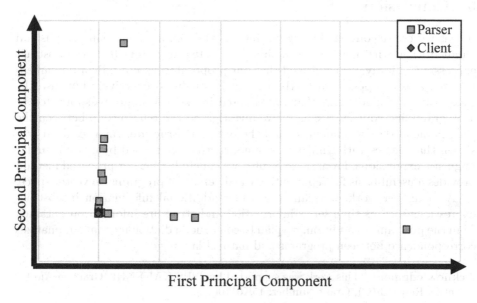

Fig. 7. Result of PCA of Parser class and Client class

Figures 5, 6 and 7 show the result of two-dimensional plots by PCA to the programs among the classes. From these figures, we can confirm that the Server and the Client have a similar trend, the Parser and the Token have the trend of the inclusion relation. Also, since the Parser and the Client are programs with different purposes, we can confirm that program diversity can be expressed.

4.4 Discussion

We performed classification in programs with the classification of programs in the same functions, the same task, and the same class name. As a result, we could confirm the usefulness of the vector representation of the proposed method. That is, by using the proposed method, we could confirm that programs for the same purpose are classified into the same cluster. Therefore, we think that it is necessary to examine the type of data structure and functional pattern from open sources and so on. Also, we think that it can be improved by adding time series information of programs and structure information when those are made into a syntax tree.

From experiment in general program, we confirm that there is some correspondence between some part of programs and natural language. We think that there is a possibility that a reference document described in natural language can be retrieved from the program. For that purpose, it is necessary to investigate correspondence between programs and natural languages, and construct a dictionary.

5 Conclusion

In this paper, we converted programs into abstract expressions using graphs and handled them with mathematical similarity between vectors from the abstract expressions. Firstly, we confirmed that our proposal work not only a fundamental programs as appears in textbooks can be analyzed effectively but also the classification of algorithms that are adopted in various programs submitted to a programming contest. Secondly, we investigated the relationship between the class name of the Java language and the program structure. As a result, it was shown that classes with similar program structure are named by certain similar linguistic expressions. From these results, we conclude that our proposed method provides a useful basis for representing the diversity of programs on vector space.

As a further work, we plan to consider additional information in abstract expressions such as time series information and structure information obtained from the program. In addition, we plan to construct a dictionary that summarizes correspondence between programs and natural language.

Acknowledgment. This work was supported by JSPS KAKENHI (Grant-in-Aid for Scientific Research(C), Grant Number: 15K01100).

References

1. Mechtaev, S., Yi, J., Roychoudhury, A.: Directfix: looking for simple program repairs. In: Proceedings of the 37th International Conference on Software Engineering, vol. 1, pp. 448–458. IEEE Press (2015)
2. Nguyen, H.D.T., Qi, D., Roychoudhury, A., Chandra, S.: Semfix: program repair via semantic analysis. In: Proceedings of the 2013 International Conference on Software Engineering, pp. 772–781. IEEE Press (2013)

3. Le Goues, C., Nguyen, T., Forrest, S., Weimer, W.: Genprog: a generic method for automatic software repair. IEEE Trans. Softw. Eng. **38**(1), 54–72 (2012)
4. Murakami, H., Hotta, K., Higo, Y., Kusumoto, S.: Predicting next changes at the fine-grained level. In: 2014 21st Asia-Pacific Software Engineering Conference (APSEC), vol. 1, pp. 119–126. IEEE (2014)
5. Mou, L., Li, G., Liu, Y., Peng, H., Jin, Z., Xu, Y., Zhang, L.: Building program vector representations for deep learning. arXiv preprint arXiv:1409.3358 (2014)
6. Gvero, T., Kuncak, V.: Synthesizing Java expressions from free-form queries. ACM SIGPLAN Not. **50**(10), 416–432 (2015)
7. EclipseJDT: Eclipsejdt. http://www.eclipse.org/jdt. Accessed 27 Sept 2017
8. Allamanis, M., Sutton, C.: Mining source code repositories at massive scale using language modeling. In: The 10th Working Conference on Mining Software Repositories, pp. 207–216. IEEE (2013)
9. Lafore, R.: Java de manabu algorithm to data kouzou Java で学ぶアルゴリズムとデータ構造 (Japanese). SB Creative (1999)
10. Kondou, Y.: Teihon Java programmer no tameno algorithm to data kouzou 定本 Java プログラマのためのアルゴリズムとデータ構造 (Japanese). SB Creative (2011)
11. Igarashi, T.: Data kouzou to algorithm データ構造とアルゴリズム (Japanese). Suurikougaku-Sha (2007)
12. Shibata, B.: Meikai Java niyoru algorithm to data kouzou 明解 Java によるアルゴリズムとデータ構造 (Japanese). SB Creative (2007)
13. Watanobe, Y.: Aizu online judge. http://judge.u-aizu.ac.jp/onlinejudge/. Accessed 27 Sept 2017

Comparing Sparse Autoencoders for Acquisition of More Robust Bases in Handwritten Characters

Takuya Okada[(✉)] and Kazuhiro Takeuchi

Graduate School of Engineering, Osaka Electro-Communication University,
Osaka, Japan
mi16a002@oecu.jp

Abstract. Autoencoders that acquire specific feature space models from unsupervised data have become an important technique for designing systems based on neural networks. In this paper, we focuses on the reusability of sparse encoder for handwritten characters. In existing studies, the training bias of sparse autoencoders is generally more constrained in the aspect of the number of the activated intermediate units than the other autoencoders. We investigate the role that trained units play as another direction of training bias for more reusable autoencoder. As a basis of the investigation, we manually selected three autoencoders and compare the reusability of them in two experiments. One is a letter identification experiment for a character whose character faded or blurred and the structure of the original character collapsed. The other is the experiment to distinguish the lines that form letters from that the line segments in subparts other than the text part constituting the sentences such as figures and tables. As a result, we found that the role that the intermediate units of the most reusable autoencoder in our experiments plays is regarded as binary functions.

Keywords: Autoencoder · Sparse coding · Pre-training
Feature extraction

1 Introduction

Pre-training that processes features of learning objects including autoencoders is important to utilize the neural network [1]. In recent years, studies on the autoencoder, which is core module acquired in pre-training, have been actively conducted and brought about improvement in performance in various tasks using them [2,3]. On the other hand, there is a problem that autoencoder modeled for such specific task is overfitted to the task excessively, which makes it difficult to reuse it in another task. For example, there is a problem that characters with high discrimination performance in the task of recognizing characters, but figures other than characters not used for modeling autoencoder are sometimes misrecognized as specific characters. This is because the target components that an autoencoder is trained do not directly correspond to the lines that the characters are drawn (we refer such

© Springer International Publishing AG, part of Springer Nature 2018
V.-N. Huynh et al. (Eds.): IUKM 2018, LNAI 10758, pp. 138–149, 2018.
https://doi.org/10.1007/978-3-319-75429-1_12

lines to as strokes in this paper). Therefore, networks built with these methods are very weak against certain types of noise, giving suddenly erroneous answers in images that look the same to human eyes [4]. In other words, determining the granularity (the length and bold size) of the strokes is not easy. For example, if they are too fine, the trained autoencoder will not distinguish strokes from noises. On the other hand, if they are is too rough, the trained basis of autoencoder will not reflect the strokes but the trained character themselves. In order to acquire more reusable bases, we have to make an appropriate criteria for the autoencoder with that the training examples can be divided into lines.

In order to address the problem in this paper, we construct several sparse autoencoders that are given with different constraint and from them we select three autoencoders that are different in reusable characteristics. Then, we investigate the functions that intermediate units realized in the autoencoders in detail. In particular, we evaluate the reusability of autoencoder by their ability to distinguish parts that characters consist of from ones that are contained in figures, tables and the other. In other words, we assume that obtaining reusable basis for character recognition by the autoencoder is ability to extract the strokes that make up the character by it. As a condition of our experiments, we use the autoencoders that are trained only with handwritten characters. This evaluation task can be also regarded as a preprocess in photographed document analysis.

2 Related Works

An autoencoder [5] is one of the architectures of neural network trained the target data is its input data itself. In other words, an autoencoder is regarded as a machine learning module which is trained from unlabeled data.

Assuming that the output of the network is y and the teacher value (input) is d, autoencoder minimizes the loss function $E(w)$ defined by Eq. (1).

$$E(w) = \sum_k (y_k - d_k)^2 \tag{1}$$

The intermediate layer of the learned autoencoder can be work as input feature extraction. Since it is sufficient to reproduce the training data, the feature obtained is rarely to learn features that successfully express those structures. Therefore, it is weak against certain noise [4,6], and it reacts equally to the target task even for inputs with completely different characteristics.

The sparse autoencoder [7,8] adopts the constraints from the target structure of input constraint into their intermediate layer of network. Specifically, a sparse autoencoder defines the loss function $E(w)$ by the Eq. (2) so that the output value of the intermediate layer approaches 0 when input is given.

$$E(w) = \sum_k (y_k - d_k)^2 + \beta \sum_j KL(\rho||\hat{\rho}_j) \tag{2}$$

In the Eq. (2), β is a parameter that determines the degree of extent regularization by the trained examples. $KL(x||y)$ in the second term is the distance function between x and y, and the definition is given by the Eq. (3).

$$KL(x||y) = x \log\left(\frac{y}{x}\right) + (1-x)\log\left(\frac{1-x}{1-y}\right) \tag{3}$$

when input is given, the average of outputs of the jth element of the intermediate layer $\hat{\rho}_j$ is expressed by the Eq. (4).

$$\hat{\rho}_j = \frac{1}{N}\sum_{n=1}^{N} h_j(\boldsymbol{x}_n) \tag{4}$$

With this regularization, the Eq. (5), which is referred to as the parameter update formula, is defined with the parameter $\boldsymbol{w}^t$ at the next time t. Where ϵ is the learning rate and z_i^{l-1} is the input from the ith element of the previous layer. δ_j^l is given by Eqs. (6) and (7) in the output layer and the intermediate layer, respectively.

$$w_{ji}^{t+1} = w_{ji}^{t} - \epsilon\delta_j^l z_i^{l-1} \tag{5}$$

$$\delta_j^{out} = y_j - d_j \tag{6}$$

$$\delta_j^{hidden} = \sum_k \delta_j^{out} w_{kj} + \beta\left(-\frac{\rho}{\hat{\rho}_j} + \frac{1-\rho}{1-\hat{\rho}_j}\right) \tag{7}$$

The second term on the right side of Eq. (7) acts as a constraint on the degree of sparseness. As long as each unit of the intermediate layer is affected by almost all inputs, the weight of the intermediate layer is suppressed toward 0. In other words, this bias limits the number of the bases constructed with repetition of training.

Therefore, the intermediate layer can represent the input by these limited bases. In this way, the sparse autoencoder automatically learns the bases reflecting the constituent units of the input. However, in order to obtain a reusable bases, adjustment of hyper-parameters is important and difficult [9].

Bases responds only to the specific letters learned, so the reusability will decrease. Conversely, if the basis does not correspond to a single character, it is a feature of the pixel unit. Therefore, this basis can't distinguish whether the input is a character or not, and it has poor reusability. For these reasons, suitable roughness of basis is on the order of stroke, but it is not easy to add directly as a constraint. On the other hand, there are attempts to reduce the number of hyper-parameters [10,11], but the criteria for choosing a network after training is not clear.

3   Reusable Bases of Handwritten Characters

Assuming that the basis of the reusable autoencoder is a stroke, it must satisfy the following three conditions (Fig. 1).

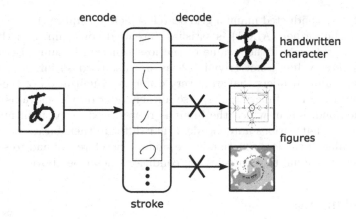

Fig. 1. Condition of stroke basis

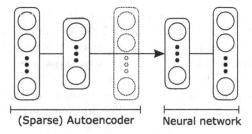

Fig. 2. Configuration of each model

- A stroke can not be divided into more strokes.
- A character is constructed by several strokes.
- Combining several strokes does not make a figure or table.

In order to examine what kind of autoencoder satisfies the conditions, we prepare the following three autoencoders that have different characteristics. Figure 2 shows the structure how each autoencoder works in a certain classification task. In the network, training data was given to the pre-trained autoencoder, and the output of the intermediate layer at that time was given as a new input to the neural network which becomes a classifier. However, the gradient from the classifier was not propagate to the sparse autoencoder, it was used as a feature extractor that does not know the target task.

- An autoencoder learned by giving output of the intermediate layer of a sparse autoencoder with intended constraint. (hereinafter referred to as ALG1)
- An autoencoder learned by giving output of the intermediate layer of a sparse autoencoder that was not subject to the intended constraint. (hereinafter referred to as ALG2)
- An autoencoder learned by giving output of the intermediate layer of ordinary autoencoder. (hereinafter referred to as ALG3)

Actually, we conducted manual grid search select to acquire the two autoencoders that the former (ALG1) is satisfies 'intended constrain' and the latter (ALG2) does not even if their value of the loss function is same. Because the hyper-parameter value is not trivial [12] as we discussed so far, we manually selected those autoencoders that were very differ in visualization of their bases. The bias with the loss function of sparse autoencoder only constrains to let it satisfies the condition in which the input is reproduced on a small number of basis. It means that various autoencoders that differ in their bases are acquired even if the value of the loss function is the same. Therefore, we had to select the autoencoder in manual that satisfies the conditions describe above.

4   Experiments

4.1   Data

The character data used in this experiment is the JIS 1st level handwritten Chinese character database ETL9B [13]. ETL 9B contains 3036 types of JIS 1st level Kanji 2965 character types and 71 types of Hiragana (including dakuten and handakuten), 200 patterns of each character type are recorded. From this time, we used 71 kinds of Hiragana.

Hiragana is a Japanese syllabary, one component of the Japanese writing system, along with katakana and kanji. The hiragana syllabary consists of 71 base characters:

- 5 singular vowels
- 40 consonant-vowel unions
- 1 singular consonant
- 20 consonant-vowel unions with dakuten
- 5 consonant-vowel unions with handakuten

These hiragana are shown in Table 1. For the data in the figure, we prepared 200 samples of figures trimmed manually. Each sample is resized to (32×32) and binarized. After that, the image was input as a 1024 dimensional vector.

4.2   Experiment Setting

In this experiment, each autoencoder with 100 nodes was trained 400 times using character data that 50 subjects wrote. The initial value was set according to [14]. As an example of the learned bases at this time, Fig. 3(a) shows a case where constraint is applied (ALG1), Fig. 3(b) shows a case where no constraints as intended are applied (ALG2), and Fig. 3(c) shows a case where a constraint is not used (ALG3).

The left hand side of Fig. 3 shows the weights between the input layer and the intermediate layer, where the positive weight is white, 0 is gray, and the negative weight is black, which is associated with each pixel of the input image. The order of this image is ordered by the element number of the intermediate

Table 1. Hiragana syllabograms

gojuon					
	a	i	u	e	o
φ	あ	い	う	え	お
K	か	き	く	け	こ
S	さ	し	す	せ	そ
T	た	ち	つ	て	と
N	な	に	ぬ	ね	の
H	は	ひ	ふ	へ	ほ
M	ま	み	む	め	も
Y	や		ゆ		よ
R	ら	り	る	れ	ろ
W	わ				を
*	ん				
gojuon with (han) dakuon					
G	が	ぎ	ぐ	げ	ご
Z	ざ	じ	ず	ぜ	ぞ
D	だ	ぢ	づ	で	ど
B	ば	び	ぶ	べ	ぼ
P	ぱ	ぴ	ぷ	ぺ	ぽ

layer. It can be seen from the figures that the normal autoencoder ALG2 has many bases to respond to all areas of the inputs, but in the case of ALG1, there are many bases that react only to a part of the inputs.

The right hand side of Fig. 3 shows a histogram of intermediate layer output values of the when test data is given as input to each model. Figure 3(a) shows the intermediate layer output values of ALG1. We seen many output values around 0 or 1, and almost not seen elsewhere. As shown in Fig. 3(b), the intermediate layer output values of ALG2 extends to all the regions from 0 to 1, and in particular, much is seen in the vicinity of 0. As shown in Fig. 3(c), the intermediate layer output values of the ALG3 is distributed throughout the range from 0 to 1 like the ALG2, but in contrast to the ALG2, the output values in the vicinity of 1 is particularly large.

From these results, when the bases is a stroke, the input seems to be reproduced only by the overlapping of a few bases, so the output of the intermediate layer seems to appear intensively at 0 or 1 when the correct constraint is applied. In contrast, in the case where it is not an appropriate constraint, the output of each element is suppressed to 0, but the input can be reproduced by strengthening or destroying multiple bases, so the output of the intermediate layer is relatively gentle distribution. On the other hand, when the constraint is not used, it is considered that the input was reproduced with a plurality of bases and never held down to 0, so that the distribution became a gentle distribution relatively concentrated around 1. Since the loss function of the sparse autoencoder uses the average value of the output of each intermediate layer element for

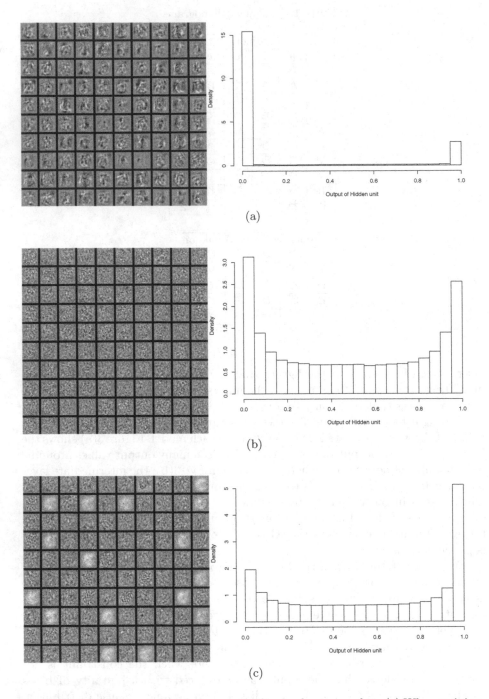

Fig. 3. Bases and intermediate layer output of trained autoencoders. (a) When training with ALG1. (b) When training with ALG2. (c) When training with ALG3

evaluation, the distribution difference like ALG1 and ALG2 can not be known from that value. ALG1 is strongly influenced by the penalty of input whose several output values are 1, and it is greatly different from the bases of ALG2 which frequently outputs a small value which is not 0 even though the average is the same degree. Also, the distribution of the intermediate layer output values of ALG1 can be represented by a binary function.

We used the feedforward neural network as the classifier, the input layer size was 100, the output of the intermediate layer was used as the input of the classifier, so it was 100 and the size of the classifier of each model the size of the intermediate layer was 50.

We conducted three experiments to evaluate each model. First, as a baseline, we perform discrimination task of character type using each trained model (hereinafter referred to as EX1). In the second experiment, we examine whether stroke-based decomposition is robust against noise. To do this, add noise to the test data that takes into account blurring of characters, and perform a classification task (hereinafter referred to as EX2). This experiment is done to make sure that it has the same capability as Denoising autoencoder [15]. However, in contrast to the denoising autoencoder, the noise is not added to the training data but only added to the test data. This indicates that the decomposition by stroke is strong against noise, and also ensures that the network is not learning noise. In the third experiment, in order to confirm the robustness against inputs other than characters, we perform figure and character classification task (hereinafter referred to as EX3).

5   Results and Discussion

5.1   The Way of Evaluation

For the evaluation of classification of character type, the order of prediction of the correct character type is used. The ranking of the forecast should be within 10th place, and the ratio correctly answered for all the test cases is taken as accuracy. That is, when the prediction rank of the ith sample is denoted by r_i, it is defined by the Eq. (8).

$$Accuracy = \frac{\sum^N f(r_i)}{N}, \ f(r_i) = \begin{cases} 1 \ (r_i \leq 10) \\ 0 \ (otherwise) \end{cases} \tag{8}$$

For evaluation of figure and character discrimination, the correct answer rate of correctly categorized sample number for all test data numbers shown in Eq. (9) is used for evaluation.

$$Accuracy = \frac{\textbf{Number of correctly classified samples}}{\textbf{Number of all test samples}} \tag{9}$$

5.2   Classification of Character Type

As an evaluation, we identify which hiragana letter each given input in the test data belongs to. We divide the data that we did not use as training data for autoencoder into the training data for neural network and the test data. Both data are written by 50 people each.

To objectively evaluate whether character characteristics can be successfully acquired with each autoencoder, we artificially modified handwritten characters to be as our test data. Artificial character modification is based on the character structure. Specifically, we randomly select one direction to draw a stroke with respect to the preset eight directions and blurred it weakening the black region. The degree of weakening is randomly selected in the range from 0 to 1. An example of artificial modification of character is shown in Fig. 4.

Figures 5 and 6 show the results of EX1 and EX2. From Fig. 5, ALG3 predicts more accurately than the ALG1, but in Fig. 6 the relationship is reversed. This shows that autoencoder is weak against noise. On the other hand, ALG1 stably maintains higher accuracy than ALG3, and it seems that ALG1 has some degree of robustness against noise.

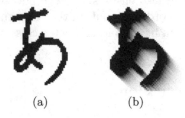

(a)   (b)

Fig. 4. Examples of blurred letters (a) Image before adding blur (b) Image in the lower right direction with an intensity of 0.7

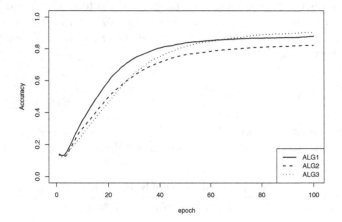

Fig. 5. Result of EX1

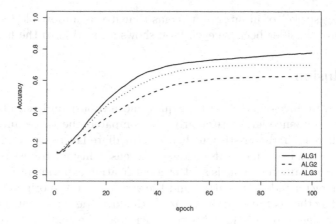

Fig. 6. Result of EX2

5.3   Discrimination Between Figures and Letters

In discrimination between figures and letters, we classify whether a given input is a figure or a character. From the character data, 10 persons not used for training in autoencoder and 150 samples from the data in the figure were used as training data. However, if this training data is used as it is, the number of data belonging to each class will be biased, so oversampling is performed until 150 samples of data in the figure are equal to the number of characters. After that, a test was carried out using another 10 persons from the character data and the remaining 50 samples of the data in the figure.

Figure 7 shows the results of EX3. From Fig. 7, it is understood that ALG1 is superior to ALG3 in terms of drawing and character classification. As a cause, in the case of autoencoder without constraint, we can not capture the continuous

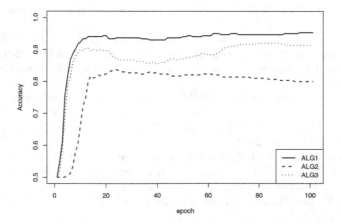

Fig. 7. Result of EX3

characteristic (stroke) of input and it seems that it was impossible to construct an appropriate classifier because each base shows a reaction to the figure.

6   Conclusion

In this paper, we investigate bias to acquire more robust and reusable strokes with a sparse autoencoder. Particularly, we compared the three autoencoders that have different characteristics in their intermediate layer. We conducted two experiments to confirm which autoencoder has reusability for robust recognition of hand-written characters: One is a letter identification experiment for a character whose character faded or blurred and the structure of the original character collapsed. The other is the experiment to distinguish the lines that form letters from that the line segments in subparts other than the text part constituting the sentences such as figures and tables. From the further analysis of the most robust autoencoder, we found that the intermediate layer of the highly reusable autoencoder is like a binary function. As a further work, we would like to develop a method to automatically select a reusable network from a trained autoencoder using this property.

References

1. Hinton, G.E., Salakhutdinov, R.R.: Reducing the dimensionality of data with neural networks. Science **313**(5786), 504–507 (2006)
2. Mikolov, T., Chen, K., Corrado, G., Dean, J.: Efficient estimation of word representations in vector space. arXiv preprint arXiv:1301.3781 (2013)
3. Vincent, P., Larochelle, H., Lajoie, I., Bengio, Y., Manzagol, P.-A.: Stacked denoising autoencoders: learning useful representations in a deep network with a local denoising criterion. J. Mach. Learn. Res. **11**, 3371–3408 (2010)
4. Goodfellow, I.J., Shlens, J., Szegedy, C.: Explaining and harnessing adversarial examples. In: ICLR (2014)
5. Bishop, C.M.: Pattern Recognition and Machine Learning. Springer, New York (2006)
6. Nguyen, A., Yosinski, J., Clune, J.: Deep neural networks are easily fooled: high confidence predictions for unrecognizable images. arXiv:1412.1897 (2014)
7. Goodfellow, I.J., Bengio, Y., Courville, A.: Deep Learning. MIT Press (2016). Book in preparation
8. Ng, A.: Sparse autoencoder. CS294A Lecture notes, Stanford University, p. 72 (2011)
9. Tsuboi, Y., Unno, Y., Suzuki, J.: Natural Language Processing by Deep Learning. Kodansha, New York (2017)
10. Makhzani, A., Frey, B.: k-Sparse autoencoders. arXiv:1312.5663 (2013)
11. Makhzani, A., Frey, B.J.: Winner-take-all autoencoders. In: Advances in Neural Information Processing Systems, pp. 2773–2781 (2015)
12. Okatani, T.: Deep Learning. Kodansha, New York (2015)

13. Electrotechnical Laboratory, Japanese Technical Committee for Optical Character Recognition, ETL Character Database (1973–1984)
14. Glorot, X., Bengio, Y.: Understanding the difficulty of training deep feedforward neural networks. In: Proceedings of AISTATS, vol. 9, pp. 249–256 (2010)
15. Vincent, P., Larochelle, H., Bengio, Y., Manzagol, P.-A.: Extracting and composing robust features with denoising autoencoders. In: ICML (2008)

Machine Learning Applications

fMKL-DR: A Fast Multiple Kernel Learning Framework with Dimensionality Reduction

Thanh Trung Giang[1(✉)], Thanh Phuong Nguyen[2,3(✉)], Tran Quoc Vinh Nguyen[4(✉)], and Dang Hung Tran[5(✉)]

[1] Tay Bac University, Son La, Vietnam
trunggt@utb.edu.vn
[2] Life Sciences Research Unit, University of Luxembourg, Luxembourg, Luxembourg
[3] Analytics and Modelling, ArcelorMittal Europe - Flat Products, Luxembourg, Luxembourg
phuong.nguyen@arcelormittal.com
[4] Da Nang University of Education, Da Nang, Vietnam
ntquocvinh@ued.udn.vn
[5] Hanoi National University of Education, Hanoi, Vietnam
hungtd@hnue.edu.vn

Abstract. Exploring and integrating data from heterogeneous sources have attracted much interest in recent years. However, one of the greatest challenges is that a lot of data are highly dimensional and diverse. In order to effectively combine multiple data sources, it is essential to reduce the number of dimensions and boost computational performance. This could be accomplished by combining multiple kernel learning with dimensionality reduction. In this paper, we propose an improved multiple kernel learning framework, referred to as fMKL-DR, that optimize equations to calculate matrix chain multiplication. To reach this conclusion, we performed several comparative evaluations on various biomedical data sets. The results demonstrate that, compared to previous work, the fMKL-DR remarkably improves computational cost. Therefore, the proposed framework is beneficial to the manipulation and integration of huge and complex datasets.

Keywords: Multiple kernel learning · Dimensionality reduction
Computational cost

1 Introduction

Data are measured and collected by a number of means; therefore the data types and data formats are highly heterogeneous. It is very useful to integrate multiple types of data to obtain a comprehensive view. However, it is challenging to develop efficient methods for dealing with data combination. There have been several research efforts to solve this problem [1–3]. In most cases, the methods addressed the problem of data representation which is in various forms such as bag-of-features and matrices because different data were extracted by typical measurements. Furthermore, the datasets were often high-dimensional and required a significant amount of calculating time. Dimensionality reduction (DR) was critical to perform with or after integrating data [2].

© Springer International Publishing AG, part of Springer Nature 2018
V.-N. Huynh et al. (Eds.): IUKM 2018, LNAI 10758, pp. 153–165, 2018.
https://doi.org/10.1007/978-3-319-75429-1_13

Lin *et al.* [4] proposed an efficient framework, integrating multiple data and reducing their dimensions, called MKL-DR. MKL-DR not only used multiple kernels learning but also applied a dimensionality reduction method based on graph embedding. This framework had two major advantages. Firstly, it could be applied to almost all data types thanks to multiple kernel learning. Secondly, any dimensionality reduction method using graph embedding could be successfully applied on MKL-DR. Recently, Spacher *et al.* [5] improved MKL-DR by adding a constraint to avoid overfitting, namely rMKL-DR. The experimental results of rMKL-DR were better than the ones of MKL-DR on cancer patient datasets. Nevertheless, both MKL-DR and rMKL-DR were likely time-consuming on large datasets due to the repetition of matrix chain multiplication. This limitation makes it difficult for other application and does not comply with one of three evaluating criteria in data analysis recommended by John Shawe-Taylor in [8].

In this paper, we have proposed a fast multiple kernel learning combined with dimensionality reduction (fMKL-DR), which improves the MKL-DR method. fMKL-DR optimizes the equations to calculate matrix chain multiplication in order to enhance performance in terms of computational cost. We performed a number of experiments on cancer patient dataset, which included three data types: Gene Expression, DNA Methylation, and Protein Expression. Comparisons of computational costs of the three methods, MKL-DR, rMKL-DR, fMKL-DR were then carried out. The results show that our method runs faster than MKL-DR and rMKL-DR did.

2 Related Work

2.1 Kernel Method

Kernel method has been proposed to solve the problems of non-linear input data. There are two important points in kernel method. Firstly, this method transforms original data into a feature space, called Hilbert space, with the assumption that the data are linearly separated in the new feature space. Secondly, linear relations in feature space are unraveled by applying linear algebra, geometry, and statistical algorithms. The data transformation is performed implicitly.

Given an input dataset $\mathcal{X} = \{x_1, x_2, \ldots, x_N\}$ and a feature space $\mathcal{F}$, we have a mapping Φ as follows:

$$\Phi : \mathcal{X} \to \mathcal{F} \tag{1}$$

$$x \mapsto \Phi(x) \tag{2}$$

$$(x_i, x_j) \to \langle \Phi(x_i), \Phi(x_j) \rangle \tag{3}$$

Kernel method does not define mapping Φ but uses the *kernel function*, which returns inner product between a pair of data points in $\mathcal{F}$. A kernel function $k(x_i, x_j) = \langle \Phi(x_i), \Phi(x_j) \rangle$ with $x_i, x_j \in \mathcal{X}$.

Kernel matrix, also known as Gram matrix, is a square matrix with elements which are inner products between a pair of data points in $\mathcal{F}$. Kernel matrix $K \in \mathbb{R}^{N \times N}$ with $K_{ij} = k(x_i, x_j)$.

In the next step, the linear relations were found on kernel matrix rather than on original data. Support Vector Machines (SVMs), which finds a linear separator, defines an affine function in $\mathcal{F}$:

$$f(x) = \langle w, \Phi(x) \rangle + b \tag{4}$$

with weight vector $w \in \mathcal{F}$. Kernel weight vector is defined by a linear combination of data points based on kernel function: $w = \sum_{i=1}^{N} \alpha_i \Phi(x_i)$. The function f is then redefined as follow:

$$f(x) = \sum_{i=1}^{N} \alpha_i k(x_i, x) + b \tag{5}$$

2.2 Multiple Kernel Learning

Kernel method presented above only applies to a single data type whereas in actuality, there are a number of data types collected for one problem. Moreover, data are presented in different forms and various measurements are employed. As a result, multiple kernel learning (MKL) is proposed to integrate multiple data into a unified dataset. Multiple kernel learning is a learning process which finds a unified kernel matrix as a linear combination of base kernel matrices.

Given a base kernel function set $\{k_m\}_{m=1}^{M}$ (or a base kernel matrix set $\{K_m\}_{m=1}^{M}$) with M is a number of data types. A unified kernel function (or a unified kernel matrix) is defined as:

$$k(x_i, x_j) = \sum_{m=1}^{M} \beta_m k_m(x_i, x_j) \tag{6}$$

$$K = \sum_{m=1}^{M} \beta_m K_m \tag{7}$$

Consequently, the binary classification model based on multiple kernel learning from data $\{(x_i, y_i \in \pm 1)\}_{i=1}^{N}$ is:

$$f(x) = \sum_{i=1}^{N} \alpha_i y_i k(x_i, x_j) + b \tag{8}$$

$$= \sum_{i=1}^{N} \alpha_i y_i \sum_{m=1}^{M} \beta_m k_m(x_i, x) + b \tag{9}$$

Finding separation function f while optimizing both $\{\alpha_i\}_{i=1}^{N}$ and $\{\beta_m\}_{m=1}^{M}$.

2.3　Graph Embedding

Graph embedding is a dimensionality reduction (DR) framework proposed by Yan *et al.* [6]. This framework is a generalized model integrating early DR method. Yan *et al.* also proposed a new DR method based on graph embedding. Graph embedding presents input data as a graph of vertices of data sample and edges as the similarities between a pair of samples. It finds a projected vector to project original data graph into a new space, which is the best characterization the relation between pairs of vertices in the original space. The project vector is defined base on the graph-preserving criterion:

$$\text{minimize} \sum_{i \neq j}^{N} ||\alpha^T K_i - \alpha^T K_j||^2 w_{ij} \tag{10}$$

$$\text{s.t.} \sum_{i=1}^{N} ||\alpha^T K_i||^2 d_{ii} = \text{const}, \tag{11}$$

$$\text{or} \sum_{i,j=1}^{N} ||\alpha^T K_i - \alpha^T K_j||^2 w'_{ij} = \text{const} \tag{12}$$

where N is a number of samples, K is a kernel matrix, K_i is ith column of K, W is a similar matrix with elements w_{ij}, D is a diagonal matrix with entries d_{ii} and W' are constrained matrices avoiding the trivial solution of the objective function (10). Many DR methods applying this framework have been developed, such as Linear Discriminant Analysis (LDA), Local Discriminant Embedding (LDE), Locality Preserving Projections (LPP), Semi-supervised Discriminant Analysis (SDA), etc. Defining W, W', and D matrices and constraint (11) or constraint (12) depends on DR method (see details in [6]).

3　Method

3.1　Multiple Kernel Learning Combined with Dimensionality Reduction

Lin *et al.* [4] proposed a multiple kernel learning combined with dimensionality reduction framework (MKL-DR) to better combine data from different sources. MKL-DR has two running processes: integrating data and dimensionality reduction.

Given a unified kernel matrix K defined by a linear combination of the base kernel matrices $\{K_m\}_{m=1}^{M}$ as (7), DR process is employed using graph embedding. MKL-DR transforms the original data ($\mathcal{X}$) into the feature space ($\mathcal{F}$) and later projects from the feature space $\mathcal{F}$ to a reduced dimension space ($\mathbb{R}^P$ with P is number of dimensions after reducing). The next steps of analyzing data are based on $\mathbb{R}^P$.

Given v is projected vector from $\mathcal{F}$ to $\mathbb{R}^P$:

$$v = \sum_{n=1}^{N} \alpha_n \phi(\mathrm{x}_n) \tag{13}$$

Associating with (2) we have a data point x_i in $\mathbb{R}^P$ as follows:

$$v^{\mathsf{T}} \phi(\mathrm{x}_i) = \sum_{n=1}^{N} \sum_{m=1}^{M} \alpha_n \beta_m k_m(\mathrm{x}_n, \mathrm{x}_i) = \alpha^{\mathsf{T}} \mathbb{K}^{(i)} \beta \tag{14}$$

with:

$$\alpha = [\alpha_1 \cdots \alpha_N]^{\mathsf{T}} \in \mathbb{R}^N,$$
$$\beta = [\beta_1 \cdots \beta_M]^{\mathsf{T}} \in \mathbb{R}^M,$$

$$\mathbb{K}^{(i)} = \begin{bmatrix} K_1(1, i) & \cdots & K_M(1, i) \\ \vdots & \ddots & \vdots \\ K_1(N, i) & \cdots & K_M(N, i) \end{bmatrix} \in \mathbb{R}^{N \times M} \tag{15}$$

From (10) and (14), MKL-DR problem defines the constrained optimization problem in 1-D space by:

$$\min_{\alpha, \beta} \sum_{i,j=1}^{N} ||\alpha^{\mathsf{T}} \mathbb{K}^{(i)} \beta - \alpha^{\mathsf{T}} \mathbb{K}^{(j)} \beta||^2 w_{ij} \tag{16}$$

$$\text{s.t.} \sum_{i,j=1}^{N} ||\alpha^{\mathsf{T}} \mathbb{K}^{(i)} \beta - \alpha^{\mathsf{T}} \mathbb{K}^{(j)} \beta||^2 w'_{ij} = \text{const} \tag{17}$$

$$\beta_m \geq 0, m = 1, .., M. \tag{18}$$

Problem (16) was defined in 1-D space. However, in fact, the number of dimensions P in $\mathbb{R}^P$ is often greater than one. In a multiple dimension case ($P > 1$), the sample coefficient vector α is extended to sample coefficient matrix A as follows:

$$A = \begin{bmatrix} \alpha_1 \alpha_2 \cdots \alpha_P \end{bmatrix} \tag{19}$$

Projected vector v is extended to $V = \begin{bmatrix} v_1 v_2 \cdots v_P \end{bmatrix}$. Equation (14) is transformed to:

$$V^{\mathsf{T}} \phi(x_i) = A^{\mathsf{T}} \mathbb{K}^{(i)} \beta \in \mathbb{R}^P \tag{20}$$

Optimization problem (16) is extended in a general case as follows:

$$\min_{A, \beta} \sum_{i,j=1}^{N} ||A^{\mathsf{T}} \mathbb{K}^{(i)} \beta - A^{\mathsf{T}} \mathbb{K}^{(j)} \beta||^2 w_{ij} \tag{21}$$

$$\text{s.t. } \sum_{i,j=1}^{N} ||A^\top \mathbb{K}^{(i)}\beta - A^\top \mathbb{K}^{(j)}\beta||^2 w'_{ij} = \text{const} \tag{22}$$

$$\beta_m \geq 0, m = 1, .., M. \tag{23}$$

A direct solution to the problem (21) base on both parameters A and β is difficult. A two-step strategy is used to alternately optimize A and β. At each iteration, either A or β is optimized while the other is fixed, and vice versa. Problem (21) is reduced to two optimization problems as follows:

Fixing A, optimize β. We have a new problem defined by:

$$\min_{\beta} \quad \beta^\top S_W^A \beta$$
$$\text{s.t. } \beta^\top S_{W'}^A \beta = 1 \quad \text{and} \quad \beta \geq 0, \tag{24}$$

with:

$$S_W^A = \sum_{i,j=1}^{N} w_{ij}(\mathbb{K}^{(i)} - \mathbb{K}^{(j)})^\top AA^\top(\mathbb{K}^{(i)} - \mathbb{K}^{(j)}), \tag{25}$$

$$S_{W'}^A = \sum_{i,j=1}^{N} w'_{ij}(\mathbb{K}^{(i)} - \mathbb{K}^{(j)})^\top AA^\top(\mathbb{K}^{(i)} - \mathbb{K}^{(j)}). \tag{26}$$

The problem (24) is nonconvex quadratically constrained quadratic programming problem that is known as a hard one. The problem can be solved by adding an auxiliary variable B:

$$\min_{\beta,B} \text{trace}(S_W^A B)$$
$$\text{s.t. } \text{trace}(S_D^A B) = 1 \text{ or trace}(S_{W'}^A B) = 1,$$
$$e_m^\top \beta \geq 0, \quad m = 1, \dots, M$$
$$\begin{bmatrix} 1 & \beta^\top \\ \beta & B \end{bmatrix} \geq 0 \tag{27}$$

with e_m is a column vector whose elements equal to 0 except its mth element equal to 1. Problem (27) is solved easily by Semidefinite Programming (SDP) [7].

Fixing β, optimize A:

$$\min_{A} \quad \text{trace}(A^\top S_W^\beta A)$$
$$\text{s.t. } \text{trace}(A^\top S_{W'}^\beta A) = 1, \tag{28}$$

with:

$$S_W^\beta = \sum_{i,j=1}^{N} w_{ij} (\mathbb{K}^{(i)} - \mathbb{K}^{(j)}) \beta \beta^\mathsf{T} (\mathbb{K}^{(i)} - \mathbb{K}^{(j)})^\mathsf{T}, \tag{29}$$

$$S_{W'}^\beta = \sum_{i,j=1}^{N} w'_{ij} (\mathbb{K}^{(i)} - \mathbb{K}^{(j)}) \beta \beta^\mathsf{T} (\mathbb{K}^{(i)} - \mathbb{K}^{(j)})^\mathsf{T}. \tag{30}$$

The problem (28) is transformed to a *ratio trace* problem, solving the below-generalized eigenvalue problem to optimize A:

$$S_W^\beta \alpha = \lambda S_{W'}^\beta \alpha \tag{31}$$

We get A after problem (31) being solved, which is a matrix with columns are P eigenvectors corresponding P smallest eigenvalues ($A = [\alpha_1 \ \alpha_2 \ \cdots \ \alpha_P]$).

Limitation of MKL-DR. Algorithm 1 (Fig. 1) shows how to use the MKL-DR model. A major part of the computational time for training MKL-DR is to calculate $S_W^A, S_{W'}^A, S_W^\beta, S_{W'}^\beta$ (at steps 3 and 5 in Algorithm 1). The $S_W^A, S_{W'}^A, S_W^\beta, S_{W'}^\beta$ values are computed by Eqs. (25), (26), (29) and (30) correspondingly, which are repeatedly calculated T times. Each equation also repeats matrix chain multiplication N times. Therefore, if the size of the samples is large, the computation cost will be drastically increased because the matrices in (25), (26), (29) and (30) are depended on the number of samples. In our experiment of 541 samples, MKL-DR required thousands of seconds for each. This limitation violates one of three criteria in evaluating a data analysis method. In the next section, we present an improved MKL-DR to reduce computational time, called fMKL-DR.

Algorithm 1. *The MKL-DR Training Procedure*

Input	The DR method specified by W and W' matrices (by (10));	
	M base kernel matrices $\{K_m\}_{m=1}^{M}$ corresponding to the M dataset;	
Output	Sample coefficient matrix $A = [a_1 \, a_2 \cdots a_P]$;	
	Kernel weight vector β;	

1   *Step 1*: Initialization value for A or β

2   *Step 2*: Repeat

3   Compute S_W^A by (25) and $S_{W'}^A$ by (26);

4   β is optimized by solving optimization problem (27) via SDP;

5   Compute S_W^β by (29) and $S_{W'}^\beta$ by (30);

6   A is optimized by solving the generalized eigenvalue problem (31);
        Until converge or reach maximum number of iteration (T)

7   *Step 3*: Return A and β;

Fig. 1. MKL-DR algorithm

3.2   Improvement of MKL-DR

There are three parameters that affect the performance of the MKL-DR, i.e., the number of samples (N), the number of data types (M), the dimensions after being reduced (P). In case where the value of M is small, often between 3–10, the number of dimensions after being reduced is small (in our experiment, we chose $P = 5$). Therefore, the computation complexity is $\mathcal{O}(N^3)$, which is polynomial time. The MKL-DR training algorithm calculates iterative Eqs. (25), (26), (29) and (30) that calculate the sum of the matrix chain multiplication. As a result, the MKL-DR will become extremely time-consuming when increasing number of samples.

Matrix chain multiplication has combinatory property, meaning that changing calculation order between a pair of the matrices will affect the number of product operations without modifying the multiplication results. Therefore, we propose to use a dynamic-programming-based procedure to find the calculation sequence that gives minimum product operations. If the number of product operations is minimum, the computation time of the equations will be reduced.

The minimum product operation order problem based on dynamic programming

Given N matrices $A_1, A_2, \ldots, A_N$ with the size of A_i is $d_{i-1} \times d_i$.

Problem: Find the order of product matrix chain $A_1 \times A_2 \times \ldots \times A_N$ to minimize the number of product operations.

Solution: Construct matrix F is a matrix $N \times N$, with the element $F(i,j)$ is the total product operations to calculate matrix chain multiplication from A_i to A_j ($A_i \times A_{i+1} \times \ldots \times A_j$). The formula to calculate $F(i,j)$ defined by:

- $F(i,i) = 0$,
- $F(i, i+1) = d_{i-1} \times d_i \times d_{i+1}$,
- $F(i,j) = \min (F(i,t) + F(t+1,j) + d_{i-1} \times d_i \times d_{i+1})$
  with $t = i+1, i+2, \ldots, j-1$. In other words, t is a midpoint to insert the parentheses to change the calculation order so that the number of product operations is minimum:

$$A_i \times A_{i+1} \times \ldots \times A_j = (A_i \times A_{i+1} \times \ldots \times A_t)(A_{t+1} \times A_{t+2} \times \ldots \times A_j)$$

Based on the above assumption, we developed an improved algorithm to matrix chain multiplication to minimize product operations shown in Fig. 2.

Algorithm 2. *The Matrix Chain Multiplication Ordering Procedure (MCMO)*

    Input N matrices sizes $d_1, d_2, ..., d_N$;

    Output   The multiplication ordering $O = [o_1, o_2, ..., o_q]$ so the number of product

                   operations is smallest

1    $F(i,i) = 0$ with $i = 1, ..., N$;

2    $F(i,i+1) \leftarrow d_{i-1} \times d_i \times d_{i+1}$ with $i = 1, ..., N-1$;

3    $O = []$;

4   for $i \leftarrow N$

5   for $j \leftarrow N$

6   if $(i \neq j)$

7              $F(i,j) = \min(F(i,t) + F(t+1,j) + d_{i-1} \times d_i \times d_{i+1})$

8   if $(i = 0)$

9   Add t to O;

10   Return O;

Fig. 2. Matrix Chain Multiplication Ordering Procedure

The MCMO complexity is $\mathcal{O}(N^3)$. However, in the above equations, N (the number of matrices in chain) is small and equals to 4. Consequently, the time consumption of MCMO is trivial. Moreover, MCMO only call 2 times in fMKL-DR, which is built as demonstrated in Fig. 3 (with input and output similar as MKL-DR).

Algorithm 3. *The fMKL-DR Training Procedure*

1   *Step 1:* Initialization value for A or β

2   *Step 2:* Compute the best ordering:

3    $O_A = \text{MCMO}(\text{size}(\mathbb{K}^{(i)} - \mathbb{K}^{(j)}), \text{size}(\beta), \text{size}(\beta^{\top}), \text{size}((\mathbb{K}^{(i)} - \mathbb{K}^{(j)})^{\top}))$

4    $O_\beta = \text{MCMO}(\text{size}((\mathbb{K}^{(i)} - \mathbb{K}^{(j)})^{\top}), \text{size}(A^{\top}), \text{size}(A), \text{size}(\mathbb{K}^{(i)} - \mathbb{K}^{(j)}))$

5   *Step 3:* Repeat

6   Compute S_W^A by (25) and $S_{W'}^A$ by (26) *based on the order* O_A;

7   Solve the optimization problem (27) to get β ;

8   Compute S_W^β by (28) and $S_{W'}^\beta$ by (29) *based on the order* O_β ;

9   Solve the generalized eigenvalue problem (31) to get A;

10 Until converge or reach maximum number of iteration (T)

11 *Step 4:* Return A and β

Fig. 3. fMKL-DR algorithm

The matrices S_W^A, $S_{W'}^A$, S_W^β, $S_{W'}^\beta$ are calculated in steps 6 and 8 of Algorithm 3 based on the ordering of O_A and O_β. These matrices have the same values as MKL-DR ones. Therefore, in our experiments, we evaluated the proposed method with the aspect of calculation time because the integrating results did no changes.

4   Experiments and Discussions

4.1   Materials

We investigated four cancer patient datasets from TCGA (The Cancer Genome Atlas, 2017) [9] including Glioblastoma Multiforme (GBM), Ovarian Serous Cystadenocarcinoma (OV), Squamous Cell Lung Carcinoma (LUNG), and Breast Invasive Carcinoma (BREAST). In each dataset, we used three data types: gene expression, protein expression, DNA methylation (see details in Table 1). The four datasets were selected because each data type had a different samples size. Hence, the comparison results between methods will be more objective and reliable.

Table 1. Cancer patients data

Cancer	Number of samples	Number of features (dimensions) per data type		
		Gene expression	DNA methylation	Protein expression
LUNG	106	12,042	23,074	218
GBM	275	12,042	22,896	218
BREAST	435	12,042	24,978	218
OV	541	12,042	21,825	218

4.2   Experimental Design

We performed three MKL method with dimensionality reduction methods, i.e., MKL-DR, rMKL-DR, fMKL-DR on the same datasets. We evaluated the results based on time consumption of each method.

Experiments were designed as follows.

- **Step 1.** Generate three base kernel matrices according to three data types (gene expression, DNA methylation, protein expression).
- **Step 2.** Set the MKL-DR, rMKL-DR, fMKL-DR methods with dimensionality reduction algorithm as LDA.
- **Step 3.** Input three base kernel matrices obtained from Step 1 into the three methods. We alternatively executed the methods with maximum iterations of 5, 10, and 20.
- **Step 4.** Evaluate results based on time consumption for each dataset and the iterations. We set up the parameters for the methods as follow.
  - Reduced dimensions $P = 5$. If the number of dimensions is high, the integrated data could be sparser and scattered.
  - For each data type, we used a Gaussian kernel function to generate kernel matrices in Step 1.

4.3   Results and Discussion

These methods were employed in Matlab R2016a and run on Windows 10 operating system with an Intel Core i7 2.6 GHz with 8 GB main memory. Table 2 shows the results of the three methods with the different iterations and datasets.

Table 2. Comparison time consumption among methods with various iterations

Cancer	Computation time (seconds)								
	5 iterations			10 iterations			20 iterations		
	MKL-DR	rMKL-DR	fMKL-DR	MKL-DR	rMKL-DR	fMKL-DR	MKL-DR	rMKL-DR	fMKL-DR
LUNG	4	4	**3**	9	9	**6**	18	19	**13**
GBM	69	69	**41**	138	139	**82**	276	279	**165**
BREAST	1064	1101	**714**	2,130	2,203	**1,428**	4,262	4,409	**2,857**
OV	1,716	1,750	**1,237**	3,433	3,502	**2,475**	6,867	7,005	**4,952**

Table 2 shows three crucial improvements of fMKL.

- Firstly, in all of the iteration cases, fMKL-DR always obtained the smallest computation time among the three methods. When applying to BREAST dataset with 5 iterations, MKL-DR required 1064 s, rMKL-DR required 1101 s. Outstandingly fMKL-DR only required 714 s, significantly improving computational cost. The results on the other datasets with different iteration demonstrated the robustness and scalability of our proposed method.
- Secondly, the time consumption was approximately proportional to the number of iterations. Applying fMKL-DR on same BREAST dataset, the computation time were 714, 1428, and 2857 s for 5, 10, and 20 respectively.
- The computation time of rMKL-DR method is largest when compared to the time-consumption of MKL-DR and fMKL-DR. Compare with MKL-DR method, rMKL-DR added a constraint to the optimization problem. Hence, time-consumption of rMKL-DR is greater than MKL-DR ones.

Figure 4 shows that the computation time of these methods on each dataset with the iterations is 10. It turns out that, if the number of samples is small, the computation time is negligible. For details, for the LUNG dataset with 106 samples, the largest computation time was just 9 s. For the GBM dataset with 275 samples, the largest computation time increased, but was still a reasonable 139 s. However, when the sample size was large, such as BREAST and OV datasets, the time-consumption rose significantly. Time histograms are noticeable when the number of samples increases to 435 and 541 and with largest time of 2203 and 3502 s respectively. Figure 4 also shows the remarkable difference between the computation time of fMKL-DR and two other methods, which can speed up to thousands of seconds.

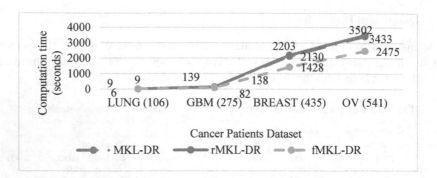

Fig. 4. Computation time in the different sizes of samples (with the same number of iterations equals 10)

5 Conclusion

In this paper, we proposed a fast multiple kernel learning method combined with dimensionality reduction. This framework is a crucial improvement of MKL-DR by optimizing the equations to reduce time-consumption of the matrix chain calculation. The experimental results showed that fMKL-DR could immensely reduce the computation time in comparison with the original frameworks. When the number of samples is large, the reduced time is especially more significant. The proposed method could be applied not only in biomedicine but also other fields. The robustness of the methods will be advantageous in the new era of Big Data, when the number of observed and analyzed data is huge and growing rapidly.

References

1. Kwon, Y.-J.: Genome-wide analysis of DNA methylation and the gene expression change in lung cancer. J. Thorac. Oncol. **7**(1), 20–33 (2012)
2. Liang, M., Li, Z., Chen, T., Zeng, J.: Integrative data analysis of multi-platform cancer data with a multimodal deep learning approach. IEEE/ACM Trans. Comput. Biol. Bioinform. (TCBB) **12**(4), 928–937 (2015)
3. Wang, B.: Similarity network fusion for aggregating data types on a genomic scale. Nat. Methods **11**, 333–337 (2014). https://doi.org/10.1038/nmeth.2810
4. Lin, Y.Y., Liu, T.L., Fuh, C.S.: Multiple kernel learning for dimensionality reduction. IEEE Trans. Pattern Anal. Mach. Intell. **33**(6), 1147–1160 (2011)
5. Speicher, N.K., Pfeifer, N.: Integrating different data types by regularized unsupervised multiple kernel learning with application to cancer subtype discovery. Bioinformatics **31**(12), i268–i275 (2015). https://doi.org/10.1093/bioinformatics/btv244
6. Yan, S., Xu, D., Zhang, B., Zhang, H.J., Yang, Q., Lin, S.: Graph embedding and extensions: a General framework for dimensionality reduction. IEEE Trans. Pattern Anal. Mach. Intell. **29**(1), 40–51 (2007)
7. Vandenberghe, L., Boyd, S.: Semidefinite programming. SIAM Rev. **38**(1), 49–95 (1996)

8. Shawe-Taylor, J., Mello, C.: Kernel method for Pattern Analysis. Cambridge University Press, Cambridge (2004)
9. The Cancer Genome Atlas. http://cancergenome.nih.gov

An Attention-Based Long-Short-Term-Memory Model for Paraphrase Generation

Khuong Nguyen-Ngoc[1], Anh-Cuong Le[2(✉)], and Viet-Ha Nguyen[1]

[1] VNU University of Engineering and Technology, Ha Noi City, Vietnam
khuongnn@dhhp.edu.vn, hanv@vnu.edu.vn
[2] NLP-KD Lab, Faculty of Information and Technology, Ton Duc Thang University,
Ho Chi Minh City, Vietnam
leanhcuong@tdt.edu.vn

Abstract. Neural network based sequence-to-sequence models have shown to be the effective approach for paraphrase generation. In the problem of paraphrase generation, there are some words which should be ignored in the target text generation. The current models do not pay enough attention to this problem. To overcome this limitation, in this paper we propose a new model which is a penalty coefficient attention-based Residual Long-Short-Term-Memory (PCA-RLSTM) neural network for forming an end-to-end paraphrase generation model. Extensive experiments on the two most popular corpora (**PPDB** and **WikiAnswers**) show that our proposed model's performance is better than the state-of-the-art models for paragraph generation problem.

1 Introduction

Paraphrase Generation (PG) is the task of re-statement of a text or passage that conveys the same meaning but with different words from the source. It recently extends to such plenty Natural Language Processing (NLP) tasks as question answering, summarization and other natural language inference [17]. There are some traditional methods for paraphrase generation such as [36] using automatically learned complex paraphrase patterns, [18] using hand-crafted rules, use thesaurus-based, [13] using semantic-driven natural language generation, or [35] using a leverage statistical machine learning (SML) method.

Recently, the prevalent approach to sequence to sequence (seq2seq) learning [31] have been used to solve various tasks in NLP with effective results: language modeling [32], machine translation [4], speech recognition [15], and dialogue systems [26]. The essence of PG can be formulated machine translation task that is performed within a single language and totally can be considered as a seq2seq learning task. However, not much work has been done in this area with regard to state-of-the-art deep neural networks like Neural Machine Translation [19]. More recently, Neural Attention Mechanism (NAM) has been used for textual entailment generation [12] which has given a very impressed result. Actually PG is also a type of bi-directional semantic relation between text fragments and thus

© Springer International Publishing AG, part of Springer Nature 2018
V.-N. Huynh et al. (Eds.): IUKM 2018, LNAI 10758, pp. 166–178, 2018.
https://doi.org/10.1007/978-3-319-75429-1_14

the methods of using NAM in Neural Paraphrase Generation (NPG) [8,24] have improved significantly performance over previous works.

Bahdanau et al. [1] proposed a prevailing attention mechanism that achieved very impressive success in neural machine translation. Prakash et al. [24] has proposed using the prevailing attention mechanism in stacked residual Long-Short-Term-Memory(LSTM) networks architecture for paraphrase generation. It then outperformed state-of-the-art models for seq2seq learning but the authors also stated that the attention-based neural network for machine translation is not suitable for paraphrase generation because not all words in the source sequence should be substituted for paraphrasing.

To overcome this limitation we will apply the prevailing attention based residual LSTM model for PG problem. We propose a new variant attention (called PCA) mechanism which is inspired by the ability to ignore some words in paraphrasing. We will combine PCA, Residual LSTM [24] and deep LSTM [34] to build a specialized encoder-decoder model for PG. Consequently this paper will provide the following contributions:

- Present a novel approach by casting the role of attention mechanism in the encoder-decoder framework for paraphrase generation.
- Propose an new end-to-end neural networks model based on a penalty coefficient attention-based deep residual long-short-term memory (PCA-RLSTM) architecture for performing PG.
- Explore the potential of PCA-RLSTM with word-level modeling for paraphrase generation. Overall, the performance of our model is competitive with relative the state-of-the-art attention-based NPG.

2   Related Work

Paraphrase generation is increasingly drawing the attention of researchers in recent years. There have been some different approaches to address the PG, typically using knowledge-driven approaches [17], utilize hand-crafted rules [18], automatically learned complex paraphrase patterns [36], thesaurus-based, semantic analysis-driven [13]. Wubben et al. [35] have crawl news headlines from Google News to built a large aligned monolingual corpus that were used in a Phrase-Based Machine Translation framework for sentential PG. By combining multiple resources, corresponding feature functions, Zhao et al. [37] propose learning phrase-based paraphrase tables and devising a log-linear SMT model. Besides that, there are other models generate paraphrases [36] for a specific application as: bilingual parallel corpora [2], determine candidate paraphrases by applying a multi-pivot approach [38].

Up to now, as far as our understanding is concerned, almost no research has investigated the effectiveness of the size of the dataset on the performance of the paraphrase generation. In this work, we also utilized several sizes of the potential dataset. Recently, Weiting [33] PPDB 2.0 (called Annotated-PPDB) which exist in 6 common sizes (S, M, L, XL, XXL, XXXL). This new version of this dataset based on PPDB 1.0's paraphrasability. Besides that, they also published another paraphrase dataset (called ML-Paraphrase) for the purpose of evaluating performance of bigram PG task. Weiting also shows that, the smaller dataset contains only better-scoring, high-precision paraphrases, while the larger ones aim for high coverage. However, for comparison with baseline methods, we only experimented with PPDB 2.0 dataset with size L for training and testing on our proposed model.

Very recently, there are some works on using residual connections with LSTM neural networks. In their works, DenseNet et al. [10] used dense connections over every layer in image recognition or Wu et al. [34] proposed deep residual LSTM with attention mechanism for machine translation. Especially, Prakash et al. [24] have used residual connections with stack LSTMs for paraphrase generation tasks which achieved promised results in PPDB, WikiAnswer and MSCOCO datasets. However, to the best of our knowledge, this is the first work on exploiting the role of the attention mechanism in paraphrase generation.

3   Model Description

3.1   Neural Encoder-Decoder Model

Sutskever et al. [31] proposed a neural sequence to sequence modeling approach which combines the encoder and decoder components as Fig. 1. In which the LSTM Encoder takes the source sequence and maps it in to an encoded low-dimensional representation. The encoded low-dimensional representation is then used by the LSTM Decoder to generate a high-dimensional target sequence. In paraphrase generation, the LSTM Encoder encodes meaning of the source sentence into a low-dimensional vector then the LSTM Decoder can use that low-dimensional vector and generate a paraphrase sentence. While generating the paraphrase sentence, the generation of each new word depends on the current word in source sentence, features of the model and the textual of preceding generated words.

Like other monolingual machine translation tasks, the main objective of training is determining model parameters for maximizing the log probability of the

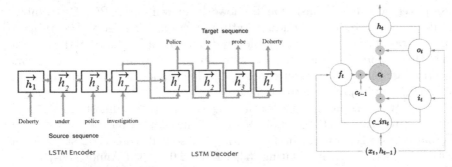

Fig. 1. Neural Encoder-Decoder framework.   **Fig. 2.** LSTM cell [24]

paraphrase sequences given the source sequences. For that reason, the desire decoded paraphrase sequence is selected in the set of all candidate sequences that has the biggest score. To find best paraphrase sequence, the definitely of beam size is used as a small set of hypotheses (candidate set). The total score for all and each of these hypotheses need to be computed. In their work, Sutskever et al. [31] have showed that although a beam size of 1, it usually achieves good results while with a higher beam size it is always better.

A crucial element of the recent success of some natural language processing (NLP) systems is the use of deep architectures, which are able to build up progressively higher level representations of data. In the next section, we will describe deep long short-term memory model for paraphrase generation.

3.2 Deep Long-Short-Term-Memory

The basic architecture of the Long-Short-Term-Memory (LSTM) cell is shown in (Fig. 2). This cell is composed of five main elements: an input gate i, a forget gate f, an output gate o, a recurring cell state c and hidden state output h. In this variant of LSTM cell, Apaszke[1] used a different approach for computing h_t. In the concrete, at every time step t, an internal memory cell $c_t \in R^n$ is added for computing h_t. Particularly, at each time step t, an LSTM cell uses the previous hidden state h_{t-1}, the input state x_t to produce the temp internal memory state c_in_t, then using the internal memory state c_{t-1} and the temp internal memory state c_in_t to produce the internal memory state c_t. By using the addition of gradient in these LSTM cells to minimize the gradient explosion (GE). Sundermeyer et al. [30] showed that LSTM outperforms vanilla RNN in almost NLP tasks. In our proposed model we will use the LSTM cell as a basic unit for both encoder and decoder modules. According to [24] the following describes the basic computations in an LSTM cell:

1. Gates
$$i_t = \sigma(W_{xi}x_t + W_{hi}h_{t-1} + b_i)$$
$$f_t = \sigma(W_{xf}x_t + W_{hf}h_{t-1} + b_f)$$
$$o_t = \sigma(W_{xo}x_t + W_{ho}h_{t-1} + b_o)$$
2. Input transform
$$c\_in_t = tanh(W_{xc}x_t + W_{hc}h_{t-1} + b_{c\_in})$$
3. State Update
$$c_t = f_t \odot c_{t-1} + i_t \odot c\_in_t$$
$$h_t = o_t \odot tanh(c_t)$$

The training stage aims to learn the parameters W_{x*}, W_{h*} for x and h respectively; $\sigma(.)$ is usually the *sigmoid* function; $tanh(.)$ denotes hyperbolic tangent functions; $\odot$ is the multiplication operator; and b denotes the bias.

Recently, deep LSTMs have been shown as the best models for text generation [6] or sentiment analysis [20]. In addition, Pascanu et al. [22] surveyed several

---

[1] https://apaszke.github.io/lstm-explained.html.

ways of incorporating multiple LSTM layers and analyzed the main challenges in training deep neural network. In our work, we employ stacked LSTM layers on the vertical direction where input of the current LSTM layer uses output of the previous layer. Note that, different from [3], in our model we will replace traditional RNN by LSTM units. Except for the first layer, our model performs passing the hidden state of the previous layer $h_t^{(l)}$ to the input of the current layer, where l denotes the layer. Therefore the activation of layer l is defined as:

$$h_t^{(l)} = f_h^{(l)}(h_t^{(l-1)}, h_{t-1}^{(l)})$$

3.3 Deep Residual LSTM

The essential idea of residual connection is to add a shortcut path from the output of prior layer to the output of the next layer. With theoretical reasoning, He et al. [9] showed that one of the most common challenges encountered during deep network training is degradation problem and residual connections can resolve this issue. In this work, we apply two very successful innovations in

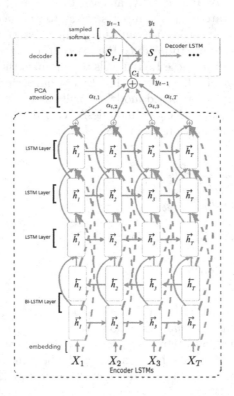

Fig. 3. PCA-LSTM architecture for paraphrase generation.

deep residual LSTM neural network, stacked residual LSTMs [24] and Google's Neural Machine Translation [34].

In our experiments, we employed four LSTM layers which stacked to each others. In addition, we also used residual connections (arrow dashed lines) as the point-wise additions which are added at the second layer (see Fig. 3) as equal:

$$\hat{h}_t^{(l)} = f_h^l(h_t^{(l-1)}, h_{t-1}^{(l)}) + x_{l-n}$$

where $\hat{h}_t^{(l)}$ is a new hidden state at layer l and is added by residual value x_{l-n} (the input to layer $l - n + 1$). This means that the input x and the output h must be in the same dimension. Prakash et al. [24] show that, when $n = 1$ the value of learn function $\hat{h}$ depends on the input x and is computed by using the standard LSTM with bias. So that, the residual connection is added after every layer that is not necessary. On the other hand, for deep LSTM, the terms of computation is very expensive when $n > 3$. That is why we choose $n = 2$ (the residual connection is added after each two layer) for this work.

3.4 Deep Attention-Based Stochastic Residual LSTM

Our proposal is inspired from Google neural machine translation model [34]. The Fig. 3 presents of our paraphrase generation architectural model. It has three components: an encoder network, a decoder network, and an attention network. In our PCA-LSTM model, the encoder uses stacked LSTM layers which include one bi-directional LSTM layer and three uni-directional LSTM layers.

At the Bi-directional LSTM encoder, the information required to paraphrase certain words on the output side can appear anywhere on the source side. Often the source side information is approximately left-to-right, similar to the target side, but depending on the language pair the information for a particular output word can be distributed and even be split up in certain regions of the input side. Then, the final hidden state h_i for each input unit x_i is the concatenation of the forward and backward hidden states.

The decoder in our propose model uses only a standard LSTM that is used for generating paraphrase sequence $y = y_0, ..., y_L$ by computing sequence of hidden state $(\overrightarrow{s_0}, ..., \overrightarrow{s_{L-1}})$ where the context of the current generated paraphrase units are encoded in s_{L-1}. Ideally, tradition attention mechanism (see Fig. 4) is modeled for computing the relevance score α_{ti} for each hidden state h_i which are used for computing the context vector c_t as follows:

$$c_t = \sum_{i=0}^{T} \alpha_{ti} h_i$$

The value of the hidden state relevance score α_{ti} denotes the most relevant source unit to focus on and is computed as:

$$\alpha_{ti} = \frac{exp(e_{ti})}{\sum_{k=0}^{T} exp(e_{tk})}$$

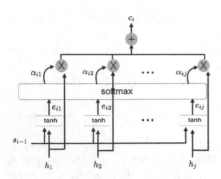

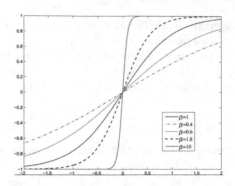

Fig. 4. Attention mechanisms.

Fig. 5. The graph of *tanh* function in neural network

where e_{ti} is called as the alignment model and calculated by using a neural network f, as follow:

$$e_{ti} = f(\beta * (W_a * s_{t-1} + U_a * h_i))$$

where f usually uses *tanh* function with the default value of β is 1. In this paper, we propose to use and evaluate the role of β parameter in generating paraphrase. Figure 5 shows that, if we consider the word corresponding to the minimum value of the *tanh* function to have no role in generating paraphrase then almost words having a role in the traditional attention model. In fact, some words have no role in paraphrasing of the PG problem (it is different from neural machine translation). To address this issue, we add a new parameter into the model. We can see the role of β in *tanh* function (see Fig. 5). We call β is the penalty coefficient (PC) on the alignment model. The goal of β is to suppress the role of a word (corresponding to the value of *tanh* function is -1) in the source sequence for the current word in the paraphrase sequence. This directly changes the value of attention weight so we call this method is Penalty Coefficient Attention (PCA). Our experiments show that our model outperforms better when the value of *beta* is estimated from 1.5 to 2 (we used *beta* = 1.8). Therefore, by using the previous hidden state s_{t-1}, the most relevant source context c_t, and previously generated textual units y_{t-1} to computed the hidden state s_t of the decoder:

$$s_t = g(s_{t-1}, y_{t-1}, c_t)$$

where g is the GRU unit. In our encoder-decoder framework based on PCA-LSTM model, the computing of generating paraphrase sequence $y = y_0, ..., y_L$ is also based on conditional distribution P, as follows:

$$P(y_L|y_1, ..., y_{L-1}, X) = f(y_{L-1}, s_L, c_T)$$

4 Experimental Settings

4.1 Dataset

We conduct our experiments on two different datasets PPDB 2.0 [23], and WikiAnswers. They are the latest benchmark and a well known paraphrase datasets.

PPDB 2.0 comes pre-packaged in 6 sizes: S to XXXL. However, for the purpose of conducting experiments to compare with one of the state-of-the-art results of the previous works we only use the paraphrase dataset PPDB 2.0 with size L [23], this dataset contains about 18 million paraphrases. According to that, we have also selected 5.3 million paraphrases (mainly short phrases) that makes it suitable for paraphrase generation [17].

The paraphrases in WikiAnswers [5] have existed as different questions which were tagged by the users as similar questions. Although WikiAnswers comes with over 29 million instances, we have randomly selected 5.3 million paraphrases (mainly long phrases) for both training and testing.

In the these datasets, we have used K-Fold Cross-Validation method to evaluation effectiveness of the proposed model. In K-fold cross-validation, the original corpus is randomly partitioned into K sections. Of the K sections, a single section is retained as the validation data for testing the model, and the remaining $K-1$ sections are used as training data. The cross-validation process is then repeated K times (the folds), with each of the K sections used exactly once as the validation data. The K results from the folds be averaged to final results (or otherwise combined). In our experiments, we choose $K = 10$ [11].

4.2 Models

Table 1 describes five different models which are used in our experiment for comparison. For each model, both the encoder and decoder were used two and four LSTM layers. Model 1 is a sequence to sequence model from Sutskever et al. [31]. Model 2 from Bahdanau et al. [1] is the model of sequence to sequence with attention mechanisms for bilingual machine translation which was used for generating paraphrase and achieved the state-of-the-art results in this problem. Model 3 from Graves et al. [7] uses stacked LSTMs. Model 4 from Prakash et al. [24]

Table 1. Models.

Models	Reference
Sequence to Sequence	(Sutskever et al. 2014)
Seq2Seq with Attention	(Bahdanau et al. 2015)
Bi-directional LSTM	(Graves et al. 2013)
Residual LSTM	(Prakash et al. 2016)
PCA-RLSTM	**(Our proposed model)**

using residual connections between LSTM layers. Finally, model 5 is our proposal.

In encoder-decoder models, the computational complexity is inversely proportional to the size of the beam search algorithm. We here use the most common beam size (5 and 10) which were used in the literature [31]. We train all models using the implementation of the stochastic gradient descent (SGD) algorithm with the one-hot vector of words as input and the learning rate begins at 1.0. Each LSTM layer has exactly 512 LSTM units and a standard dropout [29] of 50% is applied after every LSTM layer.

5   Evaluation

5.1   Metrics

Prakash et al. [24] shows that some paraphrase generation evaluation metrics, such as PEM (Paraphrase Evaluation Metric) [16] and PINC (Paraphrase In Ngram Changes) [28] have certain limitations. Furthermore, in terms of the essence of paraphrase generation is monolingual Machine Translation. So that, to evaluate the performance of models in this work, we only choose the well-known automatic evaluation metrics: BLEU [21], METEOR [14], Translation Error Rate (TER) [27] and Emb Greedy [25].

5.2   Results

Figures 6 and 7 shows the final results of five models conducted on PPDB 2.0 and WikiAnswer paraphrase datasets. The performance of our proposed model is better than other models on two main measurements (BLEU and TER). On METEOR metric, our proposed model outperforms almost other models; however, the sequence to sequence is better with beam size is 5 and four layer LSTMs. This is explained by the PPDB 2.0 dataset in our experimented contains mainly short phrases and the METEOR metric is generally appropriate with long phrases on WikiAnswer dataset for comparing parallel corpora. On Emb Greedy metric, the PCA-LSTM model outperforms almost other models

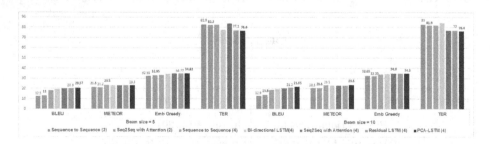

Fig. 6. Evaluation results on PPDB (best results are in **bold**).

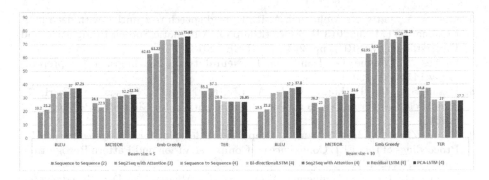

Fig. 7. Evaluation results on WikiAnswer (best results are in **bold**).

except the Attention model when beam size is 10 on **PPDB** 2.0 dataset. In addition, experimental results also show that deeper LSTMs architecture improves performance over shallow architectures.

6   Conclusion and Future Work

In this work, we have proposed a new model called PCA-RLSTM which is an extension of Residual Long Short Term Memory (RLSTM) architecture by adding penalty coefficient attention factor to RLSTM. Our model is applied to the problem of Paraphrase Generation and the experiments have shown better results in comparison with the state-of-the-art methods. This model is hoped to be useful for other NLP tasks, especially for sequence to sequence learning problems.

Acknowledgement. This paper is supported by The Vietnam National Foundation for Science and Technology Development (NAFOSTED) under grant number 102.01-2014.22.

References

1. Bahdanau, D., Cho, K., Bengio, Y.: Neural machine translation by jointly learning to align and translate. arXiv preprint arXiv:1409.0473 (2014)
2. Bannard, C., Callison-Burch, C.: Paraphrasing with bilingual parallel corpora. In: Proceedings of the 43rd Annual Meeting on Association for Computational Linguistics, pp. 597–604. Association for Computational Linguistics (2005)
3. Bengio, Y., Simard, P., Frasconi, P.: Learning long-term dependencies with gradient descent is difficult. IEEE Trans. Neural Networks 5(2), 157–166 (1994)
4. Chorowski, J.K., Bahdanau, D., Serdyuk, D., Cho, K., Bengio, Y.: Attention-based models for speech recognition. In: Advances in Neural Information Processing Systems, pp. 577–585 (2015)
5. Fader, A., Zettlemoyer, L., Etzioni, O.: Paraphrase-driven learning for open question answering. In: Proceedings of the 51st Annual Meeting of the Association for Computational Linguistics (Volume 1: Long Papers), vol. 1, pp. 1608–1618 (2013)

6. Graves, A., Jaitly, N., Mohamed, A.-R.: Hybrid speech recognition with deep bidirectional LSTM. In: 2013 IEEE Workshop on Automatic Speech Recognition and Understanding (ASRU), pp. 273–278. IEEE (2013)
7. Graves, A., Wayne, G., Danihelka, I.: Neural turing machines. arXiv preprint arXiv:1410.5401 (2014)
8. Hasan, S.A., Liu, B., Liu, J., Qadir, A., Lee, K., Datla, V., Prakash, A., Farri, O.: Neural clinical paraphrase generation with attention. In: ClinicalNLP 2016, p. 42 (2016)
9. He, K., Zhang, X., Ren, S., Sun, J.: Deep residual learning for image recognition. In: Proceedings of the IEEE Conference on Computer Vision and Pattern Recognition, pp. 770–778 (2016)
10. Huang, G., Liu, Z., Weinberger, K.Q., van der Maaten, L.: Densely connected convolutional networks. arXiv preprint arXiv:1608.06993 (2016)
11. Kohavi, R.: A study of cross-validation and bootstrap for accuracy estimation and model selection, pp. 1137–1143. Morgan Kaufmann (1995)
12. Kolesnyk, V., Rocktäschel, T., Riedel, S.: Generating natural language inference chains. arXiv preprint arXiv:1606.01404 (2016)
13. Kozlowski, R., McCoy, K.F., Vijay-Shanker, K.: Generation of single-sentence paraphrases from predicate/argument structure using lexico-grammatical resources. In: Proceedings of the Second International Workshop on Paraphrasing, vol. 16, pp. 1–8. Association for Computational Linguistics (2003)
14. Lavie, A., Agarwal, A.: Meteor: an automatic metric for MT evaluation with high levels of correlation with human judgments. In: Proceedings of the Second Workshop on Statistical Machine Translation, pp. 228–231. Association for Computational Linguistics (2007)
15. Li, X., Wu, X.: Constructing long short-term memory based deep recurrent neural networks for large vocabulary speech recognition. In: 2015 IEEE International Conference on Acoustics, Speech and Signal Processing (ICASSP), pp. 4520–4524. IEEE (2015)
16. Liu, C., Dahlmeier, D., Ng, H.T.: PEM: a paraphrase evaluation metric exploiting parallel texts. In: Proceedings of the 2010 Conference on Empirical Methods in Natural Language Processing, pp. 923–932. Association for Computational Linguistics (2010)
17. Madnani, N., Dorr, B.J.: Generating phrasal and sentential paraphrases: a survey of data-driven methods. Comput. Linguist. **36**(3), 341–387 (2010)
18. McKeown, K.R.: Paraphrasing questions using given and new information. Comput. Linguist. **9**(1), 1–10 (1983)
19. Manning, C.D., Luong, M.-T., Pham, H.: Effective approaches to attention-based neural machine translation. CoRR abs/1508.0402 (2017)
20. Nguyen, N.K., Le, A.-C., Pham, H.T.: Deep bi-directional long short-term memory neural networks for sentiment analysis of social data. In: Huynh, V.-N., Inuiguchi, M., Le, B., Le, B.N., Denoeux, T. (eds.) IUKM 2016. LNCS (LNAI), vol. 9978, pp. 255–268. Springer, Cham (2016). https://doi.org/10.1007/978-3-319-49046-5_22
21. Papineni, K., Roukos, S., Ward, T., Zhu, W.-J.: Bleu: a method for automatic evaluation of machine translation. In: Proceedings of the 40th Annual Meeting on Association for Computational Linguistics, pp. 311–318. Association for Computational Linguistics (2002)

22. Pascanu, R., Gulcehre, C., Cho, K., Bengio, Y.: How to construct deep recurrent neural networks. arXiv preprint arXiv:1312.6026 (2013)
23. Pavlick, E., Rastogi, P., Ganitkevitch, J., Van Durme, B., Callison-Burch, C.: PPDB 2.0: Better paraphrase ranking, fine-grained entailment relations, word embeddings, and style classification (2015)
24. Prakash, A., Hasan, S.A., Lee, K., Datla, V., Qadir, A., Liu, J., Farri, O.: Neural paraphrase generation with stacked residual LSTM networks. arXiv preprint arXiv:1610.03098 (2016)
25. Rus, V., Lintean, M.: A comparison of greedy and optimal assessment of natural language student input using word-to-word similarity metrics. In: Proceedings of the Seventh Workshop on Building Educational Applications Using NLP, pp. 157–162. Association for Computational Linguistics (2012)
26. Serban, I.V., Klinger, T., Tesauro, G., Talamadupula, K., Zhou, B., Bengio, Y., Courville, A.: Multiresolution recurrent neural networks: An application to dialogue response generation. arXiv preprint arXiv:1606.00776 (2016)
27. Snover, M., Dorr, B., Schwartz, R., Micciulla, L., Makhoul, J.: A study of translation edit rate with targeted human annotation. In: Proceedings of Association for Machine Translation in the Americas, vol. 200 (2006)
28. Socher, R., Huang, E.H., Pennington, J., Ng, A.Y., Manning, C.D.: Dynamic pooling and unfolding recursive autoencoders for paraphrase detection. In: NIPS, vol. 24, pp. 801–809 (2011)
29. Srivastava, N., Hinton, G., Krizhevsky, A., Sutskever, I., Salakhutdinov, R.: Dropout: a simple way to prevent neural networks from overfitting. J. Mach. Learn. Res. **15**(1), 1929–1958 (2014)
30. Sundermeyer, M., Schlüter, R., Ney, H.: LSTM neural networks for language modeling. In: Interspeech, pp. 194–197 (2012)
31. Sutskever, I., Vinyals, O., Le, Q.V.: Sequence to sequence learning with neural networks. In: Advances in Neural Information Processing Systems, pp. 3104–3112 (2014)
32. Vinyals, O., Kaiser, Ł., Koo, T., Petrov, S., Sutskever, I., Hinton, G.: Grammar as a foreign language. In: Advances in Neural Information Processing Systems, pp. 2773–2781 (2015)
33. Wieting, J., Bansal, M., Gimpel, K., Livescu, K., Roth, D.: From paraphrase database to compositional paraphrase model and back. arXiv preprint arXiv:1506.03487 (2015)
34. Wu, Y., Schuster, M., Chen, Z., Le, Q.V., Norouzi, M., Macherey, W., Krikun, M., Cao, Y., Gao, Q., Macherey, K., et al.: Google's neural machine translation system: Bridging the gap between human and machine translation. arXiv preprint arXiv:1609.08144 (2016)
35. Wubben, S., Van Den Bosch, A., Krahmer, E.: Paraphrase generation as monolingual translation: data and evaluation. In: Proceedings of the 6th International Natural Language Generation Conference, pp. 203–207. Association for Computational Linguistics (2010)
36. Zhao, S., Lan, X., Liu, T., Li, S.: Application-driven statistical paraphrase generation. In: Proceedings of the Joint Conference of the 47th Annual Meeting of the ACL and the 4th International Joint Conference on Natural Language Processing of the AFNLP, vol. 2, pp. 834–842. Association for Computational Linguistics (2009)

37. Zhao, S., Niu, C., Zhou, M., Liu, T., Li, S.: Combining multiple resources to improve SMT-based paraphrasing model. In: ACL, pp. 1021–1029 (2008)
38. Zhao, S., Wang, H., Lan, X., Liu, T.: Leveraging multiple MT engines for paraphrase generation. In: Proceedings of the 23rd International Conference on Computational Linguistics, pp. 1326–1334. Association for Computational Linguistics (2010)

Deep Neural Network-Based Models for Ranking Question - Answering Pairs in Community Question Answering Systems

Van-Tu Nguyen[1] and Anh-Cuong Le[2(✉)]

[1] VNU University of Engineering and Technology, Ha Noi City, Vietnam
tuspttb@gmail.com
[2] NLP-KD Lab, Faculty of Information Technology, Ton Duc Thang University,
Ho Chi Minh City, Vietnam
leanhcuong@tdt.edu.vn

Abstract. Ranking question-answering pairs according to their similarities to each input question is very important for any real-world community Question Answering system. To address this problem we will propose the models which use Convolutional Neural Network and Bi-Directional Long Short Term Memory. The proposed models are formulated for both representation learning and question similarity score detection. Especially in this paper we will utilize various feature kinds including both abstract features (i.e. high level representation) and conventional features. We test our proposed model on the dataset SemEval 2016 and obtain the results with the *Accuracy* and *MAP* of 82.86% and 78.43% respectively, which are best in comparison with previous studies.

Keywords: Community question answering · Deep neural network
Feature extraction · Ranking question

1 Introduction

The Question Answering (cQA) websites with a large number of real questions and human answers have become a very useful resource for users. On cQA websites, users can freely submit questions and wait to receive answers by other users. There are many cQA sites which are becoming more and more popular in the real world such as StackOverflow[1] and Quora[2]. The questions and answers on these cQA sites are varied and facilitate different users to directly request answers from complex and heterogeneous information. In a cQA website when an user submit an input question the cQA should have a mechanism to return the most related question-answering pairs in relationship to the input question. The problem here is how to rank these question-answering pairs according to their similarities to the input question so that the closest question-answer pairs are put on the top for the seeker.

---

[1] https://stackoverflow.com/.
[2] https://www.quora.com/.

© Springer International Publishing AG, part of Springer Nature 2018
V.-N. Huynh et al. (Eds.): IUKM 2018, LNAI 10758, pp. 179–190, 2018.
https://doi.org/10.1007/978-3-319-75429-1_15

For measuring the similarity between the input question and existed question-answering pairs in the database, how to represent questions is very important. The features as n-grams are so common beside some richer linguistic information such as using syntactic parsing in [1–3]. However, although using syntactic and semantic information has been shown to improve performance, it may take a lot of time to compute and require a large number of extension tools (syntactic parser, lexicons, knowledge bases, etc.). Furthermore, adapting it to new domains requires not only additional efforts to tune feature extraction pipelines but also some necessary adding new resources that may not even exist.

Recently, representation learning and deep neural models have been shown to be effective for many NLP tasks such as semantic parsing [4], search query retrieval [5], sentence modeling [6], and sentence classification [7]. Therefore in this paper we would like to exploit these advanced deep learning models for the problem of ranking related question-answering pairs for each input question. Concretely the problem will be formulated to be used in the Convolutional Neural Network (CNN) model and Long Short Term Memory (LSTM) model. Especially in this paper we will combine the representation learning and similarity measuring into the unique model. The CNN and LSTM models are used as deep neural modules in the model. The Multi-Layer Perceptron (MLP) module is used as the end part of the model for prediction. Actually we will design two models, one uses CNN and the other uses Bi-directional LSTM.

In the proposed models, we will use word embeddings as input for representation learning module of question representation. Through CNN module or LSTM module we will generate vector representations for the input question and related questions. These vectors and other conventional features are concatenated for being used as the input to the fully connected MLP. Note that all these modules are integrated into an unique model for both learning representation and scoring similarity between input question and related question-answering pairs. Consequently, we obtained similar scores between the input question and related questions for the task of ranking question-answering pairs.

2   Related Work

Ranking question-answering pairs related to the input question in cQA has become the important task for designing any cQA system. Many studies have explored the common ways for measuring the similarity between an input question and related questions, between a question and its answers by using cosine measurement between words. Moreover, some other studies proposed more advanced features and models. Cao et al. [8] classified questions into different topics and using this information as feature for building a recommendation system. Duan et al. [9] extracted questions focus and use it for the similarity measuring. Some other researchers such as [10,11] proposed a method of using topic modeling. Interestingly, Zhou et al. [12] and Jeon et al. [13] followed an approach of translation for question and answer pairs. There are also some other studies using syntactic information. Wang et al. [3] used some sub-structures of the parsed tree as features for measuring the similarity between questions. Authors in [1,2] used syntactic information, but with parsed tree kernel based on the KeLP platform from [14]. Franco-Salvador et al. [15] used SVM rank method on the distributed representation of words for the task on the ranking question at the SemEval 2016 Task 3.

Recently, deep neural network based learning methods have been shown to be useful for machine learning [16]. They have been especially successful in image processing and speech processing. More recently, such methods have begun to overtake traditional sparse, linear models for NLP [6,17]. Recent works have shown the effectiveness of neural models for sequence labeling [18], answer selection [19], answer sentence selection [20] and ranking question [21] in cQA. In [21] the authors used CNN and bag-of-words (BOW) representations of input and related questions to compute cosine similarity scores. [22] presented a neural attention model for machine translation and showed that the attention mechanism is useful for solving long sentences. [23] presented a model based on LSTM and BOW to estimate the relevance of the question and its answers.

Different from previous studies, in this study we propose novel models to calculate the similarity between the input question and the related question and then use this similarity to the ranking question-answering pairs. The models we build include traditional NLP model (based on feature extraction) and deep neural network-based models. In the NLP model, we consider various aspects of questions as well as answers to build rich knowledge vector for representing the similarity between an input question and the related question-answering pairs in the database. In the deep neural network-based models, we use CNN and BLSTM to generate the vector representation for measuring this similarity.

3   Feature Extraction for Ranking Related Question-Answering Pairs

3.1   Conventional Features

We use some common features extracted from the surface forms of the question and answers, including: the ratio of the number of words between the input question and the related question; the ratio of the number of words between the input question and the answer of the related question; the ratio of the number of sentences between the input question and the related question; the ratio of the number of sentences between the input question and the answer of the related question; bag of word, word overlap, noun overlap, name entities overlap.

3.2   Word Embeddings

Word embeddings are the result of a word representation learning model which represents words by real-valued vectors whose relative similarities correlate with semantic similarity. Word embedding models based on statistics of word occurrences in a corpus to encode semantic information which expresses how meaning is generated from these statistics, and how the resulting word vectors might represent that meaning.

In this work, we use the continuous Skip-gram model [24] of the word2vec toolkit[3] to generate vector representations of the words. First, all the sentences in input questions, related questions and answers are tokenized and the words are then converted to vectors using the pre-trained word2vec model. In order to construct the question vector and answer vector we implement the following steps:

– Each question or answer with length t is represented by a word vector $(w_1, w_2, ..., w_t)$, where w_i is word vector representation of ith word. Suppose that we need to calculate the similarity between the input question q^* and the answer a_i. Where question q^* and answer a_i are represented as follows: $q^* = (w_1, w_2, ..., w_n)$ and $a_i = (v_1, v_2, ..., v_h)$.
– For each word vector w_i in q^*, we will find the most similarity word vector v_j in a_i according to cosine measurement as in the formula 1 as below.

$$score(w_i) = \max_{1 \leq j \leq h} (cosin\_similarity(w_i, v_j)) \qquad (1)$$

where: h: the number of words in answer a_i, w_i: word vector representation of ith word in question q^*, v_j: word vector representation of jth word in answer a_i, $cosin\_similarity(w_i, v_j)$: is the cosine similarity of two vector representations of ith word in question q^* with the jth word in answer a_i. Finally, the similarity score between input question q^* and answer a_i is calculated as follows:

$$similarity(q^*, a_i) = \frac{\sum_{k=1}^{n} score(w_i)}{n} \qquad (2)$$

where n is the number of words in question q^*.

Similarly we can also calculate the similarity between q^* and q_i by the same way.

3.3 Question Categories as Features

We use the distributed semantic representation of word (i.e. word2vec) to measure the relationship between the input question category and related question category. Note that we use the term "question category" to represent the set of questions which belong to the same category. The input question categories obtained by using a question categorization module. We are given the dataset Q includes question - answer pairs extracted from cQA sites, in which each question is assigned to a category label. The question categorization module aims to classify the input question q^* into one of the question categories in the dataset Q. To this end, we implement the following steps:

1. We prepare a training dataset including questions in dataset Q, they are assigned with category labels (the label here is question category).

---

[3] https://code.google.com/p/word2vec.

2. The questions are represented as vectors.
3. A machine learning method is used (here we choose SVM) to learn the classifier.
4. For each input question, we first represent it by feature vectors and use the classifier obtained at the third step to predict the label (i.e. question category).

where u and v are two n-dimensional vectors, u_i is the ith element of u vector.

Suppose that we are given the input question q^*, we align each word in the related question of the same category to the word in q^* that has the highest vector cosine similarity, as shown in the formula 1. Finally, the similarity score between the related question category and input q^* is calculated by the formula 2.

4   CNN-Based Model for Ranking Related Question - Answering Pairs

CNN utilize layers with convolving filters that are applied to local features [25]. CNN models were originally invented for computer vision, and then proved to be effective for NLP tasks and achieved excellent results in semantic analysis [4], search query retrieval [5], sentence modeling [6], and other traditional NLP tasks. Our reason for using a CNN for measuring the similarity between the input question and the question - answering pairs in the database is that it can capture both features of n-grams and long-range dependencies [20]. These traits make CNN useful for dealing with long questions.

We now propose a new approach for ranking related question-answering pairs. We construct a CNN-based model to calculate the similarity score for each input question and related question pair and then ranking related question-answering pairs based on this similarity scores. Our models are illustrated in Figs. 1 and 2.

We used the continuous Skip-gram model [24] to generate vector representations of the words in the dataset. First, all the sentences in questions are tokenized and the words are then converted to vectors using the pre-trained word2vec model. Then each question is transformed into a question matrix. In which, each row of the question matrix is a word vector representations of each

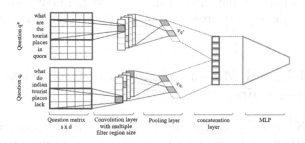

Fig. 1. The architecture of a CNN-based model for ranking related question - answering pairs

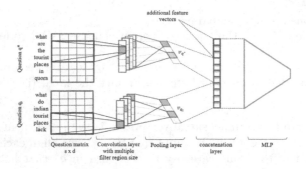

Fig. 2. The architecture of a CNN-based model adds feature vectors for ranking related question - answering pairs

token. We assume the dimension of the word vectors is d. If the length of a given question is s, then the dimensionality of the question matrix is $s \times d$.

In the question matrix, by rows representing discrete symbols (i.e. words in this case), it would be reasonable to use filters with a "width" is equal to the width of the word vectors (i.e. d). Therefore, we simply change the "height" of the filter, meaning that the number of adjacent rows is considered together. We refer to the "height" of the filter as the region size of the filter. For each region size, we use multiple filters. We then use the CNN convolution and pooling layers to generate the input question and related question vector representations. The concatenation layer connecting input question and related question vectors (as illustrated in Fig. 1) into a single vector and use as input for the fully connected MLP. Also, our CNN-based model also adds different feature kinds which are extracted from questions and answers. That is, the concatenation layer connects the input question - related question vector and the feature vector (as illustrated in Fig. 2) for the unique representation. Note that the feature vectors are extracted using NLP tools (as shown in Sect. 3). Our MLP will have two hidden layers, in which the first is a fully connected hidden layer with the same number of neurons as input vector size. The output layer contains a single neuron in order to calculate the similarity scores between an input question and corresponding related questions. This layer uses the sigmoid activation function in order to produce a probability output in the range of 0 to 1. Finally, we use the logarithmic loss function (binary cross-entropy) at the training stage for parameter estimation.

5   BLSTM-Based Model for Ranking Related Question - Answering Pairs

In case of CNN, concatenating words with various filter region size, works as n-gram models do not capture long-distance word dependencies with short filter region size. A larger filter region size can be used, but this may lead to data sparsity problem. In order to encode long-distance word dependencies, we use

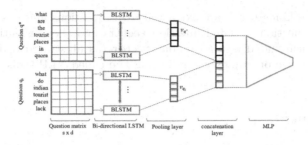

Fig. 3. The architecture of a BLSTM-based model for ranking related question-answering pairs

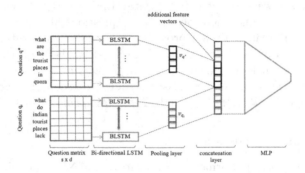

Fig. 4. The architecture of a BLSTM-based model adds feature vectors for related question-answering pairs

Long Short Term Memory (LSTM) networks, which are a special kind of Recurrent Neural Networks (RNN) capable of learning long-distance dependencies. LSTM were introduced by Hochreiter and Schmidhuber [26]. The LSTM models have been successful applied to many different NLP tasks, including: sequence labeling [18], answer selection [19], answer sentence selection [20].

In this paper, we propose a new method for ranking related question-answering pairs based on BLSTM. We build the BLSTM-based model to calculate the similarity score for each input question-related question pair and then rank the related question-answering pairs based on obtained similarity scores. Our models is illustrated in Figs. 3 and 4. The procedure of BLSTM-based model is similar to that of CNN-based model. The words of input questions and related questions are first converted into vectors by using the pre-trained word2vec model. Then they are input sequences and sequentially read by BLSTM from both directions. By this way, the contextual information across words in both input question and related question is modeled by employing temporal recurrence in BLSTM. The BLSTM then generate the input question and related question vector representations. The concatenation layer connecting input question and related question vectors (as illustrated in Fig. 3) into a single vector and used as input for a fully connected MLP. Our BLSTM-based model also add different feature kinds which are extracted from questions and answers. Therefore all input question - related question vectors and the conventional feature vectors are concatenated into an unique vector (as illustrated in Fig. 4). Note that the feature vectors are extraction using the called NLP tools (as shown in Sect. 3).

6   Experiments

6.1   Dataset and Evaluation Metrics

In order to setup our experiments, we use the cQA dataset provided by SemEval 2016[4]. The dataset was extracted from Qatar Living (http://www.qatarliving. com/forum), a web forum where people pose questions about multiple aspects of their daily life in Qatar. Table 1 provides statistics for this dataset.

We used several measures to evaluate our models, they consist of: the classification measures include: *Accuracy (Acc)*, *Precision (P)*, *Recall (R)*, and *F1 − measure (F1)*; the ranking measures include: *Mean Average Precision (MAP)*, *Average Recall (AvgRec)* and *Mean Reciprocal Rank (MRR)*.

6.2   Results and Discussion

Our experiments were performed using traditional NLP model, deep neural network-based models, and deep neural network-based models with full features (for simplicity we call these model combined models). Table 2 shows the results of our experiments. The first row of Table 2 shows our experiment results when using the traditional NLP model. We extracted many different feature kinds as shows in Sect. 3 and combined these feature kinds as input for the classifier. In this paper, we use a SVM classifier to classify related question-answering pairs into one of two classes "relevant" or "irrelevant". We then use the predicted scores to rank related question-answering pairs according to their similarities to the input question. This model has achieved very good results with *Accuracy* and *MAP* of 81.86% and 78.27% respectively.

The second and third rows of Table 2 show our experimental results when using deep neural network-based models. In these models, to represent words in the form of word embedding we use 300-dimensional vectors that were trained and provided by word2vec using a part of the Google News dataset. In the CNN-based model, we use filters with different sizes of 3, 4, 5 (i.e. filter region size is

Table 1. Statistics of the dataset

	Train	Test	Total
Input questions	267	70	337
Related question-answer pairs	2669–26690	700–7000	3369–33690

---

[4] http://alt.qcri.org/semeval2016/task3/index.php?id=data-and-tools.

Table 2. Our experimental results using different models for the task of ranking question-answering pairs in cQA

Models	Classification measures				Ranking measures		
	Acc	P	R	F1	MAP	AveRec	MRR
NLP	81.86	73.25	71.67	72.45	78.27	92.14	85.64
CNN-based	73.71	53.65	62.19	57.60	72.95	87.87	78.29
BLSTM-based	74.14	53.22	63.27	57.81	73.86	87.08	80.51
CNN-based + NLP	82.43	70.82	75.00	72.85	78.37	91.97	86.23
BLSTM-based + NLP	82.86	72.10	75.34	73.68	78.43	92.07	86.23

3, 4 and 5). Each filter region size includes 128 filter window. In the BLSTM-based model, we use BLSTM with 100 hidden units, followed by a pooling and a dropout layer. In these experiments, we observed that BLSTM-based model performed better than the CNN-based model. The BLSTM-based model gives better results than the CNN-based model in both classification (*Accuracy*) and ranking (*MAP*) measures.

The fourth and fifth rows of Table 2 show our experiment results when using deep neural network-based models combined with the traditional NLP model (i.e. adding traditional feature vector into the deep neural models). We achieved the best results when using the BLSTM-based model and traditional feature vector with the *Accuracy* and *MAP* of 82.86% and 78.43% respectively. It shows that the combination of the BLSTM-based model and traditional NLP models achieves the best performance. Figure 5 compares the performance of our different models for the task on the ranking question-answering pairs in the database related to the input question.

We also implemented a comparison of our proposed methods with previous studies which are well-known and use the same evaluation metrics as well as the same dataset. Results of this comparison are presented in Table 3.

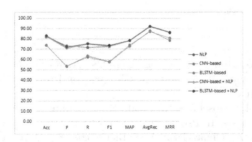

Fig. 5. Compare the performance of different models for ranking question - answering pairs task

Table 3. Comparison with previous studies for the same task and the same dataset

Models	Classification measures				Ranking measures		
	Acc	P	R	F1	MAP	AveRec	MRR
UH-PRHLT-primary [15]	76.57	63.53	69.53	66.39	76.70	90.31	83.02
ConvKN-primary [1]	78.71	68.58	66.52	67.54	76.02	90.70	84.64
Kelp-primary [2]	79.43	66.79	75.97	71.08	75.83	91.02	82.71
SLS-primary [23]	79.43	76.33	55.36	64.18	75.55	90.65	84.64
Our NLP	81.86	73.25	71.67	72.45	78.27	92.14	85.64
Our CNN-based + NLP	82.43	70.82	75.00	72.85	78.37	91.97	86.23
Our BLSTM-based + NLP	82.86	72.10	75.34	73.68	78.43	92.07	86.23

Table 3 shows that our proposal achieves the *Accuracy* and *MAP* measures of 82.86% and 78.43% respectively, which are the best results in comparison with the others.

7   Conclusion

In this paper, we have focused on the problem of ranking question-answering pairs related to the input questions in cQA systems. We have formulated this problem under CNN and BLSTM based deep neural network models for both question representation learning and question similarity score prediction. We also propose using various rich feature kinds and integrating all the features into these deep neural models. The experimental results on the dataset provided by SemEval 2016 show that our proposal models achieve the best results in comparison with the previous recent studies.

Acknowledgement. This paper is supported by The Vietnam National Foundation for Science and Technology Development (NAFOSTED) under grant number 102.01-2014.22.

References

1. Alberto, B.C., Bonadiman, D., Martino, G.D.S.: Answer and question selection for question answering on Arabic and English Fora. In: Proceedings of SemEval-2016, pp. 896–903 (2016)
2. Filice, S., Croce, D., Moschitti, A., Basili, R.: Learning semantic relations between questions and answers. In Proceedings of SemEval-2016, pp. 1116–1123 (2016)
3. Wang, K., Ming, Z., Chua, T.S.: A syntactic tree matching approach to finding similar questions in community-based QA services. In: SIGIR, pp. 187–194 (2009)

4. Yih, W.T., He, X., Meek, C.: Semantic parsing for single-relation question answering. In: Proceedings of ACL, pp. 643–648 (2014)
5. Shen, Y., He, X., Gao, J., Deng, L., Mesnil, G.: Learning semantic representations using convolutional neural networks for web search. In: Proceedings of WWW, pp. 373–374 (2014)
6. Kalchbrenner, N., Grefenstette, E., Blunsom, P.: A convolutional neural network for modelling sentences. In: Proceedings of ACL, pp. 655–665 (2014)
7. Zhang, Y., Byron, C.W.: A sensitivity analysis of (and practitioners guide to) convolutional neural networks for sentence classification (2016)
8. Cao, Y., Duan, H., Lin, C.Y., Yu, Y., Hon, H.W.: Recommending questions using the MDL-based tree cut model. In: Proceedings of WWW, pp. 81–90 (2008)
9. Duan, H., Cao, Y., Lin, C.Y., Yu, Y.: Searching questions by identifying question topic and question focus. In: Proceedings of ACL 2008: HLT, pp. 156–164 (2008)
10. Ji, Z., Xu, F., Wang, B., He, B.: Question-answer topic model for question retrieval in community question answering, pp. 2471–2474. ACM (2012)
11. Zhang, K., Wu, W., Wu, H., Li, Z., Zhou, M.: Question retrieval with high quality answers in community question answering. In: Proceedings of ACM, pp. 371–380 (2014)
12. Zhou, G., Cai, L., Zhao, J., Liu, K.: Phrase-based translation model for question retrieval in community question answer archives. In: Proceedings of ACL, pp. 653–662 (2011)
13. Jeon, J., Croft, W.B., Lee, J.H.: Finding similar questions in large question and answer archives. In: Proceedings of ACM, pp. 84–90 (2005)
14. Filice, S., Castellucci, G., Croce, D., Martino, G.D.S., Moschitti, A., Basili, R.: KeLP: a Kernel-based learning platform for natural language processing. In: Proceedings of ACL-IJCNLP, pp. 19–24 (2015)
15. Salvador, M.F., Kar, S., Solorio, T., Rosso, P.: Combining lexical and semantic-based features for community question answering. In: Proceedings of SemEval-2016, pp. 814–821 (2016)
16. LeCun, Y., Bengio, Y., Hinton, G.: Deep learning. Nature 521(7553), 436–444 (2015)
17. Goldberg, Y.: A primer on neural network models for natural language processing. arXiv preprint arXiv:1510.00726 (2015)
18. Graves, A.: Supervised Sequence Labelling with Recurrent Neural Networks. SCI, vol. 385. Springer, Heidelberg (2012). https://doi.org/10.1007/978-3-642-24797-2
19. Feng, M., Xiang, B., Glass, M.R., Wang, L., Zhou, B.: Applying deep learning to answer selection: a study and an open task. CoRR, abs/1508.01585 (2015)
20. Yu, L., Hermann, K.M., Blunsom, P., Pulman, S.: Deep learning for answer sentence selection. CoRR, abs/1412.1632 (2014)
21. Santos, C.D., Barbosa, L., Bogdanova, D., Zadrozny, B.: Learning hybrid representations to retrieve semantically equivalent questions. In: Proceedings of ACL, pp. 694–699 (2015)
22. Bahdanau, D., Cho, K., Bengio, Y.: Neural machine translation by jointly learning to align and translate. In: Proceedings of International Conference of Learning Representations (2015). arXiv:1409.0473 [cs.CL]
23. Mohtarami, M., Belinkov, Y., Hsu, W.N., Zhang, Y., Lei, T., Bar, K., Cyphers, S., Glass, J.: Neural-based approaches for ranking in community question answering. In: Proceedings of SemEval-2016, pp. 828–835 (2016)

24. Mikolov, T., Chen, K., Corrado, G., Dean, J.: Efficient estimation of word representations in vector space. In: Proceedings of Workshop at ICLR (2013)
25. LeCun, Y., Bottou, L., Bengio, Y., Haffner, P.: Gradient-based learning applied to document recognition. Proc. IEEE **86**(11), 2278–2324 (1998)
26. Hochreiter, S., Schmidhuber, J.: Long short-term memory. Neural Comput. **9**, 1681–1726 (1997)

A Hybrid Approach to Answer Selection in Question Answering Systems

Phuc H. Duong[1], Hien T. Nguyen[1(✉)], Duy D. Nguyen[2], and Hao T. Do[2]

[1] Artificial Intelligence Laboratory, Faculty of Information Technology,
Ton Duc Thang University, Ho Chi Minh City, Vietnam
{duonghuuphuc,hien}@tdt.edu.vn
[2] NewAI Research, Ho Chi Minh City, Vietnam
{duy.nguyen,hao.do}@newai.vn

Abstract. In this paper, we present a hybrid model for answer selection in question answering systems by representing multiple kinds of features, *i.e.*, lexical-based, word-alignment, and word-embedding. The model employs convolutional neural network, multilayer perceptron, and support vector machines to train the classifiers. We evaluate our model on the two popular QA datasets, *SemEval-2016 Task 3* and *TREC QA*. The experimental results show that our system outperforms the top-5 proposed systems in *SemEval-2016* workshop, and also achieves the-state-of-art results on *TREC QA* dataset.

Keywords: Question answering · Answer selection

1 Introduction

Measuring the semantic relatedness between pair of sentences is a fundamental task in many complex natural language processing systems (*e.g.*, information retrieval, paraphrase identification, question answering). In QA task, by developing intelligent QA systems, especially community-based QA, we can overcome some drawbacks in current community QA sites. For instance, the data on these sites is growing rapidly, since there are a number of questions created on everyday, and many of them have been answered in previously similar questions. Therefore, by taking advantage of natural language processing techniques, the QA systems can quickly retrieve and suggest a list of possible questions and/or answers to users. Ferrucci *et al.* [1] develop a very successful open-domain QA system - IBM's Watson, and the proposed architecture has become a standard. The IBM's Watson is a combination model of many modules, *e.g.*, answer selection, document retrieval, question analysis. In this paper, we will focus on QA task, and based on the IBM's Watson architecture, we will present a solution for the answer selection module by applying semantic measurements in order to classify a list of answers for given question in predefined labels. After the document retrieval engine crawled a list of possible answers, the answer selection module has to rank or classify them to return the relevant ones. Let's consider the following example:

© Springer International Publishing AG, part of Springer Nature 2018
V.-N. Huynh et al. (Eds.): IUKM 2018, LNAI 10758, pp. 191–202, 2018.
https://doi.org/10.1007/978-3-319-75429-1_16

- Q: "Where I can buy good oil for massage?"
- A_1: "I have never been for a massage! Can you believe it? Maybe I should go and try too! :)"
- A_2: "body shop."

In the example above, extracted from the *SemEval-2016 Task 3* dataset[1], the possible answer for Q is A_2. Although A_1 implies much information, it is not an expected answer. Therefore, many studies have concentrated to exploit multiple features from question and candidate answers to find the relatedness between them, *e.g.*, surface form of word, semantic and syntactic information; or learn deep features by applying deep learning approaches.

In this paper, first, we present a hybrid model towards the answer selection task, which exploits both surface form and the semantic of words. Besides that, with the emergence of neural-based methods recently, we also consider them in representing the distributional of word and designing learning models. Then, we conduct the experiments on *SemEval-2016 Task 3* and *TREC QA*[2] datasets, which are the popular datasets in community-based QA studies.

The rest of this paper is organized as follows. Section 2 presents related work. Section 3 presents our proposed model. Section 4 presents our experiments. And our conclusion is in Sect. 5.

2   Related Work

There have been many studies in the task of answer selection, from features exploitation to learning model. For instance of features exploitation, **lexical-based** methods concentrate on the syntactic and semantic information for measuring the relatedness between question and answer sentences, *e.g.*, TF-IDF, bag-of-word, n-gram. Jijkoun and Rijke [2] propose a lexical similarity measurement to identify whether the text t entails the hypothesis h by combining the term-weight (TF-IDF) and word-to-word similarity metrics. For measuring the similarity of words, they apply *Lin*98 and *HS*98 methods [3], which exploit WordNet[3] lexical database. Mollá [4] evaluates three features: (1) word dependencies, (2) grammatical relations, (3) logical forms. The evaluation from [4] shows that the logical forms feature gains better results than the others, since it plays an important role in considering the meaning of words in the absence of context and also in QA tasks [5,6]. Papineni *et al.* [7] propose a word-overlap method by comparing the n-grams between two sentences. The proposed method is, then, evaluated on machine translation task and yields the state-of-art performance. In [8], for lexical features, the method employs 4 metrics: (1) cosine similarity based on word n-gram, character n-gram, TF-IDF; (2) word overlap; (3) noun overlap; and (4) n-gram overlap.

---

[1] http://alt.qcri.org/semeval2016/task3/.
[2] http://trec.nist.gov/data/qa.html.
[3] http://wordnet.princeton.edu/.

In lexical-based approach, the main drawback of word-overlap measures is that it does not take into account the semantic information between words and phrases [9]. To overcome this, Duong *et al.* [10] exploit multiple features to measure the semantic similarity between two sentences, *i.e.*, named entity recognition (NER), knowledge-based and corpus-based measures, word-embedding. Madnani *et al.* [11] re-examine machine translation metrics to identify the two given texts whether paraphrase.

Another approach in answer selection is applying **tree edit distance** (TED). Yao *et al.* [12] propose a cost function, based on TED, to compute the steps of edit sequences for transforming question tree to answer tree, including three basic operations (*i.e.*, insert, delete, and rename). Each node in the dependency parse tree is represented by lemma, part-of-speech, NER, the dependency's type between current and parent nodes. These are then input to linear-chain conditional random fields model to represent the associations between question and answer types. In [13], the proposed method takes into account the complete dependency parse tree of two given sentences to compute the TED score between them. Besides that, method in [13] also evaluates holistic metrics (*i.e.*, Levenshtein distance, BLEU [7]) and unstructured word-order metric.

Instead of directly computing the similarity score, if we can exploit the constraints between question and answer sentences, the system will better understand which answers are appropriate for given question. Li and Roth [14] propose a **question classification** model to categorize given questions into 6 coarse classes (*i.e.*, abbreviation, description, entity, human, location, numeric). In [15], the proposed approach exploits bag-of-word and bag-of-*n*grams features, then, evaluated by 5 machine learning algorithms (*i.e.*, nearest neighbors, Naïve Bayes, decision tree, SNoW, and Support vector machine (SVM)). The number of question categorizes in [15] is as proposed in [14].

With the emergence of artificial intelligence, recent studies have focused on applying **artificial neural network** models, and the performance have also been improved. Yu *et al.* [16] present a deep learning approach in answer selection, including two distributional sentence models: (1) bag-of-word, (2) bi-gram model based on convolution neural network (CNN). In [17], after representing a sentence in d-dimension word-embedding, then, adding the 5-dimension of feature-embedding which is the relation between words of two given sentences. He *et al.* [18] combine sentence model and similarity measurement layer for identifying the paraphrase level between sentences. For sentence model, the method represents each word into d-dimension word-embedding by applying multiple perspectives of convolution. For similarity measurement layer, after representing input sentences, first, flatten the convolutional matrix to vector by performing comparisons on local regions, then, apply cosine metric to compute the similarity between two vectors. In [19], the proposed method exploits different level features to measure short text similarity by combining character-level CNN, word-level CNN and Long Short Term Memory (LSTM). In [20], the proposed *Sent2Vec* model is an unsupervised approach, combining word vectors and *n*-gram embeddings to compose sentence embedding model. Zhang *et al.* [21] propose an

AI-NN model, which includes two steps: (1) represent question and answer by CNN model, (2) apply the attentive model to take into account the useful segments and calculate the interaction between each pair of representations.

3 Proposed Method

In Fig. 1, we present our proposed model, including training and testing phases, to classify a list of answers for a given question (*i.e.*, good, bad, potentially-useful). First, the labeled $<Q, A_{1 \to n}>$ set, where n is the list of possible answers for a given question, goes through the preprocessing step. Next, representing them as feature vectors and input to the learning algorithms to learn the classifier. Finally, in testing phase, we input a new unlabeled $<Q', A'_{1 \to n}>$ set to label each given answers.

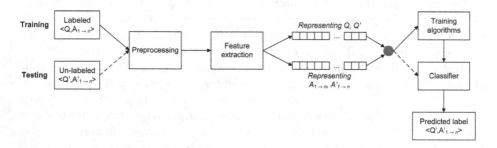

Fig. 1. Proposed model

3.1 Preprocessing

This step is essential for our model. In community question answering sites, the sentences often contain some special characters (*e.g.*, colon, dollar sign, emoticon) and website addresses (URL), but they do not contribute much information in measurement. Therefore, in preprocessing step, we propose to skip these special characters and URL but still preserve the sentence structure. We also skip all stop-word in both question and answer sentences. Since these common words do not convey significant semantic information to sentences they appear in.

3.2 Feature Extraction

In this step, we represent given pair of question and answer as feature vector by exploiting: (1) lexical-based metrics, (2) word-alignment metrics, and (3) word-embedding.

Lexical-Based Metrics. We apply 9 lexical-based metrics to compute the similarity between a pair of question and answer sentences. The first score, denoted by F_{LB1}, is computed by counting the number of overlap words between two sentences, and normalized by the length in range of $[0, 1]$. Before computing, we need to normalize each word in given sentences to the base form by applying Porter stemming algorithm. Next, we perform part-of-speech tagging and count the noun overlap, as F_{LB2}. We also compute the overlapping between named entities (NE), as F_{LB3}, by performing NER. We observe NE in three classes (*i.e.*, location, organization, person) to clearly determine not only the overlapping NE, but also the type of them. Considering this case, both question and answer sentences have NE but do not share any type in common, thus, if we only consider the overlapping, it does not reflect the right information. Therefore, we need to take into account the type of recognized NE when counting the overlapping to avoid this problem. For F_{LB4} score, we compute the n-gram cosine similarity between two sentences, for $n = 1, 2, 3$.

Before applying the other lexical-based metrics, we need to represent the two sentences as feature vectors. First, we create a joint word set W containing all distinct words of two sentences (S_1 and S_2). Since W is derived from input sentences, we use it as a standard vector to normalize the length of S_1, S_2. Next, each sentence will be represented by a feature vector whose length is $||W||$, denoted by V_1 and V_2. Elements in V are assigned by the following rules:

- **Rule 1**: if a word (w_i) appears in both sentences, assign 1 to respective position in current V.
- **Rule 2**: unless, compute the similarity score between w_i and each word in T. If computed score (s) exceed a preset threshold, assign s to respective position in current V, otherwise, assign 0.

After forming two feature vectors, V_1 and V_2, we can compute the similarity score between them by applying cosine coefficient, as presented in Eq. 1.

$$Score(V_1, V_2) = \cos(\theta) = \frac{V_1 \cdot V_2}{||V_1|| \cdot ||V_2||} \tag{1}$$

In order to calculate the similarity between two words, as presented in **Rule 2**, we exploit the lexical structure of them by applying multiple string similarity metrics, as follows:

- F_{LB5} — Levenshtein distance is a string edit-distance metric, reflecting the minimum number of deletions, insertions, or substitutions to transform source string to target string.
- F_{LB6} — Damerau-Levenshtein distance is an optimal metric of Levenshtein distance, beside considering three basic operations, Damerau-Levenshtein distance also computes the transposition of two adjacent characters.
- F_{LB7} — Jacard distance, also known as *dis*-similarity metric, between two given work is computed by dividing the size of intersection by the size of union sets, then, subtracting from 1. To unify the overall representation, we

only compute the rate between intersection and union set, that means, the similarity score, as presented in Eq. 2.

$$F_{LB7}(w_1, w_2) = \frac{|w_1 \cap w_2|}{|w_1 \cup w_2|} \tag{2}$$

- F_{LB8} — Sørensen-Dice coefficient reflects the rate between presence and absence characters, as presented in Eq. 3.

$$F_{LB8}(w_1, w_2) = \frac{2 \cdot |w_1 \cap w_2|}{|w_1| + |w_2|} \tag{3}$$

- F_{LB9} — Jaro-Winkler distance measures the minimum number of single-character transpositions to change source string to target string. This metric differs from Damerau-Levenshtein distance, since it considers the scaling factor to adjust the score to have common prefixes. Equation 4 presents F_{LB9} metric, where: d_J is Jaro distance of w_1 and w_2; l is the length of common prefix; p is scaling factor. In practice, we set up $l = 1, 2, 3$ and $p = 0.1$.

$$F_{LB9}(w_1, w_2) = d_J + (lp(1 - d_J)) \tag{4}$$

Finally, appending all lexical-based scores, F_{LB_i} ($i \in [1, 9]$), into a feature vector, which represents the lexical relations between two given sentences.

Word-Alignment Metrics. We apply 3 word-alignment metrics by identifying the relationships between words in two sentences, then forming the bipartite graph represents these relationships.

The first word-alignment score, denoted by F_{WA1}, is based on BLEU (Bi-Lingual Evaluation Understudy) method [7]. We compute the overlap of n-gram ($n = 1, 2, 3$), since [7] points out that the larger n is, the less precision will be. By practice, we find out that $n = 4$ will generate noise, since BLEU is based on geometric mean, which is calculated by multiplying all the n-gram counts. That means, if we have a 0 value, it will turn the geometric mean to 0, and it frequently happens when $n \geq 4$.

To compute F_{WA2} score, we apply TER-Plus [22] method. Beside a heuristic algorithm computes the number of edits over the average number of reference words, TER-Plus includes additional operations, i.e., paraphrase, stemming, and synonym. In practice, we take WordNet for synonym data, Porter algorithm for stemming, and pre-trained phrase table[4].

For F_{WA3} score, we compute the similarity between sentences by constructing a matcher system, considering both precision and recall measures, as proposed in [23]. This system includes four matchers: (1) exact match, (2) stemming match, (3) synonym match, and (4) paraphrase match. For (1), it means two surface form of words are identical. For (2) and (3), we apply Porter algorithm and

---

[4] https://github.com/snover/terp.

WordNet, respectively. For (4), we use the pre-trained paraphrase tables[5], as presented by the authors.

Finally, plugging all word-alignment scores together, we form a feature vector of two given sentences, including three elements, F_{WA_i} ($i \in [1, 3]$).

Word-Embedding. To represent word embeddings in both question and answer sentences, we take advantage of the continuous bag-of-words (CBOW) approach from *word2vec*[6] model [24]. In this metric, after generating all vector representation of words in the sentence, we sum these vectors and normalize by the length of sentence. In Eq. 5, $|v|$ is the total number of words, and v_i is vector representation of each word in sentence.

$$V = \frac{1}{|v|} \sum_{i=1}^{|v|} v_i.$$ (5)

4   Experiments

4.1   Dataset

We conduct experiments on the *SemEval-2016 Task 3* and *TREC QA* datasets. With *SemEval-2016 Task 3* dataset, which is a subtask in the series of evaluation of computational semantic analysis systems. This dataset is question-external comment similarity and concentrates on classification model, including 26,690 and 7,000 pairs of questions and answers for training and testing, respectively; described as follows:

- Given a question Q and 100 possible answers A_i ($i \in [1, 100]$), which are extracted from 10 related questions to the original Q.
- The output is labels of these 100 answers in three types: *Good, PotentiallyUseful,* and *Bad*.

With *TREC QA*, the task is to rank the candidate answers for each question. We adopt the dataset constructed in [25], based on the Text REtrieval Conference (TREC) QA track, which removed all answer sentences over 40 words and questions with only positive/negative words. Therefore, this clean version of *TREC QA* dataset includes 1,162 questions in the training set, 65 questions in the development set and 68 questions in the test set.

4.2   Training Model

After forming all representation vectors of given sentences in three metrics: (1) lexical-based metrics, (2) word-alignment metrics, and (3) bag-of-word; we now input them to the training process. We propose training the classifiers by applying: CNN, Multilayer perceptron (MLP), and SVM.

---

[5] http://www.cs.cmu.edu/~alavie/METEOR/.
[6] https://code.google.com/archive/p/word2vec/.

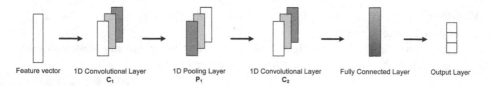

Fig. 2. Proposed CNN model

- The original CNN model is familiar with working on 2-dimension data, thus we construct a 2-layer-1-dimension-CNN, which input a vector V having d variable dimension ($V \in \mathbb{R}^d$), as presented in Fig. 2. In practice, we do not consider the *padding* parameter, since our representation vector is always in fixed-length. We apply the *categorical_crossentropy*, as a multi-class logarithmic loss function during model training, and stochastic gradient descent (SGD) as an optimizer.
- Though CNN model was proved to be useful in text processing [26], we want to evaluate another neural-based model which does not have some strengths of CNN model, *e.g.*, convolution operation, pooling layer. Therefore, we construct the original neural model, a MLP, which is a class of feed-forward neural network to compare the performance with CNN model. Same with the CNN model, we apply *categorical_crossentropy* as loss function, and SGD as optimizer.
- Beside neural-based models, we apply SVM to train a multi-label classifier. In practice, we use the most common optimization algorithm to train SVM, which is sequential minimal optimization (SMO) algorithm.

Table 1 presents the hyper-parameters of CNN and MLP models. We apply the same hyper-parameters for training on *SemEval-2016 Task 3* and *TREC QA* datasets, since we want to construct a general model, which can adapt to multiple QA datasets.

Table 1. Hyper-parameters of CNN and MLP models

Hyper-parameters	CNN	MLP
Embedding	*word2vec*, augmented features	*word2vec*, augmented features
Hidden dimension	600, 800	1000, 1200, 1400
Filter width	5	N/A
Activation function	ReLU, Softmax	ReLU, Softmax
Optimizer	SGD	SGD
Dropout	0.4	N/A
Learning rate	0.0001	0.0001

Table 2. Experimental results on *SemEval-2016 Task 3*

Method	MLP		CNN		SVM	
	Accuracy	F1	Accuracy	F1	Accuracy	F1
WE	92.26	44.58	89.47	**43.25**	84.90	39.06
WE + F_{LB}	**92.97**	44.38	**90.91**	42.03	**85.17**	**39.06**
WE + F_{WA}	92.64	44.20	85.68	41.54	84.33	39.06
WE + F_{LB} + F_{WA}	92.01	**44.62**	85.32	41.76	84.71	39.06

Table 3. Experimental results on *TREC QA*

Method	MLP		CNN		SVM	
	MAP	MRR	MAP	MRR	MAP	MRR
WE	**0.783**	**0.835**	0.770	**0.848**	0.763	0.830
WE + F_{LB}	0.777	0.818	0.774	0.822	**0.768**	0.830
WE + F_{WA}	0.771	0.831	0.761	0.839	0.765	**0.834**
WE + F_{LB} + F_{WA}	0.764	0.827	**0.779**	0.846	0.765	0.833

Table 4. Our system and the top-5 systems in *SemEval-2016 Task 3* workshop

Method	Accuracy	F1
ECNU-primary	91.07	15.88
SLS-primary	90.54	21.19
ConvKN-primary	90.51	14.65
UH-PRHLT-primary	88.56	35.87
Kelp-primary	84.79	44.21
This paper	**92.97**	**44.38**

4.3 Experimental Results

With *SemEval-2016 Task 3*, we report the *accuracy* and *F1* score in Table 2.
With *TREC QA*, we report the MAP and MRR scores in Table 3. We perform
experiments in 4 cases, as follows:

– Consider word-embedding only, denoted by WE.
– Combine WE and lexical-based, denoted by $(WE + F_{LB})$.
– Combine WE and word-alignment, denoted by $(WE + F_{WA})$.
– Combine all features, denoted by $(WE + F_{LB} + F_{WA})$.

The experimental results indicate that the neural-based approaches achieve
a better performance than SVM, especially the MLP model. That means, with
the data represented like our proposed method, the original neural-based model
can achieve a good performance. And, in most cases, if we consider only F_{LB}

Table 5. Our system and the state-of-the-art systems on *TREC QA* dataset

Method	MAP	MRR
He *et al.* [18]	0.777	0.836
Wang *et al.* [25]	0.771	0.845
Santos *et al.* [28]	0.753	0.851
Rao *et al.* [29]	0.801	**0.877**
Wang *et al.* [30]	**0.802**	0.875
This paper	0.783	0.835

or F_{WA}, the performance gains better results than combining both F_{LB} and F_{WA}. In Table 4, we show our best result in comparison to the top-5 proposed systems in *SemEval-2016 Task 3* workshop [27], and our approach improves performance than the others. In Table 5, we compare our experimental results with some state-of-the-art baselines on *TREC QA* dataset.

5   Conclusion

In this paper, we have introduced a hybrid approach to answer selection in question answering systems in order to label a list of possible answers for a given question. Our proposed model represents given questions and answers in multiple features, *i.e.*, lexical-based, word-alignment, and word-embedding. We have applied convolutional neural network, multilayer perceptron, and support vector machines to train the classifiers. The experimental results show that the proposed system achieves the state-of-the-art performance, in comparison to top-5 systems in *SemEval-2016 Task 3* workshop and baseline systems on *TREC QA* dataset.

References

1. Ferrucci, D., Brown, E., Chu-Carroll, J., Fan, J., Gondek, D., Kalyanpur, A.A., Lally, A., Murdock, J.W., Nyberg, E., Prager, J., Schlaefer, N., Welty, C.: Building Watson: An overview of the DeepQA project. AI Mag. **31**(3), 59–79 (2010)
2. Jijkoun, V., Rijke, M.: Recognizing textual entailment using lexical similarity. In: Proceedings Pascal 2005 Textual Entailment Challenge Workshop (2005)
3. Duong, P.H., Nguyen, H.T., Nguyen, V.P.: Evaluating semantic relatedness between concepts. In: IMCOM, pp. 20:1–20:8. ACM (2016)
4. Mollá, D.: Towards semantic-based overlap measures for question answering. In: Proceedings of the First Australasian Language Technology Workshop (ALTW 2003). University of Melbourne, Melbourne (2003)
5. Harabagiu, S., Moldovan, D., Pasca, M., Surdeanu, M., Mihalcea, R., Girju, R., Rus, V., Lactusu, F., Morarescu, P., Bunescu, R.: Answering complex, list and context questions with LCC's question-answering server. In: Text REtrieval Conference (TREC) TREC 2001 Proceedings. Department of Commerce, National Institute of Standards and Technology, pp. 355–362 (2001)

6. Pasupat, P., Liang, P.: Inferring logical forms from denotations. In: ACL (1). The Association for Computer Linguistics (2016)
7. Papineni, K., Roukos, S., Ward, T., Zhu, W.J.: BLEU: a method for automatic evaluation of machine translation. In: ACL, pp. 311–318. ACL (2002)
8. Franco-Salvador, M., Kar, S., Solorio, T., Rosso, P.: UH-PRHLT at SemEval-2016 task 3: combining lexical and semantic-based features for community question answering. In: Bethard, S., Cer, D.M., Carpuat, M., Jurgens, D., Nakov, P., Zesch, T. (eds.) Proceedings of the 10th International Workshop on Semantic Evaluation, SemEval@NAACL-HLT 2016, San Diego, CA, USA, 16–17 June 2016, pp. 814–821. The Association for Computer Linguistics (2016)
9. Wang, M., Manning, C.D.: Probabilistic tree-edit models with structured latent variables for textual entailment and question answering. In: Huang, C.R., Jurafsky, D. (eds.) COLING, pp. 1164–1172. Tsinghua University Press (2010)
10. Duong, P.H., Nguyen, H.T., Huynh, N.-T.: Measuring similarity for short texts on social media. In: Nguyen, H.T.T., Snasel, V. (eds.) CSoNet 2016. LNCS, vol. 9795, pp. 249–259. Springer, Cham (2016). https://doi.org/10.1007/978-3-319-42345-6_22
11. Madnani, N., Tetreault, J.R., Chodorow, M.: Re-examining machine translation metrics for paraphrase identification. In: HLT-NAACL, pp. 182–190. The Association for Computational Linguistics (2012)
12. Yao, X., Durme, B.V., Callison-Burch, C., Clark, P.: Answer extraction as sequence tagging with tree edit distance. In: Vanderwende, L., Daumé III, H., Kirchhoff, K. (eds.) HLT-NAACL, pp. 858–867. The Association for Computational Linguistics (2013)
13. McCaffery, M., Nederhof, M.J.: DTED: evaluation of machine translation structure using dependency parsing and tree edit distance. In: WMT, pp. 491–498. The Association for Computer Linguistics (2016)
14. Li, X., Roth, D.: Learning question classifiers. In: Proceedings of the 19th International Conference on Computational Linguistics, vol. 1. Association for Computational Linguistics (2002)
15. Zhang, D., Lee, W.S.: Question classification using support vector machines. In: Proceedings of the 26th Annual International ACM SIGIR Conference on Research and Development in Information Retrieval, pp. 26–32. ACM (2003)
16. Yu, L., Hermann, K.M., Blunsom, P., Pulman, S.: Deep learning for answer sentence selection. In: NIPS Deep Learning Workshop (2014)
17. Bonadiman, D., Uva, A.E., Moschitti, A.: Multitask learning with deep neural networks for community question answering. CoRR abs/1702.03706 (2017)
18. He, H., Gimpel, K., Lin, J.J.: Multi-perspective sentence similarity modeling with convolutional neural networks. In: Màrquez, L., Callison-Burch, C., Su, J., Pighin, D., Marton, Y. (eds.) Proceedings of the 2015 Conference on Empirical Methods in Natural Language Processing, EMNLP 2015, Lisbon, Portugal, 17–21 September 2015, pp. 1576–1586. The Association for Computational Linguistics (2015)
19. Huang, J., Yao, S., Lyu, C., Ji, D.: Multi-granularity neural sentence model for measuring short text similarity. In: Candan, S., Chen, L., Pedersen, T.B., Chang, L., Hua, W. (eds.) DASFAA 2017. LNCS, vol. 10177, pp. 439–455. Springer, Cham (2017). https://doi.org/10.1007/978-3-319-55753-3_28
20. Pagliardini, M., Gupta, P., Jaggi, M.: Unsupervised learning of sentence embeddings using compositional n-gram features. CoRR abs/1703.02507 (2017)

21. Zhang, X., Li, S., Sha, L., Wang, H.: Attentive interactive neural networks for answer selection in community question answering. In: Singh, S.P., Markovitch, S. (eds.) Proceedings of the Thirty-First AAAI Conference on Artificial Intelligence, San Francisco, California, USA, pp. 3525–3531. AAAI Press (2017)

22. Snover, M.G., Madnani, N., Dorr, B.J., Schwartz, R.M.: Ter-plus: paraphrase, semantic, and alignment enhancements to translation edit rate. Mach. Transl. **23**(2–3), 117–127 (2009)

23. Denkowski, M., Lavie, A.: Meteor 1.3: automatic metric for reliable optimization and evaluation of machine translation systems. In: Proceedings of the Sixth Workshop on Statistical Machine Translation, pp. 85–91. Association for Computational Linguistics (2011)

24. Mikolov, T., Chen, K., Corrado, G., Dean, J.: Efficient estimation of word representations in vector space. CoRR abs/1301.3781 (2013)

25. Wang, D., Nyberg, E.: A long short-term memory model for answer sentence selection in question answering. In: ACL (2), pp. 707–712. The Association for Computer Linguistics (2015)

26. Kim, Y.: Convolutional neural networks for sentence classification. In: Moschitti, A., Pang, B., Daelemans, W. (eds.) Proceedings of the 2014 Conference on Empirical Methods in Natural Language Processing, EMNLP 2014, Doha, Qatar. A Meeting of SIGDAT a Special Interest Group of the ACL, 25–29 October 2014, pp. 1746–1751. ACL (2014)

27. Nakov, P., Màrquez, L., Moschitti, A., Magdy, W., Mubarak, H., Freihat, A.A., Glass, J., Randeree, B.: SemEval-2016 task 3: community question answering. In: Proceedings of the 10th International Workshop on Semantic Evaluation, SemEval@NAACL-HLT 2016, San Diego, CA, USA, 16–17 June 2016, pp. 525–545. The Association for Computer Linguistics (2016)

28. dos Santos, C.N., Tan, M., Xiang, B., Zhou, B.: Attentive pooling networks. CoRR abs/1602.03609 (2016)

29. Rao, J., He, H., Lin, J.J.: Noise-contrastive estimation for answer selection with deep neural networks. In Mukhopadhyay, S., Zhai, C., Bertino, E., Crestani, F., Mostafa, J., Tang, J., Si, L., Zhou, X., Chang, Y., Li, Y., Sondhi, P. (eds.) Proceedings of the 25th ACM International Conference on Information and Knowledge Management, CIKM 2016, Indianapolis, IN, USA, 24–28 October 2016, pp. 1913–1916. ACM (2016)

30. Wang, Z., Hamza, W., Florian, R.: Bilateral multi-perspective matching for natural language sentences. CoRR abs/1702.03814 (2017)

Sensory Quality Assessment of Food Using Active Learning

Nhat-Vinh Lu[1]([✉]), Van-Nam Huynh[1], Takaya Yuizono[1],
and Trung-Ky Nguyen[2]

[1] Japan Advanced Institute of Science and Technology, Nomi, Japan
lnvinh@jaist.ac.jp
[2] University of Grenoble Alpes, Grenoble, France

Abstract. The correctness of sensory assessment of food quality based on machine learning approach is significantly growing in the food industry. It contributes to an improvement of the food composition and develops the new food products. However, this process requires human intervention. And thus, it is costly, time consuming and easily be biased. In this paper, we propose the Active learning method based on Sequential minimizing optimization in order to evaluate sensory of red wine quality. The general idea is letting the algorithm choose the most uncertain products and asking experts for their opinions. This scheme greatly reduces the number of labels needed for the training process, and, consequently leads to the reduction on the cost of the sensory evaluation process. Experimental results show that the prospect of this method can be widely applied in the optimization of food ingredient and consumer tastes from food consumption markets.

Keywords: Sensory · Active learning
Sequential minimizing optimization

1 Introduction

Food quality is a gathering of food attributes that meet the needs of the user, provide the user with the necessary nutrients and energy for living processes. Food quality has attributes that play a major role in selecting and accepting food products of the consumers, e.g., the nutritional value and the sensory attribute. The nutritional value is the basic attribute of food. And the sensory attribute is the main factor in determining the food acceptance of consumers, and food safety is an important factor to health of human. Though preferences of consumers for food products can be influenced by many factors such as trademark image, price and rival position, and the sensory character still are considered as the most important factor in ascertaining quality of food product quality that consumers seek in the food market because sensory characteristics are perceived by the feeling of human. Food quality can be considered as the sum of partial quality such as quality of raw materials, production technology, and sensory quality:

$$Q = Q_1 + Q_2 + ... + Q_{sensory} + ... + Q_i$$

© Springer International Publishing AG, part of Springer Nature 2018
V.-N. Huynh et al. (Eds.): IUKM 2018, LNAI 10758, pp. 203–213, 2018.
https://doi.org/10.1007/978-3-319-75429-1_17

Sensory quality consists of specifications such as color, shape, size, smell, taste, and structure, which can be valued easily by the senses of human [27]. The sensory attributes of food are assessed by sensory analysis method; the sensory tests are developed as a formalized and structured method that used by trained experts to assess specific food products. At present, the food industry is one of the strongest industries, thus ensuring product quality as well as identifying the sensory attributes of food products is extremely important. Sensory evaluation should be conducted for the first time select, raw material, during the producing process and until making final product, aim to make a food product with desired quality. In addition to methods such as physical, chemical and microbiological, sensory analysis method must be conducted to control the quality of food products [29].

Most traditional methods for sensory analysis are heavily based on human perceptions, which are highly biased by host subjects, costly and time consuming. Recently, data mining and machine learning methods such as data synthesis [10], clustering [2], classification [22] have been applied as additional automatic ways to evaluate the sensory quality of products. Compared to traditional techniques, data mining methods are more robust, less biased, faster and cheaper. However, despite their potentials, human perceptions are quite complicated and therefore are hardly to be replaced completely by machines. Combining human efforts and computers is therefore an interesting target to pursue.

In this work, we propose a co-operation approach between humans and computers for the sensory analysis problem. In particular, we focus on the wine quality analysis problem introduced in [3]. Cortez et al. [3] proposes to use Support Vector Machine (SVM) for assessing the quality of different kinds of wine by their chemical characteristics. Their approach is interesting and can classify wines very well. However, it also requires huge human efforts for building the training data. How can we have the same accuracy while spending fewer human attempts?

Our solution is our algorithm start with a small set of available training data provided by human analyzers. Then it iteratively classifies the wines and chooses one which is the most uncertain among others. After that, it asks the human experts for their opinion. The acquired result is used as an additional resource for the algorithm for enhancing its classification accuracy. By this way, it can reduce the number of samples to be labeled an expert need while retaining the same accuracy. Consequently, it can reduce the cost for analyzing wine as well as the biases caused by different opinions of experts.

Our paper is organized as follow: Sect. 2 briefly reviews the related works; Sect. 3 presents our proposed technique such as active learning method, sequential minimizing optimization and our proposed algorithm; Sect. 4 describes the experimental design and analyzes the results; finally, conclusions based on results of this study constitute Sect. 5.

2   Related Works

In the past decades, sensory quality assessment of food was developed and became one of the general techniques include of sensory study, food and drink consumer. Moreover, sensory assessment was broadly used in quality evaluation, product design and marketing in food companies. Traditional method likes as Principal Component and Analysis [11] is used to analyze sensory data of food that is supplied by experts. This method can effectively solve for some specific datasets, however sometimes it will lead to the loss of important information. In this case, the others methods were used such as random forest, neural network, fuzzy logic and support vector machine to deal with uncertainty of sensory assessment. For instance, [9] used random forest, [1] applied neural network, [5, 12, 18–20, 25, 26, 31] fuzzy logic was successfully implemented to sensory quality assessment of vital food products.

In recent years, some studies have approached data mining techniques for evaluating the sensory quality of food products by base on their physical and chemical composition. Cortez et al. [3] proposed a data mining method to predict the taste white and red "Vinho Verde" wine of Portugal by support vector machine, multiple regression and neural network, the result shows that the support vector machine method gives higher accuracy than the multiple regression and neural network, this model is useful to support experts in making wine and the same techniques can help market target by modeling consumer tastes from the suitable markets. Debska et al. [6] proposed to the artificial neural network to build Poland's beer quality classification model based on its chemical characteristics, and the result shows the great potential of using the artificial neural network in distinguishing between the quality of beer samples. Ghasemi-Varnamkhasti et al. [8] proposed to apply neural network model to assess the sensory properties of commercial non-alcoholic beer brands, and the result shows that radial basis function give the successful grading rate is about 97%, the results obtained in this study can be used as a reference dossier for electronic noses and electronic blades in beer quality control. Sensory evaluation plays a very important role in controlling produced beer quality, Dong et al. [7] studied the application of nonlinear models such as partial least squares, genetic algorithm back-propagation neural network, support vector machine to determine the relationship between flavor compound and sensory evaluation of beer, which the result shows that support vector machine model obtained more effective than other models and its accuracy is over 94%, support vector machine is a powerful multivariate statistical measure and has great potential for evaluating beer quality.

The above researchers have demonstrated the effectiveness of accessing data mining techniques in sensory evaluation and quality control of food products. Data mining techniques apply in sensory evaluation based on physical and chemical properties of food products have great potential to become tools to assist organizations in controlling the quality of food products. However, the cost of money and time for labeling data is very high. (for example, a study on the sensory quality of lamb was reported from November 2002 to November 2003 only with 81 samples for each animal took a significant cost, about 6 euros per kilogram plus with labor costs in the laboratory [4]), so an approach according to Active learning method to evaluate by sensory quality way based on physiological characteristics of food products will help to reduce the cost during sensory evaluation process because the goal of Active Learning method is to classify accurately with minimal cost.

3   Our Proposed Technique

In this Section, we start by introducing the notion of active learning in Sect. 3.1. Next, we talk about the Sequential Minimizing Optimization (SMO) algorithm. And the last Sect. 3.3 is dedicated to our algorithm.

3.1   Active Learning

Active learning is a special case of machine learning in which a learning algorithm can perform better with less training, it is allowed to choose the data from which it learns. An active learner may give queries, usually in the form of unlabeled data instances to be labeled by an oracle (e.g., a human annotator) that already understands the nature of the problem [23]. Active learning has been widely used in many fields [13, 15–17, 28, 30]. A typical active learning process is shown in the Fig. 1.

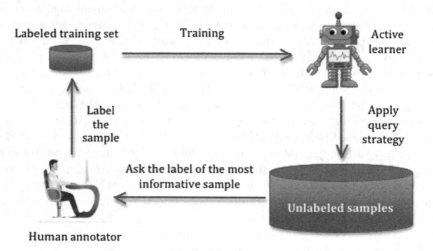

Fig. 1. The active learning process

3.2   The Sequential Minimizing Optimization

The Sequential Minimizing Optimization (SMO) algorithm was introduced by Platt [21]. This algorithm leverages the training of Support Vector Machine by solving quickly the very large quadratic programming (QP) optimization problem. In fact, the SMO divides the large QP to a series of smallest QP optimization problems. An advantage of the SMO is not need to store large matrix and QP sub-problem calling recursively while quickly settle the SVM QP. The key idea of SMO is using a selection to deal with optimization processes. At each step, the SMO chooses two Lagrange multipliers to optimize at the same time, then finds optimal values for the multipliers and update SVM to obtain the new optimal values. As a consequence, the SMO performs two main tasks: optimization solution for two Lagrange multipliers; and heuristic method for two multipliers selection. Therefore, the SMO absolutely avoids the high computational costs and memory for solving the large QP optimization problems of Support Vector Machine. This will lead to the SMO can handle with the very large training sets. Another advantage of the SMO can speed up a linear SVM by only storing a single weight vector. From these advantages of SMO algorithm, we implemented this algorithm in our approach to minimize the training time-consuming.

3.3   Our Proposed Algorithm

We introduce a general frame work for the wine classification problem. Our frame work uses SMO as the main classification technique and the pool-based uncertainty sampling algorithm for active query selection. The pseudocode is shown in Algorithm 1.

Generally, our frame work operates as follows. First, it performs classification using SMO. Then, based on the acquired results, the active learning algorithm will be used for picking x^* most uncertainty objects. These objects are then given to user for labeling them. And, they are put back into the training set for subsequence classification tasks. The whole process is repeated until we run out of query budget N.

In order to identify the most uncertainty sampling or the most informative instance, we propose the Entropy computation (Eq. 1) was introduced by Shannon [24] to calculate the class probability. We realized that the most uncertainty sampling which its class has lower probability.

$$x_H^* = \underset{x}{argmax} - \sum_y P_\theta(y|x) log P_\theta(y|x) \qquad (1)$$

In the next section we will describe about the empirical results that we obtain from the experiments and the Sect. 5 will discuss and conclude about the result.

4   Empirical Results

In this paper, we use the Wine dataset [14] for demonstrating the performance of our algorithm.

Algorithm 1. Our classification algorithm based on SMO and Uncertainty sampling

---

Input:
    L = Set of initial labeled instances
    U = A pool of unlabeled instances
    N = Number of query
Output:
    Accuracy of classification
Begin
  for i = 1, 2, 3, ..., N **do**
      θ = SMO_Train(L)
      Select x* $\in$ U, the most uncertain instance according to model θ
      Query the oracle to obtain label y*
      Add (x*, y*) to L
      Remove x* from U
  end for
  θ = SMO_Train(L)
  Return Accuracy of model θ
End

4.1 Wine Data

Data of sensory quality of Portugal wine was contributed by Cortez et al. [3] from May 2004 to February 2007 and was tested at official certification entity. The red wine dataset was composed of 1559 samples and was published in the UCI archives [14].

Input variables (based on physiochemical tests): fixed acidity, volatile acidity, citric acid, residual sugar, chlorides, free sulfur dioxide, total sulfur dioxide, density, pH, sulphates, alcohol. And the output variable (based on sensory data) is the quality (scored between 0 and 10 with 10 is the best).

Table 1 presents statistical on physicochemical components in the data, each sample is evaluated by at least three experts. In addition, sensory quality is classified from score 0 (very bad) to 10 (excellent), the final score is calculated by the mean of evaluation. Figure 2 plots histograms of the target variables, denoting a typical normal shape distribution.

4.2 Results

In this study, we carried out experiments on active learning based on SMO SVM (RBF kernel) classification algorithm with Pool based sampling model and Uncertainty sampling strategy (Algorithm 1) using Entropy. We select the number of init instances is 10, the number of query is 100, and the test mode is 5-fold cross validation.

Figure 3 shows the accuracy of both sampling techniques. In our case, the uncertainty sampling model outperforms the random sampling one. The reason is that the uncertainty model chooses objects with most uncertain labels for

Table 1. The physicochemical data statistics per wine type

Attribute (units)	Min	Max	Mean	StdDev
Fixed acidity (g(tartaric acid)/dm3)	4.6	15.9	8.31963	1.74109
Volatile acidity (g(acetic acid)/dm3)	0.12	1.58	0.52782	0.17905
Citric acid (g/dm3)	0.0	1.0	0.27097	0.19480
Residual sugar (g/dm3)	0.9	15.5	2.53880	1.40992
Chlorides (g(sodium chloride)/dm3)	0.012	0.611	0.08746	0.04706
Free sulfur dioxide (mg/dm3)	1.0	72.0	15.87492	10.46015
Total sulfur dioxide (mg/dm3)	6.0	289.0	46.46779	32.89532
Density (g/cm3)	0.99	1.00369	0.99674	0.00188
pH	2.74	4.01	3.31111	0.15438
Sulphates (g(potassium sulphate)/dm3)	0.33	2.0	0.65814	0.16950
Alcohol (vol.%)	8.4	14.9	10.42298	1.06566

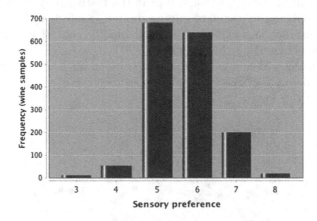

Fig. 2. The histograms for the red sensory preferences

asking the users. Thus, as a result, it requires less labeled objects to acquire the same accuracy as the random sampling technique.

In the wine industry, the sensory analysis has to be performed by human tasters. However, the evaluations are based in the experience and knowledge of the experts, which are prone to subjective factors. Thus, Cortez et al. [3] presented the definition of the tolerance concept, which uses a predefined tolerance threshold T for modifying the results. The former tolerance rounds the response into the nearest class, while the latter accepts a response that is correct within one of the two closest classes (e.g. a 3.1 value can be interpreted as grade 3 or 4 but not 2 or 5). The most practical tolerance values are $T = 0.5$ and $T = 1.0$. This approach is based on objective tests and thus it can be integrated into a

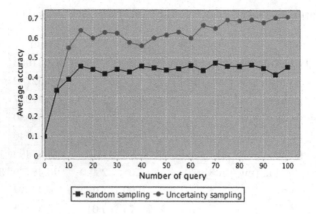

Fig. 3. Classification accuracy

decision support system, aiding the speed and quality of the oenologist performance. For instance, the expert could repeat the tasting only if his grade is far
from the one predicted by the data mining model. In effect, within this domain
the $T = 1.0$ distance is accepted as a good quality control process and high
accuracies were achieved for this tolerance.

Our experimental result with $T = 1.0$ present an accuracy classification
higher than 90% (c.f. Fig. 4).

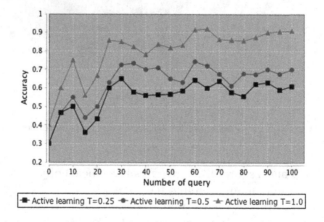

Fig. 4. Classification accuracy

5   Conclusions

Nowadays, beside the high nutritional requirements and the safety of food products, sensory evaluations such as appetite, aroma and good structure play an
important role for consumers. For many years, sensory quality assessment of

foods was performed by the company experts, who have accumulated experience. It could describe company products and sets standards of quality by which raw materials should be purchased and the product is manufactured and marketed. Recently, with the rapid development of computer science and its application in all areas of life, the need of a decision support system for food quality assessment will help to minimize costs and human biases during the sensory evaluation process. Providing a standardized environment for evaluating food products is becoming apparent and more urgent.

However, the cost is too expensive to get label for create the data of the decision support system for food quality assessment because of sensory panel. The results of this study offer a better understanding of active learning in sensory assessment of food quality. On the wine data, we demonstrate that active learning based on SMO can obtain better performance than random sampling (about 71% vs. 47%, respectively). In the case of using the concept of tolerance, within this domain the $T = 1.0$ distance is accepted as a good quality control process, our algorithm acquires high accuracy (bigger than 90%). Moreover, we know that the problem of evaluating the quality of wine is that unlabeled data is less time consuming and less expensive because of physicochemical laboratory tests routinely, while labeling costs are very expensive because of sensory panel of human experts. In our research, active learning technology is first used to evaluate the sensory quality of wine to minimize the cost while achieving the same accuracy. This approach is therefore a promising way to support the traditional sensory assessment.

We also found that the results of this research depend on ability of experts. Therefore, our future research aims at a more generic technique that can incorporate different opinions of experts into the evaluation process.

References

1. Boccorh, R.K., Paterson, A.: An artificial neural network model for predicting flavour intensity in blackcurrant concentrates. Food Qual. Prefer. **13**(2), 117–128 (2002)
2. Capozzoli, A., Lauro, F., Khan, I.: Fault detection analysis using data mining techniques for a cluster of smart office buildings. Expert Syst. Appl. **42**(9), 4324–4338 (2015)
3. Cortez, P., Cerdeira, A., Almeida, F., Matos, T., Reis, J.: Modeling wine preferences by data mining from physicochemical properties. Decis. Support Syst. **47**(4), 547–553 (2009)
4. Cortez, P., Portelinha, M., Rodrigues, S., Cadavez, V., Teixeira, A.: Lamb meat quality assessment by support vector machines. Neural Process. Lett. **24**(1), 41–51 (2006)
5. Debjani, C., Das, S., Das, H.: Aggregation of sensory data using fuzzy logic for sensory quality evaluation of food. J. Food Sci. Technol. **50**(6), 1088–1096 (2013)
6. Dębska, B., Guzowska-Świder, B.: Application of artificial neural network in food classification. Anal. Chim. Acta **705**(1), 283–291 (2011)

7. Dong, J.J., Li, Q.L., Yin, H., Zhong, C., Hao, J.G., Yang, P.F., Tian, Y.H., Jia, S.R.: Predictive analysis of beer quality by correlating sensory evaluation with higher alcohol and ester production using multivariate statistics methods. Food Chem. **161**, 376–382 (2014)
8. Ghasemi-Varnamkhasti, M., Mohtasebi, S.S., Rodriguez-Mendez, M.L., Lozano, J., Razavi, S.H., Ahmadi, H., Apetrei, C.: Classification of non-alcoholic beer based on aftertaste sensory evaluation by chemometric tools. Expert Syst. Appl. **39**(4), 4315–4327 (2012)
9. Granitto, P.M., Gasperi, F., Biasioli, F., Trainotti, E., Furlanello, C.: Modern data mining tools in descriptive sensory analysis: A case study with a random forest approach. Food Qual. Prefer. **18**(4), 681–689 (2007)
10. He, H., Wang, D., Xu, Y., Tan, J.: Data synthesis in the community land model for ecosystem simulation. J. Comput. Sci. **13**, 83–95 (2016)
11. Lawless, H.T., Heymann, H.: Sensory Evaluation of Food: Principles and Practices. Springer, New York (2010). https://doi.org/10.1007/978-1-4419-6488-5
12. Lee, S.J., Kwon, Y.A.: Study on fuzzy reasoning application for sensory evaluation of sausages. Food Control **18**(7), 811–816 (2007)
13. Li, J., Sander, J., Campello, R., Zimek, A.: Active learning strategies for semi-supervised DBSCAN. In: Sokolova, M., van Beek, P. (eds.) AI 2014. LNCS (LNAI), vol. 8436, pp. 179–190. Springer, Cham (2014). https://doi.org/10.1007/978-3-319-06483-3_16
14. Lichman, M.: UCI machine learning repository. University of California, School of Information and Computer Science, Irvine, CA (2013). http://archive.ics.uci.edu/ml
15. Mai, S.T., Assent, I., Storgaard, M.: AnyDBC: An efficient anytime density-based clustering algorithm for very large complex datasets. In: Proceedings of the 22nd ACM SIGKDD International Conference on Knowledge Discovery and Data Mining, pp. 1025–1034. ACM (2016)
16. Mai, S.T., Dieu, M.S., Assent, I., Jacobsen, J., Kristensen, J., Birk, M.: Scalable and interactive graph clustering algorithm on multicore CPUs. In: 2017 IEEE 33rd International Conference on Data Engineering (ICDE), pp. 349–360. IEEE (2017)
17. Mai, S.T., He, X., Hubig, N., Plant, C., Bohm, C.: Active density-based clustering. In: 2013 IEEE 13th International Conference on Data Mining (ICDM), pp. 508–517. IEEE (2013)
18. Martínez, L.: Sensory evaluation based on linguistic decision analysis. Int. J. Approximate Reasoning **44**(2), 148–164 (2007)
19. Martínez, L., Pérez, L.G., Liu, J., Espinilla, M.: A fuzzy model for olive oil sensory evaluation. In: Melin, P., Castillo, O., Aguilar, L.T., Kacprzyk, J., Pedrycz, W. (eds.) IFSA 2007. LNCS (LNAI), vol. 4529, pp. 615–624. Springer, Heidelberg (2007). https://doi.org/10.1007/978-3-540-72950-1_61
20. Mukhopadhyay, S., Majumdar, G., Goswami, T., Mishra, H.: Fuzzy logic (similarity analysis) approach for sensory evaluation of chhana podo. LWT-Food Sci. Technol. **53**(1), 204–210 (2013)
21. Platt, J.: Sequential minimal optimization: A fast algorithm for training support vector machines (1998)
22. Ropodi, A., Panagou, E., Nychas, G.J.: Data mining derived from food analyses using non-invasive/non-destructive analytical techniques; determination of food authenticity, quality & safety in tandem with computer science disciplines. Trends Food Sci. Technol. **50**, 11–25 (2016)
23. Settles, B.: Active learning. Synth. Lect. Artif. Intell. Mach. Learn. **6**(1), 1–114 (2012)

24. Shannon, C.E.: A mathematical theory of communication. ACM SIGMOBILE Mob. Comput. Commun. Rev. **5**(1), 3–55 (2001)
25. Singh, K., Mishra, A., Mishra, H.: Fuzzy analysis of sensory attributes of bread prepared from millet-based composite flours. LWT-Food Sci. Technol. **48**(2), 276–282 (2012)
26. Sinija, V., Mishra, H.: Fuzzy analysis of sensory data for quality evaluation and ranking of instant green tea powder and granules. Food Bioprocess Technol. **4**(3), 408–416 (2011)
27. Stone, H., Sidel, J.: Sensory Evaluation Practices. Elsevier Academic Press, California (2004)
28. Tuia, D., Muñoz-Marí, J., Camps-Valls, G.: Remote sensing image segmentation by active queries. Pattern Recogn. **45**(6), 2180–2192 (2012)
29. Varzakas, T., Tzia, C.: Handbook of Food Processing: Food Safety, Quality, and Manufacturing Processes, vol. 35. CRC Press, Boca Raton (2015)
30. Zhao, W., He, Q., Ma, H., Shi, Z.: Effective semi-supervised document clustering via active learning with instance-level constraints. Knowl. Inf. Syst. **30**(3), 569–587 (2012)
31. Zolfaghari, Z.S., Mohebbi, M., Najariyan, M.: Application of fuzzy linear regression method for sensory evaluation of fried donut. Appl. Soft Comput. **22**, 417–423 (2014)

Big Data Driven Architecture for Medical Knowledge Management Systems in Intracranial Hemorrhage Diagnosis

Thi-Hoang-Yen Le[1]([✉]) [iD], Thuong-Cang Phan[2] [iD], and Anh-Cang Phan[1] [iD]

[1] Vinh Long University of Technology Education, Vinh Long, Vietnam
yenlth@vlute.edu.vn
[2] Can Tho University, Can Tho, Vietnam

Abstract. Stroke is the most common and dangerous cerebrovascular disease. According to the statistics from World Health Organization (WHO), only following heart attack, stroke is one of the two leading causes of human deaths. In addition, in Vietnam, a shortage of specialized equipment and qualified professionals is becoming a significant problem for not only accurate diagnosis but also timely and effective treatment of stroke, especially intracranial hemorrhage (ICH), an acute case of stroke. This research will analyze challenges and show solutions for constructing an effective knowledge system in ICH diagnosis and treatment that helps to shorten professional gap among hospitals and regions. We suggest a service-oriented architecture for the big data driven knowledge system based on medical imaging of ICH. The architecture ensures the development of knowledge obeying a systematic and complete process including the exploration and exploitation of knowledge from medical imaging. Besides, the architecture adapts to modern trends in knowledge service modeling.

Keywords: Medical Knowledge Management System · Big data · SOA · Stroke
Intracranial Hemorrhage (ICH) · Medical imaging

1   Introduction

Stroke is the second leading cause of global human deaths, after ischemic heart disease. According to statistic of World Health Organization (WHO), not only their position had been kept stably from 2000 to 2015 but also the number of deaths with stroke had increased notably (from 5.41 to 6.24 million deaths) [1].

WHO define brain attack (stroke) as "rapidly developing clinical signs of focal (or global) disturbance of cerebral function, with symptoms lasting 24 h or longer or leading to death, with no apparent cause other than of vascular origin" [2]. Stoke is the most common and dangerous cerebrovascular disease with two primary types: cerebral infarction (blockage or narrowing in the arteries) occupying 85% and the remaining percent accounted for intracranial hemorrhage (ruptured cerebrovascular). The latter is considered as the acute form of stroke because of its sudden occurrence and the dramatic progression of symptoms. Based on the position sand the shape of hemorrhagic zone,

© Springer International Publishing AG, part of Springer Nature 2018
V.-N. Huynh et al. (Eds.): IUKM 2018, LNAI 10758, pp. 214–225, 2018.
https://doi.org/10.1007/978-3-319-75429-1_18

specialists divide intracranial hemorrhage (ICH) into four types: epidural hematoma, subdural hematoma, subarachnoid hemorrhage and intracerebral hemorrhage [3, 4].

The diagnosis of stroke, especially ICH (acute type), has relied on both neurological imaging and popular medical image acquisition facilities such as MRI, CT scanner and DSA [4]. As a result, a major essential in the treatment of ICH is that the doctors must be proficient in almost of ordinary medical imaging techniques. Nowadays, although the medical imaging field has recorded many impressive improvements, the stroke diagnosis is still manually. That shows the high subjectivity because it depends on not only capability of specialists in observing and analyzing images but also their expertise. A more important problem of health sector in Vietnam is the uneven distribution and the poverty force of the stroke experts at hospitals, especially local hospitals which are the first places receiving stroke patients. The actual situation illustrates the urgency of developing a knowledge system to potently support for diagnosis, monitor and treatment of stroke.

According to American Stroke Association, for every 45 s passed, the world has recorded one more patient with ICH [5]. The annual patient flow at hospitals in Vietnam is significant. A part of them are asked for CT/MRI scanning of the head to support in identification of ICH. Furthermore, the stored amount of image data at one scanning is from several hundred megabytes to few gigabytes in size. As a result, the dramatic increase in the rate and amount of data. Another important aspect is the assurance of continuity, accuracy, timeliness and effectiveness of real-time queries on database storing images, information and knowledge. These lead to the necessary of conducting research on methods of storing, processing, querying and especially developing new knowledge on the big neurological image data in a system supporting for ICH diagnosis.

Consequently, the present work aims to (*i*) analyze challenges of development and deployment knowledge-based supporting system in ICH diagnosis and treatment, (*ii*) propose a solution for big data driven and service-oriented knowledge architecture on medical neurological images. The rest of this paper is structured as follows. Section 2 describes challenges and motivations of medical image knowledge-based ICH diagnosis systems. In the next section, we propose a respective solution illustrated with a new approach for the systems. Section 4 outlines some applications of the knowledge systems in taking care community health.

2 Motivations of Research

2.1 Medical Knowledge Management Systems (KMS) in Big Data Era

From knowledge objects (KO) classified with their development level (data, information, knowledge and wisdom), knowledge is generated systematically [6]. The definition of knowledge management is organizational activities related to knowledge artifacts in which the training has occurred and intellectual capital has been accumulated and developed. A knowledge management system is a form of a knowledge system to process organizational knowledge. It contains four activities such as knowledge capture, knowledge organization, knowledge transfer, and knowledge application. Therefore, its architecture should completely support the knowledge activities. This also means that the architecture assures the comprehensive knowledge development consisting of *knowledge exploration* and *knowledge exploitation*.

Figure 1 describes the knowledge development levels. The knowledge management process comprises knowledge exploration and knowledge exploitation. Capturing and organizing knowledge are the processes in the exploration, while the exploitation refers to transfer and apply organizational knowledge [7]. A knowledge object is considered as a collect of knowledge components stratified from lowest to highest: cognitive (know-what), condition (know-when), situational (know-where), applied (know-how), and rule-of-thumb (know-why) [7]. Knowledge components of KO can be used and shared by using different processes of the knowledge conversion.

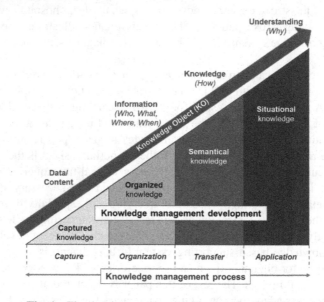

Fig. 1. The development levels of knowledge in KMSs.

According to the behavioral view, the life cycle of KO consists of captured knowledge, organized knowledge, semantical knowledge, situational knowledge. Firstly, raw and simple data exists in any form. The state of KO is captured knowledge if its data is collected and stored in knowledge repository. Next, relation connection makes meaningful data called information. Organizing data of KO according to its knowledge components corresponding to structure of knowledge puts it into the organized state. At this state, KO can help to answer the simple questions such as "What", "Who", "Where", "When". The third state, knowledge is the proper manipulation of information to create organized activities. A KO is in the state of semantical knowledge if its knowledge components corresponding to the structure of knowledge are linked to the knowledge components corresponding to the transition of knowledge [8]. KO at this state can provide or guide answer for "How" question. Finally, understanding is process by which an organization can take knowledge and synthesize new knowledge form original raw data organized to make business decisions.

Nowadays, KMSs play important role in business decision and processes of the modern and network organizations whose model is like the current system of hospitals. KMS renders more competition in grasping business opportunities between companies. However, the dramatic challenge for current KMSs is the explosion of information. A variety and unprecedented amount of data has been the result of different business activities, IT-based services, kinds of large-scale images and videos, called "big data". Computing on huge data creates a noticeable barrier for organizations to exploit efficiently the value of big data. In our research context, data consists of traditional data and big data transmitted massively from hospitals (including stored data and real-time data). They are large-scale medical images, documents related to patients. The original knowledge in this research is data and information (related to ICH), collected from CT/MRI scanner, available systems, Internet, and communication networks. It is necessary for the development of ICH knowledge in KMS to provide new knowledge in different levels suitable to particular purposes of use.

In respect of knowledge systems, researchers are being exceedingly interested in the development of KMS in healthcare industry. Nonetheless, their studies have showed the lack of both actual implement and experience for deploying knowledge management based on IT [9]. Moreover, most of proposed medical KMSs only supports for clinical diagnosis and ignore the specialized diagnosis. In fact, the support of image-based medical diagnosis is still contingent on the information analyzed by specialists and tagged in images [10, 11]. This is one of causes leading our research to propose and deploy a new architecture of KMS supporting the medical automatic diagnostic based on big data (CT/MRI image data and related) in general, diagnosis of ICH in particular.

2.2   Service-Oriented Architecture for KMSs

Although recent big data studies have shown rapid improvements, they have lacked of frameworks and architecture enable to organizations capturing value of big data systematically. As a result, one of important challenges for current KMSs is how can processing big dataset required the continuously updated contents. Therefore, the next KMS generation effectively computing on big data is need of organization, notably data-driven modern organizations.

An advanced tendency for KMS is service orientation. Service Oriented Architecture (SOA) is collection of services linked flexibly (that means an application ignoring the internal technique details when it "talks" to another application). The communication (to invoke service functions) is defined clearly and independently from system platform, and can be reused. SOA, the higher of application development, focuses on the business process and uses the communication standards to cover the inside technical complication. The crux of SOA is its design separating the service's interface from its implementation. The service functions is announced as a standard interface. The installing detail of functions is covered and not necessary for the interest of users. Users need only interact to methods displayed on interface to use the functions of service. Service-oriented KMS with its services would enable organizations to exploit knowledge effectively [12]. With advantages, SOA is a potential approach for architecture of KMSs, especially medical KMSs support to diagnose automatically.

2.3   Medical Data Lakes and a System of ICH Identification and Classification

Today, a plenty of CT/MRI image data, medical records and medical attachments (called meta-data) are being stored in hospitals serving treatment of diseases. They combined with other resources on the Internet provide valuable material for clinical diagnosis and predictive results. Unfortunately, these "data lakes" are usually in closed state, not shared or forbidden between hospitals or doctors when they would like to refer to comparable cases. In several cases, even though the doctors have treated for their patients, they cannot remember patient's name or the other information related to the disease. The major reason of this situation is the manually stored data. This storage also leads to the difficulties, delay and ineffectuality of treatment.

DICOM (Digital Imaging and Communications in Medicine), the international standard for medical images and related information, meets demand to store, transmit, retrieve and process medical imaging information. According to the standard, a DICOM file format includes not only image data but also the information of patient, image-acquisition device, etc. It is the difference from other formats. The DICOM implementation is compatible with almost every radiology, cardiology imaging, and radiotherapy device (e.g. X-ray, CT and MRI) [13]. Along with the development of medical imaging technique, the brain image data significantly increase not only the quantity but also the complexity. The storage, management and retrieval of big data quickly, accurately and efficiently is important.

One of the challenges posed for the development of medical image based knowledge system supporting for the ICH diagnosis is processing and extracting important features from images. More precisely, it is the challenge of constituting the data and knowledge repositories related to ICH in KMS to support for the automatic diagnosis. It helps to establish foundation for exploring and exploiting knowledge in the medical diagnosis. According to recent studies [14–16], a typical system for identifying and classifying ICH is shown in Fig. 2. It includes main steps of preprocessing input image, extracting features, determining major features, and identification. The extracted features used for identifying types of ICH is such as position, size, shape, color and texture of hemorrhagic zone.

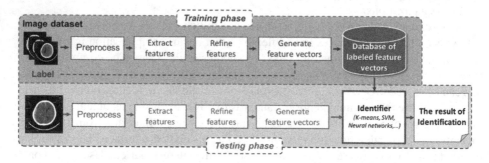

Fig. 2.   A typical system supporting for ICH diagnosis based on CT/MRI image.

In this study, we focus on the identification and classification of main ICH types including epidural hematoma, subdural hematoma, subarachnoid hemorrhage and intracerebral hemorrhage, as in Fig. 3.

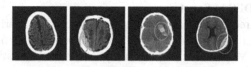

Fig. 3. Normal brain and three of the four types of ICH including subdural hematoma, intracerebral hemorrhage and epidural hematoma.

2.4   Big Data Technologies

MapReduce

Google introduced MapReduce for the first time in 2004. It is known as a programming model and associated implementation help to decrease the time of processing and generating large datasets with their capacity computed in terabytes or petabytes [17]. In MapReduce, users define *map* and *reduce* functions. The parallelization and computation of programs written in this functional style is automatic on large-scale cluster of independent machines (workers or nodes) [17]. Besides, when executing with the MapReduce model, a program code (resources of job) is copied to storage of nodes instead of replicating data (in big size) to nodes. This is one of MapReduce advantages because transmission of the program code is more economical and efficient than the latter. A MapReduce framework consists of one master and multi-workers. Master is a special node to coordinate operations among workers. The first step of MapReduce job is a process of splitting input data into small chunks and then integrating them to independent nodes (mappers executing map functions). Therefore, the number and the size of fragments depend on the number of mappers [18]. Then, each operation of map and reduce is performed in parallel and separation on key/value pairs. As a result, the program is divided into distinct phases called *map phase* and *reduce phase*.

HDFS (Hadoop Distributed File System)

Apache Hadoop is an open source framework used for distributed storage and processing of the massive amounts of data using the MapReduce model. It consists of computer cluster built from commodity hardware. When stored amount of data overcomes storage capacity of a single physical machine, it becomes necessary to partition data across a network of separate machines. In other words, it requires a distributed file system. Hadoop offers a distributed file system, called HDFS. HDFS is a file system designed for distributed storage of massive files with the highly scalable, flexible, reliable and fault-tolerant capability [19]. HDFS is organized with the architecture of namenode (master) and datanodes (slaves). Master manages file system namespace and coordinates all storage operations on Hadoop, including read and write in HDFS. It maintains a file system tree and metadata for all files and folders in the tree. Namenode identifies datanodes on which all blocks of a file are distributed. Meanwhile, datanodes store actual data. Based on assignment of namenode, datanodes are also responsible for creating, deleting and replicating data blocks.

MapReduce and HDFS in Spark

Spark is an open source framework for computing on cluster. Spark started in 2009 as a research project in the UC Berkeley RAD Lab. Spark runs on memory that allows processing data with MapReduce model more quickly than alternative approaches like Hadoop's MapReduce. Spark links to HDFS of Hadoop for its storage option. Notably, Spark manages memory by using Resilient Distributed Datasets (RDDs), a concept at the heart of Spark. The fault-tolerance and efficiency of RDD is demonstrated when its design supports in-memory data storage distributed across a cluster. The former is achieved, in part, by tracking the lineage of transformations applied to on fault-tolerant input dataset. The latter is reached by parallel process on multi-nodes in a cluster, and minimization of data replication between those nodes [20].

The mentioned viewpoints pose challenges to develop the KMS in medical diagnosis as follows: (1) *Building and querying big medical data-lake*, (2) *Transforming image data into knowledge systematically*, (3) *Identifying, classifying, and supporting to make decision in medical diagnosis*, (4) *Building an architecture for the KMS in big data era*.

3   Architecture for Medical KMSs in ICH Diagnosis

A medical KMS architecture is designed and constructed according to design science method [21]. It is suggested for KMSs to support automatic ICH identification and classification based on CT/MRI images. The architecture contains components as follows: a set of constructs, which is different conceptions related to KO of ICH; a model depicts the relation of knowledge conceptions; a method is the set of activities supporting knowledge management in KMS; finally, an instantiation is an illustration for system operation.

Data used for our medical KMS is both traditional data and big data, including CT/MRI images and related attachments. Knowledge is structured from KOs classified by the development levels. They can be data, information, knowledge and wisdom. The implemented system should include capabilities to implement steps classifying and identifying ICH automatically such as preprocessing input images, extracting features, refining main features, and identifying hemorrhagic zone. It is also necessary for supporting better in retrieving information from image content, ranking the results, classifying ICH types and making decision in ICH diagnosis. Moreover, the system can provide knowledge at different developing levels as services.

3.1   Overall Architecture

The basic idea of the architecture is to divide the design into service layers as in Fig. 4:

- *Data source* consists of batch data and real-time data related to ICH, especially CT/MRI images.

- **Data-as-a-Service (DaaS)**, the lowest main layer, its components perform the capture and storage of skull CT/MRI images, related attachment as batch data and real-time streaming data. Data services is deployed at this layer to provide raw data.
- **Information-as-a-Service (IaaS)** in which the batch and real-time data will be put in processing components respectively to be transformed into ICH information. Thence, the information is stored and supplied to user through information services.
- **Knowledge-as-a-Service (KaaS)** where ICH knowledge is created via machine learning's activities (ICH identification and classification) and a recommendation system on the available information of the IaaS layer. This is a motivation for the present research to put forward solutions for big data usage in the diagnosis and treatment of ICH. This layer includes distributed knowledge base, structure of knowledge, comments. Its functions are packed and launched as knowledge services.
- **Business Process-as-a-Service (BPaaS)** is the highest layer in the architecture. It enables users to visualize and analyze query results. BPaaS provide services for applications by integrating them with business process of ICH diagnosis.

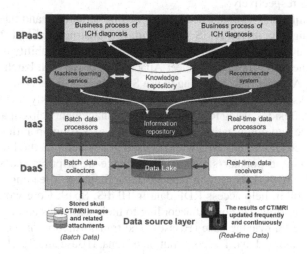

Fig. 4. The suggested architecture

3.2 Solutions for the Challenges

As mentioned above, there are the four challenges of building the KMS in ICH diagnosis. This section describes our respective solutions for layers of the architecture.

Solution 1: Building and querying the big medical data-lake

The recommended solution is related to the data sources and the data acquisition and storage at DaaS. At data source layer, the captured batch data is files and skull CT/MRI images, which are stored in hospitals and attached meta-data (information related to patient and medical reports). The real-time data is also brain CT/MRI images but it is dispatched continuously during the examination and treatment.

At DaaS, we choose HDFS to build our data-lake storing the variety of data kinds. HDFS is used because it is scalable, reliable, portable and distributed file system, especially its high fault-tolerance suitable to the data sources in our case. However, putting a plenty of data sources (real-time and batch data) into HDFS is impossible for traditional techniques. Therefore, we consider for using more feasible capture tools. With the structured data such as relational database, we use Apache Sqoop because of its efficiency in exporting and importing bulk data between HDFS and structured database [22]. Besides, ETL (Extract, Load and Transform) tools as Hadoop FS shell commands or written Java client API is used for loading bulk data from text file or NoSQL to HDFS. On the other hand, we also think about Apache Kafka, the tendency of stream processing used widely in big companies, for real-time data acquisition with its features: distributed capability, reliable, high-throughput and low latency [23].

Solution 2: Transforming image data into knowledge systematically

Using Apache HBase and Spark to implement the IaaS and KaaS layers can solve this problem. Firstly, HBase is a column-oriented NoSQL database built on the top of HDFS. It supports random, real-time access (read/write) to HDFS and the feature loading bulk [23]. As a result, we use HBase to establish information and knowledge repositories at IaaS and KaaS respectively.

Secondly, unlike conventional KMSs designed for structured and internal data, the architecture of big data driven KMS works with raw unstructured and structured data as well as internal and external data sources. Besides, it requires ability to handle batch and near real-time data. Hence, we select Apache Spark for both batch and real-time data processing. Another reason leading to use Spark is that while Hadoop operates on disk, Spark works with data in memory, supporting more quickly and better special computing tasks. Besides, using RDD for memory management helps Spark warrant for the fault tolerance and optimization of importing and exporting operations [20, 24].

RDF (Resource Describer Framework) is used as a general method for information model because it is a popular standard for resource description and semantic data published on the Web. Libraries in Apache Jena Elephas, a leading semantic web programmers' toolkit, help access RDF data in HDFS. Spark Core works with batch data from HDFS to organize content according to their semantics and to create as well as to maintain the knowledge base (HBase). The transformation of unstructured data into RDF is executed by Jena. The contribution of Jena, HBase and Spark constructs the scalable, efficient and distributed repository of RDF. With real-time data, Spark Streaming performs continuously transforming input data received from Kafka into real-time knowledge.

Solution 3: Identifying, classifying, and supporting to make decision in medical diagnosis

The CT/MRI image-based classification and identification of ICH are implemented according to the overall model (mentioned in Sect. 2.3). To identify types of ICH, we can use the extracted features such as position, size, shape, color, Hounsfield (HU) value (with DICOM images), and texture of hemorrhagic zone. However, we improve the training and testing phases (K-means, SVM, Neural network, etc.) in the parallel MapReduce environment with Spark mentioned in Sect. 2.4. For an instance, Fig. 5 shows the training phase in parallel. In the training process, the skull CT/MRI images stored on HDFS data-lake at DaaS is partitioned across mappers to preprocessing, extract features and label. After shuffling and sorting, the output of the map phase is the list of labeled feature vectors corresponding to the set of input images. The next step is the dimension decrease of feature vectors with Principal Component Analysis (PCA) executed on reducers. The output of reduce functions, the list of labels and feature vectors, is stored in information repository of IaaS to serve the matching and identifying ICH types by machine learning algorithms and the recommendation system in KaaS. All of operations of the training phase is executed in Spark Core (batch data). Likewise, matching and identifying process can be also proceeded with MapReduce in Spark by Spark streaming (real-time or query data).

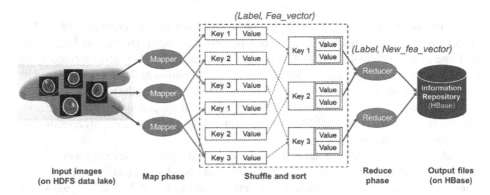

Fig. 5. The training phase in parallel with the MapReduce model in Spark.

The business process implemented at BPaaS help users apply and integrate their ICH diagnosis knowledge. It is important to model business process and then use the work-flow application to support for the automatic diagnosis of ICH as well as to facilitate organizational learning and intelligent capital development.

Solution 4: Building the architecture for the KMS in big data era

The required data by specialists and doctors in their researches, disease examine and treatment is different. Therefore, a requirement for the big data driven KMS in ICH diagnosis is that it can provide the "ICH knowledge" at different levels based on partic-ular using purpose such as raw ICH images, important features on a skull CT/MRI images, classified results of queried images, and supporting decision to diagnose ICH. Therefore, we propose the service-oriented architecture for the ICH diagnosis KMS. At each layer, users can query analyzed and stored results (data, information, knowledge) through services.

Firstly, we install WebHDFS for Hadoop to implement data services at DaaS. It enables users to access data in HDFS using an industry-standard mechanism. Moreover, the integration of Hive-server2 package provides query like SQL and return the results to clients.

At the next layer, IaaS, HBase is used for building repositories. Besides, HBase not only provides a thrift and RESTful gateway but also web service gateways for integrating and accessing HBase. Therefore, we can benefit from this to implement the information services.

Finally, at the KaaS layer, with the technique view, a system built according to SOA mechanism is structured from services. These services are defined in description languages with interfaces to support business process and implemented in different programming languages. Consequently, we select Web service, the most popular and well-known technique for the implement of SOA architecture, to install knowledge services.

4 Conclusion

In the era of big data, many challenges are posed for the development of KMSs supporting the ICH diagnosis and treatment. The first challenge is building and accessing the large-scale medical data lake (ICH images and related attachments) captured from different sources. Transforming original data into information and knowledge stored in respective repositories systematically is necessary to consider as the second challenge. The next problem is the classification, determination and support in medical diagnosis. Finally, it is important to establish the big data driven medical KMS architecture in automatic ICH diagnosis.

The proposed architecture assures a systematic and complete development of ICH knowledge (including exploration and exploitation of knowledge). It is a medical image-based KMS architecture supporting effectively and on time for disease diagnosis and treatment. As a result, it helps to rise to a better quality of ICH diagnosis in particular and in medical diagnosis in general.

The scalable capability of the architecture helps to develop a more full knowledge system serving for medical diagnosis and education. According to the architecture, the orientation of the KMS system is open and enable to share with contribution of specialists, doctors, hospitals and research institutes in order to better support community health services in Vietnam.

References

1. WHO: "The top 10 causes of death". http://www.who.int/mediacentre/factsheets/fs310/en/. Accessed 2 Mar 2017
2. Coupland, A.P., Thapar, A., Qureshi, M.I., Jenkins, H., Davies, A.H.: The definition of stroke. J. R. Soc. Med. **110**(1), 9–12 (2017)
3. Nguyen Van, D.: Cerebrovascular Accident (Stroke). Medical Publishing House, Hanoi (2006)
4. Heit, J.J., Iv, M., Wintermark, M.: Imaging of intracranial hemorrhage. J. Stroke **19**(1), 11–27 (2017)
5. Vu Hong, V.: Stroke/Cerebrovascular accident: the most dangerous cerebrovascular disease. http://noitonghop.org/dot-quy-tai-bien-mach-nao-benh-ly-mach-mau-nao-nguy-hiem-nhat/. Accessed 5 Jan 2017
6. Bierly III, P.E., Kessler, E.H., Christensen, E.W.: Organizational learning, knowledge and wisdom. J. Organ. Change Manage. **13**(6), 595–618 (2000)

7. Le Dinh, T., Rickenberg, T.A., Fill, H.-G., Breitner, M.H.: Enterprise content management systems as a knowledge infrastructure: the knowledge-based content management framework. Int. J. e-Collab. **11**(3), 49–70 (2015)
8. Le Dinh, T., Ho Van, T., Moreau, E.: A knowledge management framework for knowledge-intensive SMEs. In: Proceeding of 16th International Conference on Enterprise Information Systems, Lisbon, Portugal, pp. 435–440 (2014)
9. Chen, E.T.: An observation of healthcare knowledge management. Commun. IIMA **13**(3), 95–106 (2013). Article no. 7
10. Demigha, S., Balleyguier, C.: KMSS: a knowledge management system for senology. In: ECKM 2014, pp. 268–277 (2014)
11. Baigorri, A., Villadangos, J., Astrain, J., Córdoba, A.: A medical knowledge management system based on expert tagging (MKMST). In: Data Management and Security: Applications in Medicine, Sciences and Engineering, pp. 221–231 (2013)
12. Le Dinh, T., Phan Thuong, C., Bui, T.: Towards an architecture for big data-driven knowledge management systems. In: AMCIS 2016, San Diego (2016)
13. NEMA's DICOM Homepage. http://www.dicomstandard.org/. Accessed 2 Dec 2017
14. Al-Ayyoub, M., Alawad, D., Al-Darabsah, K., Aljarrah, I.: Automatic detection and classification of brain hemorrhages. WSEAS Trans. Comput. **12**(10), 395–405 (2013)
15. Hingene, M.C., Matkar, S.B., Mane, A.B., Shirsat, A.M.: Classification of MRI brain image using SVM classifier. LISTE Int. J. Sci. Technol. Eng. **1**(9), 24–28 (2015)
16. Fatima, S.M., Naza, S., Anjum, K.: Diagnosis and classification of brain hemorrhage using CAD system. Proc. NCRIET 2015 Indian J. Sci. Res. **12**(1), 121–125 (2015)
17. White, T.: Hadoop: The Definitive Guide, Storage and Analysis at Internet Scale, 4th edn, pp. 185–279. O'Reilly Media, Sebastopol (2015)
18. Dean, J., Ghemawat, S.: MapReduce: simplified data processing on large clusters. In: Proceedings of MSST 2004. USENIX Association, San Francisco (2004)
19. Shvachko, K., Kuang, H., Radia, S., Chansler, R.: The Hadoop distributed file system. In: Proceedings of MSST 2010, pp. 1–10. IEEE Computer Society, Washington, DC (2010)
20. Scott, J.A.: Getting Started with Apache Spark, from Inception to Production, pp. 15–20. MapR Technologies, Inc., San Jose (2015)
21. Hevner, A.R., March, S.T., Park, J., Ram, S.: Design science in information systems research. MIS Q. **28**(1), 75–105 (2004)
22. Pippal, S., Singh, S.P., Kushwaha, D.S.: Data transfer from MySQL to Hadoop: implementers' perspective. In: Proceedings of ICTCS 2014, India, pp. 79:1–79:5 (2014)
23. Assunção, M.D., Calheiros, R.N., Bianchi, S., Netto, M.A.S., Buyya, R.: Big Data computing and clouds: Trends and future directions. J. Parallel Distrib. Comput. **79–80**, 3–15 (2015). Special Issue on Scalable Systems for Big Data Management and Analytics
24. Shanahan, J.G., Dai, L.: Large scale distributed data science using Apache Spark. In: Proceedings of the 21th ACM SIGKDD, pp. 2323–2324. ACM, New York (2015)

Building Modelling Methodology Combined to Robust Identification for the Temperature Prediction of a Thermal Zone in a Multi-zone Building

Van-Binh Dinh[1(✉)] , Benoit Delinchant[1] , Frederic Wurtz[1], and Hoang-Anh Dang[2]

[1] Univ. Grenoble Alpes, CNRS, Grenoble INP, G2Elab, 38000 Grenoble, France
Van-Binh.Dinh@g2elab.grenoble-inp.fr
[2] Hanoi University of Industry, Hanoi, Vietnam

Abstract. Building thermal modelling plays an important role in managing the thermal comfort and the energy consumption of buildings. A major challenge for modellers is how to deal with uncertainty problems in order to have a robust model with an acceptable computational time for the improvement of predictive control. This paper presents a methodology which allows obtaining the good model of a controllable thermal zone able to adapt regularly to the measurements by a robust identification procedure. Its input data are achieved by the modelling simplification of adjacent zones under uncontrollable uncertainties. This method is applied for a multi-zone positive energy building in south of France to validate our approach.

Keywords: Energy in buildings · Robust identification · Predictive control
Thermal envelope modelling

1 Introduction

Faced with the climate change and the limit of fossil fuels, the modelling and controlling building energy consumption are essential because buildings use 40% of the world's primary energy [1]. Among the energy consumers in buildings, space heating and/or cooling systems are important consumers assuring the human thermal comfort.

The energy consumption of heating and air conditioning systems depends on not only the climate conditions but also thermal characteristics of building envelope and occupant behaviour. Indeed, a building is considered as a complex system made up of different entities like walls, energy systems, and occupants under the external conditions such as weather conditions, sun or wind. The internal volume of the building and the external environment are separated by the surface of walls, also called building envelope. It is through this envelope that thermal exchanges operate. The characteristics of the envelope as well as the thermal interaction around it directly influence the inside air temperature and also the consumption of heating and cooling systems. Therefore, having a reliable building thermal envelope model performing well the simulated results compared to the measured ones, becomes an important mission for the model-based predictive control and the building energy management. However, due to the complexity

© Springer International Publishing AG, part of Springer Nature 2018
V.-N. Huynh et al. (Eds.): IUKM 2018, LNAI 10758, pp. 226–237, 2018.
https://doi.org/10.1007/978-3-319-75429-1_19

of the built environment and uncertainty problems, it is not easy to achieve an accurate representation of real-world building operation.

The sources of uncertainty in building thermal modelling are generally associated to the inputs including dynamic and static types. It can be distinguished by 3 sources of uncertainty in building's operating phase:

- Uncertainties arising from static parameters as materials and dimensions of building envelope, window characteristics, infiltration rate etc. These uncertainties are due to the lack of knowledge about physical and geometrical parameters, defects during construction, or the aging of materials etc.
- Uncertainties related to dynamic parameters or solicitation variables such as weather data and occupant scenarios. Indeed, the measurement sensors of temperature, solar irradiation, and wind speed are not always available so that one does not sufficiently have input data for modelling. In addition, it is hard to estimate the number of people in a room, their behaviour and equipment use (turn on/off an air conditioning, open/ close windows ...). The prediction of these data for the next day is even far more difficult to obtain with good confidence.
- Uncertainties due to the modelling assumptions. It's about the simplification of building geometry or the limit of number of thermal zones for instance without taking into account the major effects that these approximations can induce.

As a consequence, a calibration can be done in order to match the simulation outputs with measured data, and then compensate all these uncertainties. However in prediction, it is important to keep the physical meaning of model parameters which is a trade-off to find with calibration.

Amara et al. [2] introduced building modelling approaches and devised them into three categories: white-box, black-box and grey-box. The white-box approach (sometimes called knowledge modelling) is based on the physical laws to describe the set of phenomena of building. This approach allows extrapolating the models to other situations which have been not present previously. Nonetheless, it requires a significant amount of building knowledge which is not compatible with real uncertainties. While the black-box models (or universal models), such as polynomial models (ARX, ARMAX ...), are built from the observations without using a priori physical knowledge. It is very hard to extrapolate these models for the accurate prediction while the quality of these models strongly depends on the training data used as input. Finally, the authors indicated that the grey-box approach (semi-physical approach), based on both physical knowledge and observations, is the best alternative for the predictive control. By getting all strengthen from white-box and black-box approaches, the final approach permits to obtain lightweight models which can be rapidly tuned in conserving the physical meaning. The equivalent electrical circuit [3] using the thermal-electrical analogy is well known as a grey-box approach for the building thermal model.

The main purpose of this paper is to propose a modelling methodology using the equivalent electrical circuit to control a thermal zone in a multi-zone building. Adjacent zones, that one has very little information about, and that cannot be controlled, will be simplified. The controllable thermal zone model is then regularly calibrated by an automatic and robust identification procedure using our meta-optimization approach based on a scatting analysis.

The remainder of the paper is organized as follows: Sect. 2 reviews the methods and works associated to uncertainties for the building energy models. Then, our methodology is detailed and applied to positive energy building in Sect. 3. Section 4 summarizes the main conclusions of the paper and proposes some future works.

2   A Brief Overview of Methods and Works Associated to Uncertainties for Building Energy Models

2.1   Sensitivity Analysis

The sensitivity analysis is the technique studying the effect of input parameters x on the uncertainty of the output $y = f(x) = f(x_1, x_2, \ldots, x_i, \ldots, x_n)$. The technique allows to: (i) identify the input parameters having a significant or negligible influence on the output; (ii) determine the interactions between parameters, permitting a concentration on a group of parameters rather than separated parameters.

Saltelli et al. [4] has given a broad overview on methods of sensitivity analysis applied in many fields. In our article, popular methods used in the building science are briefly presented.

Screening Methods
Screening methods are generally used for the computationally expensive models with a large number of inputs. It is based on the "One At a Time" OAT design which modifies one by one each input in fixing others and recording the results. In particular, the method of Morris [5] discretizes the space of input factors and constructs a series of trajectories in this space in moving inputs randomly one-parameter-at-a-time. This method permits to classify the inputs into 3 groups according to their effects: inputs having negligible effects, inputs having linear effects without interactions, and ones having nonlinear effects and/or interactions.

As a result, the screening methods allow obtaining qualitative information on the non-influential and influential inputs with a limited number of simulations but they do not quantify exactly the relative importance of a parameter compared to another, as well as the interaction between them. An application example of these methods was given by Heiselberg in [6] that measured the influence of design parameters on total energy demand of an office building in Denmark.

Local Sensitivity Methods
Based on the OAT approach too, local sensitivity methods evaluate the output variability in terms of the variation of one parameter at a time around its reference value, while all other parameters are held constant. In this case, a sensitivity index estimates the partial derivative of the output y with respect to the parameter x_i in the neighbourhood of its nominal value:

$$S_i = x_i \cdot \frac{\partial y}{\partial x_i} \tag{1}$$

Local sensitivity methods require a reasonable number of simulations to obtain the quantitative information for the analysis of the influence of inputs on the output of the models with an important number of parameters. Nevertheless, these methods only take into account the local effect of input parameters but neither their variation range nor the correlation between them. Westphal and Lamberts [7] used the local sensitivity analysis to specify thermal loads for the calibration of a public office building. Mejri [8] applied this method for the identification of dynamic models for the performance evaluation and energy diagnosis of existing buildings. It has also applied to illustrate the importance of occupancy and indoor temperature of the residential building in Czech Republic [9].

Global Sensitivity Methods

Global sensitivity methods help studying the influence of input parameters in their all range of variation. This approach takes into account the probability distribution of each input and in many cases all parameters can be simultaneously varied for observing the interactions between them. According to Lavin et al. [10], global sensitivity analysis methods can be categorized into three groups: Monte Carlo based methods, variance based methods, and graphical methods.

Monte Carlo based methods aim at making a large number of evaluations with randomly selected model inputs, and then using regression-based measures [11, 12] to analyse the contribution of input factors to the output uncertainty.

Variance based methods, such as Sobol [13, 14] and FAST [15, 16], study how the variance of output is due to the variation of each parameter and its related interactions via sensitivity indexes of different orders.

Graphical methods estimate the qualitative measures of sensitivity using the graphical assessment with charts, graphs, or surfaces of pairs of inputs-corresponding outputs. These methods can bring us the complementary visual information about the meaning of numerical sensitivity indices and the enhancement of results of other quantitative methods. Scatterplot [17] is one of the most common forms of graphical sensitivity methods.

2.2 Model Calibration

Model calibration aims at tuning the model parameters so that simulation outputs match closely with the measured data.

According to Clarke et al. [18], calibration methodologies can be grouped into 4 principal categories: calibration based on manual, iterative and pragmatic intervention; calibration based on a suite of informative graphical comparative displays; calibration based on special tests and analytical procedures; and calibration based on analytical and mathematical methods. Reddy [19] revised these methods and agreed that the first three categories of the group involves tuning and refining the initial simulation input parameters in a heuristic manner, which depends on the experience and expertise of the user. On the contrary, the final one based the analytical and mathematical formulation is associated to an optimization problem which identifies automatically multiple solutions within a parameter space to minimize an objective function. Reddy also indicated that sensitivity analysis should be used in the first stage of calibration process for reducing the number of parameters to be calibrated so that the numerical optimization is more efficient. Such a systematic and automated way is applied by O'Neill and Eisenhower [20] who used a sensitivity analysis for identifying the most important parameters for tuning an office building energy model. An overview of applications of the building model calibration can be found in [21, 22].

2.3 Our Proposals for the Improvement of Building Thermal Models Under Uncertainties

In order to contribute to the uncertainty management in modelling, we have proposed some approaches for improving the reliability of building energy and thermal models.

Grandjacques et al. [23] developed sensitivity analysis methods for dependent and dynamic inputs based on the method of Pick and Freeze for the estimation of Sobol's indexes. These methods were applied for an existing platform building in France to take into account both the dynamic aspect of the inputs (room temperatures, heating system, presence of occupancy, inertia of heat exchange) and the one of the output.

Dang et al. [24] presented a thermal envelope model based on the equivalent electrical circuit for the building optimal management. From 16 input parameters including thermal resistances and capacitances, the method of Morris was used to determine 8 non-influential inputs which were then fixed by their analytic values. The rest of input factors which have a significant influence on the output were calibrated with one data set and then validated with only another data set. It is notable that such a model is not still robust since the model may provide poor performances when applied to environmental conditions that significantly differ from those which calibration referred to. That is why the model recalibration, and even the regular and automatic recalibration, becomes more and more important in real time predictive control. However, it is essential to understand that more parameters of model there are, less robust the calibration process is, because of the convergence problems in particular.

With the aim of improving the robustness of calibration for predictive control, we have developed in Nguyen Hong et al. [25] a calibration procedure of building energy models using a called meta-optimisation approach with the help of scattering parameters analysis. In this study, the scattering analysis aims to reach a reduced number of parameters to be identified. The parameter estimation is then performed by an automatic and regular optimization process to obtain the more robust prediction relative to the environment change. Through different tests, the authors also demonstrated the performance of this approach compared to the classical methods in terms of computational cost, prediction errors. Nevertheless, it is still a big challenge to apply this approach for real complex buildings with an important number of zones, such as office, commercial or apartment buildings. Indeed, modelling the overall system is a really great challenge, and the convergence issues of the optimization process applied to multi-zone buildings is still present.

In the context of studying the thermal model for predictive control of a zone in a multi-zone building, the uncertainties significantly increase when ones do not have any knowledge and information about the adjacent zones that directly influence the temperature of considered zone. Therefore, modelling whole building with the same level of detail for each zone is called into question. Considering our past studies, the next part is addressing the issue of robust calibration for the predictive control of a single zone in a multi-zone building.

3   Building Modelling Methodology Based on the Robust Identification for the Temperature Prediction of One Thermal Zone in the Building

3.1   Methodology: Assumptions for Simplification

Our methodology aims to improve the modelling efficiency permitting a more robust calibration process for the temperature prediction of one thermal zone located in a multi-zone building as described in the Fig. 1.

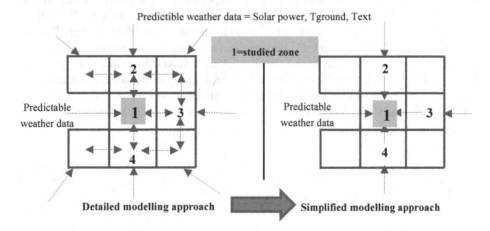

Fig. 1. Modelling assumptions for the temperature prediction of a studied zone "1"

On the left of this figure, the 2-way arrows represent the 2-way interactions between the building zones in prediction. In order to predict the temperature of the studied zone, called "zone 1", it is necessary to know weather conditions and the temperatures of adjacent zones (2, 3 and 4). However, the temperatures of adjacent zones, themselves, are also unknown and have to be predicted depending on the temperature of "zone 1", and other boundaries. Consequently, the crossed interaction between building zones requires the model of whole building for the prediction of only one zone.

As previously discussed, the grey-box model in the form of equivalent electrical circuit is a good choice for prediction. Nevertheless, using this semi-physical approach for modelling the whole building can bring a complex model with a large number of parameters which are difficult to calibrate regularly and robustly. Regarding the lack of information about adjacent zones (scenario of use, materials …), our methodology is as following:

- simplify the modelling for adjacent zones by the assumption that their temperatures in prediction depend only on the exterior environment (such as solar power, exterior temperature and ground temperature which are predictable by weather forecast services) and neither on the zone of interest (zone 1) nor on other boundaries. This can be illustrated on the right of the Fig. 1.
- use simple approximate models (simplified structure of an equivalent electrical circuit) to predict the temperatures of the adjacent zones and then take the results obtained as input data for the prediction model of zone 1.

3.2   Methodology: Electrical Equivalent Models for the Zone of Interest and Adjacent Zones

Generically, the equivalent electrical circuit of zone 1 can be seen as Fig. 2. This circuit considers that electrical components like voltage sources, current sources, resistors and capacitors are respectively corresponding to temperatures, heat gains, thermal resistances and capacitances. The 2R1C-structure used for each wall (contact between 2 zones) is considered as a good trade-off between the accuracy and the robust prediction. Particularly, R_{ext_1}, R_{ext_2} and C_{ext} represent respectively the external resistance, internal resistance and capacitance of wall of zone 1 linked to exterior. R_{adj2_1} and R_{adj2_2} and C_{adj2} represent the external resistance, internal resistance and capacitance of wall of zone 1 linked to adjacent zone 2. R_{vent} introduces the resistance linked to the ventilation. The analytic values of resistances and capacitances are calculated from the envelope characteristics.

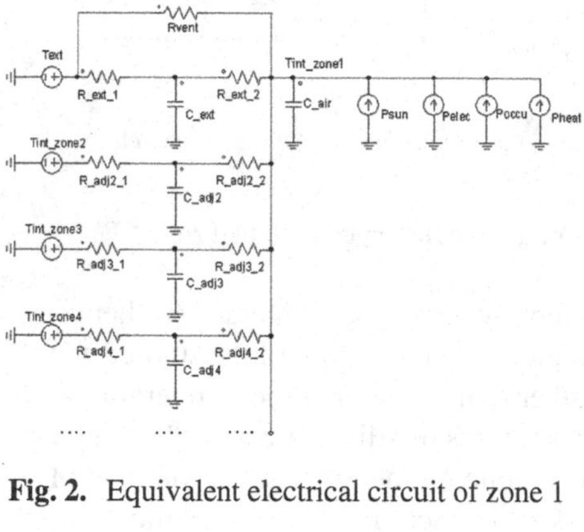

Fig. 2. Equivalent electrical circuit of zone 1 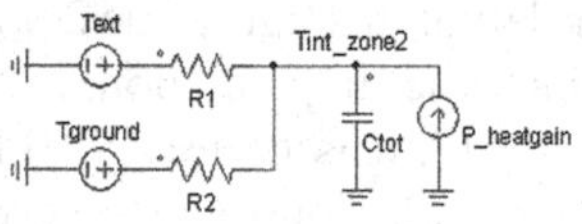   **Fig. 3.** Approximate model of adjacent zones (e.g. zone 2)

P_{sun}, P_{elec}, P_{occu} and P_{heat} are heat gains inside the zone 1, which are associated to sun, electrical appliances, occupancy and heating system. T_{ext} and T_{int_zone1} are external temperature and internal air temperature of zone 1. T_{int_zone2}, T_{int_zone3} and T_{int_zone4} are respectively the air temperatures of adjacent zones.

Meanwhile, with the proposed modelling assumptions, the temperature of adjacent zones, for example T_{int_zone2} of zone 2, is calculated by a much simpler circuit (Fig. 3). Here only the heat exchange between zone 2 and exterior (and maybe zone 2 and ground if this zone is close to the ground) is taken into account by one resistance and one capacitance while the thermal effect of other boundaries on zone 2 is neglected.

The model of an adjacent zone (e.g. zone 2) can be expressed by the equation:

$$T'_{int\_zone2} = \frac{1}{C_{tot}} \cdot \left(\left(\frac{1}{R_1} + \frac{1}{R_2} \right) \cdot T_{int\_zone2} + \frac{1}{R_1} \cdot T_{ext} + \frac{1}{R_2} \cdot T_{ground} + P_{\_heatgain} \right) \quad (2)$$

It remains some physical meaning in parameters R_1, R_2, C_{tot}, but less than those of the zone of interest. Thus, they can be obtained during calibration process from a simple temperature sensor or a simulation.

The simplification in modelling the adjacent zones is acceptable because the uncontrollable sources of uncertainty are very high so that a detailed modelling approach is not suited. Moreover, we are mainly interested in the behavior of zone 1 which will be frequently calibrated with measured data to fit with the environmental changes.

3.3 Methodology: Calibration

A periodic recalibration process is using a meta-optimisation approach combined with scattering parameters analysis, as presented in Fig. 4.

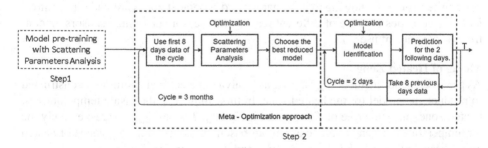

Fig. 4. Robust calibration procedure for temperature prediction of zone 1

The first step aims at pre-training the model with the scattering parameters analysis to determine how many parameters among n parameters of initial model should be fixed. The detailed description of this method can be seen in [25].

The step 2 is an automatic process during real time control, in which the scattering parameters analysis is one more time used with multi-start optimization to define which k parameters have to be fixed in taking into account the data of first 8 days of the season. This is done each three-month to adapt to the season changes. This model is re-calibrated every two days to improve the prediction for the next days. Such a process is running continuously and automatically during one year. The application of this methodology is presented in the coming section, using the simplified model of a building based on adjacent zone approximations.

3.4 Application for a Positive Energy Building

Case Study

Our case study is a positive energy building located in south of France with a floor area of more than 200 m². This household includes one main zone, called heated zone which is regulated by a heating system, one garage zone and two basements (room basement and office basement). It has been built, with high performance materials to reduce heat losses and to ensure a summer thermal comfort without cooling system (Fig. 5).

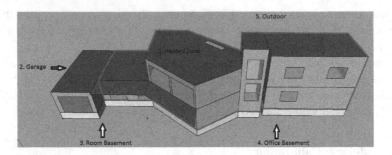

Fig. 5. Overview of studied building

The purpose of this study is to apply our methodology for the prediction of heated zone air temperature knowing that the garage and basement zones are under uncontrollable uncertainties because of different reasons (no sensors or limited sensors, uncontrolled scenarios of use…).

Model of Heated Zone

As proposed in our methodology above, an equivalent electrical circuit was constructed to produce the model for the heated zone. In this study, T_{int}, the inside temperature of heated zone, plays the role of T_{int_zone1} in Fig. 2. T_{gar}, T_{off} and T_{room} are respectively the air temperatures of garage, office and room basement zones (adjacent zones of heated zone), playing the role of T_{int_zone2}, T_{int_zone3} and T_{int_zone4} in Fig. 2.

Model of Adjacent Zones

The garage zone is thermally linked with outdoor, room basement and heated zone. As discussed in our methodology for modelling simplification, we neglect the thermal effect of heated zone on the garage zone and replace the heat transfer to the garage from the room basement by the ground in assuming that the thermal insulation between the room basement and the ground is so weak. In this case, the heat gain inside the garage comes mainly from the sun. As a result, we obtain the simplified model expressed by the equation:

$$T'_{gar} = \alpha.T_{gar} + \beta.T_{ext} + \gamma.T_{ground} + \delta.P_{sun} \tag{3}$$

With $\alpha = \dfrac{1}{C_{tot}} \cdot \left(\dfrac{1}{R_1} + \dfrac{1}{R_2} \right)$; $\beta = \dfrac{1}{C_{tot}} \cdot \dfrac{1}{R_1}$; $\gamma = \dfrac{1}{C_{tot}} \cdot \dfrac{1}{R_2}$; $\delta = \dfrac{1}{C_{tot}}$.

The parameters α, β, γ and δ were identified in our study using a detailed dynamic thermal model (EnergyPlus[1] software). It would have been done using measures. The prediction performance of our simplified model is evaluated using another dataset and shown in Fig. 6 by a Bland Altman plot. It is observable that the simplified model is strongly correlated to data. The root mean squared error (RMSE) is 1.17 °C while mean absolute error (MAE) is 1.08 °C. The coefficient of determination (r^2) is 0.9911 that is very close to 1, expresses how well the predictions of our simplified model fits data.

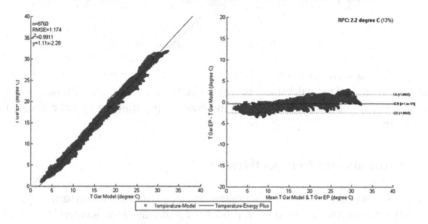

Fig. 6. Bland Atman plot between predictions of our simplified model and data.

Despite this, we are always aware that the temperature prediction of garage zone based on this simple model would be sometimes not accurate but the most important things are the prediction accuracy for the heated zone and the reduction of computation time.

A similar work was done for the modelling of office and room basement zones.

Robust Identification For Heated Zone Temperature Prediction
We focus now on results achieved by applying our methodology with the robust calibration procedure mentioned in Fig. 4 for the temperature prediction of heated zone. In fact, the pre-training in step 1 with scattering parameters analysis permitted to find the optimal degree of freedom for this model with 7 parameters to be fixed from 14 parameters of model. The automatic and continuous calibration process over one year in step 2 leads to very good predictions (Fig. 7) with a mean error of 0.88 °C and a maximal error of 2.51 °C.

---

[1] http://apps1.eere.energy.gov/buildings/energyplus.

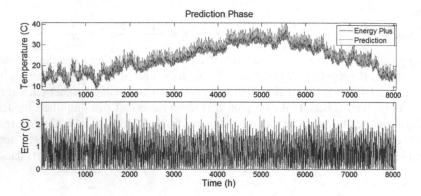

Fig. 7. Prediction results throughout one year

The prediction accuracy is very similar throughout the year, which can be explained by the regular and robust calibration process. The total computational time for whole year anticipation is about 1 h 40 min with the time for pre-training model estimated of about 20 min.

4   Conclusions and Perspectives

This paper has introduced a modelling methodology permitting to perform the robust identification to improve the prediction process of a thermal zone located in the multi-zone building. The obtained results indicate the robustness of the prediction with a reasonable computational time. It is to conclude that the simplification in modelling of adjacent zones with various uncontrollable uncertainties is a good solution to generate the data input for the temperature prediction of the studied main zone. Furthermore, this methodology should be tested for other cases study and be integrated into real time anticipative energy management system.

References

1. Costa, A., Keane, M.M.: Building operation and energy performance: monitoring, analysis and optimisation toolkit. Appl. Energ. **101**, 310–316 (2013)
2. Amara, F., Agbossou, K., Cardenas, A., Dubé, Y., Kelouwani, S.: Comparison and simulation of building thermal models for effective energy management. Smart Grid Renew. Energ. **6**, 95–112 (2015)
3. Mendes, N., Oliveira, G.H., De Araújo, H.X.: Building thermal performance analysis by using Matlab/Simulink. In: Seventh International IBPSA Conference, Rio de Janeiro, Brazil, pp. 473–480 (2001)
4. Saltelli, A., Ratto, M., Andres, T., Campolongo, F., Cariboni, J., Gatelli, D., Saisana, M., Tarantola, S.: Global Sensitivity Analysis: The Primer. Wiley, Chichester (2008)
5. Morris, M.D.: Factorial sampling plans for preliminary computational experiments. Technometrics **33**, 161–174 (1991)

6. Heiselberg, P., Brohus, H., Hesselholt, A., Rasmussen, H., Seinre, E., Thomas, S.: Application of sensitivity analysis in design of sustainable buildings. Renew. Energ. **34**, 2030–2036 (2009)
7. Westphal, F.S., Lamberts, R.: Building simulation calibration using sensitivity analysis. In: 9th IBPSA Conference, Montréal, Canada (2005)
8. Mejri, O.: Développement de méthodes de diagnostic énergétique des bâtiments. Ph.D. thesis, Universités de Tunis El Manar et Bordeaux 1 (2011)
9. Kominek, P., Tumova, E., Hirs, J.: Sensitivity analysis of residential building simulation parameters. In: 9th International Conference, High Tatras, Slovakia (2017)
10. Lavin, R.B., Rohlig, K.J., Becker, D.K.: Sensitivity analysis techniques for the performance assessment of a radioactive waste repository. In: Euradwaste - Seventh European Commission Conference, pp. 387–397 (2008). ISBN: 978-92-79-13105-9
11. Hopfe, C.J., Hensen, J.L.M.: Uncertainty analysis in building performance simulation for design support. Energ. Build. **43**, 2798–2805 (2011)
12. Nguyen, A.T., Reiter, S.: A performance comparison of sensitivity analysis methods for building energy models. Build. Simul. **8**(6), 651–664 (2015)
13. Sobol, I.M.: Sensitivity estimates for nonlinear mathematical models. MMCE **1**(4), 407–414 (1993)
14. Sobol, I.M., Tarantola, S., Gatelli, D., Kucherenko, S.S., Mauntz, W.: Estimating the approximation error when fixing unessential factors in GSA. Reliab. Syst. Saf. **92**, 957–960 (2007)
15. Cukier, R.I., Fortuin, C.M., Shuler, K.E., Petschek, A.G., Schaibly, J.K.: Study of the sensitivity of coupled reaction systems to uncertainties in rate coefficients I. Theory. J. Chem. Phys. **59**(8), 3873–3878 (1973)
16. Saltelli, A., Tarantola, S., Chan, K.: A quantitative, model independent method for global sensitivity analysis of model output. Technometrics **41**(1), 39–56 (1999)
17. Friendly, M., Denis, D.: The early origins and development of the scatterplot. J. Hist. Behav. Sci. **41**(2), 103–130 (2005)
18. Clarke, J.A., Strachan, P., Pernot, C.: An approach to the calibration of building energy simulation models. ASHRAE Trans. **99**, 917–927 (1993)
19. Reddy, T.A.: Literature review on calibration of building energy simulation programs: uses, problems, procedures, uncertainty and tools. ASHRAE Trans. **112**, 226–240 (2006)
20. O'Neill, Z., Eisenhower, B.: Leveraging the analysis of parametric uncertainty for building energy model calibration. Build. Simul. **6**, 365–377 (2013)
21. Fabrizio, E., Monetti, V.: Methodologies and advancements in the calibration of building energy models. Energies **8**, 2548–2574 (2015)
22. Coakley, D., Raftery, P., Keane, M.: A review of methods to match building energy simulation models to measured data. Renew. Sustain. Energy Rev. **37**, 123–141 (2014)
23. Grandjacques, M., Janon, A., Adrot, O., Delinchant, B.: Pick-freeze estimation of sensitivity indices for models with dependent causal processes inputs. MCQMC, KU Leuven, Belgium (2014)
24. Dang, H.A., Delinchant, B., Wurtz, F.: Toward building energy management: electric analog modeling for thermal behavior simulation. In: IEEE International Conference on Sustainable Energy Technologies (ICSET), Hanoi, pp. 246–250 (2016)
25. Nguyen Hong, Q., Le-Mounier, A., Dinh, V.B., Delinchant, B., Ploix, S., Wurtz, F.: Meta-Optimization and Scattering Parameters Analysis for Improving On Site Building Model Identification for Optimal Operation. IBPSA, Berkeley (2017)

Technology Adoption Optimization with Heterogeneous Agents and Carbon Emission Trading Mechanism

Chenhao Fang[1] and Tieju Ma[1,2(✉)]

[1] East China University of Science and Technology, Meilong Road 130, Shanghai 200237, China
forever707@live.cn, tjma@ecust.edu.cn
[2] International Institute for Applied System Analysis, Schlossplatz 1, 2361 Laxenburg, Austria

Abstract. The adoption of new technologies with high efficiency and low emission is of great importance in achieving sustainable development. Most studies of technology adoption have been criticized of idealistically assuming only one global decision agent. In this paper, a model of optimizing technology adoption with heterogeneous agents is proposed. Each agent attempts to identify the optimal solution for a portion of the entire system. The heterogeneity in agents is the different demands they face. In order to internalize the external effects of emission, a quantity-based market incentive policy instruments - Carbon Emission Trading is implemented. With two heterogeneous agents, a bargaining process is introduced to reasonably allocate the profit to them. Computational tests are conducted with different market shares and different discounting factors. Numerical results show the impact of heterogeneity and carbon emission trading mechanism on the optimal technology adoptions. It is suggested that a smaller gap of agents' market shares leads to earlier and more adoptions. Besides, adoptions remain no change when both agents have a same discounting factor. A big discounting factor of the seller will accelerate the adoptions in the buyer agent and the entire system if agents have different discounting factors.

Keywords: Technology adoption · Heterogeneous agents
Carbon emission trading · Bargaining process

1 Introduction

With the rapid development of the global economy, environmental problems become increasingly serious, especially for climate change. To realize the Global Warming of 1.5 °C [1], it is necessary to develop and adopt clean energy technologies. Compared to old technologies, new technologies have advantages on production efficiency and environmental protection. However, they need high investment in research and development (R&D) and establishment of capacities at the early stage of development. Thus, decision makers will be reluctant to change the original technical structure, called technology lock-in effect. Despite this, due to the technological learning of new technologies, the cost using them will decrease as the experience with them accumulates [2–4]. Hence, reasonable adoption decisions are particularly important for the sustainable development of the entire energy system.

© Springer International Publishing AG, part of Springer Nature 2018
V.-N. Huynh et al. (Eds.): IUKM 2018, LNAI 10758, pp. 238–249, 2018.
https://doi.org/10.1007/978-3-319-75429-1_20

Early researches on technology adoption are either based on the acceptance of individuals or from the perspective of systematic planning [5, 6]. However, most of theses models have been criticized of being too idealistic by assuming only one global decision agent. In a real energy system, there may be more than one decision agent. One purpose of this paper is to explore the optimal technology adoption in this condition.

Besides, environmental problem caused by climate change arises because the economic agent cannot bear all the costs of its emissions, which is a format of market failure [7]. Thus, reducing greenhouse gas emissions requires the introduction of government interventions such as administrative imperative policies and market incentive policies. The former includes emission standards and technology adoption standards, and the latter includes price-based policy instruments like emissions taxes or subsidies and quantity-based policy instruments like emission trading. The concept of emission trading originated from the paper published by Coase [8] in 1960. With this policy, a country having fewer carbon emissions or more carbon emission permits is able to sell the permits to make profit and a country having more carbon emissions or fewer carbon emission permits can purchase the permits to emit more. Carbon emission trading is a form of emissions trading that specifically targets carbon dioxide. Researches about technology adoptions with carbon emission trading are various, using agent-based models [9, 10], optimization models [11] or empirical investigations [12, 13]. However, trading is a behavior between individuals. Optimization models from the perspective of systematic planning always ignore detailed trading processes, especially for the formation of trading price. We intend to provide an intuitive understanding of the impact of carbon emission trading on the optimal technology adoption by incorporating trading processes into optimization model, which is another purpose of this paper.

In this paper, the optimal technology adoption with heterogeneous agents (e.g. firms) and carbon emission trading mechanism will be studied. Technology adoption is no longer decided by a global agent but by heterogeneous agents instead. They are almost the same except they face different demands. Each agent attempts to optimize technology adoption for a portion of the entire system. The government sets a national emission abatement target, and each agent is allocated some free permits according to its market share. Carbon emission trading mechanism is implemented. We intend to reveal the impact of heterogeneity and carbon emission trading mechanism on the optimization of technology adoption.

The model presented here is not intended by any means to be a "realistic" model showing technological details. Rather, the model is mainly used for exploratory modeling purposes and as a heuristic research device to examine in depth the impacts of heterogeneity and carbon emission trading on the optimal technology adoption.

The rest of this paper is organized as follows. Section 2 introduces the model of technology adoption considering heterogeneous agents and carbon emission trading mechanism. Section 3 presents the computational tests and analyzes the impact of heterogeneity and carbon emission trading mechanism on the optimal adoptions of technologies. A bargaining process to allocate the profit of trade between agents is proposed. Section 4 gives the conclusions.

2 Formulation

2.1 The Simplified Energy System

The simplified energy system is consisted of three parts: resources, demands, and technologies. For example, in coal-electricity system, coal is resource and the demand is electricity. Technologies that can generate electricity with coal link resources to demands. The demands increase over time.

There are three technologies that can be used to produce goods from resources, namely, existing-T1, incremental-T2, and revolutionary-T3. The existing technology is assumed to be entirely mature. It has low efficiency and low initial investment cost, and the carbon emission caused by using it is high. The incremental technology is a new technology, with higher efficiency and higher initial investment cost, and it produces less carbon emission than the existing technology. The revolutionary technology is also a new technology, with much higher efficiency and much higher initial investment cost, and it produces almost no air pollution. The incremental and revolutionary technologies have learning potential, which means their initial investment costs will decrease as the experience to them increases.

2.2 Models with Carbon Emission Trading Mechanism

Suppose that there are N decision agents in the system, for agent $n(n = 1, 2, \ldots, N)$, let $y^t_{i,n}$ denote its newly installed capacity of technology $i(i = 1, 2, 3)$ at time t, so its total installed capacity can be calculated as

$$C^t_{i,n} = \begin{cases} \sum\limits_{j=t-\tau_i}^{t} y^j_{i,n} & t > \tau_i \\ \sum\limits_{j=1}^{t} y^j_{i,n} + \dfrac{\tau_i - t}{\tau_i} C^0_{i,n} & t \leq \tau_i \end{cases} \tag{1}$$

where τ_i denotes the plant life of technology i, $C^0_{i,n}$ denotes agent n's initial installed capacities of technology i. Since our model assumes that the initial demand is satisfied by the existing technology, $\dfrac{\tau_i - t}{\tau_i} C^0_{i,n}$ denotes the remaining part of agent n's initial installed capacities of technology i.

At each decision interval, agent n's own cumulative experience with technology i at time t can be represented as its cumulative production,

$$E^t_{i,n} = E^0_{i,n} + \sum_{j=1}^{t} x^j_{i,n} \tag{2}$$

where $E^0_{i,n}$ denotes agent n's initial cumulative experience with technology i, $x^j_{i,n}$ denotes agent n's production using technology i at time j.

Since T1 has no learning potential, its unit investment cost is fixed. T2 and T3 have learning potential, so their unit investment costs will decrease as their total experience increase. The unit investment cost can be written as

$$CF_{i,n}^t = \begin{cases} CF_{i,n}^0 & i = 1 \\ CF_{i,n}^0 \cdot \left(E_{i,n}^{t-1}\right)^{-b_i} & i = 2, 3 \end{cases} \tag{3}$$

where $CF_{i,n}^0$ denotes agent n's initial unit investment cost for technology i, $1 - 2^{b_i}$ is the learning rate of technology i, which means the percentage reduction of future unit investment cost for every doubled total experience.

Agent n's total resource consumption at time t is the function of its production,

$$R_n^t = \sum_{i=1}^{3} \frac{x_{i,n}^t}{\eta_i} \tag{4}$$

where η_i denotes the efficiency of technology i. Then agent n's cumulative resource consumption by time t can be presented as

$$\overline{R}_n^t = \sum_{j=1}^{t} R_n^j \tag{5}$$

since the cost of extracting each unit resource is linear correlation with the cumulative resource consumption, then we have

$$CE_n^t = CE_n^0 + k\overline{R}_n^t \tag{6}$$

where CE_n^0 denotes agent n's initial cost of extracting each unit resource, k denotes the sensitivity of the extraction cost to cumulative resource consumption.

The demand is exogenous and increases over time with an annual growth rate,

$$D^t = D^0(1 + \alpha)^t \tag{7}$$

where D^0 denotes the initial total demand, α denotes the annual growth rate of total demand.

Besides, different agents are assumed to face different demands. Let w_n denote agent n's market share, which satisfies $\sum_n w_n = 1$.

With carbon emission trading, for every agent n in year t, the optimization model can be written as follows:

$$\min \sum_{i=1}^{3} \frac{1}{(1+\delta)^t} CF_{i,n}^t \cdot y_{i,n}^t + \frac{1}{(1+\delta)^t} CE_n^t \cdot R_n^t + \sum_{i=1}^{3} \frac{1}{(1+\delta)^t} COM_i \cdot x_{i,n}^t - \frac{1}{(1+\delta)^t} p^t Q_n^t \tag{8}$$

$$\text{s.t. } w_n D^t \leq \sum_{i=1}^{3} x_{i,n}^t \tag{9}$$

$$x_{i,n}^t \leq C_{i,n}^t, \forall i \tag{10}$$

$$\sum_{i=1}^{3} \frac{\lambda_i}{\eta_i} x_{i,n}^t + Q_n^t \leq Cap_n^t \tag{11}$$

$$x_{i,n}^t \geq 0, \forall i \tag{12}$$

$$y_{i,n}^t \geq 0, \forall i \tag{13}$$

where δ denotes the discount rate; COM_i denotes the operation and maintenance cost of technology i; p^t denotes the price of emission permits at time t; Q_n^t denotes the volume of emission permits traded by agent n at time t, when Q_n^t is positive, agent n is a seller, otherwise is a buyer; Cap_n^t denotes agent n's emission cap at time t, which is decided by the reduction target of government and allocated freely.

The optimization model can be solved for a sufficiently large sample M, where the size of M is determined through successive experiments. Several successive model runs with the same sample size M are compared. If no major changes in the solution structure and the objective function can be observed, M is considered sufficiently large (For more detail, see in [14]).

3   Computational Experiments

In this section, we will investigate the impact of heterogeneity and carbon emission trading mechanism on the optimal technology adoption. For simplicity and being comparable, only two agents ($N = 2$) are considered. The weights w_1 and w_2 are assumed to be 0.8 and 0.2 respectively. The reduction target is assumed to be 5% every 10 years. The cap of each agent is set according to its market share.

3.1   The Initial Value of Parameters

To explore the optimal technology adoption, we assume that all agents are cooperative with same resource consumption function, same demand dynamics and same technological learning rate, and all adoption strategies are made simultaneously. Table 1 lists the initial value of parameters that will be used in our simulations.

3.2   Numerical Results

Optimal Technology Adoptions without Cap or Trading, Only with Cap and with Cap & Trading. Our experiment will first show the differences of optimal technology adoptions among models without cap or trading, only with cap and with cap & trading.

It is assumed that a carbon abatement plan will be implemented since 2020, together with carbon emission trading policy. The price of unit carbon emission permit is assumed to be $20/t. The carbon cap in 2020 is assumed to be 278.48 kt. This simulation does not aim to replicate any reality. The main purpose is to demonstrate the difference to take carbon emission trading into account.

Table 1. Initial value of parameters

	Existing technology	Incremental technology	Revolutionary technology
Parameters			
Initial investment cost (US$/kw)	$CF^0_{1,n} = 1000$	$CF^0_{2,n} = 2000$	$CF^0_{3,n} = 40000$
Efficiency (%)	$\eta_1 = 30$	$\eta_2 = 40$	$\eta_3 = 90$
Plant life (year)	$\tau_1 = 30$	$\tau_2 = 30$	$\tau_3 = 30$
Initial installed capacity (kw)	$C^0_{1,n} = 1000$	$C^0_{2,n} = 0$	$C^0_{3,n} = 0$
Learning rate (%)	$1 - 2^{-b_1} = 0$	$1 - 2^{-b_2} = 20$	$1 - 2^{-b_3} = 30$
O+M cost (US$/kw)	$COM_1 = 30$	$COM_2 = 50$	$COM_3 = 50$
Carbon emission coefficient	$\lambda_1 = 0.8$	$\lambda_2 = 0.6$	$\lambda_3 = 0.1$
Other parameters			
Initial total demand	$D^0 = 62500$	Annual growth rate of demand (%)	$\alpha = 2.6$
Initial extraction cost (US$/kw)	$CE^0_n = 20$	Extraction cost coefficient	$k = 10^{-9}$
Discount rate (%)	$\delta = 5$	Scale of problem (year)	$T = 100$, decision interval is 10 years

Figure 1 plots the optimal adoptions of T3 in all three cases. As we can see, without cap or trading, T3 will not be adopted at all. However, with carbon cap both agents will have to adopt T3, no matter it is profitable or not. With carbon emission trading T3 diffuses more quickly 20 years later after the implementation of carbon abatement plan, which means carbon emission trading mechanism do help to promote the diffusion of new technology. However, with more T3 being adopted, unit abatement cost decreases. When it is lower than the price of unit carbon emission permit, the effect of carbon emission trading mechanism is not obvious.

Optimal Technology Adoptions with Heterogeneity in Market Share. In the last simulation, the permit price is assumed to be $20/t. However, the real price depends on agents' dynamic unit abatement costs. Reasonable price should be in the interval of the lowest price at which the seller is willing to sell and the highest price at which the buyer is willing to buy [15, 16]. Both agents will be glad to join the collaboration voluntarily if trade is profitable. Since the collaboration is based on mutual benefits shared between the participating agents through negotiation, the cooperation is also a bargaining process to properly allocate the profits between them [17]. To describe this bargaining process, the Rubinstein game theory [18] is applied, in which an indefinite bargaining game with

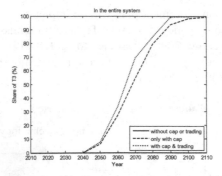

Fig. 1. The optimal adoptions of T3 without cap or trading, only with cap and with cap & trading

two participants is characterized: Two players have to reach an agreement on the partition of a pie of size 1. Each has to make in turn, a proposal as to how it should be divided. After one player has made an offer, the other must decide either to accept it, or to reject it and continue the bargaining. Let δ_1 and δ_2 be the fixed discounting factors of agent 1 and agent 2 respectively. A perfect equilibrium partition is expressed as

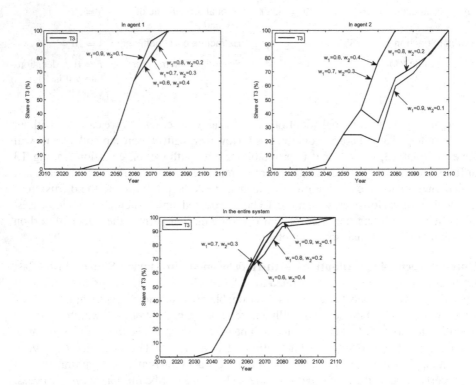

Fig. 2. The optimal adoptions of T3 in agent 1, agent 2 and the entire system with heterogeneous market shares of agents

$M = (1 - \delta_2)/(1 - \delta_1\delta_2)$. Here, no short supply is allowed, which can be avoided by raising the equilibrium price or cancelling the trade.

To explore how heterogeneous market shares influence the adoptions of T3, we run simulations with different weights when $\delta_1 = \delta_2 = 0.8$. The carbon cap in 2020 is assumed to be 215.44 kt. Figure 2 shows the optimal adoptions of T3. As observed, the adoption of T3 increases as market share increases, especially in agent 2. This means a higher demand contributes to the adoption of T3. In the entire system, when the gap of the market shares between agents gets closer, more T3 will be adopted. Besides, a big gap postpones the year when T3 is totally adopted. We can infer that with a small market share agent 2 prefers to buy permits rather than adopt T3. Although it will promote technological change in agent 1 in a certain extent, at the same time it strengthens the lock-in effect of T2 and postpones the adoption of T3 greatly.

Results of trading prices and volumes show that the trading price increases as the market share of seller (agent 1) increases. With a small and decreasing market share, agent 2 tends to buy more permits, so its unit abatement cost increases. Although agent 2's desire to buy will induce agent 1 to adopt more, the decrease in the unit abatement cost of agent 1 is less than the increase in the unit abatement cost of agent 2. As a result, the equilibrium price increases. Besides, with an increasing gap of market share, agent 2 strengthens the lock-in effect while agent 1 prefers to adopt T3 earlier. A high buying price and a low selling price contribute to a high possibility of a successful trade. On the contrary, a decreasing gap of market shares leads to a lower possibility and a fewer volume of a successful trade.

Optimal Technology Adoptions with Different Discounting Factors. In the last simulations, δ is assumed to be a fixed value. A small discounting factor indicates agent's impatience with negotiation while a big one indicates agent's patience with negotiation. In this subsection, we intend to analyze the impact of different discounting factors on the optimal technology adoptions. The demands in agent 1 and agent 2 are assumed to be 50000 and 12500 respectively.

First, it is assumed that both agents have the same discounting factors $(\delta_1 = \delta_2 = \delta)$, ranging from 0.1 to 0.9. Results show that no matter how the discounting factor changes, the adoptions of T3 are almost the same in all cases. Although in agent 2 differences exist, the gap is less than 3.6%. Besides, the demand in agent 1 is 4 times as that in agent 2, so the difference in the entire system is less than 1%. Therefore, we can hold the idea that the optimal adoptions of T3 remain no change when both agents have the same discounting factors. As for trading price, a small discounting factor leads to a high equilibrium price. Because at most trading points, agent 1 is the seller. A small discounting factor indicates that agent 1 gets more shares of the profit, leading to a higher equilibrium price. In addition, the largest trading volume occurs in 2070. The decrease in the adoption of T3 in 2070 in agent 2 and the increasing demand can account for this.

Second, we explore the cases with δ_1 changing from 0.1 to 0.9 and δ_2 being 0.8. Results show that the adoption of T3 increases as the discounting factor of agent 1 increases. Besides, the adoptions of T3 are almost the same in agent 1 no matter how the discounting factor changes. This is because agent 1's high demand contributes to its early adoption of T3, so the influence of carbon emission trading is very small. The adoptions of T3 in agent 2 suggest that it prefers to adopt more T3 rather than buy permits from agent 1 when the discounting factor of agent 1 increases. As we know, the increase in the discounting factor of agent 1 leads to a higher equilibrium price and a less profitable trade for agent 2. As for trading price, the highest price increases as the discounting factor of agent 1 increases. However, results are different in 2080 and 2090, especially in 2090 the situation is almost the opposite. In addition, except some special cases, the total trading volumes go down in most cases when the discounting factor of agent 1 increases. Due to this increase, the equilibrium price increases, making trade less profitable. Thus, the total trading volumes decrease.

Third, cases with δ_2 changing from 0.1 to 0.9 and δ_1 being 0.8 are explored. Results show that the adoptions of T3 in agent 1 are almost the same. Although in agent 2 different result exists when $\delta_2 = 0.9$, the adoptions in the entire system are still almost the same, only with a gap less than 2.5%. This means when agent 1 faces high demand and agent 2 faces low demand changes in the discounting factor of agent 2 will almost not influence the optimal adoption of T3. Besides, trading price decreases as the discounting factor of agent 2 increases. This is because the increase in the discounting factor of agent 2 means more shares of the profit it can get, so the equilibrium price decreases. In addition, except some special cases, the total trading volume increases in most cases as the discounting factor of agent 2 increases. Due to the increase in the discounting factor of agent 2, the equilibrium price decreases, making trade more profitable. Thus, the total trading volumes increase.

Carbon Emission Paths of Different Simulations. There are two factors contributing to the carbon emission paths. One is the consumption and the other is the technology used to satisfy the demand. Figure 3 shows the different carbon emission paths of different simulations. As we can see, without carbon cap the total carbon emissions in the entire system increase as demand increases. In this case, although T2 will be adopted when demand increases, the reduction in carbon emissions using T2 cannot compensate the increase in carbon emission caused by the increasing demand. Hence, the total carbon emissions increase as demand increases.

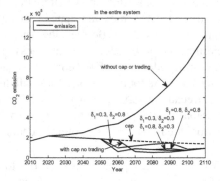

Fig. 3. Different carbon paths of different simulations

With carbon cap being implemented, carbon abatement plan is realized. Carbon emission is controlled below the cap 40 years later since the implementation of carbon cap. However, with carbon emission trading mechanism it takes only 30 years. This means that carbon emission trading mechanism contributes to an earlier transition into a low-carbon economy.

Besides, except the case with $\delta_1 = 0.3, \delta_2 = 0.8$, the total emissions in the cases with carbon emission trading are less than that in the case only with cap, although more carbon will be emitted from 2070 to 2100. It is suggested that in the market where a seller has more power to manipulate the price carbon emission trading is an effective policy used to reduce greenhouse gas emissions.

4   Conclusion

This paper presents a model of optimizing technology adoption with heterogeneous agents and carbon emission trading mechanism. Although the simulations are highly stylized, they can still provide a good idea as to how the heterogeneity in agents and the carbon emission trading mechanism influence the diffusions of technologies. By comparing the adoptions in the cases without cap or trading, only with cap and with cap & trading, the paper first shows the difference to take carbon emission trading into account. Incorporating the carbon emission trading mechanism with a price negotiation process, this paper then explores how the differences in the discounting factors of agents influence the adoptions. The main findings what we can learn from the simulations are summarized:

1. The simulations demonstrate that the heterogeneity in agents influences the optimal technology adoptions. A higher demand leads to more adoptions in each agent. However, the technology adoption in the entire system is influenced by the gap of market shares between agents. A smaller gap leads to more adoptions and less time for the revolutionary technology to be totally adopted.
2. When carbon emission trading policy is implemented, with a same discounting factor of both agents, the optimal adoptions of revolutionary technology remain no change no matter how the discounting factor changes.
3. When the buyer agent has a big discounting factor, a big discounting factor of the seller accelerates the adoptions in the buyer agent and in the entire system because it leads to a higher trading price, meaning trade less profitable for the buyer.
4. When the seller agent has a big discounting factor, a big discounting factor of the buyer will almost not influence the technology adoptions in the entire system because of its small market share.

This paper combines the idea of agent-based modeling with optimization modeling by incorporating the trading process into optimization model and provides an intuitive understanding of the impact of carbon emission trading on optimal technology adoptions. However, the study has some limitations.

In terms of the trading process, only two heterogeneous agents are considered. In the real world, more than two will be in a system. Besides, the buyers or sellers may form coalitions to ensure that they can get more profit. Thus, the bargaining process will be more complicated. In our future work, we will explore the optimal technology adoptions with more heterogeneous decision agents.

In addition, in this paper the remaining emission permits after trade in every single period cannot be used in later periods, which means no banking is allowed. However, banking may affect agents' technology adoption decisions and permit trading behavior. In our future work, we will investigate how different the design of banking will influence the optimal technology adoptions.

Acknowledgments. This work was supported by the Ministry of Education, China under Grant 222201718006; National Natural Science Foundation of China under Grant 71571069.

References

1. Intergovernmental Panel on Climate Change (IPCC): Chair's Vision Paper, AR6 Scoping Meeting, Addis Ababa, Ethiopia (2017)
2. Ma, T., Grubler, A., Nakamori, Y.: Modeling technology adoptions for sustainable development under increasing returns, uncertainty, and heterogeneous agents. Eur. J. Oper. Res. **195**, 296–306 (2009)
3. Ma, T.: Coping with uncertainties in technological learning. Manage. Sci. **56**(1), 192–201 (2010)
4. Lin, B., Li, J.: Analyzing cost of grid-connection of renewable energy development in China. Renew. Sustain. Energy Rev. **50**, 1373–1382 (2015)
5. Chen, H., Ma, T.: Technology adoption with limited foresight and uncertain technological learning. Eur. J. Oper. Res. **239**(1), 266–275 (2014)
6. Chen, H., Ma, T.: Optimizing systematic technology adoption with heterogeneous agents. Eur. J. Oper. Res. **257**(1), 287–296 (2017)
7. Fan, Y., Mo, J.-L., Zhu, L.: Carbon Trading in China: Policy Design and Social-Economic Impact. Science Press, Beijing (2016)
8. Coase, R.H.: The problem of social cost. J. Law Econ. **3**(4), 1–44 (1960)
9. Tang, L., Wu, J., Yu, L., Bao, Q.: Carbon emissions trading scheme exploration in China: a multi-agent-based model. Energy Policy **81**, 152–169 (2015)
10. Zhang, H., Cao, L., Zhang, B.: Emissions trading and technology adoption: an adaptive agent-based analysis of thermal power plants in China. Resour. Conserv. Recycl. **121**, 23–32 (2016)
11. Zhao, J., Hobbs, B.F., Pang, J.-S.: Long-run equilibrium modeling of emissions allowance allocation systems in electric power markets. Oper. Res. **58**(3), 529–548 (2009)
12. Camacho-Cuena, E., Requate, T., Waichman, I.: Investment incentives under emission trading: an experimental study. Environ. Resour. Econ. **53**(2), 229–249 (2012)
13. Liu, X., Fan, Y., Li, C.: Carbon pricing for low carbon technology diffusion: a survey analysis of China's cement industry. Energy **106**, 73–86 (2016)
14. Messner, S., Golodnikov, A., Gritsevskii, A.: A stochastic version of the dynamic linear programming model MESSAGE III. Energy **21**(9), 775–784 (1996)
15. Stańczak, J., Bartoszczuk, P.: CO2 emission trading model with trading prices. Clim. Change **103**(1–2), 291–301 (2010)
16. Sabzevar, N., Enns, S.T., Bergerson, J., Kettunen, J.: Modeling competitive firms' performance under price-sensitive demand and cap-and-trade emissions constraints. Int. J. Prod. Econ. **184**, 193–209 (2017)

17. Ding, H., Zhao, Q., An, Z., Tang, O.: Collaborative mechanism of a sustainable supply chain with environmental constraints and carbon caps. Int. J. Prod. Econ. **181**, 191–207 (2016)
18. Rubinstein, A.: Perfect equilibrium in a bargaining model. Econometrica **50**(50), 97–109 (1982)

Statistical Methods

On a Generalized Method of Combining Predictive Distributions for Stock Market Index

Son Phuc Nguyen[1(✉)], Uyen Hoang Pham[1], and Thien Dinh Nguyen[2]

[1] Department of Mathematics, University of Economics and Law, VNU-HCM,
Ho Chi Minh city, Vietnam
sonnp@uel.edu.vn
[2] CEFR, University of Economics and Law, VNU-HCM,
Ho Chi Minh city, Vietnam

Abstract. Forecast combinations have frequently been found in empirical studies to produce better forecasts on average than methods based on the ex ante best individual forecasting model. In this paper, we study a generalized method of aggregation in the form of a nonlinear transformation of a linear mixture model. The major advantage of the nonlinear transformation is an excellent flexibility to calibrate predictive cumulative distributions. This method proves to be particularly useful to accommodate complex volatility in the stock market.

As for applications, we study two stock market indices, namely the Vietnamese VN30 index and the Thai SET50 index. The forecasts are in the form of empirical densities estimated by Bayesian inference.

Keywords: Beta calibration · Finite mixture models
Stock market index · Density forecast · Bayesian inference · Stan

1 Introduction

In order to construct effective portfolios, investors have a need for decent predictions of stock market data. Nevertheless, prediction of any individual asset is notoriously difficult since its returns can be decomposed into two components: systematic and idiosyncratic. The systematic component is driven by factors that are common to other securities, such as sector, macroeconomic, and asset class effects. Meanwhile, the idiosyncratic components of returns is particular to a single asset, such as the company's competitive position, individual deals and perceived management capabilities, and are not related to industry or other factor trends. When many assets are bundled together to form an index, the idiosyncratic return components tend to cancel each other out, as winners and loser are both embedded in the index. On the other hand, systematic return components do not cancel each other out because they affect all members of the index. This feature of portfolio diversification is desirable from the point of view of investors because the idiosyncratic components add unnecessary risk to their portfolios. Instead, investors can focus on a few systematic factors.

© Springer International Publishing AG, part of Springer Nature 2018
V.-N. Huynh et al. (Eds.): IUKM 2018, LNAI 10758, pp. 253–263, 2018.
https://doi.org/10.1007/978-3-319-75429-1_21

Based on these investors' insights, we study index returns with a goal of extracting the systematic information for better predictions. However, decision makers often deal with multiple forecasts of the same variable like index return. This could reflect differences in forecasters subjective judgements due to heterogeneity in their information sets in the presence of private information or due to differences in modeling approaches. In the latter case, two forecasters may well arrive at very different views depending on the maintained assumptions underlying their forecasting models, e.g., constant versus time-varying parameters, linear versus nonlinear forecasting models, etc.

Faced with multiple forecasts of the same variable, an issue that immediately arises is how best to exploit information in the individual forecasts. In particular, should a single dominant forecast be identified or should a combination of the underlying forecasts be used to produce a pooled summary measure? From a theoretical perspective, unless one can identify ex ante a particular forecasting model that generates smaller forecast errors than its competitors (and whose forecast errors cannot be hedged by other models forecast errors), forecast combinations offer diversification gains that make it attractive to combine individual forecasts rather than relying on forecasts from a single model. Even if the best model could be identified at each point in time, combination may still be an attractive strategy due to diversification gains, although its success will depend on how well the combination weights can be determined.

In this paper, each stock which is used to construct a particular market index is treated as a forecaster for the index itself. Hence, we first predict individual stocks by classical tools like Bayesian GARCH model. Then, we propose two combination methods of individual stock predictions to form a joint prediction for the index. The first method is the well-known linear mixture model, and the second is the new method of calibrating the mixture model by a beta transformation.

For applications, we are interested in the Vietnam VN30 index where the market is a frontier market. For comparison, we have selected a neighboring country's market with the Thailand SET 50 index. Note that the Thai market is an emerging market with a much longer history and with the total value traded more than 10 times larger than the value traded in Vietnam.

The plan of the paper is as follows. Section 2 introduces a finite linear mixture of GARCH models together with a beta calibration process. We then apply Bayesian inference to find optimal weights for the mixture models as well as best-fit parameters for the beta calibration. Section 3 deploys the framework to study the Vietnamese and Thai stock market data. This section also contains a careful evaluation of the predictive results. The paper ends with a conclusion in Sect. 4.

2   Finite Mixtures of GARCH Models and Calibration

2.1   GARCH Model (Generalized Autoregressive Conditional Heteroscedasticity)

Over the three decades since its introduction, the GARCH model (see [3,9]) and numerous variants have improved volatility estimation and risk management.

Perhaps, the most important characteristic of the GARCH model family is that it describes the return volatility at time t in a deterministic manner, conditional on information that is available at time $t - 1$. In their strong form, GARCH models comprise innovations that are independent and identically distributed (i.i.d.), with zero mean and unit variance.

In this paper, normal and student t distributions have been experimented as innovations for GARCH models on individual stock returns. However, there is no clear advantage of student t innovation compared to the normal one. We also try the GARCH with a threshold to account for the asymmetric behavior of individual stock returns (i.e. returns are more sensitive to bad news than good news). Once again, the results of threshold GARCH are not better than those of a simple GARCH with normal innovations.

Therefore, each individual stock return in both the Vietnamese VN30 and Thai SET50 markets is fitted to a Bayesian GARCH(1,1) with normal innovation. Then, one-day-ahead forecasts are carried out for the 22 out-of-sample days (about one month of trading).

Let $r_{i,t}$ be the return of stock i on day t, and μ_i, $\sigma_{i,t}$ be the mean and standard deviation of the return $r_{i,t}$. Below is the GARCH(1,1) formula used to predict volatility and return for every stock in this research:

$$
\begin{aligned}
r_{i,t} &= \mu_i + \sigma_{i,t} z_{i,t} \\
\sigma_{i,t}^2 &= \alpha_0 + \alpha_1 \cdot (r_{i,(t-1)} - \mu_i)^2 + \beta_1 \cdot \sigma_{i,t-1}^2
\end{aligned} \tag{1}
$$

where $z_{i,t} \sim N(0, 1)$ is the standard white noise process.

Note that μ_i, α_0, α_1 and β_1 are Bayesian parameters and will be estimated using historical data of each stock.

2.2   Finite Mixture Models

Probabilistic forecasts aim to provide calibrated and sharp predictive distributions for future quantities or events of interest. In many situations, complementary or competing probabilistic forecasts from dependent or independent information sources are available. For example, the individual forecasts might stem from distinct experts, organizations, or statistical models. The prevalent method for aggregating the individual predictive distributions into a single combined forecast is the linear pool (see [13]). While other methods for combining predictive distributions are available (see [6,10]), the linear pool is typically the method of choice.

In this article, each stock included in the sets of VN30 or the Thai50 index serves as an information source to predict the index itself. In detail, we utilize the GARCH(1,1) estimation for each stock as a component of the linear mixture. In particular, for the VN30 index, there are 30 GARCH components corresponding to 30 stocks that contribute to the index, and for the Thai SET50 index, there are 45 GARCH components corresponding to 45 stocks contributing to the index. See Sect. 3 for more details.

Initially, we have tried the harmonic opinion pool and logarithmic opinion pool (see [5]) but the forecasting results are mixed when compared to the simple linear pool. Thus, we have decided to keep only the linear pool and improve predictions by applying the beta transformation to the linear mixtures.

As in [11] (see page 6), define the finite mixture cumulative distribution function as follows:

$$H_t(y \mid \omega) = \sum_{i=1}^{K} \omega_i F_{it}(y) \tag{2}$$

where $\omega_i \geq 0$, $\sum_{i=1}^{K} \omega_i = 1$ and $F_{it} \sim N(\mu_i, \sigma_{it})$ is the cumulative distribution function of stock i on day t.

The corresponding density function can be derived as follows

$$h_t(y \mid \omega) = \sum_{i=1}^{K} \omega_i f_{it}(y) \tag{3}$$

where f_{it} is the density function of stock i on day t.

In Sect. 3.2, we will estimate the weights ω_i using Bayesian inference.

2.3   Beta Calibration

Extending the ideas of calibration using functionals as in [1], a transformation utilizing the beta cumulative distribution function $B_{\alpha, \beta}$ with two parameters α, β (α, $\beta > 0$) is applied to the finite mixture model in Eq. (2), namely

$$G_t(y) = B_{\alpha, \beta}(H_t(y \mid \omega)) \tag{4}$$

α and β are to be estimated by the index historical data under the Bayesian framework in Sect. 3.2.

The corresponding density function can be easily derived as follows:

$$g_t(y) = b_{\alpha, \beta}\big(H_t(y \mid \omega)\big) \cdot h_t(y \mid \omega) \tag{5}$$

where $b_{\alpha, \beta}$ is the density function of the beta distribution and h_t is the density in Eq. (3).

2.4   Density Forecast

A density forecast provides a complete description of the uncertainty associated with a prediction, and it stands in contrast to a point forecast which contains no description of the associated uncertainty. When the predictive distribution cannot be assumed to be symmetric, a density forecast plays a vital role in constructing any interval estimate.

In this article, we construct density forecasts for the two indices of the Vietnamese and Thai stock markets for an out-of-sample test set of 22 days, which provide interval and point estimates of the indices in near future.

2.5   Bayesian Inference and Stan

Doing Bayesian inference in a multidimensional parameter space is difficult due to a high computational complexity. In this paper, we use Stan, a probabilistic programming language invented by Carpenter et al. (see [4]), to carry out all parameter estimations to produce the final density forecasts. In detail, first, we estimate a GARCH model for each stock. Then, we estimate optimal weights for mixture models of individual stocks, and finally, construct density forecasts by finding parameters for a beta transformation.

3   Main Results

In this section, we proceed to construct two predictive distributions for the Vietnam VN30 market index and the Thailand SET50 market index, respectively. The main results are two beta calibrated models which produce better predictions compared to the classic linear mixture models.

Recall that we consider each stock in the VN30 or the SET50 index as "an opinion" in predicting the market indices. Next, we combine them into a mixture; then, calibrate the mixture using a beta transformation.

3.1   Data

In this paper, originally, we selected only the VN30 index from our home country, Vietnam, to study. However, since the Vietnam stock market, classified as a frontier market, is still in its infancy, it is sometimes unstable, and is easily affected by so many different socio-economic factors. Therefore, for comparison, we have decided to investigate the index of a neighboring emerging market, the Thailand stock market which shares some characteristics with the Vietnam market, but has a much longer history with more than ten times capitalization, and hence, is more stable.

First are historical datasets for individual stocks in the Thai and Vietnamese markets. In particular, the training set for Thai stock market consists of daily returns for 45 tickers in about two years of trading from May 26th, 2015 to June 8th, 2017 while the training set for Vietnamese stock market consists of daily returns for 30 tickers in about 1.5 years of trading from July 20th, 2015 to February 13th, 2017.

Second, to train the beta calibrated and the mixture models, the historical data for index returns in the corresponding time frame are obtained, i.e., for the Vietnamese market, we collect the VN30 daily returns from May 26th, 2015 to June 8th, 2017, and use those to do inference on the parameters, and for the Thai market, the SET 50 index returns from July 20th, 2015 to February 13th, 2017 are gathered.

Third, in each market, the next period of 22 days serves as the out-of-sample test set for the weight and calibration parameters of the linear mixture and the beta calibrated predictive distributions.

3.2   Procedure to Construct Predictive Distributions

There are 3 steps in the procedure: First, a Bayesian GARCH(1,1) estimation for each stock is carried out; second, a probability weight vector ω is estimated to combine every stock in each index; and third, a beta calibration is applied to the linear mixture.

Individual GARCH(1,1). For any particular stock of interest, we collect its historical data to estimate the parameters in formula (1). Then, using those estimations, we generate 22 one-day forecasts for the mean and standard deviation of its return.

In detail, the first group consists of 30 stocks comprising the VN30 index; we obtain 30 different GARCH(1,1) estimations, and the second group of 45 stocks in the SET50 index generates 45 individual GARCH(1,1) estimations.

Combining Individual Forecasts into a Linear Mixture. In the next step, 30 stocks in the VN30 index are combined using formulas (2) and (3) with a parametric weight vector $\omega^1 = (w_1^1, \ldots, w_{30}^1)$, $\sum_{i=1}^{30} w_i = 1$, i.e.

$$H_t^1 = \sum_{i=1}^{30} \omega_i^1 F_{it}^1 \tag{6}$$

where $F_{it}^1 \sim N(\mu_i, \sigma_{it})$ is the normal cumulative distribution function with parameters μ_i, σ_{it} being the maximum likelihood estimator of the Bayesian GARCH(1,1) model of stock i on day t.

Similarly, with a parametric weight vector $\omega^2 = (w_1^2, \ldots, w_{45}^2)$, 45 stocks in the SET50 index are combined into

$$H_t^2 = \sum_{i=1}^{45} \omega_i^2 F_{it}^2 \tag{7}$$

H_t^1 and H_t^2 are the linear-mixture predictive distributions on day t for the VN30 and the SET50 indices, respectively. Historical data of the two indices are used to estimate the posterior distributions for the weight vectors ω^1 and ω^2.

Beta Calibration of the Linear Mixture. Suppose α and β are two positive real parameters. Let $B_{\alpha, \beta}$ and $b_{\alpha, \beta}$ be the cumulative distribution function and density functions of the beta distribution. In order to improve predictions of the two indices, we apply a beta calibration process by applying $B_{\alpha, \beta}$ to H_t^1 and to H_t^2 as in formulas (4) and (5).

In particular, the calibrated VN30 predictive cumulative distribution and density functions are

$$G_t^1(y) = B_{\alpha, \beta}(H_t^1(y)) \tag{8}$$

$$g_t^1(y) = b_{\alpha, \beta}(H_t^1(y)) \cdot h_t^1(y) \tag{9}$$

where α and β are parameters to be estimated from the historical VN30 returns. Similarly, the final SET50 predictive distribution and density functions are

$$G_t^2(y) = B_{\alpha, \beta}(H_t^2(y)) \tag{10}$$

$$g_t^2(y) = b_{\alpha, \beta}(H_t^2(y)) \cdot h_t^2(y) \tag{11}$$

In this case, α and β are parameters to be estimated from the historical SET50 returns.

3.3   Forecasting Results

By implementing Bayesian inferences with historical VN30 data on formula (6), we obtain a posterior distribution for the parametric vector ω^1. Let $\hat{\omega}^1$ be the maximum likelihood estimator (MLE) of ω^1 from the posterior distribution, and let $\hat{F}_{it}$ be the predicted cumulative distribution of each stock with MLE parameters $\hat{\mu}_i$, $\hat{\sigma}_{it}$. For each day t, $1 \leq t \leq 22$, we get a linear-mixture density forecast $\hat{H}_t^1 = \sum_{i=1}^{30} \hat{F}_{it}^1$ for the VN30 index on that day. Moreover, by calibrating the mixture model, we also get a calibrated density forecast $\hat{G}^1 = B_{\hat{\alpha}, \hat{\beta}}(\hat{H}_t^1)$ with MLE parameters $\hat{\alpha}$ and $\hat{\beta}$.

In a similar manner, we obtain two forecasts $\hat{H}_t^2 = \sum_{i=1}^{45} \hat{F}_{it}^2$ and $\hat{G}^2 = B_{\hat{\alpha}, \hat{\beta}}(\hat{H}_t^2)$ for the SET50 index.

For illustration purposes, Fig. 1 shows the density plots of the predictive densities on the fifth day of the 22 out-of-sample test days.

On each of the 22 days, we create a point estimate $\hat{r}_t$ for the index return by finding the mode of the daily predictive density function.

On the other hand, using a standard error (se) equal to 80% the average absolute change of the index returns (from historical data), we construct an interval forecast $[\hat{r}_t - se, \hat{r}_t + se]$ for day t. If the true index return in the test set belongs to this interval, we record a correct prediction.

Table 1 summarizes the accuracy rates of the interval forecasts over the 22-day periods of test data.

3.4   Result Evaluations

Prediction occupies a distinguished position in econometrics; hence, evaluating predictive ability is a fundamental concern. Until recently, most attention has been paid to evaluating point and interval forecasts. The literature on evaluating density forecasts is very little due to various reasons. Please see [8] and [12] for a survey.

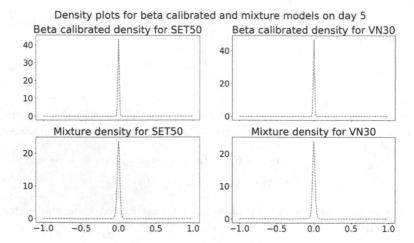

Fig. 1. Density forecasts for the fifth day in the out of sample test set

Table 1. Accuracy rates of interval predictions for daily returns

	SET50 index	VN30 index
Standard error	0.0028	0.004
Beta calibrated model	73%	61%
Linear mixture model	68%	60%

In this paper, we utilize a standard approach of calculating the probability integral transform (PIT) values of the outcomes in the forecast distributions. Assessment then rests on the question of whether such a sequence "looks like" a random sample from the uniform distribution on the interval $[0, 1]$ (see [7], p. 281): if so, the forecasts are said to be well calibrated.

In this article, the PIT values are computed for the 22-day out-of-sample test data i.e., let

$$q_t^1 = \hat{G}_t^1(y_t^1), \quad 1 \leq t \leq 22 \tag{12}$$

where $\hat{G}_t^1$ is the estimated beta-calibrated cumulative distribution as in formula (8) and y_t is the real VN30 return in the test set.

Similarly, we can define

$$q_t^2 = \hat{G}_t^2(y_t^2), \quad 1 \leq t \leq 22 \tag{13}$$

for the SET50 return test set.

The more q_t^1 and q_t^2 resemble random samples from the uniform distribution on $[0, 1]$ the better the predictive distribution G_t^1 and G_t^2 are.

To check the sequences q_t^1 and q_t^2, first, we present the probability-probability plots (PP-plots) to compare q_t^1 and q_t^2 with the uniform cumulative distribution function on the interval $[0, 1]$. Note that the uniform distribution on $[0, 1]$ is

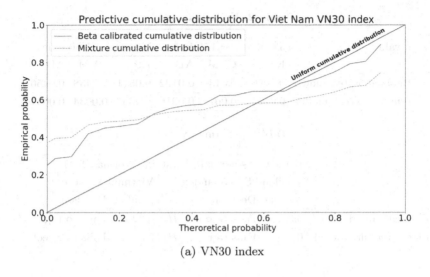

(a) VN30 index

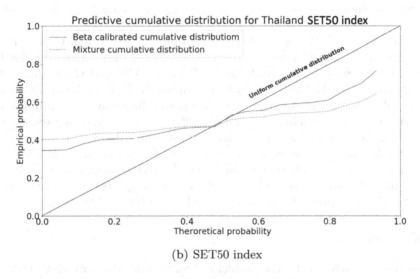

(b) SET50 index

Fig. 2. P-P plots for PIT values

the probability integral transform of the unknown true distribution. Please see Fig. 2a and b.

In the two figures, the solid curves are created from q_t^1 and q_t^2, and for comparison, we also plot the PIT values of the linear mixture models as the dashed curves. It can be realized that, when compared with the dashed curve representing the simple linear mixture, the solid curve representing the beta calibrated distribution lies closer to the 45° straight line representing the uniform distribution, especially in the upper and lower tail area. The plots suggests that the beta calibrated model produces better predictions, notably in the tail area which

Table 2. Uniformity tests

p values	Thailand SET50 index			Vietnam VN30 index		
	KS	CVM	AD	KS	CVM	AD
Beta-calibrated model	0.0076	0.0135	0.0122	0.0511	0.0438	0.0456
Linear mixture model	0.01224	0.0011	0.0016	0.0067	0.0034	0.0041

Table 3. Normality test

	Berkowitz's normality test H_0: Normal(0, 1)			
	Thailand SET50 index		Vietnam VN30 index	
	p value	Decision	p value	Decision
Beta-calibrated model	0.3325	Fails to reject H_0	0.3253	Fails to reject H_0
Linear mixture model	0.3212	Fails to reject H_0	0.3124	Fails to reject H_0

considerably affects the calculations of certain risk measures like value at risk or expected shortfall.

Second, following [2], we make use of a number of statistical tests to check uniformity for all the PIT values above. Calculations are carried out for the Kolmogorov-Smirnov (KS), the Anderson-Darling (AD) and the Cramer von-Mises (CVM) methods. Moreover, by transforming the PIT results into standard normal values using the standard normal quantile function, we can apply the Berkowitz test for normality. A summary of uniformity tests is presented in Table 2 and the summary for the Berkowitz test is in Table 3.

With a 5% significance level, the predictive PIT values for both VN30 and SET50 indices pass the uniform test. However, the SET50 index has much smaller p values in all statistical tests which, to some extent, suggests that the predictions of the SET50 are stabler, hence, more reliable than those of the VN30.

4   Conclusion

In this paper, motivated by the problem of forecasting stock market index, we propose a generalized combination method based on the classical linear mixture model and a beta transformation. Each stock in a market index serves as a forecaster for the index itself. Evidently, the composite index does not behave like any single stock; but, an appropriate combination of individual stocks will capture the systematic information of the market, therefore, predict the index well.

As for applications, the predictions for an emerging market like the Thailand stock index seem to be more reliable and stable while the results for a frontier market like the Vietnam stock index is not up to expectation yet. Hence, in subsequent research, we are exploring some new sources of market information to incorporate into our combined models. Meanwhile, we are also experimenting with some other transformations besides the beta one.

Acknowledgement. We would like to express our gratitude to professor Nguyen Trung Hung of New Mexico State University and Chiang Mai University for his generous support, guidance, encouragements and discussions over many years.

References

1. Bassetti, F., Casarin, R., Ravazzolo, F.: Bayesian nonparametric calibration and combination of predictive distributions. J. Am. Stat. Assoc. (2017)
2. Berkowitz, J.: Testing density forecasts, with applications to risk management. J. Bus. Econ. Stat. **19**(4), 465–474 (2001)
3. Bollerslev, T.: Generalized autoregressive conditional heteroskedasticity. J. Econ. **31**(3), 307–327 (1986)
4. Carpenter, B., Gelman, A., Hoffman, M., Lee, D., Goodrich, B., Betancourt, M., Brubaker, M.A., Guo, J., Li, P., Riddell, A.: Stan: a probabilistic programming language. J. Stat. Softw. **20**, 1–37 (2016)
5. Casarin, R., Mantoan, G., Ravazzolo, F.: Bayesian calibration of generalized pools of predictive distributions. Econometrics **4**(1), 17 (2016)
6. Clemen, R.T., Winkler, R.L.: Aggregating probability distributions. na (2005)
7. Dawid, A.P.: Present position and potential developments: some personal views: statistical theory: the prequential approach. J. Royal Stat. Soc. Ser. A (General) **147**, 278–292 (1984)
8. Diebold, F.X., Gunther, T.A., Tay, A.S.: Evaluating density forecasts (1997)
9. Engle, R.F.: Autoregressive conditional heteroscedasticity with estimates of the variance of United Kingdom inflation. Econ. J. Econ. Soc. **50**, 987–1007 (1982)
10. Genest, C., Zidek, J.V.: Combining probability distributions: a critique and an annotated bibliography. Stat. Sci. **1**, 114–135 (1986)
11. McLachlan, G., Peel, D.: Finite Mixture Models. Wiley, Hoboken (2004)
12. Mitchell, J., Wallis, K.F.: Evaluating density forecasts: forecast combinations, model mixtures, calibration and sharpness. J. Appl. Econ. **26**(6), 1023–1040 (2011)
13. Stone, M., et al.: The opinion pool. Ann. Math. Stat. **32**(4), 1339–1342 (1961)

Quantum Econometrics: How to Explain Its Quantitative Successes and How the Resulting Formulas Are Related to Scale Invariance, Entropy, and Fuzziness

Kittawit Autchariyapanitkul[1], Olga Kosheleva[2], Vladik Kreinovich[2](✉),
and Songsak Sriboonchitta[3]

[1] Faculty of Economics, Maijo University, Chiang Mai, Thailand
kittawit_a@mju.ac.th
[2] University of Texas at El Paso, El Paso, TX 79968, USA
{olgak,vladik}@utep.edu
[3] Faculty of Economics, Chiang Mai University, Chiang Mai 50200, Thailand
songsajecon@gmail.com

Abstract. Many aspects of human behavior seem to be well-described by formulas of quantum physics. In this paper, we explain this phenomenon by showing that the corresponding quantum-looking formulas can be derived from the general ideas of scale invariance and fuzziness. We also use these ideas to derive a general family of formulas that include non-quantum and quantum probabilities as particular cases – formulas that may be more adequate for describing human behavior than purely non-quantum or purely quantum ones.

1 Formulation of the Problem

Somewhat surprisingly, quantum models describe several aspects of human behavior. The main issue with which we deal in this paper is that many aspects of human behavior seem to be well-described by formulas from quantum physics; see, e.g., [1,3,12].

This is understandable on the qualitative level. The success of quantum models in describing several aspects of human behavior is understandable on the qualitative level (see, e.g., [8]): similar to quantum physics, every time we gain new knowledge we inevitably change the system. For example, once we learn a new dependence between the economic variables, we can make better predictions of economic phenomena and thus, change the behavior of decision makers.

But how can we explain this success on the quantitative level? The above qualitative arguments do not explain why not only *ideas* but also *formulas* from quantum physics are helpful in describing human behavior (see, e.g., [1]). In other words, while the *qualitative* success of quantum models is reasonable, their *quantitative* success remains a mystery.

© Springer International Publishing AG, part of Springer Nature 2018
V.-N. Huynh et al. (Eds.): IUKM 2018, LNAI 10758, pp. 264–275, 2018.
https://doi.org/10.1007/978-3-319-75429-1_22

What we do in this paper. In this paper, we show that the analysis of different types of uncertainty leads to the desired quantitative explanation of the success of quantum models in describing human behavior. To be more precise, we show that this analysis leads to a general formula that includes non-quantum and quantum probabilities as particular cases. We hope that some intermediate cases of this general formula will be even more accurate in describing human behavior than the currently used formulas based on non-quantum and quantum probabilities.

The structure of the paper is as follows. In Sect. 2, we briefly recall the main ideas behind quantum physics, and recall how the corresponding models help in describing human behavior. In Sect. 3, we show how to explain the quantum formulas, and how to derive a general formula that contains non-quantum and quantum formulas as particular cases. In Sects. 4 and 5, we describe how this formula is related to entropy and fuzzy.

2 Quantum Models and How They Describe Human Behavior: A Brief Reminder

The main difference between non-quantum and quantum probabilities. According to [4], the main difference between non-quantum and quantum probabilities can be explained on the example of the following two-slot experiment.

This experiment is about particle propagation. We have a particle generator – e.g., a light source or a radio source that generates photons, or a radioactive element that generates electrons or alpha-particles. There is an array of sensors at some distance from this generator. By detecting the particles, these sensors helps us estimate the probability that the original particle goes to the location x of the sensor. To be more precise, what we estimate is the probability density $\rho(x)$ corresponding to the sensor's location x.

There is a barrier between the source of the signals and the sensors. In this barrier, there are two slots that can be open or close. If both slots are closed, no particles come through, so the sensors do not detect anything. If one or both slots are open, detectors detect the particles.

We assume that the particles do not interact with each other; this is a reasonable assumption for electromagnetic waves (photos), and even for electrons – as long as their density is not too high.

Let us assume that first we open the first slot and leave the second slot closed. Let $\rho_1(x)$ denote the resulting probability density. Now, we can close the first slot, open the second slot, and measure the new probability density. We will denote this new probability density by $\rho_2(x)$. What will happen if we open both slots? What will then be the resulting probability density $\rho(x)$?

In non-quantum physics, the answer is simple. Indeed, in the new experiment, a particle reaches the sensor if it either went through the first slot or it went through the second slot. We set up the experiment in such a way that the particle cannot go through both slots. Thus, the probability that a particle passes via one of the two slots is equal to the sum of the probability of passing through the first slot and the probability of passing through the second slot: $\rho(x) = \rho_1(x) + \rho_2(x)$.

However, this is *not* what we observe in quantum physics. In quantum physics, to properly describe uncertainty, it is not sufficient to describe the corresponding probabilities, we also need to describe a complex-valued function $\psi(x)$ – called *wave function* – for which the probability density is equal to $\rho(x) = |\psi(x)|^2$.

In this case, if we know the wave function $\psi_1(x)$ corresponding to the first-slot-open case and the wave function $\psi_2(x)$ corresponding to the second-slot-open case, then the wave function $\psi(x)$ corresponding to the case when both slots are open is equal to $\psi(x) = \psi_1(x) + \psi_2(x)$. In this case, in general,

$$\rho(x) = |\psi(x)|^2 = |\psi_1(x) + \psi_2(x)|^2 \neq |\psi_1(x)|^2 + |\psi_2(x)|^2 = \rho_1(x) + \rho_2(x).$$

For example, in situations in which the values of all wave functions are positive real numbers, we get $\psi_1(x) = \sqrt{\rho_1(x)}$, $\psi_2(x) = \sqrt{\rho_2(x)}$, and thus, $\rho(x) = \psi^2(x)$ has the form

$$\rho(x) = (\sqrt{\rho_1(x)} + \sqrt{\rho_2(x)})^2 = \rho_1(x) + \rho_2(x) + 2\sqrt{\rho_1(x) \cdot \rho_2(x)} \neq \rho_1(x) + \rho_2(x).$$

Comment. In general, the values of $\psi_i(x)$ may be complex. In this case, by using triangle inequality, the only thing that we can conclude about $|\psi(x)| = |\psi_1(x) + \psi_2(x)|$ is that

$$||\psi_1(x)| - |\psi_2(x)|| \leq |\psi(x)| \leq |\psi_1(x)| + |\psi_2(x)|.$$

By squaring all three parts of this double inequality, we can conclude that the probability density function $\rho(x) = |\psi(x)|^2$ satisfies the inequality

$$(\sqrt{\rho_1(x)} - \sqrt{\rho_2(x)})^2 \leq \rho(x) \leq (\sqrt{\rho_1(x)} + \sqrt{\rho_2(x)})^2,$$

i.e., that

$$\rho_1(x) + \rho_2(x) - 2\sqrt{\rho_1(x) \cdot \rho_2(x)} \leq \rho(x) \leq \rho_1(x) + \rho_2(x) + 2\sqrt{\rho_1(x) \cdot \rho_2(x)}.$$

How is this related to human behavior. In the early 1980s, a group of researchers from the Republic of Georgia observed the behavior of kids in a two-door room; see, e.g., [13,14]. In some cases, both doors were open, in other cases, only one door was open. On the other side, boxes with treats were placed, and the researchers measured how frequently kids pick up treats from the box located at spatial location x.

It turns out that the older kids, after walking through the door, mostly went to the box which was the closest to this door. For these kids, the frequency $f(x)$ of selecting the box x when both doors were open was (approximately) equal to the sum $f_1(x) + f_2(x)$ of the frequencies $f_i(x)$ corresponding to the cases when the i-th door was open and the other door was closed. In other words, older kids exhibited non-quantum behavior.

Somewhat surprising, for younger kids (3–4 years old), the frequency $f(x)$ was different from the sum $f_1(x) + f_2(x)$: not only that, it was close to the quantum formula $f(x) = (\sqrt{f_1(x)} + \sqrt{f_2(x)})^2$.

For the adults – just like for the older kids – a large part of their decision making behavior is described by the traditional (non-quantum) probabilities. However, surprisingly, some aspects of their behavior are better described by the quantum formulas; see, e.g., [1,3].

Remaining questions. How can we explain the usability of quantum formulas? And how can we come up with formulas that take into account some similarity to quantum phenomena without requiring that all formulas are quantum ones?

In this paper, we use several uncertainty-related approaches to come up with such a more general formula.

3   How to Explain the Quantum Formulas and to Get a General Expression Containing Non-quantum and Quantum Formulas as Particular Cases

Formulation of the problem: reminder. For each sensor location x, we know the values of the probability densities $\rho_1(x)$ and $\rho_2(x)$ corresponding to situations when one of the slots is open. Based on these values, we need to estimate the value $\rho(x)$.

Let $f(a, b)$ denote the algorithm that transforms the known values $a = \rho_1(x)$ and $b = \rho_2(x)$ into the estimate for $\rho(x)$. In terms of this algorithm, the desired estimate has the form $\rho(x) = f(\rho_1(x), \rho_2(x))$.

The problem is: which function $f(a, b)$ is the most appropriate?

Comment. The two-slot experiment is just an example. The algorithm $f(a, b)$ can be used not only for the two-slot experiment, but for all possible situations when we need to combine the known probabilities $a = P(A)$ and $b = P(B)$ of two events A and B into an estimate $f(a, b)$ for the probability $P(A \vee B)$ of the disjunction $A \vee B$.

First natural requirement: commutativity. The estimate for $\rho(x)$ should not depend on which door we call the first one and which one we call the second one. Thus, we must have $f(\rho_1(x), \rho_2(x)) = f(\rho_2(x), \rho_1(x))$ for all possible values $\rho_1(x)$ and $\rho_2(x)$.

In other words, we must have $f(a, b) = f(b, a)$ for all possible values a and b, i.e., the function $f(a, b)$ must be commutative.

Second natural requirement: continuity. In practice, all probability values are estimated only approximately, as frequencies $\widetilde{\rho}_i(x) \approx \rho_i(x)$. The more observations we have, the more accurate these approximations, i.e., the closer the estimates to the actual probabilities. It is therefore reasonable to require that as the approximate values $\widetilde{\rho}_1(x)$ and $\widetilde{\rho}_2(x)$ tend to the actual values $\rho_1(x)$ and $\rho_2(x)$, the resulting estimate $f(\widetilde{\rho}_1(x), \widetilde{\rho}_2(x))$ should tend to the estimate

$f(\rho_1(x), \rho_2(x))$ based on the actual values. In other words, it is reasonable to require that the function $f(a, b)$ be *continuous*.

Third natural requirement: monotonicity. It is also reasonable to require that if one of the probabilities $\rho_i(x)$ increases, the resulting overall probability $\rho(x)$ should increase as well. In other words, it is reasonable to require that the function $f(a, b)$ is a (non-strictly) *increasing* function of each of its variables.

Fourth natural requirement: associativity. If we have three doors instead of two, then we can estimate the probability $\rho(x)$ corresponding to the case when all 3 doors are open in two different ways:

- We can first use the algorithm $f(a, b)$ to estimate the probability density function $\rho_{12}(x) = f(\rho_1(x), \rho_2(x))$ corresponding to the case when the first two doors are open, and then again apply the same algorithm $f(a, b)$ to combine the probability density function $\rho_{12}(x)$ with the probability density function $\rho_3(x)$, resulting in the value $f(f(\rho_1(x), \rho_2(x)), \rho_3(x))$.
- Alternatively, we can first combine the probability density functions corresponding to doors 2 and 3, resulting in $\rho_{23}(x) = f(\rho_2(x), \rho_3(x))$, and then combine the resulting probability density function with $\rho_1(x)$, resulting in

$$f(\rho_1(x), \rho_{23}(x)) = f(\rho_1(x), f(\rho_2(x), \rho_3(x))).$$

It is reasonable to require that these two estimates are equal for all possible values $\rho_i(x)$, i.e., that $f(f(a, b), c) = f(a, f(b, c))$ for all a, b, and c – in other words, that the operation $f(a, b)$ is *associative*.

Fifth natural requirement: scale-invariance. By definition, the probability density is probability divided by the length or area (or volume). In principle, we can use different units for measuring length, and thus, different units for measuring area or volume. If we replace the original measuring unit with the one which is λ times smaller, the numerical values of the probability density gets multiplied by λ.

It is reasonable to require that the estimating function $f(a, b)$ should not change if we thus re-scale all the values of the probability density, that $\rho(x) = f(\rho_1(x), \rho_2(x))$ should imply $\lambda \cdot \rho(x) = f(\lambda \cdot \rho_1(x), \lambda \cdot \rho_2(x))$. Thus, we require that $f(\lambda \cdot a, \lambda \cdot b) = \lambda \cdot f(a, b)$ for all possible values a, b, and λ.

Now, we are ready for formulate our main result.

Definition 1. *We say that a function $f(a, b) : \mathbb{R}_0^+ \times \mathbb{R}_0^+ \to \mathbb{R}_0^+$ from non-negative real numbers to non-negative real numbers is a* scale-invariant estimation function *if it is commutative, associative, continuous, (non-strictly) increasing, and $f(\lambda \cdot a, \lambda \cdot b) = \lambda \cdot f(a, b)$ for all a, b, and λ.*

Proposition 1. *The only scale-invariant estimation functions are $f(a, b) = 0$, $f(a, b) = \min(a, b)$, $f(a, b) = \max(a, b)$, and $f(a, b) = (a^\alpha + b^\alpha)^{1/\alpha}$ for some α.*

4 Proof

Comment. In principle, we could shorten our proof if we take into account the known general structure of 1-D semigroups [6,10], but, for pedagogical purposes, we decided to have a longer "from scratch" proof, since in this proof, all steps are clear and it is, thus more convincing to non-mathematical readers.

1°. Depending on whether the value $f(1,1)$ is equal to 1 or not, we have two possible cases: $f(1,1) = 1$ and when $f(1,1) \neq 1$. Let us consider these two cases one by one.

2°. Let us first consider the case when $f(1,1) = 1$. In this case, the value $f(0,1)$ can be either equal to 0 or different from 0. Let us consider both subcases.

2.1°. Let us first consider the first subcase, when $f(0,1) = 0$.

In this case, for every $b > 0$, scale invariance with $\lambda = b$ implies that

$$f(b \cdot 0, b \cdot 1) = b \cdot 0,$$

i.e., that $f(0,b) = 0$. By taking $b \to 0$ and using continuity, we also get $f(0,0) = 0$. Thus, $f(0,b) = 0$ for all b.

By commutativity, we have $f(a,0) = 0$ for all a. So, to fully describe the operation $f(a,b)$, it is sufficient to consider the cases when $a > 0$ and $b > 0$.

2.1.1°. Let us prove, by contradiction, that in this subcase, we have $f(1,a) \leq 1$ for all a.

Indeed, let us assume that for some a, we have $b \overset{\text{def}}{=} f(1,a) > 1$. Then, due to associativity and $f(1,1) = 1$, we have $f(1,b) = f(1, f(1,a)) = f(f(1,1),a) = f(1,a) = b$.

Due to scale-invariance with $\lambda = b$, the equality $f(1,b) = b$ implies that $f(b,b^2) = b^2$. Thus, $f(1,b^2) = f(1, f(b,b^2)) = f(f(1,b),b^2) = f(b,b^2) = b^2$.

Similarly, from $f(1,b^2) = b^2$, we conclude that for $b^4 = (b^2)^2$, we have $f(1,b^4) = b^4$, and, in general, that $f(1,b^{2^n}) = b^{2^n}$ for every n.

Scale invariance with $\lambda = b^{-2^n}$ implies that $f(b^{-2^n},1) = 1$. In the limit $n \to \infty$, we get $f(0,1) = 1$, which contradicts to our assumption that $f(0,1) = 0$. This contradiction shows that indeed, $f(1,a) \leq 1$.

2.1.2°. For $a \geq 1$, monotonicity implies $1 = f(1,1) \leq f(1,a)$, so $f(1,a) \leq 1$ implies that $f(1,a) = 1$.

Now, for any a' and b' for which $0 < a' \leq b'$, if we denote $r \overset{\text{def}}{=} \dfrac{b'}{a'} \geq 1$, then scale-invariance with $\lambda = a'$ implies that $a' \cdot f(1,r) = f(a' \cdot 1, a' \cdot r) = f(a',b')$. Here, $f(1,r) = 1$, thus $f(a',b') = a' \cdot 1 = a'$, i.e., $f(a',b') = \min(a',b')$. Due to commutativity, the same formula also holds when $a' \geq b'$. So, in this case, $f(a,b) = \min(a,b)$ for all a and b.

2.2°. Let us now consider the second subcase of the first case, when $f(0,1) > 0$.

2.2.1°. Let us first show that in this subcase, we have $f(0,0) = 0$.

Indeed, scale-invariance with $\lambda = 2$ implies that from $f(0,0) = a$, we can conclude that $f(2 \cdot 0, 2 \cdot 0) = f(0,0) = 2 \cdot a$. Thus $a = 2 \cdot a$, hence $a = 0$. The statement is proven.

2.2.2°. Let us now prove that in this subcase, $f(0,1) = 1$.

Indeed, in this case, for $a \overset{\text{def}}{=} f(0,1)$, we have, due to $f(0,0) = 0$ and associativity, that $f(0,a) = f(0, f(0,1)) = f(f(0,0),1) = f(0,1) = a$. Here, $a > 0$, so by applying scale invariance with $\lambda = a^{-1}$, we conclude that $f(0,1) = 1$.

2.2.3°. Let us now prove that for every $a \leq b$, we have $f(a,b) = b$. So, due to commutativity, we have $f(a,b) = \max(a,b)$ for all a and b.

Indeed, from $f(1,1) = 1$ and $f(0,1) = 1$, due to scale invariance with $\lambda = b$, we conclude that $f(0,b) = b$ and $f(b,b) = b$. Due to monotonicity, $0 \leq a \leq b$ implies that $b = f(0,b) \leq f(a,b) \leq f(b,b) = b$, thus $f(a,b) = b$. The statement is proven.

3°. Let us now consider the remaining case when $f(1,1) \neq 1$.

3.1°. Let us denote $v(k) \overset{\text{def}}{=} f(1, f(\ldots, 1) \ldots)$ (k times). Then, due to associativity, for every m and n, the value $v(m \cdot n) = f(1, f(\ldots, 1) \ldots)$ ($m \cdot n$ times) can be represented as

$$f(f(1, f(\ldots, 1) \ldots), \ldots, f(1, f(\ldots, 1) \ldots)),$$

where we divide the 1s into m groups with n 1s in each. For each group, we have $f(1, f(\ldots, 1) \ldots) = v(n)$. Thus, $v(m \cdot n) = f(v(n), f(\ldots, v(n)) \ldots)$ (m times).

We know that $f(1, f(\ldots, 1) \ldots)$ (m times) $= v(m)$. Thus, by using scale-invariance with $\lambda = v(n)$, we conclude that $v(m \cdot n) = v(m) \cdot v(n)$, i.e., that that function $v(n)$ is multiplicative. In particular, this means that for every number p and for every positive integer n, we have $v(p^n) = (v(p))^n$.

3.2°. If $v(2) = f(1,1) > 1$, then by monotonicity, we get $v(3) = f(1, v(2)) \geq f(1,1) = v(2)$, and, in general, $v(n+1) \geq v(n)$. Thus, in this case, the sequence $v(n)$ is (non-strictly) increasing.

Similarly, if $v(2) = f(1,1) < 1$, then we get $v(3) \leq v(2)$ and, in general, $v(n+1) \leq v(n)$, i.e., in this case, the sequence $v(n)$ is strictly decreasing.

Let us consider these two cases one by one.

3.2.1°. Let us first consider the case when the sequence $v(n)$ is increasing. In this case, for every three integers m, n, and p, if $2^m \leq p^n$, then $v(2^m) \leq v(p^n)$, i.e., $(v(2))^m \leq (v(p))^n$.

For all m, n, and p, the inequality $2^m \leq p^n$ is equivalent to $m \cdot \ln(2) \leq n \cdot \ln(p)$, i.e., to $\dfrac{m}{n} \leq \dfrac{\ln(p)}{\ln(2)}$. Similarly, the inequality $(v(2))^m \geq (v(p))^n$ is equivalent to $\dfrac{m}{n} \leq \dfrac{\ln(v(p))}{\ln(v(2))}$. Thus, the above conclusion "if $2^m \leq p^n$ then $(v(2))^m \leq (v(p))^n$" takes the following form:

for every rational number $\dfrac{m}{n}$, if $\dfrac{m}{n} \leq \dfrac{\ln(p)}{\ln(2)}$ then $\dfrac{m}{n} \leq \dfrac{\ln(v(p))}{\ln(v(2))}$.

Similarly, for all m', n', and p, if $p^{n'} \leq 2^{m'}$, then $v(p^{n'}) \leq v(2^{m'})$, i.e., $(v(p))^{n'} \leq (v(2))^{m'}$. The inequality $p^{n'} \leq 2^{m'}$ is equivalent to $n' \cdot \ln(p) \leq m' \cdot \ln(2)$, i.e., to $\dfrac{\ln(p)}{\ln(2)} \leq \dfrac{m'}{n'}$. Also, the inequality $(v(p))^{n'} \leq (v(2))^{m'}$ is equivalent to $\dfrac{\ln(v(p))}{\ln(v(2))} \leq \dfrac{m'}{n'}$. Thus, the conclusion "if $p^{n'} \leq 2^{m'}$ then $(v(p))^{n'} \leq (v(2))^{m'}$" takes the following form:

for every rational number $\dfrac{m'}{n'}$, if $\dfrac{\ln(p)}{\ln(2)} \leq \dfrac{m'}{n'}$ then $\dfrac{\ln(v(p))}{\ln(v(2))} \leq \dfrac{m'}{n'}$.

Let us denote $\gamma \overset{\text{def}}{=} \dfrac{\ln(p)}{\ln(2)}$ and $\beta \overset{\text{def}}{=} \dfrac{\ln(v(p))}{\ln(v(2))}$. For every $\varepsilon > 0$, there exist rational numbers $\dfrac{m}{n}$ and $\dfrac{m'}{n'}$ for which $\gamma - \varepsilon \leq \dfrac{m}{n} \leq \gamma \leq \dfrac{m'}{n'} \leq \gamma + \varepsilon$. For these numbers, the above two properties imply that $\dfrac{m}{n} \leq \beta$ and $\beta \leq \dfrac{m'}{n'}$ and thus, that $\gamma - \varepsilon \leq \beta \leq \gamma + \varepsilon$, i.e., that $|\gamma - \beta| \leq \varepsilon$. This is true for all $\varepsilon > 0$, so we conclude that $\beta = \gamma$, i.e., that $\dfrac{\ln(v(p))}{\ln(v(2))} = \gamma$. Hence, $\ln(v(p)) = \gamma \cdot \ln(p)$ and thus, $v(p) = p^\gamma$ for all integers p.

3.2.2°. We can reach a similar conclusion $v(p) = p^\gamma$ when the sequence $v(n)$ is decreasing and $v(2) < 1$, and a conclusion that $v(p) = 0$ if $v(2) = 0$.

3.3°. By definition of $v(n)$, we have $f(v(m), v(m')) = v(m + m')$. Thus, we have $f(m^\gamma, (m')^\gamma) = (m + m')^\gamma$. By using scale-invariance with $\lambda = n^{-\gamma}$, we get $f\left(\dfrac{m^\gamma}{n^\gamma}, \dfrac{(m')^\gamma}{n^\gamma}\right) = \dfrac{(m + m')^\gamma}{n^\gamma}$. Thus, for $a = \dfrac{m^\gamma}{n^\gamma}$ and $b = \dfrac{(m')^\gamma}{n^\gamma}$, we get $f(a, b) = (a^\alpha + b^\alpha)^{1/\alpha}$, where $\alpha \overset{\text{def}}{=} 1/\gamma$.

Rational numbers $r = \dfrac{m}{n}$ are everywhere dense on the real line, hence the values r^γ are also everywhere dense, i.e., every real number can be approximated, with any given accuracy, by such numbers. Thus, continuity implies that $f(a, b) = (a^\alpha + b^\alpha)^{1/\alpha}$ for every two real numbers a and b.

The proposition is proven.

Discussion. For $\alpha = 1$, we get the usual formula for the probability for the event $A \vee B$ when A and B are disjoint. For $\alpha = 0.5$, we get the quantum formula. Thus, we get the desired justified general formula or which traditional probabilistic formula and the quantum formula are particular cases.

5   Relation to Entropy

What we do in this section. Let us start our analysis of the resulting general formula. In this section, we show that this formula has an interesting relation to Shannon's entropy.

Informal analysis of the problem. As we have mentioned, most aspects of human behavior and human decision making can be described by the usual probabilistic formulas. This means that while some deviations from the usual formulas are needed – to take into account some aspects of human behavior which are better described by quantum formulas – the corresponding value α should be close to the value $\alpha = 1$ corresponding to the usual probabilistic case.

Since quantum formulas seem to capture some aspects of human behavior, and quantum formulas correspond to $\alpha < 1$, this means that the actual value α is close to 1 and smaller than 1. Thus, we can conclude that $\alpha = 1 - \varepsilon$ for some small $\varepsilon > 0$.

The fact that ε is small means that we can safely ignore terms which are quadratic (or of higher order) in terms of ε, and keep only terms which are linear in ε. Let us see how the above formula can be this simplified.

Formal analysis of the problem. For every value a, we have $a^\alpha = a^{1-\varepsilon} = a \cdot a^{-\varepsilon}$. Here, since $a = \exp(\ln(a))$, we get

$$a^{-\varepsilon} = (\exp(\ln(a))^{-\varepsilon} = \exp(-\varepsilon \cdot \ln(a)) \approx 1 - \varepsilon \cdot \ln(a).$$

Thus, $a^\alpha \approx a \cdot (1 - \varepsilon \cdot \ln(a)) = a - \varepsilon \cdot a \cdot \ln(a)$, and

$$(\rho_1(x))^\alpha + (\rho_2(x))^\alpha \approx \rho_1(x) + \rho_2(x) - \varepsilon \cdot (\rho_1(x) \cdot \ln(\rho_1(x)) + \rho_2(x) \cdot \ln(\rho_2(x))).$$

Similarly, since $\dfrac{1}{\alpha} = \dfrac{1}{1-\varepsilon} \approx 1 + \varepsilon$, we get $a^{1/\alpha} = a + \varepsilon \cdot a \cdot \ln(a)$. Thus,

$$\rho(x) = ((\rho_1(x))^\alpha + (\rho_2(x))^\alpha)^{1/\alpha}$$
$$\approx (\rho_1(x) + \rho_2(x) - \varepsilon \cdot (\rho_1(x) \cdot \ln(\rho_1(x))$$
$$+ \rho_2(x) \cdot \ln(\rho_2(x))) + \varepsilon \cdot (\rho_1(x) + \rho_2(x)) \cdot \ln(\rho_1(x) + \ln(\rho_2(x))).$$

Resulting formula and its relation to Shannon's entropy. Thus, we arrive at the following formula: $\rho(x) \approx \rho_1(x) + \rho_2(x) + \varepsilon \cdot \Delta\rho(x)$, where we denoted

$$\Delta\rho(x) \stackrel{\text{def}}{=} -\rho_1(x) \cdot \ln(\rho_1(x)) - \rho_2(x) \cdot \ln(\rho_2(x))$$
$$-(-(\rho_1(x) + \rho_2(x)) \cdot \ln(\rho_1(x) + \rho_2(x))).$$

The sum of expressions of the type $-\rho(x) \cdot \ln(\rho(x))$ is exactly what we see in the formula for Shannon's entropy, so we get a (unexpected but clear) connection of quantum-type effects with Shannon's entropy.

6   Fuzzy and Probabilistic Interpretations of Our General Formula

It is not easy to interpret our formula in probabilistic terms. In general, in the traditional probability theory, the probability $P(A \vee B)$ of the disjunction

$A \vee B$ cannot exceed the sum of the two corresponding probabilities: $P(A \vee B) \leq P(A) + P(B)$. Similarly, for the probability density functions $\rho_1(x)$ and $\rho_2(x)$, we should have $\rho(x) \leq \rho_1(x) + \rho_2(x)$.

However, in the quantum case,

$$\rho(x) = \rho_1(x) + \rho_2(x) + 2\sqrt{\rho_1(x) \cdot \rho_2(x)} > \rho_1(x) + \rho_2(x).$$

This is one of the results showing that quantum formulas cannot be interpreted in terms of the traditional probabilities.

A similar inequality holds for all possible values $\alpha < 1$. So what shall we do?

Comment. The fact that some aspects of human behavior cannot be adequately described in probabilistic terms is well known; see, e.g., [5]. For example, in certain situations, people estimate the probability that a person X is a professional *and* a feminist as higher than the probability that this person is a feminist – an inequality which is impossible for probabilities, since the probability of a sub-event cannot exceed the probability of the original super-event.

First option: fuzzy interpretation. A natural first option is to take into account that, since here, $P(A \vee B) > P(A) + P(B)$, the values $P(A)$, $P(B)$, and $P(A \vee B)$ are not real probabilities, they are non-probabilistic degrees of certainty. Thus, to describe these degrees, it is reasonable to consider the most well-known non-probabilistic uncertainty formalism: the formalism of fuzzy logic; see, e.g., [2,7,9,11,15].

Second option: probabilistic interpretation. Another option is to explicitly take into account that, e.g., $\rho_1(x)$ is the probability that the particle passed through the first slot *and* did not pass through the second slot.

This is a known possible interpretation of the above feminist paradox – that when people are asked to compare the probability that X is a feminist and the probability that X is a professional and a feminist, they interpret the first option as saying that X is a feminist but *not* a professional.

If we denote by A_1 the event that the particle passed through the first slot, this means that $\rho_1(x)$ is *not* the probability of this event A_1, but rather the probability of a composite event $A_1' \stackrel{\text{def}}{=} A_1 \,\&\, (\neg A_2)$, i.e., the value $P(A_1) - P(A \,\&\, A_2)$ Similarly, $\rho_2(x)$ is the probability of $A_2' \stackrel{\text{def}}{=} A_2 \,\&\, (\neg A_1)$, i.e., $P(A_2) - P(A_1 \,\&\, A_2)$.

Here, $\rho(x)$ is the probability of $A_1 \vee A_2$, i.e., the probability $P(A_1 \vee A_2) = P(A_1) + P(A_2) - P(A_1 \,\&\, A_2)$. Thus, our formula $(\rho(x))^\alpha = (\rho_1(x))^\alpha + (\rho_2(x))^\alpha$ can be interpreted as the following indirect formula for determining a function that described the probability $C(u, v) \stackrel{\text{def}}{=} P(A_1 \,\&\, A_2)$ in terms of the probabilities $u \stackrel{\text{def}}{=} P(A_1)$ and $v \stackrel{\text{def}}{=} P(A_2)$:

$$(u + v - C(u, v))^\alpha = (u - C(u, v))^\alpha + (v - C(u, v))^\alpha.$$

Comment. This interpretation is related to a fuzzy one. Indeed, in the probabilistic interpretation, we started with the assumption that the particle cannot go through both slots, i.e., that going through the second slot is equivalent to the negation of going through the first slot. We ended up by realizing that, to make sense of the corresponding formulas for the probabilities, we need to allow a non-zero probability that the particle goes through both slots.

This is exactly what fuzzy does: instead of assuming that a person is either young or not young, it takes into account that the same person can be to some extent young and to some extent not young.

In the quantum case, we have an explicit expression for the corresponding function $C(u, v)$. In general, from the above formula, we cannot extract an explicit expression for $C \overset{\text{def}}{=} C(u, v)$, but it is possible in the quantum case, when $\alpha = 1/2$. In this case, by squaring both sides of the above equation, we conclude that

$$u + v - C = u - C + v - C + 2\sqrt{(u - C) \cdot (v - C)}.$$

By cancelling equal terms on both sides and moving C to the left-hand side, we conclude that $C = 2\sqrt{(u - C) \cdot (v - C)}$. Squaring both sides of this equality, we get $C^2 = 4u \cdot v - 4(u + v) \cdot C + 4C^2$, i.e., the quadratic equation

$$3C^2 - 4(u + v) \cdot C + 4u \cdot v = 0,$$

with an explicit solution

$$C(u, v) = \frac{2(u + v) \pm \sqrt{4(u + v)^2 - 48u \cdot v}}{6} = \frac{u + v \pm \sqrt{(u + v)^2 - 12u \cdot v}}{3}.$$

Comment. It is worth mentioning that, as one can check, the resulting operation $C(u, v)$ is *not* associative.

Acknowledgments. We acknowledge the partial support of the Center of Excellence in Econometrics, Faculty of Economics, Chiang Mai University, Thailand. This work was also supported in part by the US National Science Foundation grant HRD-1242122. The authors are thankful to the anonymous reviewers for valuable suggestions.

References

1. Baaquie, B.E.: Quantum Finance: Path Integrals and Hamiltonians for Options and Interest Rates. Cambridge University Press, New York (2004)
2. Belohlavek, R., Dauben, J.W., Klir, G.J.: Fuzzy Logic and Mathematics: A Historical Perspective. Oxford University Press, New York (2017)
3. Busenmeyer, J.R., Bruza, P.D.: Quantum Models of Condition and Decision. Cambridge University Press, New York (2012)
4. Feynman, R., Leighton, R., Sands, M.: The Feynman Lectures on Physics. Addison Wesley, Boston (2005)
5. Kahneman, D.: Thinking Fast and Slow. Farrar, Straus, and Giroux, New York (2011)

6. Klement, E.P., Mesiar, R., Pap, E.: Triangular Norms. Springer, Heidelberg (2000). https://doi.org/10.1007/978-94-015-9540-7
7. Klir, G., Yuan, B.: Fuzzy Sets and Fuzzy Logic. Prentice Hall, Upper Saddle River (1995)
8. Kreinovich, V., Nguyen, H.T., Sriboonchitta, S.: Quantum ideas in economics beyond quantum econometrics. In: Proceedings of the 2018 International Econometrics Conference, Hanoi, Vietnam, 15–16 January 2018 (to appear)
9. Mendel, J.M.: Uncertain Rule-Based Fuzzy Systems: Introduction and New Directions. Springer, Cham (2017). https://doi.org/10.1007/978-3-319-51370-6
10. Mostert, P.S., Shield, A.L.: On the structure of semigroups on a compact manifold with boundary. Ann. Math. **65**, 117–143 (1957)
11. Nguyen, H.T., Walker, E.A.: A First Course in Fuzzy Logic. Chapman and Hall/CRC, Boca Raton (2006)
12. Penrose, R.: The Emperor's New Mind: Concerning Computers, Minds and The Laws of Physics. Oxford University Press, Oxford (1989)
13. Proceedings of the Conference on Dialogue on Personal Computers, IVERSI 1985, Tbilisi (1985). (in Russian)
14. Proceedings of the Conference on Semiotic Aspects of Formalizing Intelligent Activity, Kutaisi 1985, Moscow (1985). (in Russian)
15. Zadeh, L.A.: Fuzzy sets. Inf. Control **8**, 338–353 (1965)

Pairs Trading via Nonlinear Autoregressive GARCH Models

Benchawanaree Chodchuangnirun[1], Kongliang Zhu[2(✉)],
and Woraphon Yamaka[2,3]

[1] Faculty of Economics, Ramkhamhaeng University, Bangkok, Thailand
[2] Faculty of Economics, Chiang Mai University, Chiang Mai, Thailand
258zkl@gmail.com
[3] Center of Excellence in Econometrics, Chiang Mai University,
Chiang Mai, Thailand

Abstract. Pairs trading is a well-established speculative investment strategy in financial markets. However, the presence of extreme structural change in economy and financial markets might cause simple pairs trading signals to be wrong. To overcome this problem in detecting the buy/sell signals, we propose the use of three non-linear models consisting of Kink, Threshold and Markov Switching models. We would like to model the return spread of potential stock pairs by these three models with GARCH effects and the upper and lower regimes in each model are used to find the trading entry and exit signals. We also identify the best fit nonlinear model using the Akaike Information Criterion (AIC) and Bayesian Information Criterion (BIC). An application to the Dow Jones Industrial Average (DJIA), New York Stock Exchange (NYSE), and NASDAQ stock markets are presented and the results show that Markov Switching model with GARCH effects can perform better than other models. Finally, the empirical results suggest that the regime-switching rule for pairs trading generates positive returns and so it offers an interesting analytical alternative to traditional pairs trading rules.

Keywords: Pairs trading · Nonlinear model · GARCH · Stock index

1 Introduction

Many investors have taken a well-known strategy, which is pairs trading and it was invented at Morgan Stanley in 1987. Pairs trading strategy works by taking the arbitrage opportunity of temporary abnormality between two related assets. When such event exists, one asset can be overvalued relative to its pair. Then the overvalued asset is sold while the undervalued asset is bought. Pairs trading is a market-neutral strategy following two-step process: first, identify two stocks whose prices have moved together historically, and second, sell the winner and buy the loser when the price relation is broken. The profit can be made and the prices of the two stocks will converge to a mean if the past is a good mirror of the future. There are a great number of different studies within

© Springer International Publishing AG, part of Springer Nature 2018
V.-N. Huynh et al. (Eds.): IUKM 2018, LNAI 10758, pp. 276–288, 2018.
https://doi.org/10.1007/978-3-319-75429-1_23

pairs trading framework, such as distance approach, co-integration approach and time series approach. These can be sorted into three main approaches. Firstly, the distance method utilizes nonparametric distance matrices to calculate the sum of squared deviations between two normalized stock prices as the criteria to form pair trading opportunities. The most cited paper was that by Gatev et al. [8] who found that the strategy provides average annualized excess returns of up to 11% based on the large sample of US equities. Later, Perlin [16] furthered the analysis to examine the profitability and risk of the pairs trading strategy for Brazilian stock market. Do and Faff [6] replicated the original methodology of Gatev et al. [8] and by the sample period extension to June 2008. They confirmed pairs trading strategy to be profitable for a long period of time, despite at a decreasing rate. Secondly, Vidyamurthy [17] developed a co-integration approach. The co-integration approach describes how to figure out co-moving stocks relying on formal co-integration testing. Applying this method to pairs trading is mostly based on Gatev et al. [8] threshold rule. Vidyamurthy [17] suggested a univariate co-integration approach, which is employed to preselect the potential co-integrated pairs, and to design the trading rule with nonparametric methods, based on statistical information. By using co-integration approach, Miao [15] provided high frequency and dynamic pairs trading system. For co-integrated assets in a continuous-time economy, Chiu and Wong originated the optimal pairs trading strategy in a closed-form solution. Thirdly, the time series approach was developed by Elliott et al. [9], which utilizes a Kalman filter for estimating a parametric model of the mean-reverting spread, in which the formation period is ignored and the spread is assumed to follow the state space model. This approach focuses on describing mean-reversion of the spread with other time series methods rather than co-integration. Do et al. [6] criticized and extended the method of Elliott et al. [9] into the stochastic residual spread method to improve the former method.

In previous literature, the pair spread exhibited a non-linear behavior and it seems to switch across the economic regimes. In addition, the deviation of the spread may also temporarily or persistently endure (Bock and Mestel [2]). Therefore, non-linear models have been proposed to deal with this behavior and they were found to provide a better fitting performance to the pair spread than linear models. Bock and Mestel [2] and Yang et al. [18], Zhu et al. [19] employed the Markov Switching to develop a trading rules for pairs trading. Chen et al. [5] proposed an alternative threshold model to capture mean and volatility asymmetries in pair spread. We found that all of those studies show the superior and the better fitting performance of the non-linear models. However, as financial time series often exhibit a volatility clustering, asymmetry in conditional mean and variance, and fat-tailed distributions [5], it is important to capture this volatility. In order to capture the dynamic volatility of the pair spread, a generalized autoregressive conditional heteroskedasticity (GARCH) model of Engle [7] and Bollerslev [3] is also proposed to the non-linear models. To find the optimal investment decisions, the spreads of pairs stock are compared with predictions from calibrated model. In this study, three non-linear models consisting of Markov Switching, threshold, and kink models are employed to find the best prediction. Thus, we propose Markov Switching AR-GARCH, threshold-AR-GARCH, and Kink-AR-GARCH and compare the prediction performance of these competing models. For the Kink-AR-GARCH, based on our best knowledge, this is the first study that employed the Kink-AR-GARCH of Boonyasana and Chinnakum [4] to develop useful trading rules for pairs trading.

This paper is organized as follows: Sect. 1 offers the introduction. The three non-linear models with GARCH effects are presented in Sect. 2. In Sect. 3, we discuss the criteria for pairs trading rules. The empirical results are shown in Sect. 4. Finally, Sect. 5 presents the conclusions.

2   Methodology

In this study, we propose three non-linear models consisting of kink, threshold and Markov Switching models. We would like to model the return spread of potential stock pairs by these three models with GARCH effects and the upper and lower regimes in each model are used to find the trading entry and exit signals. In addition, in this study, we consider only lag-one in our model because we aim to test zero serial correlations lag-one autocorrelation. If the estimated parameter at lag-one is not statistically significant, this indicates that potential pair arbitrage opportunities may not exist [5]. For the GARCH equation, our study also considers only AR(1)-GARCH(1,1) since it is able to reproduce the volatility dynamics of financial data and also bring a good fit and accurate predictions in many empirical applications.

2.1   Kink Autoregressive GARCH Model

The model has been proposed in Boonyasana and Chinnakum [4]. They extend the classical GARCH model of Bollerslev [3] to the kink model with unknown threshold of Hansen [12] and the model is called Kink Autoregressive GARCH (KAR-GARCH). The feature of the model is that it allows structural change in both mean and variance equations. The function of each equation is continuous but the slope has a discontinuity at a threshold point or kink point. It splits the lag data into two groups based on a function. KAR-GARCH model can be written as

$$y_t = \phi_0 + \phi_1^-(y_{t-1} \leq \gamma) + \phi_1^+(y_{t-1} > \gamma) + \varepsilon_t, \tag{1}$$

$$\varepsilon_t = \sqrt{h_t}\eta_t, \tag{2}$$

$$h_t = \alpha_0 + \alpha_1^- \varepsilon_{t-1}^2(y_{t-1} \leq \gamma) + \alpha_1^+ \varepsilon_{t-1}^2(y_{t-1} > \gamma) \\ + \beta_1^- h_{t-1}(y_{t-1} \leq \gamma) + \beta_1^+ h_{t-1}(y_{t-1} > \gamma), \tag{3}$$

where ϕ_0, ϕ_1^- and ϕ_1^+ are the estimated parameters for mean equation (1) while α_0, α_1^-, α_1^+, β_1^- and β_1^+ are the estimated parameters of variance equation (3). r is the threshold parameter or kink point value defining the regimes for both mean and variance equations through indicator function and $(y_{t-1} > \gamma)$ for upper regime, $(y_{t-1} \leq \gamma)$ for lower regime. The error term, ε_t, is the noise with the i.i.d. distribution and assumed to have normal distribution, a Student-t distribution, and a skewed-t distribution. Consider the GARCH equation, the estimated parameters $\alpha_1^-, \alpha_1^+ > 0$, $\beta_1^-, \beta_1^+ > 0$, $\alpha_1^+ + \beta_1^+ < 1$ and $\alpha_1^- + \beta_1^- < 1$.

2.2   Threshold Autoregressive GARCH Model

This model has been proposed by Li and Li [13]. It allows mean and the conditional variance to vary across regimes. A general two-regime threshold autoregressive GARCH model can be expressed as

$$
\begin{aligned}
y_t &= \phi_0^1 + \phi_1^1 y_{t-1} + \varepsilon_{1t} & y_{t-d} &\leq \gamma \\
y_t &= \phi_0^2 + \phi_1^2 y_{t-i} + \varepsilon_{2t} & y_{t-d} &> \gamma
\end{aligned}
\tag{4}
$$

$$
\varepsilon_{1t} = \sqrt{h_{1t}}\,\eta_{1t}, \varepsilon_{2t} = \sqrt{h_{2t}}\,\eta_{2t},
\tag{5}
$$

$$
\begin{aligned}
h_{1t} &= \alpha_0^{(1)} + \alpha_1^{(1)} \varepsilon_{1t-1}^2 + \beta_1^{(1)} h_{1t-1} & y_{t-d} &\leq \gamma \\
h_{2t} &= \alpha_0^{(2)} + \alpha_1^{(2)} \varepsilon_{2t-1}^2 + \beta_1^{(2)} h_{2t-1} & y_{t-d} &> \gamma
\end{aligned}
\tag{6}
$$

where Eqs. (4) and (6) are the conditional mean and variance equation, respectively. $\phi_i^{(j)}$, $\alpha_i^{(j)}$, $\beta_1^{(j)}$, $j = 1, 2$ are the estimated parameters of the model in 2 regimes. Here, we restrict $\alpha_1^{(j)} > 0$, $\beta_1^{(j)} > 0$ and $\alpha_1^{(j)} + \beta_1^{(j)} < 1$. ε_t is the residual term which consists of the standard variance, h_t, and the standardized residual, η_t. Note that we assumed to have a normal distribution, a Student-t distribution, and a skewed-t distribution. The movement of the observations between the regimes is controlled by threshold variable y_{t-d} with the delay parameter being a positive integer. Note that we consider only lag-one, d is set to be 1, y_{t-1}. If y_{t-1} is greater or lower than threshold parameter, γ, the separated observations can be estimated as different regressions then the model can vary across regimes.

2.3   Markov Switching Autoregressive–GARCH Model

Roughly speaking, this model consists of the Markov regime-switching model proposed by Hamilton [11] and the GARCH of Engle [7] and Bollerslev [3]. As discussed by Bauwens et al. [1], the persistence in the estimated single regime of GARCH process could be considered as resulting from the misspecification and thus they introduced a way to control it using an MS-GARCH model where the regime switches are governed by a hidden Markov chain. Our study follow the works of Haas et al. [10]; Marcucci [14]; and Bauwens et al. [1], whose are used the Markov Switching GARCH approach to gain more ability to capture some stylized facts of financial time series. The general form of the Markov Switching AR–GARCH(1,1) model can be written as

$$
y_t = \phi_{0,S_t} + \phi_{i,S_t} y_{t-1} + \varepsilon_t,
\tag{7}
$$

$$
\varepsilon_t = \sqrt{h_{t,S_t}}\,\eta_t,
\tag{8}
$$

$$
h_{t,S_t} = \alpha_{0,S_t} + \alpha_{1,S_t} \varepsilon_{t-1}^2 + \beta_{1,S_t} h_{t-1},
\tag{9}
$$

where Eqs. (7) and (9) are the mean and variance equations, respectively, and both are regime dependent. This means that these two equations are allowed to switch across regime. The estimated variance equation parameters $\alpha_{1,S_t} > 0$, $\beta_{1,S_t} > 0$ and $\alpha_{1,S_t} + \beta_{1,S_t} < 1$ are to ensure the positive conditional variance, h_{t,S_t}. S_t is the state variable which is the probabilistic structure of the switching regime indicator and is defined by first-order Markov process with constant transition probabilities Q.

$$Q = \begin{bmatrix} p_{11} & p_{12} \\ p_{21} & p_{22} \end{bmatrix}, \tag{10}$$

where $p_{11} = \Pr(S_t = 1 | S_{t-1} = 1)$, $p_{22} = \Pr(S_t = 2 | S_{t-1} = 2)$, $p_{21} = \Pr(S_t = 2 | S_{t-1} = 1)$, and $p_{12} = \Pr(S_t = 1 | S_{t-1} = 2)$. To estimate the parameter set in this model, the maximum likelihood method is used and the general form of the likelihood can be defined as

$$L(\Theta_{S_t} | y) = f(\Theta_{S_t} | y) Pr(S_t = k), \tag{11}$$

where $f(\Theta_{S_t})$ is the density function, $\Theta_{S_t} = \{\phi_{0,S_t}, \phi_{i,S_t}, \alpha_{0,S_t}, \alpha_{1,S_t}, \beta_{1,S_t}\}$ is state dependent parameter set of the model and $Pr(S_t = k)$ is the filtered probabilities in each regime (k). To estimate $Pr(S_t = k)$ we employed Hamiltons filter of Hamilton [11] which can be written as

$$\Pr(S_t = k | \Theta_{S_t}) = \frac{f(y_t | S_{t=k}, \Theta_{S_t\,t-1}) \Pr(S_{t=k} | \Theta_{S_t,t-1})}{\sum\limits_{k=1}^{2} f(y_t | S_{t=k}, \Theta_{S_t\,t-1}) \Pr(S_{t=k} | \Theta_{S_t,t-1})}, \tag{12}$$

where $f(y_t | S_{t=k}, \Theta_{S_t\,t-1})$ is the density function of each regime k.

2.4 Pairs Trading

In this study, we employ pairs trading strategy, therefore the selection of pair stock is very important. To select the appropriate pair stock, we followed the study of Chen et al. [5] where the pair stock is selected using the lowest value of the Minimum Squared Distance method (MSD). The formula of this method is given as follows.

$$MSD = \sum_{t=1}^{T} (P_t^A - P_t^B)^2, \tag{13}$$

where P_t^A and P_t^B is the normalized stock price and $P_t^i = (p_t^i - \bar{p}_t^i)/sd_i$. Here, the two stocks with the first five smallest MSD pair among the 30 stocks are selected. Then, the selected pairs are used to calculate the series of returns by taking differences of the logarithms of the daily closing price. The next step is to compute the return spread, $y_t = r_t^A - r_t^B$ and fit our non-linear model with GARCH effects to the return spread. Once the best non-linear model is fitted, we conduct one of the trading rules, called distance method, introduced in Yang et al. [18] and first proposed by Gatev et al. [8].

$$
\begin{aligned}
&sell\ A\ buy\ B = y_t \geq \mu + \delta h_t \\
&sell\ B\ buy\ A = y_t < \mu - \delta h_t'
\end{aligned}
\tag{14}
$$

where μ is the predicted value that was estimated during the pair-formation period of the best fit model, δ is set at 1.96, 1.99 and 2.3428 when the error is normal, student-t, and skewed-t distribution, respectively, at the 5% significance level. Finally, we can obtain the average return of pairs trading returns on the sell stock A and buy stock B position by

$$
r_1 = \frac{1}{D}\left[-\ln\frac{P_{sell}^A}{P_{buy}^A} + \ln\frac{P_{sell}^B}{P_{buy}^B}\right].
\tag{15}
$$

Likewise, the average trading return on the buy stock A and sell stock B position is given as follows

$$
r_2 = \frac{1}{D}\left[\ln\frac{P_{sell}^A}{P_{buy}^A} - \ln\frac{P_{sell}^B}{P_{buy}^B}\right],
\tag{16}
$$

where D is number of holding days.

3   Estimate Results

3.1   Data Description

The daily closing prices of 36 companies in the Dow Jones Industrial Average (DJIA), New York Stock Exchange (NYSE), and NASDAQ stock markets are used as an illustration. The data are obtained from Thomson Reuters data stream, Faculty of Economics, Chiang Mai University from January 3, 2005 to December 30, 2016. Before the estimation of our model, we transform all the daily data to be log-return and the Augmented Dickey Fuller (ADF) test is employed for stationary test and we found that all log-returns are stationary at the level.

In this study, we select 36 companies comprising 3M (MMM), Apple (AAPL), American Express (AXP), AT&T(T), Bank of America (BAC), Boeing (BA), Caterpillar (CAT), Chevron Corporation (CVX), Cisco Systems (CSCO), Coca-Cola (KO), Dupont (DD), ExxonMobil (XOM), General Electric (GE), Google (GOOGL), The Goldman Sachs Group (GS), HewlettPackard (HPQ), The Home Depot (HD), Intel (INTC), IBM (IBM), Johnson & Johnson (JNJ), JPMorgan Chase (JPM), Lowes Companies (LOW), McDonalds (MCD), Merck (MRK), Microsoft (MSFT), Nike (NKE), PepsiCo (PEP), Pfizer (PFE), Procter & Gamble (PG), Travelers (TRV), UnitedHealth Group Incorporated (UNH), United Technologies Corporation (UTX), Verizon Communications (VZ), Wal-Mart (WMT), Walt Disney (DIS), and Yahoo (YHOO).

Table 1. Pair selection

Pair	Stock 1	Stock 2	MSD
1	HOME	LOWE'S	113.225
2	DISNEY	TRAVELERS	161.048
3	NIKE	DISNEY	174.666
4	TRAVELERS	3M	193.21
5	PEPSI	COLA	728.665

Source: Calculation

Prior to illustrating the pairs trading strategy, we calculate the MSD between any two normalized price series for all possible pair stocks. The number of possible pairs is 630. The MSD is conducted here to select the first five stock pairs that provide the lowest MSD. We find the five pairs trading candidates as presented in Table 1.

Then we fit a non-linear model with GARCH effects to these five selected pair returns. Once the model is fitted, the upper and lower threshold values, which are calculated from the standard deviation of return spread of the stock pair, are used as trading signals. In this study, we follow a line of literatures in the pairs trading strategy by specifying that if return spread is above or below the upper or lower threshold value, we then either sell or buy one stock and either buy or sell the other stock. Once the position is open and the spread falls back to the standard deviation line, the position is closed.

3.2 Model Selection

As we mentioned before, the study would like to model the return spread of potential stock pairs by the proposed three models with GARCH effects. The study also conducted with three different error distributions, namely Normal (norm), Student-t (std), and Skew Student-t (sstd), in two-regime model. To select the best fit non-linear model and distribution for our models, the Akaike Information Criterion (AIC) and Bayesian Information Criterion (BIC) are employed to compare the performance of our models. Table 2 provides evidence that Markov Switching models is the best fit non-linear model for all pairs. We find that NIKE-DISNEY pair, TRAVELERS-3M pair, PEPSI-COLA pair prefer student-t while Skew Student-t provides the best fit to HOME-LOWE pair, DISNEY-TRAVELERS pair. Please note that the numbers between AIC and BIC of PEPSI-COLA pair are not consistent, we select student-t the best fit for this pair according to BIC.

3.3 Estimation of MS-AR-GARCH Model

Table 3 shows the estimated results of MS-AR-GARCH(1,1) when the error term has student-t distribution for NIKE-DISNEY pair, TRAVELERS-3M pair, PEPSI-COLA pair, and Skew Student-t distribution for HOME-LOWE pair,

Table 2. Model selection

HOME-LOWE'S	norm		std		sstd	
	AIC	BIC	AIC	BIC	AIC	BIC
AR-GARCH	4198.4	4228.7	3700.2	3736.4	3701.2	3743.5
Kink-AR-GARCH	4183.6	4238.1	3706.2	3766.7	3893.5	3954.0
Threshold AR-GARCH	4310.3	4401.1	3946.0	4024.7	3771.7	3862.4
MS-AR-GARCH	3700.6	3773.2	3669.4	3754.1	**3630.8**	**3727.6**
DISNEY-TRAVELERS	norm		std		sstd	
	AIC	BIC	AIC	BIC	AIC	BIC
AR-GARCH	5695.3	5725.5	5238.2	5274.4	5239.2	5281.5
Kink-AR-GARCH	5687.2	5741.6	5249.3	5309.8	5383.3	5443.8
Threshold AR-GARCH	6095.9	6186.6	5563.4	5642.1	5476.9	5567.7
MS-AR-GARCH	5281.3	5353.9	5208.8	5293.5	**5074.6**	**5171.4**
NIKE-DISNEY	norm		std		sstd	
	AIC	BIC	AIC	BIC	AIC	DIC
AR-GARCH	6391.1	6421.3	5585.4	5621.7	5586.8	5629.1
Kink-AR-GARCH	6362.2	6416.6	5637.5	5698.0	5799.9	5860.4
Threshold AR-GARCH	6697.0	6787.8	5743.4	5822.0	5732.7	5823.5
MS-AR-GARCH	5714.4	5787.0	**4922.6**	**5007.3**	5390.0	5486.8
TRAVELERS-3M	norm		std		sstd	
	AIC	BIC	AIC	BIC	AIC	BIC
AR-GARCH	5087.2	5117.4	4385.7	4422.0	4387.3	4429.6
Kink-AR-GARCH	5118.7	5173.2	4413.8	4474.3	4602.3	4662.8
Threshold AR-GARCH	5694.6	5785.4	4662.7	4741.3	4679.3	4770.0
MS-AR-GARCH	4383.2	4455.8	**3716.6**	**3801.3**	4397.9	4494.7
PEPSI-COLA	norm		std		sstd	
	AIC	BIC	AIC	BIC	AIC	BIC
AR-GARCH	2473.9	2504.2	2061.2	2097.5	2061.4	2103.7
Kink-AR-GARCH	2487.4	2541.8	2065.1	2125.6	2276.5	2337.0
Threshold AR-GARCH	2635.9	2726.6	2152.8	2231.4	2116.5	2207.2
MS-AR-GARCH	2075.0	2147.6	**2004.0**	**2088.7**	2002.2	2099.0

Source: Calculation

DISNEY-TRAVELERS pair. The model provides two equations namely, mean equation and variance equation for two regimes. Lets consider the variance equation in order to interpret the meaning of each regime. It is important to identify which of these regimes presents a high volatility and which regime presents a low volatility. To answer this question, we consider the persistence of volatility shocks for each regime. Generally, the volatility persistence can be measured by the sum and the higher value of corresponds to the higher unconditional variance

Table 3. Estimation results of MS-AR-GARCH for the five pair returns

PAR	HOME-LOWE'S	DISNEY-TRAVELERS	NIKE-DISNEY	TRAVELERS-3M	PEPSI-COLA
$\phi_{0S_t=1}$	0.0049	0.0053	−0.047	−0.0012	−0.0048
$\phi_{1S_t=1}$	−0.0658***	−0.0087	−0.1059	−0.0456***	−0.0137
$\alpha_{0S_t=1}$	0.1295***	0.1016***	0.1068	0.0566	0.0589***
$\alpha_{1S_t=1}$	0.1191***	0.0844***	0.0442	0.0414	0.0155***
$\beta_{1S_t=1}$	0.7804***	0.9143***	0.9516	0.9538***	0.9764***
$v_{S_t=1}$	4.5917***	3.5132***	3.314	3.6676***	4.5210***
$\gamma_{S_t=1}$	1.0645***	0.9864***	NaN	NaN	NaN
$\phi_{0S_t=2}$	0.0149***	0.0013***	0.00001	0.00001	−0.0019
$\phi_{1S_t=2}$	0.0012	0.001	−0.0001	0.0001	− 0.1242***
$\alpha_{0S_t=2}$	0.0016***	0.00001***	0.00001	0.0001	0.0001
$\alpha_{1S_t=2}$	0.0001	0.0011***	0.00001	0.0001	0.735
$\beta_{1S_t=2}$	0.5360***	0.4141***	0.3956	0.4074***	0.2878***
$v_{S_t=2}$	2.3422***	2.1	2.1555	2.1000***	3.9705***
$\gamma_{S_t=2}$	1.8506***	1.5393***	NaN	NaN	NaN
p_{11}	0.9604***	0.9648***	0.9673	0.9763***	0.6004***
p_{22}	0.2720***	0.4144***	0.446	0.5505***	0.5204

Source: Calculation
*** indicates 1% significiant level.

of the process. According to Table 3, we can interpret the first regime for all pairs as the high persistence of volatility shock regime and second regime as low persistence volatility regime since the intercept of regime 1, $f_{0S_t=1}$, is higher than regime 2, $f_{0S_t=2}$. This evidence is very important for investors because putting an investment in different period seems to face with a different market situation.

Moreover, the Table 3 also provides the result of the transition matrix and shows that the regimes in all pairs are persistent because the probability of staying in their own regime is larger than 96%, while the probability of switching between these regimes is less than 4%, except for PEPSI-COLA pair. This indicates that only an extreme event can switch the pair returns to change between regimes.

4   Pairs Trading Strategy

The estimated results of MS-AR-GARCH(1,1) are then extended to find the upper and lower threshold values, which are calculated from their predicted value of return spread of the stock pair and use them as trading signals. Note that we sell stock A and buy stock B when the observed spread is larger than the predicted value. In contrast, we will sell stock B and buy stock A when the observed spread is smaller l than the predicted value, see Eq. (14).

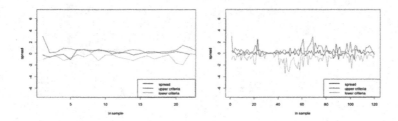

Fig. 1. HOME-LOWES pair return spread (Color figure online)

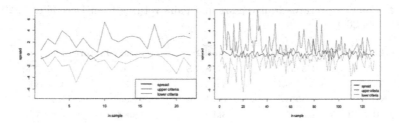

Fig. 2. DISNEY-TRAVELER pair return spread (Color figure online)

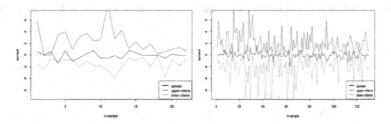

Fig. 3. NIKE-DISNEY pair return spread (Color figure online)

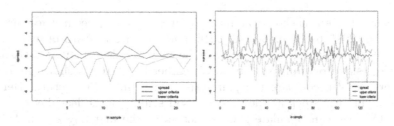

Fig. 4. TRAVELER-3M pair return spread (Color figure online)

Figures 1, 2, 3, 4 and 5 show the pairs return spread of five pairs return, during the in-sample period (1 month and 6 months), the red and green lines located at upper and lower criteria, respectively, which are employed as trading entry and exit signals. When the pairs trading strategy is used, the returns of pair trades are shown in Tables 4 and 5.

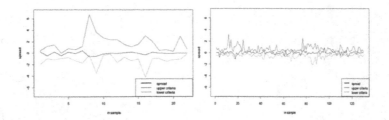

Fig. 5. PEPSI-COLA pair return spread (Color figure online)

Table 4. Company returns in five pairs and pairs returns from December 1, 2016 to December 30, 2016

Pair	Stock1	Individual return	Stock2	Individual return	Pair return
1	HOME	3.5607%	LOWE'S	0.0169%	3.0277%
2	DISNEY	5.3365%	TRAVELERS	6.0739%	1.3453%
3	NIKE	0.3553%	DISNEY	5.3365%	3.4213%
4	TRAVELERS	6.0739%	3M	3.4408%	6.9327%
5	PEPSI	5.6548%	COLA	3.2113%	No trade signal

Source: Calculation

Table 5. Company returns in five pairs and pairs returns from July 1, 2016 to December 30, 2016

Pair	Stock1	Individual return	Stock2	Individual return	Pair return
1	HOME	3.4408%	LOWE'S	−10.6507%	8.9199%
2	DISNEY	6.3143%	TRAVELERS	3.0037%	8.2369%
3	NIKE	−8.5955%	DISNEY	6.3143%	12.2944%
4	TRAVELERS	3.0037%	3M	1.7261%	5.9173%
5	PEPSI	0.9467%	COLA	−8.1117%	−2.5769%

Consider Table 4, it shows the mean returns of companies in five pairs from December 1, 2016 to December 30, 2016, and the mean return of five pairs. It is a one-month in sample result. In a similar way, we also calculate a six-month in-sample result from July 1, 2016 to December 30, 2016, as shown in Table 5. Moreover, we also compare the pair return with individual return. Here, we assume buying the stock at the first day and selling at the last day of trading period. Thus we can calculate gains or losses at the first day and last day of trading in-sample period. According to Table 4, we find that our trading signals contribute a positive return to all stock pairs during in-sample period. The profits are respectively 3.0227%, 1.3453%, 3.4213%, and 6.9327% for HOME-LOWE pair, DISNEY-TRAVELERS pair, NIKE-DISNEY pair, TRAVELERS-3M pair, PEPSI-COLA pair. For the comparison, we consider the individual return of stock and find that most individual stocks have returns less than pair stock returns.

In Table 5, the in-sample period is expanded to be six months and we find that our trading signals contribute a positive return to all stock pairs from July 1, 2016 to December 31, 2016. Lets consider the individual pair return, we find that NIKE-DISNEY pair provides the highest pair return, followed by HOME-LOWE pair. When we compare our pair returns with the single stock return, we find that the returns from our pairs trading strategy generate a higher return in all cases.

Finally, we can conclude that our pairs trading signal can generate a higher return when compared with the single mean return of individual stock. Thus, the obtained trading signal which was computed under the Markov switching approach works well in our application study.

5   Conclusions and Future Research

Movements of financial variables exhibit extreme fluctuations during turbulence period and market uncertainty. They can be affected by the institutional policies and intervention of regulatory authorities. Some studies also mentioned that news release from these institutes and government is another factor leading to structural change in the market. In this study, we aim to develop a pairs trading model that combines the non-linear approach to search for trading entry and exit signals. We have employed three nonlinear GARCH models consisting of Markov Switching GARCH, Threshold GARCH, and Kink GARCH and also compared the results with previous linear GARCH in order to confirm the structural change in our analysis. Additionally, the trading rule was applied for the investing universe of companies in the Dow Jones Industrial Average (DJIA), New York Stock Exchange (NYSE), and NASDAQ stock markets. The comparison results show that the Markov Switching model performs slightly better than other models for all pair returns.

The empirical results suggest that the regime-switching rule for pairs trading generates positive returns and so it offers an interesting analytical alternative to traditional pairs trading rules. Our pairs trading signal can generate a higher return when compared with the single mean return of individual stock. Thus, the obtained trading signal which was computed under the Markov switching approach works well in our application study.

A natural line of future research could be the extension of our framework to more than two Markov-regimes. This, however, leads to highly parameterized models which become increasingly difficult to estimate. However, other estimation procedures rather than our ML approach may be implemented, for example Bayesian Markov Chain Monte Carlo (MCMC) algorithms which have the potential to provide an alternative way of circumventing the problem of path dependence [1].

References

1. Bauwens, L., Preminger, A., Rombouts, J.: Theory and Inference for a Markov Switching GARCH Model, Center for 2009-11
2. Bock, M., Mestel, R.: A regime-switching relative value arbitrage rule. In: Fleischmann, B., Borgwardt, K.H., Klein, R., Tuma, A. (eds.) Operations Research Proceedings. Springer, Heidelberg (2009). https://doi.org/10.1007/978-3-642-00142-0_2
3. Bollerslev, T.: Generalized autoregressive conditional heteroskedasticity. J. Econom. **31**, 307–328 (1986)
4. Boonyasana, P., Chinnakum, W.: Forecasting Chinese international outbound tourists: Copula Kink AR-GARCH model. Thai J. Math. 215–229 (2016)
5. Chen, C.W.S., Chen, M., Chen, S.-Y.: Pairs trading via three-regime threshold autoregressive GARCH models. In: Huynh, V.-N., Kreinovich, V., Sriboonchitta, S. (eds.) Modeling Dependence in Econometrics. AISC, vol. 251, pp. 127–140. Springer, Cham (2014). https://doi.org/10.1007/978-3-319-03395-2_8
6. Do, B., Faff, R., Hamza, K.: A new approach to modeling and estimation for pairs trading. In: Proceedings of 2006 Financial Management Association European Conference, pp. 87–99, May 2006
7. Engle, R.F.: Autoregressive conditional heteroskedasticity with estimates of the variance of United Kingdom inflation. Econometrica **50**, 987–1007 (1982)
8. Gatev, E., Goetzmann, W.N., Rouwenhorst, K.G.: Pairs trading: performance of a relative-value arbitrage rule. Rev. Fin. Stud. **19**, 797–827 (2006)
9. Elliott, R.J., Van Der Hoek, J., Malcolm, W.P.: Pairs trading. Quant. Financ. **5**(3), 271–276 (2005)
10. Haas, M., Mittnik, S., Paolella, M.S.: A new approach to Markov-switching GARCH models. J. Financ. Econom. **2**(4), 493–530 (2004)
11. Hamilton, J.D.: A new approach to the economic analysis of nonstationary time series and the business cycle. Econometrica J. Econ. Soc. **57**, 357–384 (1989)
12. Hansen, B.E.: Regression kink with an unknown threshold. J. Bus. Econ. Stat. **35**(2), 228–240 (2017)
13. Li, C.W., Li, W.K.: On a double-threshold autoregressive heteroscedastic time series model. J. Appl. Econom. **11**, 253–274 (1996)
14. Marcucci, J.: Forecasting stock market volatility with regime-switching GARCH models. Stud. Nonlinear Dyn. Econom. **9**(4) (2005)
15. Miao, G.J.: High frequency and dynamic pairs trading based on statistical arbitrage using a two-stage correlation and cointegration approach. Int. J. Econ. Financ. **6**(3), 96 (2014)
16. Perlin, M.S.: Evaluation of pairs-trading strategy at the Brazilian financial market. J. Deriv. Hedge Funds **15**(2), 122–136 (2009)
17. Vidyamurthy, G.: Pairs Trading: Quantitative Methods and Analysis, vol. 217. Wiley, Hoboken (2004)
18. Yang, J.W., Tsai, S.Y., Shyu, S.D., Chang, C.C.: Pairs trading: the performance of a stochastic spread model with regime switching-evidence from the S&P 500. Int. Rev. Econ. Financ. **43**, 139–150 (2016)
19. Zhu, K., Yamaka, W., Sriboonchitta, S.: Pair trading rule with switching regression GARCH model. In: Huynh, V.-N., Inuiguchi, M., Le, B., Le, B.N., Denoeux, T. (eds.) IUKM 2016. LNCS (LNAI), vol. 9978, pp. 586–598. Springer, Cham (2016). https://doi.org/10.1007/978-3-319-49046-5_50

A Regime Switching for Dynamic Conditional Correlation and GARCH: Application to Agricultural Commodity Prices and Market Risks

Benchawanaree Chodchuangnirun[1], Woraphon Yamaka[2,4], and Chatchai Khiewngamdee[3,4(✉)]

[1] Faculty of Economics, Ramkhamhaeng University, Bangkok, Thailand
[2] Faculty of Economics, Chiang Mai University, Chiang Mai, Thailand
[3] Department of Agricultural Economy and Development, Faculty of Agriculture, Chiang Mai University, Chiang Mai, Thailand
[4] Center of Excellence in Econometrics, Chiang Mai University, Chiang Mai, Thailand
getiiecon@gmail.com

Abstract. Time varying correlations are often estimated with dynamic conditional correlation, generalized autoregressive conditional heteroskedasticity (DCC-GARCH) models which are based on a linear structure in both GARCH and DCC parts. In this paper, a Markov regime-switching DCC-GARCH (MS-DCC-GARCH) model is proposed in order to capture the time variations and structural breaks in both GARCH and DCC processes. The parameter estimates are driven by first order Markov chain. We provide simulation study to examine the accuracy of the model and apply it for empirical analysis of the dynamic volatility correlations between commodity prices and market risks. The proposed model is clearly preferred in terms of likelihood, Akaike information criterion (AIC), and likelihood ratio test.

Keywords: Markov switching · Credit default swap
Volatility index · Dynamic conditional correlation

1 Introduction

It is well known that there exists co-movement of financial volatilities more or less closely over time across assets and markets. Hence it is important to take into account this time varying co-movement. To date, the multivariate GARCH model are usually employed to measure the dynamic volatility and correlations between assets by estimating the covariance matrix between the assets. To extend the multivariate GARCH under the dynamic context, the dynamic copula and dynamic conditional correlation (DCC) is proposed to decompose the conditional covariance matrix into a conditional standard deviation matrix and a conditional correlation matrix, for example, dynamic conditional correlation DCC-GARCH

© Springer International Publishing AG, part of Springer Nature 2018
V.-N. Huynh et al. (Eds.): IUKM 2018, LNAI 10758, pp. 289–301, 2018.
https://doi.org/10.1007/978-3-319-75429-1_24

(Engle [1]) and dynamic copula-GARCH (Patton [2]). Although these dynamic linear approaches have been undertaken for time varying variance and correlations, they still have some limitations. For instance, it does not take into account the nonlinear structure in dynamic correlations since empirical evidences suggested that the behavior of economic and financial time series may exhibit different patterns over time (see, Billio and Caporin [3] and Pastpipatkul et al. [4,5]). Hence, instead of using the linear conditional variance, it has become more reliable to extend the linear model into the model that properly reflects those structural change patterns. As a consequence, the Markov Switching model was extended to those linear models (see, Billio and Caporin [3] and Da Silva Filho et al. [6]).

In this work, we focus on the Markov Switching dynamic correlation GARCH (MS-DCC-GARCH) model of Billio and Caporin [3]. The model has the flexibility of univariate GARCH but not the complexity of conventional multivariate GARCH. The model is estimated in two steps: firstly, a series of univariate GARCH parameters are estimated and then dynamic conditional correlation parameters are estimated in the second step. In other words, the parameters to be estimated in the correlation and GARCH processes are independent (Engle [1]). In the extension of Billio and Caporin [3], the model was modified to allow only a dynamic correlation to switch between two or more regimes. Therefore, they restricted the regime dependent structure only to the DCC parameters excluding GARCH parameters. They mentioned that the model in which all parameters are allowed to switch might lead to unstable results and difficult to reach the global maximum likelihood due to the large number of switching parameters. However, we are still concerned about the consistency and asymptotic normality of this two-step estimation. We need to have a reasonable regularity conditions to ensure that the first step will ensure consistency of the second step. The maximum of the second step will be a function of the first step parameter estimates, so these two steps need to be consistency. To overcome this problem, our study modifies the likelihood function of MS-DCC-GARCH by incorporating a likelihood of DCC to likelihood of Multivariate GARCH in order to do a one-step maximum likelihood and also modifies possible change in all parameters in both GARCH and DCC parts.

The main objective of this research is to extend MS-DCC GARCH by allowing the parameters in mean, variance and correlation equations to switch across different regimes or to be state dependent according to the first order Markov process. This means that all parameters are governed by a state variable S_t which is assumed to evolve according to S_{t-1} with transition probability. Then the one step maximum likelihood is constructed as an estimator of the model. Subsequently, we apply our model to study the dynamic volatility correlations between commodity prices and market risks.

Measuring the dynamic volatility correlations between commodity prices and market risks has an important implication for economic growth and investment. The increased integration between financial markets and the commodity markets provides an alternative way for investors to invest, diversify and hedge their portfolio's risks. To manage the portfolio's risks, investors and regulators need to take into account the correlation between assets across international financial markets. This study therefore aims to investigate the relationship between the agricultural commodity prices in the light of the risk perceptions and uncertainty that lead to the global financial crisis. We consider the indicators of the perceived global risk and global market conditions, namely, the volatility index and credit default swaps index. These two indicators are the benchmark proxy for the risk perception of investors (Gozgor, Kablamac [7]). As a consequence, in this study, we will show the performance of our model to measure the dynamic correlation and volatility among agricultural commodities including corn, wheat, and soybean; and global risk including volatility index and credit default swaps index.

The organization of this paper is as follows. Section 2 describes the methodologies used in the model. Section 3 conducts a simulation study to evaluate and illustrate our model. Section 4 employs the proposed model to investigate the dynamics between three agricultural commodities and market risks. Finally, we provide concluding remarks in Sect. 5.

2   Methodology

In this study, our concern is on the non-linear behavior of financial data which entitles the conventional DCC-GARCH model not appropriate for describing the correlation and volatility of the returns of the assets. To deal with this problem, a Markov Switching dynamic conditional correlation GARCH (MS-DCC-GARCH) is considered in this study. We generalizes the MS-DCC-GARCH model of Billio and Caporin [3], Pelletier [8] and Chen [9] in that the parameters to be estimated in the GARCH and DCC processes are dependent and allowed to vary across regimes.

2.1   Univariate GARCH (1, 1) Model

In the application study, the GARCH (1, 1) model is sufficient to capture the volatility clustering in the data (Bollerslev et al. [10]). Thus, the GARCH (1, 1) specification for the volatility spillover model follows Bollerslev [11] and is specified as

$$r_{i,t} = u_i + \varepsilon_{i,t} = u_i + \sqrt{\sigma_{i,t}^2 a_{i,t}}, \tag{1}$$

$$\sigma_{i,t}^2 = \alpha_{i,0} + \alpha_{i,1}\varepsilon_{i,t-1}^2 + \beta_{i,1}\sigma_{i,t-1}^2, \tag{2}$$

where $r_{i,t}$ is return of asset i at time t, u_i is constant term, $\varepsilon_{i,t}$ is the error term composed of a sequence of iid standardized residual with normal distribution, $a_{i,t}$, and conditional variance, $\sigma_{i,t}^2$. Bollerslev [11] provides a systematic framework for asset volatility modeling which has proved particularly valuable in time series with time-varying variance, $\sigma_{i,t}^2$ as presented in Eq. 2. Some restrictions are set in this model as follows: $\alpha_0 > 0, \alpha_1 \geq 0, \beta_1 \geq 0$, and $(\alpha_1 + \beta_1) < 1$. The latter constraint on $\alpha_1 + \beta_1$ implies that the unconditional variance of $r_{i,t}$ is finite.

2.2 DCC-GARCH (1, 1) Model

The DCC-GARCH model is the extension of GARCH model with the purpose to capture multivariate volatility (see, Engle [1]). The advantage of the DCC model is that we can estimate the time-varying correlations between multi-dimensional returns instead of a constant correlation (CCC model of Bollerslev [12]). Considering n-dimensional return series R_t we can write multivariate DCC-GARCH as:

$$R_t = U + \sqrt{H_t^2 A_t}, \tag{3}$$
$$H_t = D_t Z_t D_t, \tag{4}$$

where U is an $n \times 1$ vector of constants, A_t is an $n \times T$ matrix of standardized residuals. H_t is an $n \times T$ matrix of conditional variances from univariate GARCH model in Sect. 2.1. Z_t is the conditional correlation matrix of the standardized residuals $\varepsilon_{i,t}$. $D_t = \operatorname{diag}(\sigma_{1,t}^{1/2}, \cdots, \sigma_{n,t}^{1/2})$ is the diagonal matrix composed of the conditional variance of n returns. Z_t is given by:

$$Z_t = Q_t^{*-1/2} Q_t Q_t^{*-1/2}, \tag{5}$$

then, the DCC equation can be specified by

$$Q_t = (1 - \theta_1 - \theta_2)\overline{Q} + \theta_2 Q_{t-1} + \theta_2 \varepsilon_{t-1}\varepsilon_{t-1}', \tag{6}$$

where, $\overline{Q} = \dfrac{1}{T}\sum_{t=1}^{T} \varepsilon_{t-1}\varepsilon_{t-1}'$ is the unconditional covariance matrix of the standardized residuals, $\varepsilon_{t-1} = \{\varepsilon_{1,t-1}, \cdots, \varepsilon_{n,t-1}\}$ and Q_t^* is a diagonal matrix with the square root of the diagonal elements of Q_t. The coefficients θ_1 and θ_2 must satisfy: $0 \leq \theta_1 + \theta_2 < 1$.

2.3 Markov Switching DCC-GARCH (1, 1)

As we mentioned before, the study aims to allow all parameters to switch across states or regimes. Thus, the general form of the Markov switching DCC-GARCH (1, 1) model can be written as

$$R_t = U + \sqrt{H_{S_t}} A_{S_t}, \tag{7}$$
$$H_{S_t} = D_{S_t} Z_{S_t} D_{S_t}, \tag{8}$$
$$Z_{S_t} = Q_{S_t}^{*-1} Q_{S_t} Q_{S_t}^{*-1}, \tag{9}$$
$$Q_{S_t,t} = (1 - \theta_{1,S_t} - \theta_{2,S_t})\overline{Q}_{S_t,t} + \theta_{2,S_t} Q_{S_t,t-1} + \theta_{2,S_t} \varepsilon_{S_t,t-1}\varepsilon_{S_t,t-1}', \tag{10}$$

where Eqs. (7) and (8) are the mean and variance equations, respectively, and they are allowed to switch across regime. The feature of the Markov switching model is the estimated parameters in mean, variance and correlation equations, Eqs. (7), (8) and (9) can switch across different regimes or are state dependent according to the first order Markov process. This means that all parameters are governed by a state variable S_t which is assumed to evolve according to S_{t-1} with transition probability, p_{ij}, thus

$$P(S_j = 1| S_{t-1} = i) = p_{ij}, \sum_{j=1}^{J} p_{ij} = 1, \text{ for } i = 1, \cdots, J. \tag{11}$$

In this study, two regimes are considered thus, the first order Markov process could be written as:

$$\begin{aligned} P(S_j = 1| S_{t-1} = 1) = p_{11} \\ P(S_j = 1| S_{t-1} = 2) = p_{12} \\ P(S_j = 2| S_{t-1} = 1) = p_{21} \\ P(S_j = 2| S_{t-1} = 2) = p_{22} \end{aligned} \tag{12}$$

Let $\Theta = (U, \alpha_{S_t}, \beta_{S_t}, \theta_{1,S_t}, \theta_{2,S_t}, p_{11}, p_{22})$ be the vector of model parameters. From Eqs. (7–10), according to Engle [1], we can write the likelihood function of DCC-GARCH (1, 1) as

$$L(\Theta) = L_v(\theta) \cdots L_Z(\alpha, \beta), \tag{13}$$

where $L_v(\theta)$ and $L_Z(\alpha, \beta)$ are volatility part and a correlation part, respectively. Thus, we can rewrite our likelihood of Markov switching DCC-GARCH (1, 1) as

$$L(\Theta_{S_t}) = L_v(\theta_{S_t}) \cdots L_Z(\alpha_{S_t}, \beta_{S_t}), \tag{14}$$

where

$$L_v(\theta_{S_t}) = \prod_{t=1}^{T} \frac{1}{(2\pi)^{k/2}} \exp\{-\frac{1}{2}\varepsilon_{S_t}^T \varepsilon_{S_t}\}, \tag{15}$$

and

$$L_R(U_{S_t}, \alpha_{S_t}, \beta_{S_t}) = \prod_{t=1}^{T} \frac{1}{(2\pi)^{k/2} |Z_{S_t}|} \exp\{-\frac{1}{2} e_{S_t}^T Z_{S_t}^{-1} e_{S_t}\}. \tag{16}$$

Therefore, we will denote the full likelihood of Markov switching DCC-GARCH (1, 1) with J regimes for k assets by

$$L(\Theta_{S_t}| r_{1,t}, \cdots, r_{kt}) = \sum_{j=1}^{J} \left(\prod_{t=1}^{T} \prod_{k=1}^{K} L(\Theta_{S_t}| r_{kt}) \left(Pr(S_t = j| \Theta_{S_t})\right) \right). \tag{17}$$

2.4   Hamilton Filter

According to Eq. (17), the filtered probability $Pr(S_t = j| \Theta_{t,S_t})$ is an important process to assign estimated coefficient and variance parameters into two different regimes. The most famous filtering approach is introduced in Hamilton [13] called the Hamilton Filter. Suppose, we assume two-regime Markov switching DCC-GARCH, then the Hamilton filter is executed according to the following algorithm.

1. Given an initial guess of transition probabilities which are the probabilities P of switching between regimes, the transition probabilities of two regimes are

$$P = \begin{bmatrix} p_{11} & p_{12} \\ p_{21} & p_{22} \end{bmatrix} \tag{18}$$

2. Update the transition probabilities of each state with the past information including the parameters in the system equation, $\Theta_{S_{t-1}}$ and P, for calculating the likelihood function in each state at time t. After that, the probabilities of being in each state are to be updated by the following formula,

$$Pr(S_t = j)| \theta_{S_t}) = \frac{f(r_{1,t}, \cdots, r_{kt}| S_t = j\Theta_{S_{t-1}})Pr(S_t = j| \Theta_{S_{t-1}})}{\displaystyle\sum_{j=1}^{J} f(r_{1,t}, \cdots, r_{kt}| S_t = j\Theta_{S_{t-1}})Pr(S_t = j| \Theta_{S_{t-1}})} p_{jj}, \tag{19}$$

where $f(r_{1,t}, \cdots, r_{kt}| S_t = j\Theta_{S_{t-1}})$ is the full likelihood function in Eq. (10) and $Pr(S_t = j| \Theta_{S_{t-1}})$ is a filtered probability.

3. Iterate steps 1 and 2 for $t = 1, \cdots, T$.

3   Simulation Study

In this section, we examine the accuracy of the proposed model by conducting a simulation study. To this end, we consider a two-regime MS-DCC-GARCH with 3 variables and the true parameters are provide in the Table 1. In this simulation, the model is simulated based on a normal distribution. Moreover, we consider three sample sizes: $T = 1000, T = 2000$, and $T = 3000$. We randomly the state variable S_1 from a first-order Markov process taking values $\{1, 2\}$ and set $p_{11} = 0.90$ and $p_{22} = 0.85$.

From the simulation results in Table 1, we can see that the estimated parameters are close to their true values in the first column. In addition, when the number of sample is larger, the model performs more accurate.

4   Real Data Estimation

4.1   Data Description

We examine a systematic relationship between agricultural commodity prices, consisting of corn, wheat, and soybean futures, and the market risks, which are volatility index (VIX) and credit default swaps index (CDS), over the period from 1/4/2005 through 31/3/2017, covering 625 observations. We collect all the weekly data from Thomson Data stream database and convert into returns in a standard method as log differences.

Table 1. Simulation results

Parameter	True	Estimated		
		T = 1000	T = 2000	T = 3000
$\alpha_{01,S_t=1}$	0.001	0.0006	0.0005	0.0006
$\alpha_{11,S_t=1}$	0.02	0.001	0.0015	0.0021
$\beta_{11,S_t=1}$	0.7	0.8008	0.7998	0.8098
$\alpha_{02,S_t=1}$	0.003	0.002	0.0017	0.002
$\alpha_{12,S_t=1}$	0.005	0.0003	0.0027	0.005
$\beta_{12,S_t=1}$	0.7	0.7998	0.7994	0.8023
$\alpha_{03,S_t=1}$	0.004	0.0047	0.004	0.0042
$\alpha_{13,S_t=1}$	0.04	0.0001	0.0047	0.004
$\beta_{13,S_t=1}$	0.8	0.7997	0.7996	0.8086
$\alpha_{01,S_t=2}$	0.009	0.0257	0.0241	0.0275
$\alpha_{11,S_t=2}$	0.07	0.0538	0.0548	0.0614
$\beta_{11,S_t=2}$	0.9	0.8896	0.8878	0.8882
$\alpha_{02,S_t=2}$	0.005	0.0467	0.0481	0.0477
$\alpha_{12,S_t=2}$	0.02	0.016	0.0161	0.0155
$\beta_{12,S_t=2}$	0.95	0.8984	0.8873	0.8865
$\alpha_{03,S_t=2}$	0.006	0.0084	0.0068	0.0078
$\alpha_{13,S_t=2}$	0.15	0.043	0.206	0.1445
$\beta_{13,S_t=2}$	0.8	0.8936	0.8972	0.904
$\theta_{1,S_t=1}$	0.01	0.0078	0.0079	0.0079
$\theta_{2,S_t=1}$	0.95	0.9213	0.9203	0.9205
$\theta_{1,S_t=2}$	0.02	0.0149	0.013	0.0146
$\theta_{2,S_t=2}$	0.6	0.5005	0.5	0.5002
P_{11}	0.9	0.8943	0.8953	0.8944
P_{22}	0.85	0.8794	0.8751	0.8682

The descriptive statistics of return series are shown in Table 2. This table shows the characteristic of the data where the positive mean are found in all returns, excepting VIX return. Also, the table presents the Maximum, Minimum, and Median of the returns. We can see that the returns show similar characteristics. All returns exhibit high kurtosis but small skewness. This indicates that these returns have a small tail. Furthermore, the normality of returns are examined by the normality Jarque-Bera test and the result shows a strongly rejected of all returns. In addition, the result of the Augmented Dickey-Fuller (ADF) test suggests that the null hypothesis of the unit root test can be rejected and that all the variables are stationary.

Table 2. Descriptive statistics

	Corn	Wheat	Soybean	VIX	CDS
Mean	0.0004	0.0002	0.0003	−0.0001	0.0002
Median	0.001	−0.0002	0.0018	−0.0041	−0.0025
Maximum	0.09	0.07	0.10	0.34	0.17
Minimum	−0.11	−0.07	−0.11	−0.24	−0.13
Std. Dev.	0.02	0.02	0.02	0.06	0.03
Skewness	−0.42	0.13	−0.83	0.63	0.42
Kurtosis	6.11	3.88	8.99	5.79	5.92
Jarque-Bera	271.67	22.04	1007.61	244.51	241.08
Probability	0	0	0	0	0
ADF-Prob.	0	0	0	0	0

5 Empirical Findings

5.1 MS-DCC-GARCH (1, 1) Results

Table 3 reports two-regime MS-DCC-GARCH (1, 1) estimation results for all pairs to examine the regime dependent dynamic conditional correlation and variance. Note that, our model allows the GARCH and DCC parameters to vary across the regimes. It is clear from the table that almost all parameters are highly significant at 1%. The degree of volatility persistence for the model can be obtained by summing ARCH, α_{1,S_t}, and GARCH, β_{1,S_t}, parameters. Different from the previous MS-DCC-GARCH models, our model distinguishes the volatility persistence into two regimes. Different results have been obtained from these parameters. We find that the sum of α_{1,S_t} and β_{1,S_t} in regime 1 is not close to 1 for all returns and also lower than in regime 2, indicating that volatility is likely to be high in regime 2. The volatility persistence coefficients measured by $\alpha_{1,S_t} + \beta_{1,S_t}$ in the GARCH specification are 0.8064, 0.6433, 0.7479, 0.8350, and 0.8914 in regime 1 and 0.9372, 0.9059, 0.9999, 0.9219, and 0.9912 in regime 2 for the the returns of corn futures, wheat futures, soybean futures, VIX, and CDS, respectively. These findings lead us to interpret regime 1 as low volatility regime while regime 2 as high volatility regime. Then, considering the regime dependent conditional correlations, the parameters θ_{1,S_t} and θ_{2,S_t} are highly significant in both regimes. Therefore, there are significant correlations among the returns in both regimes. The estimates $\theta_{1,S_t} + \theta_{2,S_t}$ for across the regimes are not quite different, the low and high volatility regimes are characterized by different dynamic correlation structures. Indeed, the sums $\theta_{1,S_t} + \theta_{2,S_t}$ are 0.9602 (0.9480) for the low (high) volatility regime. The findings suggest that the low correlation is associated with the high volatility regime and vice versa regarding the relationship between agricultural commodities and market risks. Da Silva Filho [6]

Table 3. MS-DCC-GARCH parameters estimated

			Regime 1		
GARCH					
	Corn	**Wheat**	**Soybean**	**VIX**	**CDS**
$\alpha_{0,S_t=1}$	0.0001***	0.0002	0.0001***	0.0015***	0.0003***
	(0.00001)	(0.431)	(0.00001)	(0.00001)	(0.00001)
$\alpha_{1,S_t=1}$	0.0554	0.0903	0.1901***	0.0658	0.1545***
	(0.1450)	(0.184)	(0.00001)	(0.4953)	(0.00001)
$\alpha_{1,S_t=1}$	0.8818***	0.8156***	0.8098***	0.8561***	0.8367***
	(0.00001)	(0.0001)	(0.00001)	(0.00001)	(0.00001)
Unconditional variance	0.9372	0.9059	0.9999	0.9219	0.9912
			Regime 2		
GARCH					
	Corn	**Wheat**	**Soybean**	**VIX**	**CDS**
$\alpha_{0,S_t=2}$	0.0001***	0.0002	0.0001***	0.0015***	0.0003***
	(0.00001)	(0.431)	(0.00001)	(0.00001)	(0.00001)
$\alpha_{0,S_t=2}$	0.0554	0.0903	0.1901***	0.0658	0.1545***
	(0.1450)	(0.184)	(0.00001)	(0.4953)	(0.00001)
$\alpha_{0,S_t=2}$	0.8818***	0.8156***	0.8098***	0.8561***	0.8367***
	(0.00001)	(0.0001)	(0.00001)	(0.00001)	(0.00001)
Unconditional variance	0.9372	0.9059	0.9999	0.9219	0.9912

DCC Regime 1	0.1405***		0.8197***
	(0.00001)		(0.0001)
DCC Regime 2	0.0296***		0.9184***
	(0.0048)		(0.00001)

	Transition parameter	**Duration**
p_{11}	0.8513***	6.7276
	(0.0001)	
p_{22}	0.8172***	5.4689
	(0.0001)	

Criterion		**DCC-GARCH**	**MS-DCC-GARCH**	**MS-DCC-fixed GARCH**
AIC		−12016.42	−12998.54	−12132.35
log-likelihood		6025.221,df=608	6535.2703,df=589	6087.174,df=604
	1) LR-test (H0:1 Vs Ha: 2)			P-value = 0.0000
	2) LR-test (H0:1 Vs Ha: 3)			P-value = 0.0000
	3) LR-test (H0:2 Vs Ha: 3)			P-value = 1.0000

Notes: P values are in parentheses. Parameter estimates are based on the MS-DCC (1,1)-GARCH (1,1) model.

and Pastpipatkul et al. [4,5] have found that the conditional correlation during market upturns is less than that during market downturns. Thus, this confirms that the high correlation mostly exists in the market downturn or low volatility regime and the low correlation mostly exists in the market upturn or high volatility regime.

The persistence and regime properties of these returns, as captured by the estimates of the MS-DCC-GARCH parameters, show similar analogous features in terms of transition probabilities, ergodic (regime) probabilities, and duration in each regime. In the Table 3. We denote the probabilities $Pr(S_t = 1| S_{t-1} = 1)$ by p_{11} and $Pr(S_t = 2| S_{t-1} = 2)$ by p_{22}. We can notice that both of the regimes are persistent because of the high values obtained for the probabilities $p_{11} = 0.8513$ and $p_{22} = 0.8172$. We typically observe that the agricultural commodities, VIX, and CDS have a high probability to spend much of the time in the low volatility regime, resulting in higher duration estimates for the low volatility regime compared to the high volatility regime. The duration of the low volatility regime is 6.7276 weeks while that of the high volatility regime is 5.4689 weeks.

The estimated MS-DCC-GARCH (1, 1) model also produces the probabilities of two regimes for the period from 2006 to 2016. In this section, we plot only the low volatility regime probabilities for all the returns presented in Fig. 1. Based on the results, it is evident that the volatility of all returns has mostly taken place in the low volatility regime, except for the period from 2007 to 2010 (the blue-dashed line). This period corresponds to the Global financial crisis in 2007–2008. Our findings confirm the high volatility of all returns during the financial crisis. This period also coincided with the expansion of the quantitative easing of the US and the announcement of the Federal Reserve (FED). Therefore, we expect that these policies would probably put the high pressure directly to VIX and CDS and thereby pushing the high volatility in this regime.

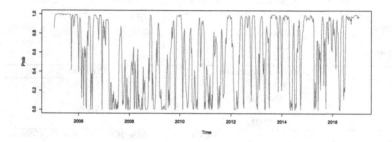

Fig. 1. Smoothed probabilities of low volatility regime

Furthermore, what is interesting is to compare the log-likelihood and AIC of the different models. In this study, we examine the performance of our model MS-DCC-GARCH (1,1) with regime switching in both GARCH and DCC parts, by comparing with two other conventional models namely MS-DCC-GARCH with regime switching only DCC part and DCC-GARCH with no regime switching in both GARCH and DCC parts. According to the results in Table 3, we observe that our model performs slightly better than the two conventional models since we obtain the lowest AIC value. However, Billio and Caporin [3] mentioned that it is not easy to compare the best fit model through log-likelihood value, but they proposed to compare the models using the Likelihood ratio test statistics. The null hypothesis of restricted model (H0) is tested against unrestricted model (Ha). The results of LR-test statistics reject the first two tests and accept the third one. This indicates that our proposed model outperforms other 2 models.

As our model is two-regime MS-DCC-GARCH (1,1), two-regime conditional volatilities are obtained from the GARCH process for each regime.

Following Gray [14], the expected conditional volatility was proposed to present the expected conditional volatility of returns. In this case, we can compute the expected conditional volatility as

$$E\left(\sigma_{i,S_t}^2\right) = \sum_{S_t=j}^{2} \left[\left(\alpha_{i,0,S_t} + \alpha_{i,1,S_t}\varepsilon_{i,t-1,S_t}^2 + \beta_{i,1,S_t}\right) \cdot \left(Pr(S_t = j \,|\, \Theta_{S_t})\right)\right] \quad (20)$$

The result is plotted in Fig. 2.

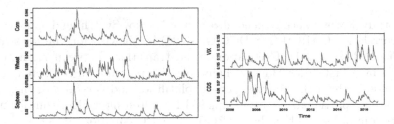

Fig. 2. Smoothed Volatilities. The expected conditional volatilities of corn, wheat, soybean, and CDS are relatively high during 2007–2008, corresponding to the period of global financial crisis.

5.2 Smoothed Correlation of Returns

Similar to the expected conditional volatility; in this section, we plot the expected dynamic conditional correlations from our model as well as the dynamic conditional correlations obtained from one-regime DCC-GARCH model (as shown in

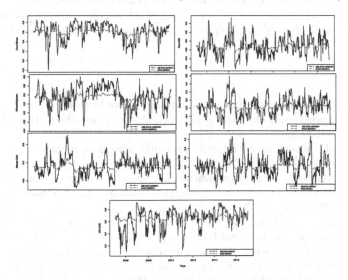

Fig. 3. Time-varying smoothed correlation between variables. The forecasting from the MS-DCC-GARCH (1, 1) (solid lines) is too noisy to represent the correlation when compared to DCC-GARCH (1, 1) (dash lines).

Fig. 3). The differences between one-regime and two-regime models allow us to recognize the advantages of using the structural change detecting method. To be clearer, Fig. 3 shows a high volatile correlation during the period 2008–2010, but a less volatile after 2012. This indicates that the Markov Switching approach yields higher volatilities and may better describe the characteristics in real empirical study.

6   Conclusions

In this study, we introduced a generalization of Markov Switching dynamic conditional correlation GARCH (MS-DCC-GARCH) by allowing for Markov switches in both GARCH and DCC equations. The transitions between regimes were governed by a first order Markov chain. We also presented a restricted version of our model where the changes across volatilities and correlations in a given regime are proportional. We also employed Hamilton filter algorithm for model estimation in order to filter the GARCH and DCC equations into two regimes. In order to break the curse of the consistency and asymptotic normality of this estimator, we constructed the full maximum likelihood of MS-DCC-GARCH and employed a one-step estimation procedure to estimate all unknown parameters of our model. We then proposed a simulation study to examine the accuracy of our model and found that our model and estimation are accurate and efficient.

An application of this model is on investigating the dynamic correlations among three major agricultural commodity prices and two market risks. The comparison of our regime switching model with the DCC model and with the MS-DCC-GARCH shows that our model has a better performance according to the AIC criteria. An interesting aspect of our regime switching model is that we obtain a weak and strong persistence in the Markov chain, which produces both high and less volatility of dynamic correlations along the sample period.

References

1. Engle, R.: Dynamic conditional correlation: A simple class of multivariate generalized autoregressive conditional heteroskedasticity models. J. Bus. Econ. Stat. **20**(3), 339–350 (2002)
2. Patton, A.J.: Modelling asymmetric exchange rate dependence. Int. Econ. Rev. **47**(2), 527–556 (2006)
3. Billio, M., Caporin, M.: Multivariate markov switching dynamic conditional correlation GARCH representations for contagion analysis. Stat. Methods Appl. **14**(2), 145–161 (2005)
4. Pastpipatkul, P., Yamaka, W., Sriboonchitta, S.: Analyzing financial risk and co-movement of gold market, and Indonesian, Philippine, and Thailand stock markets: dynamic copula with markov-switching. In: Huynh, V.-N., Kreinovich, V., Sriboonchitta, S. (eds.) Causal Inference in Econometrics. SCI, vol. 622, pp. 565–586. Springer, Cham (2016). https://doi.org/10.1007/978-3-319-27284-9_37

5. Pastpipatkul, P., Yamaka, W., Sriboonchitta, S.: Dependence structure of and co-movement between Thai currency and international currencies after introduction of quantitative easing. In: Huynh, V.-N., Kreinovich, V., Sriboonchitta, S. (eds.) Causal Inference in Econometrics. SCI, vol. 622, pp. 545–564. Springer, Cham (2016). https://doi.org/10.1007/978-3-319-27284-9_36

6. Da Silva Filho, O.C., Ziegelmann, F.A., Dueker, M.J.: Modeling dependence dynamics through copulas with regime switching. Insur. Math. Econ. **50**(3), 346–356 (2012)

7. Gozgor, G., Kablamac, B.: The linkage between oil and agricultural commodity prices in the light of the perceived global risk (2014)

8. Pelletier, D.: Regime switching for dynamic correlations. J. Econometrics **131**(1), 445–473 (2006)

9. Chen, R.: Regime switching in volatilities and correlation between stock and bond markets (2009)

10. Bollerslev, T., Chou, R.Y., Kroner, K.F.: ARCH modeling in finance: A review of the theory and empirical evidence. J. Econometrics **52**(1–2), 5–59 (1992)

11. Bollerslev, T.: Generalized autoregressive conditional heteroskedasticity. J. Econometrics **31**(3), 307–327 (1986)

12. Bollerslev, T.: Modelling the coherence in short-run nominal exchange rates: a multivariate generalized ARCH model. Rev. Econ. Stat. **72**, 498–505 (1990)

13. Hamilton, J.D.: A new approach to the economic analysis of nonstationary time series and the business cycle. Econometrica J. Econometric Soc. **57**(2), 357–384 (1989)

14. Gray, S.F.: Modeling the conditional distribution of interest rates as a regime-switching process. J. Financ. Econ. **42**(1), 27–62 (1996)

Modeling Dependence with Copulas: Are Real Estates and Tourism Associated?

Roengchai Tansuchat[1,2(✉)] and Paravee Maneejuk[1,2]

[1] Centre of Excellence in Econometrics, Chiang Mai University, Chiang Mai, Thailand
roengchaitan@gmail.com, mparavee@gmail.com
[2] Faculty of Economics, Chiang Mai University, Chiang Mai, Thailand

Abstract. Several families of copulas are considered in this study to illustrate the correlation between real estate–particularly hospitality real estate investment trust- and the tourism sector. In essence, this study uses Elliptical copulas and Archimedean copulas, and more recent classes, like extreme value copulas and mixed copulas to conduct the experiment. Under a specific data set, it is revealed that the classical classes of copulas i.e., Elliptical and Archimedean, are selected most often for illustrating the dependency, followed by the extreme value class, particularly the Husler-Reiss copula. However, surprisingly, the mixed copula is not entirely preferable for this data set.

Keywords: Dependence · Copulas · Real estates · Tourism

1 Introduction

The Real Estate Investment Trust (REIT) is a type of security that invests in real estate through property or mortgages and often traded on stock exchange. Many investors enjoy investing in properties through REITs because they can invest in the real estate sector without having to spend close to a million dollar on an entire property. Equity REITs are type of investment that allows investor to get involved in the real estate market without actually buying any real estate directly. Equity REITs own and operate income-producing real estate assets like residential properties, retail properties, office buildings, medical properties, hotels and serviced residences properties, and industrial purposed properties. Therefore, REITs can be classified as residential, retail, office, healthcare, hotels and resorts, and warehouses and manufacturing centers. In addition, the expansion of the real estate industry is in line with the expansion of the economy. Zhang et al. [15] and Hong [3] affirmed the positive effect of real estate investment on economic growth. Therefore, real estate investment is considered as one of the potential factors used when investigating short-term economic fluctuations.

"Tourism real estate" is an industry which amalgamates real estate and tourism through the integration of planning and design, construction and marketing, hotel management, and other aspects [13], and has become one of the fastest growing sectors in the entire leisure market. Tsai et al. [14] explained that tourism real estate was characterized by the development of large-scale tourism resources (e.g. resorts and theme parks) along with residential properties, under the assumption that they would increase

© Springer International Publishing AG, part of Springer Nature 2018
V.-N. Huynh et al. (Eds.): IUKM 2018, LNAI 10758, pp. 302–311, 2018.
https://doi.org/10.1007/978-3-319-75429-1_25

property value. Bardhan et al. [2] explained the connection between real estate industry and the tourism sector. Therefore, tourism real estate contributes to hospitality REITs.

Hospitality REITs is REITs that own and operate a collection of hotels and other lodging properties such as hotels or motels, lodging and serviced residences. There are many real estate investment trusts related to tourism sector such as hotel REIT, lodging/ resort REITs, local shopping REIT PLC (denoted by LSR), and HTL/RSRT/LSR REITs. Hotel & Lodging REITs own different classes of hotels based on such features as the hotels' level of service and amenities. Hotel & Lodging REITs' properties serve a wide range of customers, from business travelers to vacationers. Nowadays, there are many hospitality REITs traded in many countries both in advance economy and emerging economy such as the United States, United Kingdom, Singapore, Thailand, Japan, South Korea, Malaysia, Taiwan, Australia, and New Zealand.

International Labour Organization reported that hotel REITs investors seek benefits such as high liquidity, lower risk levels and higher returns compared to average common stock. Jackson [4] found that the hotel REITs performed better than the equity REIT portfolio but riskier than other equity REITs. However, Jackson [5] found that the performance of lodging REITs was inferior to the market portfolios as a whole and to other equity REITs except for retail and specialty REITs. Because hotels and resorts in the hospitality sector are almost completely dependent on tourist arrivals. Thus, economic downturn, political instability, major health outbreaks, natural disasters and bad news have an adverse impact on tourist arrivals by the lower visitor numbers, and expected profit and return of hotel and lodge REITs [7]. For example, Ang et al. [1] explained that the Singaporean hospitality trust highly relies on tourist arrivals from Southeast Asia and China, and Moody's Investors Service reported that Singapore's hospitality REITs exhibited weaker performance due to lower tourist arrivals [10].

Hence, this study will examine the relationship between real estates and tourism with a special focus on most visited countries in Asia, namely Singapore, Taiwan, Korea, Japan, Malaysia, Thailand, and nearby countries, specifically Australia and New Zealand. In particular, we use hotel stock index and the number of tourist arrivals to represent the real estate investment trust (REIT) and the tourism sector of each country respectively. Next, to investigate the structure and degree of dependency between these two terms, we employ one of the most popular methods for illustrating the dependence between variables that is the copula approach. This study considers both classical copula classes, i.e. Elliptical copulas and Archimedean copulas, and more recent classes, like extreme value copulas and mixed copulas. The best copula among the considered classes will be chosen to illustrate the structural relationship between hotel stock indices and the number of tourist arrivals.

The rest of this paper is structured as follows. Section 2 is about methodology, in which we show the basic concept of copulas and some brief explanations of the four classes of copula functions. Then, in Sect. 3, we will present data description. Section 4 contains results of the empirical study, including model selection, estimated results, and some discussion on the estimated parameters. Finally, the conclusion will be presented in Sect. 5.

2   Basic Concepts and Preliminaries

2.1   Basic Concepts of Copula

The basic idea of copulas refers to the Sklar's theorem [12] which stated that multivariate distribution function can be thought of as univariate marginal distributions and copula, in which the copula is described as a linkage between marginal distributions, $F_1(x_1), \ldots, F_n(x_n)$, such that

$$H(x_1, \ldots, x_n) = C(F_1(x_1), \ldots, F_n(x_n)) = C(u_1, \ldots, u_n), \tag{1}$$

where $H(x_1, \ldots, x_n)$ is a joint distribution, linking marginal distributions of each random variable, $x_1, \ldots, x_n$. Similarly, C denotes the copula, which is used to join the marginal distribution functions in unit interval $[0, 1]$, $u_1, \ldots, u_n$.

Then, to obtain the copula density, Eq. (1) is differentiated. So, the multivariate copula density function c can be written by

$$c(u_1, \ldots, u_n) = \frac{\partial^n C(F_1(x_1), \ldots, F_n(x_n))}{\partial u_1, \ldots, \partial u_n}. \tag{2}$$

However, we should explain how we could get the marginal distribution for each random variable. To address this question, the univariate ARMA (p, q)-GARCH (m, n) is employed. A specific form of this model can be written by

$$y_t = \phi_0 + \sum_{i=1}^{p} \phi_i y_{t-i} + \sum_{j=1}^{q} \theta_j \varepsilon_{t-j} + \varepsilon_t, \tag{3}$$

$$\varepsilon_t = h\eta_t, \tag{4}$$

$$h_t^2 = \alpha_0 + \sum_{i=1}^{m} \alpha_i \alpha_{t-i}^2 + \sum_{j=1}^{n} \beta_j h_{t-j}^2. \tag{5}$$

Equations (3) and (5) are the conditional mean and variance equation, respectively. ε_t is the residual, consisting of standard variance, h_t, and standardized residual, η_t. In this study, the distribution of residual is allowed to be normal distribution, student-t distribution, skew-t distribution, skew normal distribution, generalized error distribution and/or skewed generalized error distribution. The standardized residual will be given by the best-fit ARMA (p, q)-GARCH (m, n), and then transformed into a uniform distribution $[0, 1]$ (see, [9]).

2.2   Families of Copulas

This part presents copula density function of the four classes, namely elliptical copulas, Archimedean copulas, extreme value copulas, and mixed copulas. We focus on the bivariate case as there are only two major variables, i.e. real estates and tourism, that we aim to find their dependence.

(1) Elliptical copulas

- **Gaussian copula**

Consider a case of two-dimensional copula, the density of the Gaussian copula is given by [11]

$$c_G(u, v|\theta_G) = \int_{-\infty}^{\Phi^{-1}(u\,\Phi^{-1}(v))} \int_{-\infty}^{} \frac{1}{2\pi\sqrt{1-\theta_G}} \exp\left(\frac{x_1^2 + x_2^2 - 2\theta_G x_1 x_2}{2(1 - \theta_G^2)}\right) dx_1 dx_2, \tag{6}$$

where Φ is bivariate standard normal cumulative distribution and θ_G is the correlation parameter of copula lying in the interval $[-1, 1]$.

- **Student's t copula**

Student's t copula is described as a copula function with fat tail. θ_T is the parameter of the copula lying in the interval $[-1, 1]$ and v is the degree of freedom. In bivariate case, the probability density function of student's t copula is

$$c_T(u, v|\theta_T) = \int_{-\infty}^{t_v^{-1}(u)\,t_v^{-1}(v)} \int_{-\infty}^{} \frac{1}{2\pi\sqrt{1-\theta_T}} \exp\left(\frac{x_1^2 + x_2^2 - 2\theta_T x_1 x_2}{v(1 - \theta_T^2)}\right)^{-(v+2)/2} dx_1 dx_2, \tag{7}$$

where $t_d(v, 0, \theta_T)$ is the standard univariate student-t distribution with v degree of freedom, mean 0 and variance $(v + 2)/2$.

(2) Archimedean copulas

In Archimedean class, copula functions allow us to capture the asymmetric nature of the data. In the estimation, this class of copula allows modeling the dependence with only one parameter. The probability density function of the bivariate Archimedean copulas, consisting of the Clayton, Gumbel, Frank, and Joe, can be shown as follows.

- **Clayton copula**

The density function of Clayton copula is

$$c_C(u, v|\theta_C) = \left(1 + (u^{-\theta_c} - v^{-\theta_c} - 1)\right)^{-1/\theta_c}, \tag{8}$$

where θ_C is the degree of dependence on the value of $0 < \theta_C < \infty$. If $\theta_C \to \infty$, the Clayton copulas will converge to a monotonicity copula with positive dependence. But if $\theta_C = 0$, it will correspond to independence.

- **Gumbel copula**

    The density function of Gumbel copula is

    $$c_G(u, v|\theta_G) = \exp\left(-\left[(-\ln u)^{\theta_G} + (-\ln v)^{\theta_G}\right]\right)^{1/\theta_G}, \tag{9}$$

where the parameter θ_G is the degree of dependence on the value $1 < \theta_G < \infty$.

- **Frank copula**

    The density function of Frank copula is

    $$c_F(u, v|\theta_F) = -\theta_F^{-1} \log\left(1 + \frac{(e^{-\theta_F(u)} - 1)(e^{-\theta_F(v)} - 1)}{e^{-\theta_F} - 1}\right), \tag{10}$$

where θ_F is the degree of dependence on the value $-\infty < \theta_F < \infty$.

- **Joe copula**

    The density function of Joe copulas is

    $$c_J(u, v|\theta_J) = 1 - \left((1-u)^{\theta_J} + (1-v)^{\theta_J} - (1-u)^{\theta_J}(1-v)^{\theta_J}\right)^{1/\theta_J}, \tag{11}$$

where the parameter θ_J is the degree of dependence on the value $1 < \theta_J < \infty$.

(3) **Extreme value copulas**

The multivariate maxima in Extreme Value Theory can be expressed in terms of copulas. Extreme value copulas belong to a class of copulas that emerge as the natural limiting dependence structures for multivariate maxima. Following Lu et al. [6], the density functions of the EV copulas used in this study can be shown as follows.

- **E-Gumbel copula**

    $$c_{EG}(u, v|\theta_{EG}) = \exp(-((-\ln u)^{\theta_{EG}} + (-\ln v)^{\theta_{EG}})^{1/\theta_{EG}}), \tag{12}$$

where the parameter θ_{EG} is the degree of dependence on the value $1 < \theta_{EG} < \infty$.

- **Galambos copula**

    $$c_{Gal}(u, v|\theta_{Gal}) = uv \exp(-((-\ln u)^{-\theta_{Gal}} + (-\ln v)^{-\theta_{Gal}})^{-1/\theta_{Gal}}), \tag{13}$$

where the parameter θ_{Gal} is the degree of dependence on the value $0 < \theta_{EG} < \infty$.

- **Husler-Reiss copula**

    $$c_H(u, v|\theta_H) = \exp\left(-\tilde{u}\Phi\left(\frac{1}{\theta_H} + \frac{1}{2}\theta_H \ln\left(\frac{\tilde{u}}{\tilde{v}}\right)\right) - \tilde{v}\Phi\left(\frac{1}{\theta_H} + \frac{1}{2}\theta_H \ln\left(\frac{\tilde{u}}{\tilde{v}}\right)\right)\right), \tag{14}$$

where the parameter θ_H is the degree of dependence on the value $0 < \theta_H < \infty$. $\tilde{u} = -\ln(u)$, $\tilde{v} = -\ln(v)$, and Φ denotes standardized normal distribution.

(4) Mixed copulas

The conventional copula may lead to a problem associated with dependency measure -or a misspecification problem- since each family has a strong restriction on its dependence parameter (see, [8]). As a result, the mixed copulas are generated to solve this problem. This class of copula allows more flexibility for capturing almost all possible dependence structures between variables. Using mixed copula class can gain several additional advantages. First, it provides flexibility in modelling dependence as copulas allow for separate modelling of marginal distributions and corresponding dependence structures. This means, the marginal distributions can be generated separately and then combined using a specific copula. Consequently, creating complex non-normal distributions for capturing unfamiliar or uncommon dependency -as happened in financial variables and/or the co-movement between variables during market upturn and downturn- is possible under this class of copula. Second, the dependence structures obtained from mixed copulas are invariant despite the data transformations, meaning that the dependency will not change although the data is transformed into several types.

By using convex combination approach, the mixed copula can be computed by

$$c_{Mix}(u, v|\theta_1, \theta_2) = wc_{\theta_1}(u, v|\theta_1) + (1 - w)c_{\theta_2}(u, v|\theta_2), \tag{15}$$

where w is the weight parameter within $[0, 1]$. One of the advantages of the convex combination approach is the flexibility in assigning weights for calculating appropriate value between two copula functions.

3   Data Description

This study uses monthly time series (January 2008 to April 2017) of the numbers of tourist arrivals and Hotel or Hospitality REITs price index of eight countries; Singapore (SIN), Taiwan (TW), Korea (KOR), Japan (JAP), Malaysia (MAL), Thailand (THA), Australia (AUS), and New Zealand (NZ). This data set was retrieved from DataStream, and then transformed into logarithmic return or continuously compounded return. Table 1 presents the descriptive statistics of the transformed series as well as correlation values between the number of tourist arrivals and the hotel indices. In addition, the Augmented Dickey Fuller (ADF) test is conducted to show the stationarity of the data series, and the results are presented at the bottom of the table.

Table 1. Descriptive statistics

	SINT	SINR	THAT	THAR	JAPT	JAPR	KORT	KORR
Mean	0.010	0.000	0.016	0.024	0.022	0.003	0.013	0.008
Median	0.012	0.005	0.017	0.019	0.019	0.002	0.014	−0.017
Maximum	0.115	0.433	0.243	0.434	0.320	0.228	0.565	0.617
Minimum	−0.071	−0.310	−0.167	−0.208	−0.552	−0.234	−0.391	−0.437
Std. Dev.	0.041	0.093	0.074	0.096	0.103	0.083	0.093	0.151
Skewness	0.236	0.477	0.039	0.911	−1.230	0.058	1.006	0.877
Kurtosis	3.024	8.144	3.320	5.858	11.479	3.388	15.99	5.516
JB-test	1.034	126.600	0.503	53.137	360.510	0.756	800.100	43.505
Prob.	0.596	0.000	0.778	0.000	0.000	0.685	0.000	0.000
ADF-test	-10.24^a	-4.99^a	-10.55^a	-8.94^a	-9.34^a	-10.08^a	-9.52^a	-11.41^a
	MALT	MALR	TWT	TWR	AUST	AUSR	NZT	NZR
Mean	0.011	0.001	0.015	0.011	0.043	−0.005	0.034	0.002
Median	−0.016	0.002	0.016	0.002	0.041	0.000	0.037	0.003
Maximum	0.408	0.115	0.151	0.617	0.345	0.192	0.266	0.132
Minimum	−0.321	−0.149	−0.149	−0.359	−0.111	−0.217	−0.665	−0.298
Std. Dev.	0.115	0.041	0.060	0.116	0.074	0.051	0.095	0.061
Skewness	0.872	−0.511	−0.152	1.446	0.919	−0.863	−3.292	−0.991
Kurtosis	5.195	5.370	3.029	9.863	5.354	8.412	27.431	7.097
JB-test	36.342	30.818	0.432	256.494	41.260	149.234	2960.93	95.824
Prob.	0.000	0.000	0.806	0.000	0.000	0.000	0.000	0.000
ADF-test	-15.91^a	-10.12^a	-11.47^a	-10.31^a	-18.59^a	-3.49^a	-8.59^a	-11.49^a

Source: Calculation.

Note: For each variable, the first three (or two) letters correspond to country but the last letter corresponds to either the number of tourist arrivals (T) or the hotel stock indices (R) – representing the REIT; "a" denotes a significance at 1% level.

4   Results and Discussions

The main purpose of this study is to examine the dependence between real estates (REIT) and tourisms in the countries known as tourist destinations in Asia and nearby regions. This section presents the main outcomes, starting from model comparison, followed by the estimated dependence parameters and tail dependences. Four classes of copulas, particularly elliptical copulas, Archimedean copulas, extreme value copulas, and mixed copulas, are considered. Given a set of copulas, the Akaike information criterion (AIC) is used to choose the best fitting copula among these candidates in such a way that the model with minimum AIC is statistically preferable.

Table 2 presents AIC values for the comparison of models. From this table, we can see that the classical copula families, elliptical and Archimedean copulas, are selected most often for illustrating the dependency between tourist arrivals and REITs under this data set, except for the pairs of Japan and Taiwan where the Husler-Reiss copula, which is in the extreme value class, is chosen. Surprisingly, the mixed copula is not entirely preferable for this data set.

Table 2. Model comparison using AIC

Tourist arrival and REIT of

Copula	SIN	THA	JAP	KOR	MAL	TW	AUS	NZ
Gaussian	1.099	−393.148	0.253	3.578	1.964	1.971	1.987	0.649
Student-t	3.050	−390.005	1.581	1.965	4.049	4.145	3.827	2.826
Gumbel	11.688	−379.598	0.330	10.809	8.791	11.305	14.077	1.988
Clayton	1.626	−321.175	0.675	8.551	1.656	1.997	1.964	1.695
Frank	1.629	−372.376	1.551	1.748	1.916	1.988	1.742	0.071
Joe	9.693	−321.175	0.675	8.992	1.890	1.890	−0.105	1.988
E-Gumbel	2.000	−379.598	0.330	2.000	2.000	2.000	0.772	1.647
Galambos	2.000	−380.881	2.000	2.000	2.000	2.000	2.000	2.000
Husler-Reiss	2.000	−391.447	−1.544	2.000	2.000	1.726	0.979	1.647
C-G	6.435	−375.598	2.736	8.551	8.589	9.915	5.338	7.616
C-F	5.040	−368.376	5.551	2.835	5.916	5.193	5.745	4.039
F-G	5.040	−375.597	3.634	2.137	5.059	5.989	2.049	4.074
C-J	8.422	−317.171	2.166	8.551	6.669	6.577	3.895	7.604
G-J	9.693	−375.598	4.744	8.994	6.666	8.594	3.895	7.755

Source: Calculation

Note: C = Clayton, G = Gumbel, F = Frank, J = Joe.

In detail, the Gaussian copula is selected for the pairs of Singapore and Thailand as it has the smallest value of AIC, i.e. 1.099 and −393.148, respectively. Frank copula is the most preferable dependence structure for the pairs of Korea and New Zealand due to minimum AIC values, 1.748 and 0.071, respectively. However, for the relationship between tourist arrivals and REIT in Malaysia and Australia, the Clayton and Joe copulas are chosen respectively with the lowest values of AIC, 1.656 and −0.105.

The last two tables present the estimated values of dependence parameters, Kendall's tau, and tail dependence. Table 3 shows particularly the estimated degree of dependence of the copula parameter, where we can see that different copula families provide different range of copula parameters. In short, it appears that REIT in terms of the hotel stock indices and tourist arrivals in Singapore and Korea are negatively correlated. However, under the Kendall's tau correlations shown in Table 4, the dependencies are in the same range bounded in the interval [−1, 1], so we can compare them. As we can see, the highest dependency between the rate of change between tourist arrival and the hotel index is Thailand with the positive tau value 0.8888. Similarly, the remaining countries, except for Singapore and Korea, also have the positive value of tau. The positive tau indicates that the property value in these countries is likely to grow with the growth of tourism. On the other hand, negative tau implies that the rise or crash of the real estate indices is correlated negatively to the changes in tourist arrivals. Moreover, there exists also the upper tail dependence in the pairs of Japan, Malaysia, Taiwan, and Australia, meaning that we can find the dependence of extreme events between the variables in these countries. The upper tail indicates that the joint occurrence of extreme values of the two variables rises together.

Table 3. Estimated dependence parameter

Par.	SIN	THA	JAP	KOR	MAL	TW	AUS	NZ
θ	−0.0905	0.9848	0.5503	−0.3034	0.0628	0.5279	1.0930	0.9038
	(0.0972)	(0.0189)	(0.1195)	(0.5543)	(0.1029)	(0.1366)	(0.2547)	(0.6446)

Source: Calculation.

Table 4. Kendall's tau and tail dependence

	SIN	THA	JAP	KOR	MAL	TW	AUS	NZ
Kendall's Tau	−0.0577	0.8888	0.5503	−0.0336	0.0336	0.4365	0.0508	0.0996
Upper tail	0.0000	0.0000	0.0691	0.0000	0.00001	0.0219	0.1145	0.0000
Lower tail	0.0000	0.0000	0.0000	0.0000	0.0000	0.0000	0.0000	0.0000

Source: Calculation.

5 Conclusions

This study paid attention to the tourism real estate, which is considered as one of the fastest growing sectors in the entire leisure market. There were evidences showing the correlation between real estate and tourism in several countries in both advance economy and emerging economy. Therefore, this study was conducted to examine this correlation in some countries in Asia and nearby countries. We used the hotel stock indices to represent the hospitality real estate investment trust and the number of tourist arrivals to represent the tourism sector.

We examined the true dependence structure between the two variables using various copula families, namely elliptical copulas, Archimedean copulas, extreme value copulas, and mixed copulas. The results showed that the classical copula families, elliptical and Archimedean copulas, were selected for most of the countries, except for Japan and Taiwan where the extreme value copula (Husler-Reiss) was more preferable. However, estimated dependence parameters just showed a low dependency between the hospitality REITs and tourisms in all the countries. Moreover, it was found that different copula families provided different range of copula parameters, so the copula parameters were transformed into the Kendall's tau correlations. The negative tau value was found for the pair of Singapore's hospitality REIT and tourist arrival as well as Korean's hospitality REIT and its tourist arrival. On the contrary, the remaining countries of concern, which are Taiwan, Japan, Malaysia, Thailand, Australia, and New Zealand, were found to have a positive tau. This proved joint occurrences between hospitality real estate investment and tourism crashing (or rising) happened in these countries.

Acknowledgement. We are grateful for financial support from Centre of Excellence in Econometrics, Chiang Mai University and also thank to Mr. Woraphon Yamaka for helping us with code.

References

1. Ang, J., Wei, N.G., Ng, R., Feimiao, J.: Fundamental analysis department S-REITs industry report. NUS Investment Society. National University of Singapore Mochtar Riady (2014)
2. Bardhan, A., Begley, J., Kroll, C.A., George, N.: Global tourism and real estate. In: 2008 Industry Studies Conference Paper (2008)
3. Hong, L.: The dynamic relationship between real estate investment and economic growth: evidence from prefecture city panel data in China. IERI Procedia **7**, 2–7 (2014)
4. Jackson, L.A.: The structure and performance of US hotel real estate investment trusts. J. Retail Leis. Prop. **7**(4), 275–290 (2008)
5. Jackson, L.A.: Lodging REIT performance and comparison with other equity REIT returns. Int. J. Hosp. Tour. Adm. **10**(4), 296–325 (2009)
6. Lu, J., Tian, W. J., Zhang, P.: The extreme value copulas analysis of the risk dependence for the foreign exchange data. In: 2008 4th International Conference on Wireless Communications, Networking and Mobile Computing, WiCOM 2008, pp. 1–6. IEEE (2008)
7. Newell, G., Seabrook, R.: Factors influencing hotel investment decision making. J. Prop. Invest. Finance **24**(4), 279–294 (2006)
8. Nguyen, C., Bhatti, M.I., Komorníková, M., Komorník, J.: Gold price and stock markets nexus under mixed-copulas. Econ. Model. **58**, 283–292 (2016)
9. Pastpipatkul, P., Maneejuk, P., Sriboonchitt, S.: The best copula modeling of dependence structure among gold, oil prices, and U.S. currency. In: Huynh, V.-N., Inuiguchi, M., Le, B., Le, B.N., Denoeux, T. (eds.) IUKM 2016. LNCS (LNAI), vol. 9978, pp. 493–507. Springer, Cham (2016). https://doi.org/10.1007/978-3-319-49046-5_42
10. Poh, J., Lotter, P.L.: Announcement: Moody's: Singapore's hospitality REITs exhibit weaker performance due to lower tourist arrivals. Moody's Investor Service (2015)
11. Schepsmeier, U., Stöber, J.: Derivatives and Fisher information of bivariate copulas. Stat. Pap. **55**(2), 525–542 (2014)
12. Sklar, M.: Fonctions de repartition à n-dimensions et. leurs marges. Publ. Inst. Stat. Univ. Paris **8**, 229–231 (1959)
13. Sun, H.L., Wang, S.: The effective way to deal with the backlog of real estate. Commer. Res. **10**, 136–138 (2002)
14. Tsai, H., Huang, W.J., Li, Y.: The impact of tourism resources on tourism real estate value. Asia Pac. J. Tour. Res. **21**(10), 1114–1125 (2016)
15. Zhang, J., Wang, J., Zhu, A.: The relationship between real estate investment and economic growth in China: a threshold effect. Ann. Reg. Sci. **48**(1), 123–134 (2012)

A Markov-Switching Model with Mixture Distribution Regimes

Paravee Maneejuk[1,2(✉)], Woraphon Yamaka[1,2], and Songsak Sriboonchitta[1,2]

[1] Centre of Excellence in Econometrics,
Chiang Mai University, Chiang Mai, Thailand
mparavee@gmail.com, woraphon.econ@gmail.com
[2] Faculty of Economics, Chiang Mai University, Chiang Mai, Thailand

Abstract. This study proposes the mixture Markov-switching autoregressive model, which allows variation in error distribution across different regimes. This model is generalized from the ordinary MS-AR model owing to two considerations, but related to each other. First, we have concern about the mixture of distributions or populations, which often prevails in economic time series. Second, when using the MS models to analyse economic fluctuation, we doubt if each regime in the model can have distinct distribution. All of these concerns are addressed by an empirical study.

Keywords: Regime switching · Mixture of distribution
Stages of economic growth · Economic fluctuation

1 Introduction

The ability of the Markov-switching processes has been proven in various studies, also advocated in the literature. In economics, the Markov-switching model is used often to generate the stylized facts of business cycle as it is able to provide, for example, expected regime durations and amplitudes of expansions and recessions of the cycle. One outstanding feature of the Markov-switching approach, which makes it suitable for illustrating business cycle, is that this model contains several regimes and allows changes in outcomes across the regimes. The switches can occur due to a probability of regime change, which is estimated in addition to the parameters of the model. However, ordinary Markov-switching (MS) approach assumes that the parameters are normally distributed; in other words, all regimes in the MS model are assumed to have the same distribution, which is regularly a normal distribution. But in economics, time series data often involves a mixture of distributions or populations, as a result of different characteristics of the data associated with different time, for example, distinct economic behaviors during economic upturn and downturn. So, using the ordinary MS model on the data displaying a continuous signal of having a mixture of distributions may yield a poor result and false implications. This becomes our concern and leads to the question: Can different regimes in a Markov-switching model have different distributions?

© Springer International Publishing AG, part of Springer Nature 2018
V.-N. Huynh et al. (Eds.): IUKM 2018, LNAI 10758, pp. 312–323, 2018.
https://doi.org/10.1007/978-3-319-75429-1_26

There is not only an argument from the economic aspect, several empirical studies of financial time series also show that vast majority of MS-type models, when fitting to weekly returns or shorter horizons, imply quite heavy tails in the data and also skewness innovation distributions. A sizeable and growing number of candidate distributions are suggested by many studies, especially the GARCH-type studies. Haas et al. [3], Haas [4] and, recently, Ardia et al. [1] suggested that the asymmetry was usually captured by allowing different regime means. Thus, they allowed different GARCH distribution for each regime to capture the difference in variance dynamics in both high and low volatility periods.

Following this idea, we expect the distribution of regimes for example, during expansion and recession - to be different, and, therefore, introduce a Markov-switching model that allows distinct distribution for each regime. More specifically, the study generalizes a Markov-switching autoregressive model (MS-AR) of Hamilton [5] to the one with mixture distribution regimes. The conventional MS model is conducted under normal density, which, sometimes, cannot account for excess kurtosis and heavy tails as found in economic data (see, [2]). To address this, we will allow the model to have different distribution across regimes, $\{1, ..., h\}$. Generally, the MS model relates to an unobservable variable S_t interpreted as the regime of the model. In this study, we let S_t come from various distributions, for which each time t selects. So, the density function, $f_{tj}(y_t) / \sum_{j=1}^{h} f_{tj}(y_t)$, can be either the same or distributed differently. In the case that the S_t variables are independent and only one distribution is selected, this will correspond to the classical Markov-switching model.

One main advantage of this proposed approach is that it allows for heterogeneous regimes, which is what we expect to yield better results than the conventional MS-AR model. This will be proven in the empirical study, where we will apply this approach to examine the US business cycle and then compare the results with the ordinary MS-AR model. The rest of this paper is structured as follows. Section 2 is about methodology, in which we review concept and specific form of the ordinary MS-AR model, followed by the MS-AR model with mixture distribution regimes, and some explanations of how parameters change across regimes through Hamilton's filter, respectively. Section 3 contains a simulation study. Section 5 presents main results of the empirical study, as mentioned earlier, and some discussions about the comparison between the ordinary model and the alternative model. Finally, the conclusion will be presented in Sect. 5.

2 Methodology

2.1 An Ordinary Markov-Switching Autoregressive Model

The ordinary Markov-switching autoregressive (MS-AR) model can be specified as follows.

$$y_t = \beta_{0,S_t} + \sum_{i=1}^{p} \beta_{p,S_t} y_{t-i} + \varepsilon_{t,S_t}, \tag{1}$$

where $\varepsilon_{t,S_t} \sim i.i.d.\ N(0,\sigma_{S_t}^2)$. y_t is dependent variable and y_{t-i} is lag dependent variable, in which $i = 1, ..., p$. β_{0,S_t} denotes regime dependent intercept term and β_{p,S_t} denotes the regime dependent autoregressive coefficients of $AR(P)$. The regime represented by S_t is considered as an unknown parameter. Let $\{S_t\}_{t=1}^h$ be the finite Markov chain with h regimes, $\{1, ..., h\}$, the stationary (one-step) transition probability is

$$p_{ij} = P(S_{t+1} = j \mid S_t = i), \tag{2}$$

where p_{ij} is the probability of transitioning from regime i to regime j and $\sum_{j=1}^h p_{ij} = 1$. So, we can write the transition matrix Q which contains the transition probabilities p_{ij} as

$$Q = \begin{bmatrix} p_{11} & p_{12} & \cdots & p_{1j} \\ p_{21} & \ddots & & p_{2j} \\ \vdots & & \ddots & \vdots \\ p_{i1} & p_{i2} & \cdots & p_{ij} \end{bmatrix}. \tag{3}$$

In this study, we deal with the statistic estimation of unknown parameters in finite density,

$$\mathbf{f(y)} = \sum_{j=1}^h \prod_t f_j(y_t), \tag{4}$$

where f_j is density function of regime j. In the conventional approach, densities are assumed to be normally distributed despite different means and variances. According to this, the normal probability density function is written as follows:

$$f_{S_t} = \frac{1}{\sqrt{2\pi\sigma_{S_t}^2}} e^{\frac{-\varepsilon_{S_t}^2}{2\sigma_{S_t}^2}}. \tag{5}$$

Of course this density is assumed to have the same distribution across different regimes, $\{1, ..., h\}$. On the other hand, the basic assumption here is the existence of S_t comes from the same distribution which for each t selects the distributions $f_{tj}(y_t)$, where $f_{tj}(y_t)$ has the same distribution as mean μ_j and variance σ_j. For normal MS-AR model, Eq. (1), it also depends on lagged variable y_{t-p}, thus we can rewrite Eq. (4) as

$$\mathbf{f(y \mid y_{t-p})} = \sum_{j=1}^h \prod_t f_j(y_t \mid y_{t-p}). \tag{6}$$

Let $\theta_{S_t} = (\beta_{S_t}, \sigma_{S_t}, Q)$ be the vector of model parameters in Eq. (1), the complete likelihood function is shown as follows:

$$L(\theta_{S_t} \mid y) = \sum_{j=1}^h \left(\prod_{t=1}^T \left(\frac{1}{\sqrt{2\pi\sigma_{S_t=j}^2}} \times \exp\left[\frac{(\varepsilon_{t,S_t=j})^2}{2\left(\sigma_{S_t=j}^2\right)} \right] \right) (\Pr(S_t = j \mid \theta_{S_t})) \right), \tag{7}$$

where $\varepsilon_{t,S_t} = y_t - \beta_{0,S_t} - \sum_{i=1}^{p} \beta_{p,S_t} y_{t-i}$. Since the whole process of S_t is unobserved at time t, the Hamilton's filter -proposed in Hamilton [5]- is employed to estimate the filtered probability, $\Pr(S_t = j)$, where the formula of the filter is

$$\Pr(S_t = j | \theta_{S_t}) = \frac{f(y_t | S_t = j, \theta_{S_{t-1}}) \Pr(S_t = j | \theta_{S_{t-1}})}{\sum_{j=1}^{h} f(y_t | S_t = j, \theta_{S_{t-1}}) \Pr(S_t = j | \theta_{S_{t-1}})}, \tag{8}$$

where $f(y_t | S_t = j, \theta_{S_{t-1}})$ is the normal density function of each regime j. (see, [8]).

2.2 The MS-AR Model with Mixture Distribution Regimes

This study aims to relax the strong assumption in the conventional normal MS-AR model by allowing the model to have different densities distributions across different regimes, $\{1, ..., h\}$. To let the regimes S_t be governed by different distributions of y_t, we need to rewrite the complete likelihood function, Eq. (8), to be

$$L(\theta_{S_t} | y) = \sum_{j=1}^{h} \left(\prod_{t=1}^{T} \left(f(\theta_{S_t=j} | y) \right) (\Pr(S_t = j | \theta_{S_t})) \right), \tag{9}$$

where $f(\theta_{S_t=j} | y)$ is the density function which can have different distributions across regimes. $\theta_{S_t=j}$ is a set of state dependent parameter of regime j. More specifically, we consider 6 different distributions, namely normal, student-t, generalized error distribution (GED), skewed GED, skewed normal, and skewed student-t distributions. Thus, the density function $f(\theta_{S_t=j} | y)$, a part of Eq. (9), can be written in 6 different forms according to the distribution being used.

2.3 Hamilton's Filter

The Hamilton's filter, which was first introduced by Hamilton in 1989, refers to the process of unobserved variable S_t interpreted as a regime variable. As it is unobserved, we will never know which regime prevails at a certain point of time [6]. According to Eq. (8), the filter probability $(\Pr(S_t = j | \theta_{S_t}))$ is an important process. We need to filter out the estimated coefficient and variance into different regimes. Following Perlin [8], Hamilton's filter is determined using the following algorithm.

1. Give an initial guess of transition probabilities Q in Eq. (3)
2. Then update the transition probabilities of each regime with the past information, including the parameters in the system equation, θ_{S_t}, and Q for calculating the likelihood function in each state at time t, as shown in Eq. (9). The probability of each regime is updated by the following formula

$$\Pr(S_t = j | \theta_{S_t}) = \frac{f_j(y_t | S_t = j, \theta_{S_{t-1}}) \Pr(S_t = j | \theta_{S_{t-1}})}{\sum_{j=1}^{h} f_j(y_t | S_t = j, \theta_{S_{t-1}}) \Pr(S_t = j | \theta_{S_{t-1}})}, \tag{10}$$

where $f_j(y_t | S_t = j, \theta_{S_{t-1}})$ is the likelihood function of regime j which is allowed to differ across regimes, and is filtered probabilities at time $t - 1$.
3. Repeat step 1 and 2 for $t = 1, ..., T$.

3 Simulation Study

To evaluate the performance of our model and estimation procedure, we conduct an extensive set of simulations. In the simulation study, we compare the performance of the Markov-switching autoregressive model with mixture distribution regimes (denoted by MSMD) with the ordinary Markov-switching autoregressive model the one based on normal error distribution (denoted by Normal MS). We consider six distributions: the standard normal distribution (N), student-t distribution (T), generalized error distribution (GED), skew student-t distribution (ST), skew generalized error distribution (SGED), and skew normal distribution (SN). In this simulation, we consider two-regime model as a simple example, $S_t = 1, 2$. The model can be specified by

$$
\begin{aligned}
y_t &= \beta_{0,S_t=1} + \beta_{S_t=1} y_{t-1} + \sigma_{S_t=1} \varepsilon \\
y_t &= \beta_{0,S_t=2} + \beta_{S_t=2} y_{t-1} + \sigma_{S_t=2} \varepsilon_t
\end{aligned}
\tag{11}
$$

We set the value for each parameter as follows: $\sigma_{S_t=1} = 1$, $\sigma_{S_t=2} = 0.9$, $\beta_{0,S_t=1} = -1$, $\beta_{S_t=1} = 2$, $\beta_{0,S_t=2} = 1.5$, and $\beta_{S_t=2} = 3$. In addition for the skew and high kurtosis distribution, the degree of freedom for both regimes are set to be $v_{S_t=1} = v_{S_t=2} = 4$, and skew parameters are also set to be $\gamma_{S_t=1} = \gamma_{S_t=2} = 1.5$. Furthermore, the regime variable S_t is generated from

$$
S_t = 1\{U_t \geq \tau(S_{t-1})\},
\tag{12}
$$

where U_t is drawn from uniform $[0, 1]$ and $\tau(S_{t-1})$ is the cumulative probability of transition probability matrix Q in previous regime. Here, we set

$$
Q = \begin{bmatrix} 0.95 & 0.05 \\ 0.05 & 0.95 \end{bmatrix}.
\tag{13}
$$

In the Monte Carlo simulations, we consider three Scenarios:

Scenario 1. We consider MS-AR model whose error distributions are normal in regime 1 and student-t in regime 2.

Scenario 2. We consider MS-AR model whose error distributions are student-t in regime 1 and skew student-t in regime 2.

Scenario 3. We consider MS-AR model whose error distributions are skew student-t in regime 1 and GED in regime 2.

In this study, we consider sample size $n = 100, 200$ and compare the performance of the proposed model (MSMD) with the conventional model (Normal MS). The results show that when the skew distribution is assumed in the model, the small sample size brings a large bias to the estimated parameters. But when the sample size increases, the estimated parameters are close to the true values and the estimated standard errors decrease.

4 Empirical Study: The US Business Cycle

This section contains main results of the empirical study. We examine regime shift in a stochastic process of economic growth of the United States using quarterly time series (Q1 1947 to Q2 2017) of gross domestic product (GDP), measured in billions of current US dollars. This study uses two-regime MS-AR model for characterizing a common business cycle; that includes expansions and recessions. The results are divided into three parts. First, we consider a set of distributions as a choice for the error distribution for each regime. Then, the best fitting model among candidates is selected using the Akaike information criterion (AIC) in such a way that the model with minimum AIC is statistically preferable. Second, all unknown parameters are estimated by that selected model. In addition, the parameters are estimated by the ordinary MS-AR model in comparison with the proposed algorithm - the one that is selected from the first step. And third, the smooth probabilities and the filtered probabilities, interpreted as the US business cycle, are illustrated.

Firstly, we consider six distributions consisting of standard normal distribution (N), student-t distribution (T), generalized error distribution (GED), skew student-t distribution (ST), skew generalized error distribution (SGED), and skew normal distribution (SN). Table 4 shows the AIC values for each model. From this table, we can see that the best fitting model is the MS-AR model with

Table 1. Scenario 1

Parameter	True	Normal MS		MSMD (N-T)	
		$N = 100$	$N = 200$	$N = 100$	$N = 200$
$\beta_{0,S_t=1}$	-1	-0.9956	-0.9895	-0.9676	-1.0513
		(0.1296)	(0.1172)	(0.1309)	(0.1024)
$\beta_{1,S_t=1}$	2	1.7522	2.0172	2.1736	2.1221
		(0.1882)	(0.1132)	(0.1455)	(0.0949)
$\sigma_{S_t=1}$	0.9	1.0789	1.0126	0.9765	0.9481
		(0.0924)	(0.0828)	(0.0955)	(0.0739)
$\beta_{0,S_t=2}$	1.5	1.4309	1.5052	1.6323	1.4360
		(0.1127)	(0.0948)	(0.1638)	(0.1031)
$\beta_{1,S_t=2}$	3	3.0345	3.0046	3.1698	3.0067
		(0.1476)	(0.0965)	(0.1563)	(0.1023)
$\sigma_{S_t=2}$	1	0.5534	0.9215	0.9765	0.9481
		(0.1476)	(0.0677)	(0.0955)	(0.0739)
$v_{S_t=2}$	4			4.0125	4.0014
				(0.0014)	(0.0010)
p_{11}	0.95	0.9224	0.9236	0.9424	0.9554
		(0.0531)	(0.0275)	(0.0254)	(0.0199)
p_{22}	0.95	0.9695	0.9149	0.9674	0.9601
		(0.0214)	(0.0336)	(0.0155)	(0.0102)

Table 2. Scenario 2

Parameter	True	Normal MS		MSMD (T-ST)	
		N = 100	N = 200	N = 100	N = 200
$\beta_{0,S_t=1}$	−1	−1.0862 (0.1584)	−0.8846 (0.1248)	−1.2032 (0.1279)	−0.9261 (0.1154)
$\beta_{1,S_t=1}$	2	1.6299 (1.0255)	2.0516 (0.1347)	1.6306 (0.1559)	1.9226 (0.124)
$\sigma_{S_t=1}$	0.9	1.7488 (0.5253)	0.9276 (0.0873)	0.5799 (0.0946)	0.9289 (0.0805)
$v_{S_t=1}$	4			2.8391 (0.1306)	2.9273 (0.1270)
$\beta_{0,S_t=2}$	1.5	0.3995 (1.1125)	1.4838 (0.0687)	1.5938 (0.1060)	1.4705 (0.0790)
$\beta_{1,S_t=2}$	3	2.5254 (0.1819)	3.0485 (0.0691)	2.9204 (0.1051)	2.9877 (0.0542)
$\sigma_{S_t=2}$	1	1.673 (0.1168)	0.8028 (0.0487)	0.8670 (0.1925)	0.9646 (0.0637)
$v_{S_t=2}$	4			4.8511 (1.1208)	4.8909 (1.0781)
$\gamma_{S_t=2}$	1.5			0.9567 (0.1610)	0.8823 (0.0980)
p_{11}	0.95	0.0001 (0.7617)	0.9455 (0.0307)	0.863 (0.0424)	0.8984 (0.048)
p_{22}	0.95	0.9999 (0.0019)	0.9700 (0.1473)	0.9415 (0.0152)	0.9649 (0.0180)

Source: Calculations.
Note: Inside (.) is standard error.

N-T distribution regimes; that is, a normal distribution is selected for regime 1 and student-t distribution is chosen for regime 2. Moreover the AIC value noted by −* is of the model where both regimes are based on normal distribution, corresponding to the ordinary MS-AR model of Hamilton (1989). The results show that the AIC value of the best-fitting mixture MS-AR model, −1835.72, is less than that of the classical MS-AR model, which is −1816.26. This means the proposed model is more accurate and better than the conventional model based on AIC values. The finding implies that in different economic regimes the data appears to be distributed differently. And the experiment discovers that US economic growth over regime 2 has heavier tails than regime 1, meaning that it is more likely to produce values that fall far from its mean over this regime (Tables 1, 2 and 3).

Note that we re-checked our estimation by applying the same data to the original MS-AR model of Hamilton [5]. We used computer code provided by Hamilton to estimate the MS-AR(1) again, and, in addition to the estimated

Table 3. Scenario 3

Parameter	True	Normal MS		MSMD (ST-GED)	
		N = 100	N = 200	N = 100	N = 200
$\beta_{0,S_t=1}$	−1	−0.8389 (0.0921)	−1.1025 (0.1008)	−1.2265 (0.2037)	−1.1098 (0.2062)
$\beta_{1,S_t=1}$	2	2.021 (0.1234)	2.0336 (0.1152)	2.1149 (0.1270)	2.0487 (0.1793)
$\sigma_{S_t=1}$	0.9	0.7315 (0.1002)	1.0047 (0.0955)	0.7139 (0.0986)	1.0138 (0.0917)
$v_{S_t=1}$	4			3.2125 (0.1025)	3.8815 (0.0824)
$\beta_{0,S_t=2}$	1.5	1.4870 (0.1238)	1.3965 (0.1023)	1.4957 (0.1291)	1.3694 (0.1059)
$\beta_{1,S_t=2}$	3	3.0189 (0.1240)	2.9888 (0.1043)	3.1321 (0.1168)	2.9997 (0.1059)
$\sigma_{S_t=2}$	1	0.8489 (0.0921)	0.9490 (0.0857)	0.9476 (0.0944)	0.9590 (0.0781)
$v_{S_t=2}$	4			2.1055 (1.3695)	2.1000 (1.3055)
p_{11}	0.95	0.9182 (0.0410)	0.9357 (0.0275)	0.9306 (0.0388)	0.9353 (0.0294)
p_{22}	0.95	0.8918 (0.0539)	0.9557 (0.0217)	0.9148 (0.0477)	0.9552 (0.0237)

Source: Calculations.
Note: Inside (.) is standard error.

Table 4. AIC of mixture MS-AR models

Distribution	Regime 1					
Regime 2	N	SN	T	ST	GED	SGED
N	−1816.26*	−1814.51	−1835.34	−1611.08	−1812.23	−1812.59
SN	−1813.98	−1812.62	−1833.15	−1592.03	−1812.57	−1810.64
T	**−1835.72**	−1833.65	−1824.21	−1831.61	−1833.71	−1831.71
ST	−1807.33	−1810.43	−1821.52	−1600.99	−919.22	−1604.11
GED	−1817.81	−1812.23	−1833.35	−1825.91	−1806.47	−1609.63
SGED	−1667.55	7532.69	−789.77	−1831.21	−556.22	−1727.96

Source: Calculations.
Note: The processes are conducted under the first-order autoregression, AR(1); The bold value is minimum AIC; "*" is the AIC value of the mixture MS-AR model, where both regimes are based on normal distribution, corresponding to the ordinary MS-AR model of Hamilton [5].

Table 5. Estimated parameters from the mixture MS-AR model

	Regime 1 Normal	Regime 2 Student-t
Intercept	0.0291*** (0.0090)	0.0072*** (0.0008)
Coefficient	0.1774*** (0.0141)	−0.4689*** (0.0554)
Sigma	0.0112*** (0.0012)	0.0123*** (0.0034)
Degree of freedom		2.5155*** (0.4194)
Duration	188.68	34.84
Transition matrix	P_{1j}	P_{2j}
P_{i1}	0.9713	0.0053
P_{i2}	0.0287	0.9947

Source: Calculation.
Note: "***" denotes the significance at 1% level and inside (.) is
standard error.

results, we got the AIC value pretty similar to the one obtained from our
main experiment. Consider a two-regime MS-AR(1) model, the specific form is
given by

$$\Delta y_t = \beta_{0,S_t} + \beta_{1,S_t} \Delta y_{t-1} + \sigma_{S_t} \varepsilon_t \tag{14}$$

where Δy_t is the growth rate of the United States' GDP, β_{0,S_t} and β_{1,S_t} are
intercept and autoregressive coefficient, respectively. These two terms are regime
dependent as well as the error. Table 5 shows the estimated parameters obtained
from the mixture MS-AR model (the one with N-T distribution regimes), followed by Table 6 which shows the estimated results of the ordinary MS-AR
model. From Table 5, we can see a considerable difference in mean and coefficient across regimes, as well as in Table 6. One obvious thing that we can see
from these two tables is the negative sign of the coefficient in regime 2. This
means, under this regime the growing economy in previous time can cause the
falling in GDP growth at the present time. Moreover, we find that the value of
intercept in regime 2 is considerably less than in regime 1. Therefore, regime 2
is likely to conform to the recession, whilst regime 1 corresponds to expansion.

Furthermore, the transition matrix allows us to observe the asymmetry of the
business cycle in terms of duration. This is even more obvious in the result of the
proposed model. We find that expansions have a duration of approximately 189
quarters, whereas the recessions have only 35 quarters, approximately. The probability of staying in regime 1 (expansions) is 0.9713 (97.13%) but the probability
of switching from regime 1 to regime 2 (recessions) is 0.0287 (2.87%). Similarly,
the probability of moving from recession to expansion is 0.0053 (0.53%), while
the chance of remaining in the same state is 99.47%.

The US business cycle over the years 1947 to 2017 is illustrated in Figs. 1
and 2 through the filtered and smoothed probabilities, lying between 0 and 1.
We can see that the different persistence of the regimes can be observed in
these figures. The US economic growth fell certainly between 1970 and 1980s

Table 6. Estimated parameters from the ordinary MS-AR model[a]

	Regime 1	Regime 2
	Normal	Normal
Intercept	0.0092*** (0.0017)	0.0086*** (0.0013)
Coefficient	0.4960*** (0.0725)	−0.3339*** (0.0959)
Sigma	0.0121*** (0.0010)	0.0051*** (0.0010)
Duration	12.15	18.87
Transition matrix	P_{1j}	P_{2j}
P_{i1}	0.9171	0.053
P_{i2}	0.0823	0.947

Source: Calculation.
Note: [a], the ordinary MS-AR model refers to the model of Hamilton (1989); "***" denotes the significance at 1% level and inside (.) is standard error.

when the world faced economic stagnation, high unemployment, and thereby recessions. And the estimated results reveal that the probabilities of staying in recessions were almost equal to one during that time. This is even clearer in Fig. 1. Besides that, overall results suggest that the filtered probabilities and smoothed probabilities obtained from the mixture MS-AR model can, more or less, illustrate a clearer and smoother business cycle than the ones from the ordinary model. At least, the more clearly the period of 1970s recession and early 1980s recession can be captured by the proposed model.

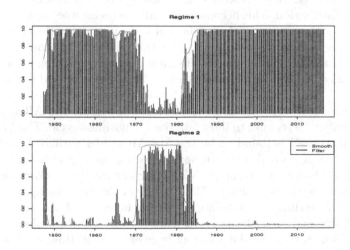

Fig. 1. The US business cycle obtained from the mixture MS-AR(1) model

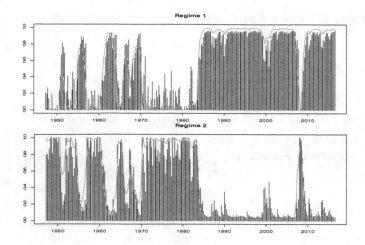

Fig. 2. The US business cycle obtained from the ordinary MS-AR(1) model

5 Conclusions

This study realizes the advantage of the Markov-switching models for illustrating trend and fluctuation in a time series data, particularly economic fluctuation or the so-called business cycle [7]. However, besides this usefulness, the MS-type models also have some limitations; that is, a normal distribution regime. To illustrate, we consider two-regime Markov-switching autoregressive (MS-AR) model to be an example. The ordinary MS-AR model assumes all parameters to be normally distributed; in other words, all regimes in the MS model are restricted to a normal distribution. This becomes our concern because economic time series appear often with a mixture of distribution or populations. Hence, this study generalizes the ordinary MS-AR model to the one with mixture distribution regimes. This model allows different distributions across regimes, so regime 1 and regime 2 can be conducted under distinct distributions, and not need to be normal.

In the empirical application, we apply the two-regime MS-AR model with mixture distribution regimes to illustrate the US business cycle. The best fitting model based on AIC is found to be the MS-AR model with N-T distribution regimes; that is, a normal distribution is selected for regime 1 and student-t distribution is chosen for regime 2. We also compare the performance of the proposed model with the ordinary MS-AR model through the application. We discover that the estimated results obtained from both models are not so different in terms of size and sign of the coefficients, except for the estimated regime durations which are considerably different from each other. However, the mixture MS-AR model can provide a superior fit compared to the ordinary model.

In essence, our finding just proves more or less the persistence of distinct distributions across regimes in the MS model. Nevertheless, this should deserve a further study. One of the things that should be improved in future study is the use of a variety of distributions. Even though this study can prove the existence of the different distributions across regimes but this is still limited to a set of specific distributions. Another thing is the model structures because the real economy is related with several factors. Therefore, the autoregressive model may not reflect well.

Acknowledgement. We are grateful for financial support from Puay Ungpakorn Centre of Excellence in Econometrics, Faculty of Economics, Chiang Mai University.

References

1. Ardia, D., Bluteau, K., Boudt, K., Catania, L., Peterson, B., Trottier, D.A.: Markov-Switching GARCH Models in R: The MSGARCH Package (2016)
2. Deschamps, P.J.: A flexible prior distribution for Markov switching autoregressions with Student-t errors. J. Econom. **133**(1), 153–190 (2006)
3. Haas, M., Mittnik, S., Paolella, M.S.: A new approach to Markov-switching GARCH models. J. Financ. Econom. **2**(4), 493–530 (2004)
4. Haas, M.: Skew-normal mixture and Markov-switching GARCH processes. Stud. Non-linear Dyn. Econom. **14**(4), 1–56 (2010)
5. Hamilton, J.D.: A new approach to the economic analysis of nonstationary time series and the business cycle. Econometrica J. Econom. Soc. **57**, 357–384 (1989)
6. Kole, E.: Regime switching models: an example for a stock market index. Econometric Institute, Erasmus School of Economics, Erasmus University Rotterdam, April 2010, Unpublished manuscript
7. Manccjuk, P., Pastpipatkul, P., Sriboonchitta, S.: Economic growth and business cycle: the case of Thailand. Int. J. Econ. Res. **14**(6), 263–274 (2017)
8. Perlin, M.: MS-Regress-the MATLAB package for Markov regime switching models (2015)

Estimation of Volatility on the Small Sample with Generalized Maximum Entropy

Quanrui Song[1]([✉]), Songsak Sriboonchitta[1,2], Somsak Chanaim[1,2], and Chongkolnee Rungruang[3]

[1] Faculty of Economics, Chiang Mai University, Chiang Mai 50200, Thailand
song_quanrui@cmu.ac.th
[2] Center of Excellence in Econometrics,
Chiang Mai University, Chiang Mai 50200, Thailand
[3] Faculty of Commerce and Management, Prince of Songkla University,
Trang Campus, Trang 92000, Thailand

Abstract. Generalized autoregressive conditional heteroscedasticity (GARCH) provides useful techniques for modeling the dynamic volatility model. Several estimation techniques have been developed over the years, for examples Maximum likelihood, Bayesian, and Entropy. Among these, entropy can be considered an efficient tool for estimating GARCH model since it does not require any distribution assumptions which must be given in Maximum likelihood and Bayesian estimators. Moreover, we address the problem of estimating GARCH model characterized by ill-posed features. We introduce a GARCH framework based on the Generalized Maximum Entropy (GME) estimation method. Finally, in order to better highlight some characteristics of the proposed method, we perform a Monte Carlo experiment and we analyze a real case study. The results show that entropy estimator is successful in estimating the parameters in GARCH model and the estimated parameters are close to the true values.

Keywords: Volatility · GARCH$(1, 1)$ model
Generalized Maximum Entropy

1 Introduction

Estimation of volatility is very important in financial economics, because volatility is a measure of uncertainty on observed time series of financial data such as stock price or stock index. Many studies realized that the volatility of financial data should not be constant overtime but invariably varying through time. The popular model to estimate the time-varying volatility is Autoregressive conditional heteroskedasticity model (ARCH) proposed by Engel [1] in 1982 and was extended by Bollerslev [2]. This paper introduces a new volatility model called Generalized autoregressive conditional heteroskedasticity model (GARCH(p,q)) in the following form

$$\varepsilon_t = \sigma_t \nu_t, \ , \nu_t \sim N(0,1), t = 1, 2 \cdots , T,$$

© Springer International Publishing AG, part of Springer Nature 2018
V.-N. Huynh et al. (Eds.): IUKM 2018, LNAI 10758, pp. 324–334, 2018.
https://doi.org/10.1007/978-3-319-75429-1_27

where

$$\sigma_t^2 = \omega_0 + \sum_{i=1}^{p} \omega_{1i}\sigma_{t-i}^2 + \sum_{j=1}^{q} \omega_{2j}\varepsilon_{t-j}^2, \ \omega_0, \omega_{1i}, \omega_j > 0 \ \forall i = 1, \cdots, p, \ j = 1, \cdots, q.$$

In this study, we consider GARCH(1,1) model because this is an extension of ARCH model and relies only on past observation and on past volatility. In general, GARCH parameters have been estimated by using Maximum Likelihood (MLE) approach which assumes normality. However, the assumption of conditional normality is not always appropriate. Maximum Entropy (ML) modeling which has a flexible functional form to use with many distributions has been applied in financial field. Park and Bera [3] applied two separate maximum entropy densites in ARCH model (MEARCH model) where moment functions are selected based on the sample to estimate NYSE stock returns. They show the MEARCH model was useful to capture the behavior of the sample.

From assumption on long-term stability, GARCH model will give a wrong answer if data is not stability trend. Financial time series data generally are characterized with a large sample size and structural change. To be consistent with this stability assumption, we suggest small sample size for estimating GARCH(1,1) model. Hwang and Valls Pereira [4] investigated ML estimation with small sample in GARCH model with non-negative Bollerslev's condition that guarantees positive conditional volatility and they showed GARCH models with small sample problems from their results that the estimated parameters are negatively biased. They also suggested the minimum size of sample needed for GARCH(1,1) model to be 500 observations.

When the data has a heavy-tailed distribution, the analysis of GARCH time series data by using quasi maximum likelihood estimation (QMLE) can lead to inconsistency in parameter noted by Lee et al. [5]. Who applied maximum entropy to estimate GARCH(1,1) model for 503 observations of S&P 500 index.

Generalized Maximum Entropy (GME) method for ARCH model can be found in the book by Golan et al. [6]. In this paper, we use GME to estimate GARCH(1,1) parameters because this method does not require the large observation and assumption about distribution function for innovation. This method has only independent assumption for all random variables and support space for each random variable. We show the result by using simulation and applying the model to estimate volatility on stock price returns with small number of observations.

2   Methodology

Let ε_t, $t = 1, \cdots, T$ is sequence random variable from time series data with mean zero. The GARCH(1,1) model is defined by

$$\varepsilon_t = \sigma_t \cdot \nu_t, \ t = 1, 2, \cdots, T, \tag{1}$$

$$\sigma_t^2 = \omega_0 + \omega_1 \sigma_{t-1}^2 + \omega_2 \varepsilon_{t-1}^2, \ t = 1, 2, \cdots, T, \tag{2}$$

where $\omega_0 > 0, \omega_1, \omega_2 \in (0,1)$ and stationary condition is $\omega_1 + \omega_2 < 1$ and $\nu_t \sim F(\cdot)$ is sequence of independent random variable or innovation. In this study, the parameters $\omega_0, \omega_1, \omega_2$ are estimated using generalized maximum entropy (GME). The basic idea of this estimator is that the entropy, which refers to an amount of the uncertainty, is maximized subject to model and data constraints. Here, we consider the Shannon's entropy measure proposed by Shannon [7]. This Shannon's entropy is represented by the amount of the uncertainty of a discrete probability distribution and the sum of all outcomes probability equals to one. The constraint primal problem for ARCH model can be written as follows:

$$\max H(P_0, P_1, P_2, W_1, W_2, \cdots, W_T) = H(P_0) + H(P_1) + H(P_2) + \sum_{t=1}^{T} H(W_t), \quad (3)$$

subject to

$$\sqrt{\left(\sum_{i=1}^{k} z_{0i}p_{0i}\right)\left(\sum_{i=1}^{k} s_i w_{1i}\right)} = \varepsilon_1,$$

$$\sqrt{\sum_{i=1}^{k} z_{0i}p_{0i} + \left(\sum_{i=1}^{k} z_{1i}p_{1i}\right)\left(\frac{\varepsilon_1}{\sum_{i=1}^{k} s_i w_{1i}}\right)^2 + \left(\sum_{i=1}^{k} z_{2i}p_{2i}\right) \cdot \varepsilon_1^2 \left(\sum_{i=1}^{k} s_i w_{2i}\right)} = \varepsilon_2,$$

$$\sqrt{\sum_{i=1}^{k} z_{0i}p_{0i} + \left(\sum_{i=1}^{k} z_{1i}p_{1i}\right)\left(\frac{\varepsilon_2}{\sum_{i=1}^{k} s_i w_{2i}}\right)^2 + \left(\sum_{i=1}^{k} z_{2i}p_{2i}\right) \cdot \varepsilon_2^2 \left(\sum_{i=1}^{k} s_i w_{2i}\right)} = \varepsilon_3,$$

$$\vdots$$

$$\sqrt{\sum_{i=1}^{k} z_{0i}p_{0i} + \left(\sum_{i=1}^{k} z_{1i}p_{1i}\right)\left(\frac{\varepsilon_{T-1}}{\sum_{i=1}^{k} s_i w_{(T-1)i}}\right)^2 + \left(\sum_{i=1}^{k} z_{2i}p_{2i}\right) \cdot \varepsilon_{T-1}^2 \left(\sum_{i=1}^{k} s_i w_{(T-1)i}\right)} = \varepsilon_T,$$

$$\frac{\sum_{i=1}^{k} z_{0i}p_{0i}}{1 - \sum_{i=1}^{k} z_{1i}p_{1i} - \sum_{i=1}^{k} z_{1i}p_{0i}} = var(\varepsilon_t),$$

$$\sum_{i=1}^{k} p_{0i}, \sum_{i=1}^{k} p_{1i}, \sum_{i=1}^{k} p_{2i}, \sum_{i=1}^{k} w_{1i}, \cdots = \sum_{i=1}^{k} w_{Ti} = 1,$$

$$p_{ji}, w_{ti} \in (0,1) \; \forall j = 0,1,2, \; i = 1, \cdots, k, \; t = 1, \cdots, T,$$

where $H(P_j) = -\sum_{i=1}^{k} p_{ji} \log(p_{ji}), j = 0, 1, 2$ and $H(W_j) = -\sum_{i=1}^{k} w_{ti} \log(w_{ti})$, $t = 1, \cdots, T$, $z_{0i}, z_{1i}, z_{2i}, s_i$ are the discrete support space, and $var(\varepsilon_t)$ is the sample variance. After optimizing this function, we can estimate $\widehat{\omega_0}, \widehat{\omega_1}, \widehat{\omega_2}, \widehat{\nu_t}$ by

$$\widehat{\omega}_0 = \sum_{i=1}^{k} z_{0i} p_{0i}, \quad \widehat{\omega}_1 = \sum_{i=1}^{k} z_{1i} p_{1i}, \quad \widehat{\omega}_2 = \sum_{i=1}^{k} z_{2i} p_{2i}, \quad \widehat{\nu}_t = \sum_{i=1}^{k} s_i w_{ti}, \ t = 1, 2, \cdots, T.$$

The standard deviation of parameters $\widehat{\omega_0}, \widehat{\omega_1}$ and $\widehat{\omega_2}$ is to be estimated by

$$\text{std. of } \widehat{\omega}_j = \sqrt{\sum_{i=1}^{k} (z_{ji} - \widehat{\omega}_j)^2 p_{ji}}, \ j = 0, 1, 2.$$

3   Simulation Study

In this section, a simulation study was conducted to evaluate performance and accuracy of GME estimation in GARCH(1,1) model with small observation $T = \{50, 100\}$. For every support space, we define 5 points support for $\omega_0, \omega_1, \omega_2$ with $z_0 = z_1 = z_2 = [0, 0.25, 0.50, 0.75, 1]$ and support space for innovation ν_t is $[-10, -5, 0, 5, 10]$ for all $t = 1, 2, \cdots, T$. We simulated the data from the GARCH(1,1) model with 1,000 paths, where the innovation is assumed to have standard normal distribution $N(0, 1)$. We consider $\omega_0, \omega_1, \omega_2$ for 5 cases.

The simulation results are provided in Tables 1, 2, 3, 4 and 5 and the similar results are obtained. According to Table 1, by the simulation we observe that the estimation mean of ω_0 is underestimated but those for ω_1, ω_2 are overestimated for both T = 50 and 100. From Table 2, we see that the estimation mean of ω_0 and ω_1 are underestimated while ω_2 value is overestimated for both T = 50 and 100. From Table 3, the different results are obtained, the means of ω_0 and ω_1 are overestimated but that for ω_2 is underestimated for both T = 50 and 100. From Table 4, we see that the means of ω_0, ω_1 and ω_2 are

Table 1. Case 1: $\omega_0 = \omega_1 = \omega_2 = 0.2$

	True	T = 50				T = 100			
		Mean	Std	Quantile 5%	Quantile 95%	Mean	Std	Quantile 5%	Quantile 95%
ω_0	0.2	0.0393	0.0215	0.0168	0.0769	0.0428	0.0202	0.0241	0.0712
ω_0	0.2	0.3533	0.0181	0.3202	0.3804	0.3468	0.0159	0.3180	0.3715
ω_0	0.2	0.3298	0.0133	0.3077	0.3505	0.3067	0.0107	0.2893	0.3228
Entropy		83.3738	0.1663	83.1607	83.6627	163.3534	0.1802	163.1272	163.6612

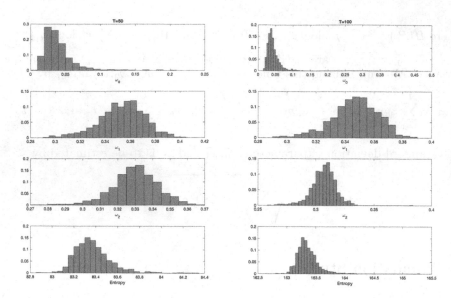

Fig. 1. Histogram of parameter estimates from simulation $T = 50$(left), $T = 100$(right)

Table 2. Case 2: $\omega_0 = .7$, $\omega_1 = .5$, $\omega_2 = 0.1$

	True	$T = 50$				$T = 100$			
		Mean	Std	Quantile 5%	Quantile 95%	Mean	Std	Quantile 5%	Quantile 95%
ω_0	0.7	0.5099	0.0287	0.4624	0.5569	0.5318	0.0228	0.4985	0.5708
ω_1	0.5	0.4363	0.0255	0.3907	0.4760	0.4605	0.0215	0.4251	0.4962
ω_2	0.1	0.3862	0.0236	0.3446	0.4235	0.3617	0.0190	0.3292	0.3930
Entropy		84.5511	0.0928	84.3883	84.6699	164.4394	0.0982	164.2645	164.5694

Table 3. Case 3: $\omega_0 = .2$, $\omega_1 = .2$, $\omega_2 = 0.7$

	True	$T = 50$				$T = 100$			
		Mean	Std	Quantile 5%	Quantile 95%	Mean	Std	Quantile 5%	Quantile 95%
ω_0	0.2	0.4048	0.1826	0.0975	0.6389	0.4405	0.1771	0.1000	0.6342
ω_1	0.2	0.3741	0.0669	0.2938	0.4846	0.3912	0.0804	0.2920	0.4946
ω_2	0.7	0.3777	0.0737	0.2920	0.5061	0.3909	0.0768	0.2898	0.5190
Entropy		84.3664	0.5714	83.5372	84.8916	164.4499	0.9400	162.1926	165.1179

overestimated for both T = 50 and 100; however, they are close to the true values. Finally, from Table 5, the estimation mean of ω_0 is underestimated and overestimated for T = 50 and T = 100 respectively; the estimation mean of ω_1 is underestimated for both T = 50 and T = 100; and the mean of ω_2 is overestimated for both T = 50 and T = 100.

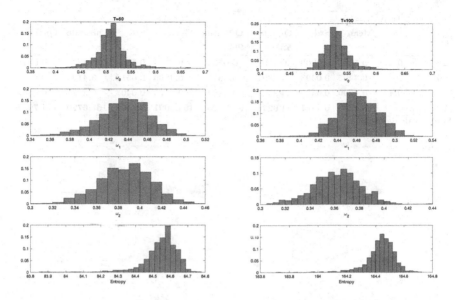

Fig. 2. Histogram of parameter estimates from simulation $T = 50$(left), $T = 100$(right)

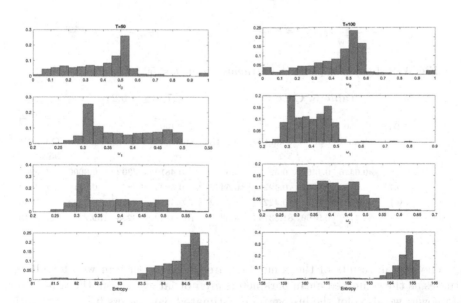

Fig. 3. Histogram of parameter estimates from $T = 50$(left), $T = 100$(right)

Table 4. Case 4: $\omega_0 = .5$, $\omega_1 = .4$, $\omega_2 = 0.4$

	True	T = 50				T = 100			
		Mean	Std	Quantile 5%	Quantile 95%	Mean	Std	Quantile 5%	Quantile 95%
ω_0	0.5	0.5365	0.0809	0.4402	0.6538	0.5389	0.1380	0.4393	0.6611
ω_1	0.4	0.4495	0.0393	0.3691	0.4975	0.4800	0.0518	0.4124	0.5203
ω_2	0.4	0.4290	0.0435	0.3457	0.4904	0.4275	0.0408	0.3620	0.4882
Entropy		84.5230	0.4421	84.0229	84.7723	164.3091	1.0100	161.6720	164.7983

Source calculation.

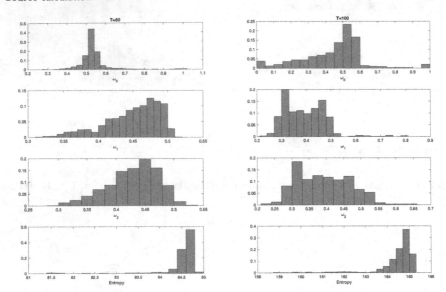

Fig. 4. Histogram of parameter estimates from $T = 50$(left), $T = 100$(right)

Table 5. Case 5: $\omega_0 = .6$, $\omega_1 = .7$, $\omega_2 = 0.2$

	True	T = 50				T = 100			
		Mean	Std	Quantile 5%	Quantile 95%	Mean	Std	Quantile 5%	Quantile 95%
ω_0	0.6	0.6476	0.1462	0.5285	0.9912	0.4819	0.3302	0.0006	0.9904
ω_1	0.7	0.5170	0.0202	0.4921	0.5442	0.5833	0.0849	0.5118	0.7500
ω_2	0.2	0.4565	0.0231	0.4227	0.4862	0.4569	0.0809	0.3933	0.5686
Entropy		83.77	0.9866	81.3374	84.5333	162.4112	1.9288	159.2213	163.3010

The overall results of the simulation are likely to perform well for all case studies as the estimated mean parameters are not far away from the true values. Moreover, we also plot the histogram of estimated parameters from 1,000 paths. We present all case studies and plot in Figs. 1, 2, 3, 4 and 5. We can observe that most estimated parameters are close to the true values and that the standard

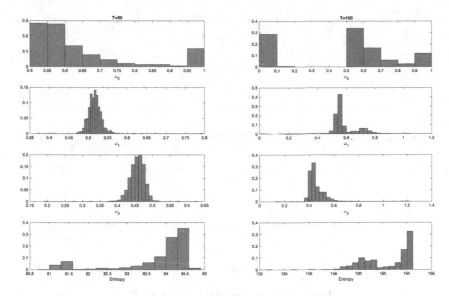

Fig. 5. Histogram of parameter estimates from $T = 50$(left), $T = 100$(right)

deviations are rather small. We may thus conclude that the proposed method estimates the GARCH(1,1) parameters quite well.

4   Real Data Application

In this section, we compare the performance of the entropy GARCH(1,1) using the stock index data. We consider closing stock price and daily return from Advanced Micro Devices, Inc. (AMD) from March 1, 2017 to July 28, 2017. The data is obtained from Thomson Reuters DataStream. We plot the daily closing price and return of AMD in Fig. 6. The summary statistics is shown in Table 6.

In this application study, we consider the simple GARCH(1,1) to estimate the volatility of the AMD return. The support spaces of the GARCH parameters are specified just like in the simulation study section. The estimated results are shown in Table 7. We can see that our estimated standard errors of $\omega_0, \omega_1, \omega_2$ in Table 7 are large and all parameters are insignificant. We suspect that our support spaces are perhaps too large and thereby leading a high standard error. Therefore, we try to estimate again but using a narrower range of support spaces for $\omega_0, \omega_1, \omega_2$. We define a new support space as $z_0 = z_1 = z_2 = [0, 0.125, 0.25, 0.375, 0.5]$ and s_i as $[-10, -5, 0, 5, 10]$. The results are shown in Table 8, and when compared to Table 7, values of parameter estimates change and standard errors decrease; however, we can still obtain the significant results. We also plot the conditional variance and innovation error from our GARCH(1,1) (Fig. 7).

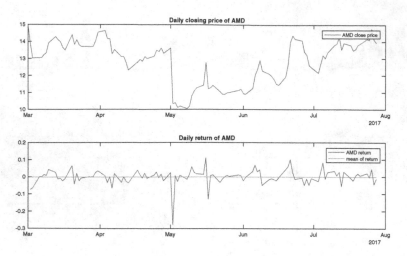

Fig. 6. Closing stock price (up) and return of AMD (down)

Table 6. Summary statistics of AMD

Mean	Median	Std	Min	Max	Skewness	Kurtosis	Obs.
−0.0007	0.0008	0.0443	−0.2775	0.1102	−2.2995	16.958	104

Table 7. Estimation of GARCH(1,1) parameter by GME

Parameter	Value	Std
ω_0	7.3761×10^{-4}	0.0137
ω_1	0.3689	0.3375
ω_2	0.2496	0.2927
Entropy	168.8636	

The support space for every parameter is the same as in the simulation part.

Table 8. Parameters estimated from GME with the support space

Parameter	Value	Std
ω_0	0.0011	0.0118
ω_1	0.2333	0.1763
ω_2	0.2079	0.1735
Entropy	169.3612	

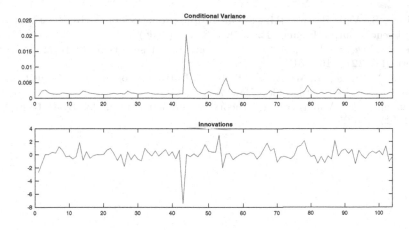

Fig. 7. Estimation of conditional variance (upper) and innovation (lower)

5   Conclusions and Future Research

It is not easy to get the big data with certain economic situation or stable environment. Thus many estimations face the ill-posed problem. In this study, GME estimator is proposed to estimate the unknown parameters in GARCH model. The simulation results show that the GME is workable well on some values of parameters since either underestimated or overestimated results are obtained in some parameters. However, the results are still acceptable in this study. From the real data analysis, we cannot find the significant result of the GARCH parameters. The problem of GME estimation in GARCH(1,1) is that the value of parameter estimates depends on the support space. We find that the narrow support space will lead a smaller standard error of the parameters.

Therefore in future research, we should find the new method to estimate a GARCH model or other volatility models that can handle the small observation problem. GME method should also be extended to GARCH(p,q) and other stochastic volatility (IGRACH, GJR-GRACH, NGARCH etc.) models.

Acknowledgement. The authors thank Mr. Woraphon Yamaka for his suggestions for GME method and applied to the GARCH(1,1) model. We would like to thank the referee for giving comments on the manuscript.

References

1. Engle, R.F.: Autoregressive conditional heteroscedasticity with estimates of the variance of United Kingdom Inflation. Econometrica **50**(4), 987–1007 (1982)
2. Bollerslev, T.: Generalized autoregressive conditional heteroskedasticity. J. Econometrics **31**(3), 307–327 (1986)
3. Park, S.Y., Bera, A.K.: Maximum entropy autoregressive conditional heteroskedasticity model. J. Econometrics **150**(2), 219–230 (2009)

4. Hwang, S., Valls Pereira, P.L.: Small sample properties of GARCH estimates and persistence. Eur. J. Finance **12**(6–7), 473–494 (2006)
5. Lee, J., Lee, S., Park, S.: Maximum entropy test for GARCH models. Stat. Methodol. **22**, 8–16 (2015)
6. Golan, A., Judge, G., Miller, D.: Maximum Entropy Econometrics: Robust Estimation with Limited Data. Wiley, New York (1997)
7. Shannon, C.E.: A mathematical theory of communication. Bell Syst. Tech. J. **27**, 379–423 (1948)

Estimating and Predicting Financial Series by Entropy-Based Inferential Model

Tanarat Rattanadamrongaksorn[1]($\boxtimes$) (iD), Duangthip Sirikanchanarak[1,3],
Jirakom Sirisrisakulchai[1,2], and Songsak Sriboonchitta[1,2]

[1] Faculty of Economics, Chiang Mai University, Chiang Mai 50200, Thailand
tanarat_ra@cmu.ac.th
[2] Center of Excellence in Econometrics, Faculty of Economics,
Chiang Mai University, Chiang Mai 50200, Thailand
[3] Bank of Thailand, Bangkok 10200, Thailand

Abstract. In this study, the non-parametric *Inferential Model* or *IM* with the entropy-based random set has been proposed for the investigation of financial data in the two statistical domains i.e. estimation and prediction. The samples from five financial markets were chosen for representing the different types of financial assets to make a conclusion about this new framework. We found that the Inferential Model performed equally well compared with the traditional method but was more robust so that it might be more appropriate for some specific uses.

Keywords: Non-parametric · Prior-free · Weak belief · Plausibility
Bayesian statistics · Dempster-Shafer · Evidence theory · Random set
P-value ban

1 Introduction

The frequency-based approach to statistics (we shall simply called it as the Frequentist) is the most widely-used among major statistical philosophies. Dominating others by its simplicity, the Frequentist has long been used and accepted by the statistical communities in general. However, in view of recent P-value ban, this approach is criticized to be invalid for statistical inference. Fortunately, the three approaches of (1) Bayesian statistics, (2) Information Theoretic statistics, and (3) Inferential Model have been addressed as solutions to this dilemma in order to avoid problems resulted by the Frequentist approach [1].

This suggestion raises our attention to these methods in order to find other possible implementations to statistical topics such as data estimation and prediction which are two of the most fundamental problems in the financial economics and any other statistics intensive areas.

In this research, we applied one of the methods namely the Inferential Model to these problems. The Inferential Model is a new framework and a descendant of Bayesian statistics equipped with the features of prior-free and frequency-calibrated properties. The following sections contain discussions on concepts and

© Springer International Publishing AG, part of Springer Nature 2018
V.-N. Huynh et al. (Eds.): IUKM 2018, LNAI 10758, pp. 335–346, 2018.
https://doi.org/10.1007/978-3-319-75429-1_28

backgrounds of the Inferential Model (Sect. 2); data, and results of demonstrated examples (Sect. 3); and conclusions (Sect. 4).

2 Inferential Model

The Inferential Model (IM) is the extension of the Dempster-Shafer Model (DSM) in searching for the prior-free inference by the distributions of random set. The IM is frequency-calibrated by utilizing the Predictive Random Set (PRS) to capture uncertainty in the model.

Let a parameter, $\theta \in \Theta$, be random variable and data, $X = \{x_1, \ldots, x_n\}$, be random samples with an unknown cumulative distribution function (cdf), $F(x)$ and $x \in \mathbb{R}$, but assumed independent and identically distributed (iid). To make an inference about the data based on the DSM is to determine the probabilities of random samples. These random sets are called the focal sets and could be obtained through a multivalued mapping,

$$M_X(\omega) = \{\theta \in \Theta | Pl_X(\theta) \geq \omega\}, \tag{1}$$

from $[0, 1]$ to 2^Θ and interpreted as parameter regions whose plausibilities (Eq. 3 below) higher than a specified probability threshold, $\omega \in [0, 1]$. The results of the DSM are belief functions which are two cases of posterior probabilities i.e. Belief (Bel) and Plausibility (Pl):

$$Bel_X(\mathcal{A}) = P(M_X(\omega) \subseteq \mathcal{A}), \tag{2}$$
$$Pl_X(\mathcal{A}) = P((M_X(\omega) \cap \mathcal{A}) \neq \emptyset), \tag{3}$$

where $\mathcal{A}$ is an assertion which is a considered subset of parameters, $\mathcal{A} \subseteq \Theta$, in a problem of interest. The belief represents a probability of evidence that definitely supports the assertion and the plausibility is a probability of evidence that does not conflict with the assertion. These belief functions are commonly normalized by the probability of a non-empty focal set, $P(M_X(\omega)) \neq \emptyset$. In general, either belief function alone is insufficient because of its sub-additivity, $Bel_X(A) + Bel_X(A^C) \leq 1$. The outputs of inferential model is more beneficial and more comprehensive in case both belief and plausibility, i.e. lower and upper probabilities, can be reported. However, the belief investigated under lower-dimensional sub-space, e.g. the assertion is singleton, will always be nought [2].

The rest of this part includes the literature review of the Inferential Model in chronological order (Sect. 2.1), the necessary concepts and backgrounds of the non-parametric Inferential Model (Sect. 2.2), and the proposed predictive random set based on entropy measurement (Sect. 2.3).

2.1 Literature Review

For historical interest, the developments of the Inferential Model will be compiled as complete as information is available to us.

2008: A type of situation-specific inference called the Weak Belief Dempster-Shafer (WBDS) method was extended from the Dempster-Shafer (DS) theory of evidence in order to answer statistical questions that models based solely on sampling approach cannot perform well [3][1]. In claiming, the standard DS inference could be improved by the weak belief and could retain the trade-off between credibility and efficiency by the Maximal Belief (MB) [4].

2009: The concept of the weak belief was described again comprehensively by the problem of statistical inference of a single observation in normal model with a known variance, $\mathcal{N}orm(\theta, 1)$. Comparisons were made and showed some difficulty in the interpretation of the objective Bayesian inference on some conflict case [5]. There was a concurrent endeavor by weakening the belief of unobserved parameters by modifying the Interval Dempster-Shafer Model with unknown-order unit intervals that the desired properties of embedding, symmetry, invariance, and neutrality have been obtained through the posterior characterized by a Dirichlet distribution [6]. The non-parametric IM had been drafted for a one-sample inference but was excluded later from the official publication [7].

2010: The idea of the Inferential Model was formulated by relaxing the "continue to believe" assumption of DSM and the unobserved variables were predicted by the so-called Predictive Random Sets [8].

2011: The 2008 version of WB and MB was officially released with the omission of some mathematical derivations [9]. In addition, a three-step procedure and how to make an inference with IM were suggested [10]. The adoption of IM to a linear regression was proposed that included the IM-based model checking and variable selection [2]. Another interesting feature was introduced by the generalized association and focused on the case that the observed data were insufficient for data generating mechanism [11].

2012: The elastic belief (EB) was proposed as a more flexible version of the WB for solving the conflict case in the restricted space by stretching out the predictive random sets until reaching the minimum random set [12].

2013: The officially and publicly available definitions of the IM have been rewritten from the old version [10] with a lot of improvements in the fundamental concepts such as credibility, validity, optimality with the abundances of mathematical proofs [13,14].

2014: Two important classes of the Inferential Model were drafted i.e. the Marginal and Conditional IMs (MIM and CIM respectively) and shown in the Bivariate normal with unknown correlation and Behrens-Fisher problem [15]. There were applied researches of multinomial inferences on high-dimensional

---

[1] These earlier versions, e.g. unpublished, pre-print or non-public versions, were revised for official releases but are worth exploring.

gnome-wide association carried out by the Generalized Inferential Model (GIM) [16] and a multiple testing for many-normal-mean problems with the newly introduced form of the PRS for predicting sorted uniforms [17].

2015: The standard IM may not be efficient in real-life application e.g. the problem of parameters in high dimensions due to an increasing number of the PRS. Two important classes of IM were developed in order to reduce the problem into lower dimensions. First, the CIM could be explored in smaller domains by the fact that some auxiliary variables were fully observed [18]. Second, the MIM could ignore a nuisance or stretch the PRS to cover all possible random sets in one dimension and project down the rest in order to sharpen the PRS [19].

2016: The GIM was revisited and shown in the advantages over several other methods i.e. Jeffreys prior Bayesian posterior, Maximum Likelihood, and the standard IM [20]. It was applied to the prediction of future observations such as the problems of quality control, system breakdown, disease count, etc. [21]. Two more applications of IM were the uses of IMs in hypothesis testing for the non-inferiority of odds ratio in matched-pairs design [22] and in the problem of change-point hypothesis testing [23]. Importantly and interestingly, the plausibility of the IM was addressed as one of the solutions to the P-value (Frequentist) crisis [1,24].

2017⁺ : The research of IM in the linear regression continued and covered more possible types of assertions i.e. basic, complex, multiple-complex assertions and evaluation of model certainties [25].

2.2   Non-parametric Inferential Model

Consider the ordered observed data, $X = \{x_1, \ldots, x_n\}$ and $x_1 \leq \cdots \leq x_n$ and their corresponding unknown cdf, $F(X) = \{F(x_1), \ldots, F(x_n)\}$, the association for the non-parametric IM takes the form of the *anti*-inverse transformation [7,17]:

$$Z =_1 F(X),   \tag{4}$$

where $Z = \{z_1, \ldots, z_n\}$ is an ordered unit vector sampled from uniform distribution, $Unif(0, 1)$. The symbol of "$=_1$" means "equal in distribution" and, that is, the uniform samples are equal in distribution to the cdf of data. However, these quantities, are actually unobservable, therefore they would be predicted by the random set [7,17]:

$$S(U) = \{Z : Z \in [0, 1]; h(Z) \geq h(U)\},   \tag{5}$$

where U is the n-sample vector randomly drawn from uniform distribution; $Unif(0, 1)$. Consequently, by Eqs. 4 and 5 the IM can be extended to [7,17]:

$$S(U)^* = \{F : F \in [0, 1] | h(F(X)) \geq h(U)\},   \tag{6}$$

where U is the n samples from uniform distribution with the boundary function calculated by the proposed random set which is presented in the next section.

2.3 Entropy-Based Random Set

In this research, an asymmetric random set has been proposed due to the nature of financial data that is sensitive to negative influences. For instance, changing the administrative team could be interpreted as either good or bad news. In case people consider this as a negative factor, the stock price of this company will abruptly fall and vice versa. In order to predict the uncertainty concerning this biased character, the proper distribution should then be right-skewed as depicted in Fig. 1 (shown in bivariate case exclusively for the purpose of comprehension). For such a reason, the random set for the non-parametric IM for financial data may take the form of:

$$h(P) = -\sum_{i=1}^{n} p_i \log p_i, \tag{7}$$

where p_i is the probability of data and $P = \{p_i, \ldots, p_n\}$. The above formula indicates a boundary of probabilities measured by the entropy.

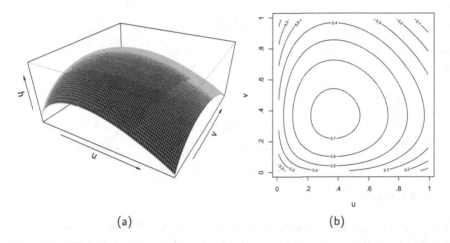

(a)   (b)

Fig. 1. (a) Perspective and (b) contour plots of the entropy-based random set.

Generally, the entropy is an artificial quantity computed by the negative summation of the products between the probabilities and their logarithms. The application of entropy is more useful in the analysis of the distribution such as purity, order, uniformity, etc. for the reason that it is known maximal at the uniform distribution that all of the values are equally probable.

This PRS has been adapted from the measurement of the uniform deviate from median [7,13,17] so that our random set (Eq. 7) also inherits its advantages as a result. Figure 1 visually confirms that the PRS is continuous and nested so that the proposed PRS is valid and admissible. The consequence of these properties have been proved [13] to be stochastically greater than the uniform distribution that is required for a frequency calibration.

For the assertion, $\mathcal{A}_h : X = U$, the plausibility can be approximated by averaging the results from the Monte-Carlo approximation algorithm [7]:

$$Pl(\mathcal{A}_h) = \frac{1}{m} \sum_{j=1}^{m} \mathbb{1}_{[h(F(X)) \geq h(\mathcal{A}_h)]}, \tag{8}$$

where m is the number of simulations and $\mathbb{1}_{h([F(X)) \geq h(\mathcal{A}_h)]}$ is the indicator function that tests whether or not the uncertainties can be captured by the proposed random set. The procedure is also summarized by the following algorithm:

Algorithm 1. Calculation of plausibility by Entropy-based Non-parametric IM

1: Initialize sum of test results: $SumT = 0$
2: From data, x_i compute cdf: F_i where $i = 1, \ldots, n$ and $x_1 \leq \ldots \leq x_n$
3: **for** each x_i: $i = 1$ **to** n **do**
4:     Compute entropy of F_i: $hF = F_i log(F_i)$
5:     **for** each simulation: $j = 1$ **to** m **do**
6:         Sample u_i from $Unif(0, 1)$
7:         Compute entropy of u_i: $hu = u_i log(u_i)$
8:         **if** $hF \geq hu$ **then**
9:             $SumT \leftarrow SumT + 1$
10:       **else**
11:           $SumT \leftarrow SumT$
12:       **end if**
13:     **end for**
14:     Compute plausibility of x_i: $Pl_i = SumT/m$
15: **end for**

---

3 Examples

This part contains brief descriptions of data used for demonstration (Sect. 3.1), methods for comparison in the standard estimation and time-series prediction (Sect. 3.2), and numerical and graphical results (Sect. 3.3).

3.1 Data

In order to conclude the behaviors of the IM for the applications of estimation and prediction; five financial samples, i.e. forex, gold, rice, foreign stock, and domestic stock; were selected aiming to cover various financial markets (currency, precious metal, commodity, foreign and domestic stock markets respectively). Although the data generally used in other financial researches were Return On Asset (ROA), it might not be a good portrayal for the prediction of time-series so that we chose price instead.

3.2 Statistical Problems

The IM was examined in the two statistical problems i.e. (1) standard estimation of probability distribution and (2) the prediction of time series. The results were compared with the commonly used approach i.e. the Kernel Distribution Estimation (KDE) method through the fast Fourier transform with linear approximation and Gaussian smoothing kernel. In the first problem, the comparisons were made between the probability and plausibility distributions. The parameters of interest were mean, variance, median, mode, 95-percent confidence interval (C.I.) and 95% credible intervals (C.R.).

In the same manner, the comparison were made in the second problem on the predicted observations by means of KDE and IM. Their results were back-tested and then evaluated by the criterion of the Mean Squared Error (MSE). The results and analyses are discussed in the next section.

3.3 Results

The numerical results for the estimations and predictions are summarized in Table 1 and their graphical results are illustrated in Figs. 2 and 3 respectively. At a first glance, the parameter values of the samples from both methods were all quite close. The estimations of forex, for instance, are (41.13 vs 41.19) for the mean, (1.18 vs 1.70) for the variance, (40.94 vs 41.19) for the median, (40.37 vs 40.66) for the mode, and [39.32, 43.30] for the confidence interval (C.I.) vs [38.86, 43.48] for the credible interval (C.R.) – the numbers in the brackets are the estimated values by the KDE and IM respectively.

Even though the tabular results could conclude that both methods produced the similar outcomes, the information obtained from the plots presented the additional insight however. Figure 2 shows that while KDE (dashed lines) gave the more fitted curves to the histograms (gray-border bars), the IM (solid lines), on the other hand, produced the less fluctuating curves. In the same asset, the distribution of forex produced two modes, i.e. there existed another mode around the price of 42–43, but the height difference in probability between the two was higher in the KDE than that in the IM. This phenomenon was consistent for every asset but more obvious in the forex and domestic stock. The reason of this behavior might be from the weaker belief produced from the random set, instead of random variable, in the prediction of uncertainty and, therefore, could cause the lower probability in posterior or less sensitive to change.

The predictions of time-series from both methods were similar as well. The MSEs for the forex are (1.14 vs 1.06), for example, and three out of five assets had the IM performing better. However, the accuracy might not be the most interesting by the current setup. Instead, the study of the behavior of the IM was more of our objective. In all cases, we may make a conclusion that the trend of prediction by the IM (solid line) is more persistent than that of the KDE (dashed line). In addition, whenever the actual observation went beyond the credible interval, the accuracy of prediction would fall immediately as could easily be seen in the sample of the domestic stock.

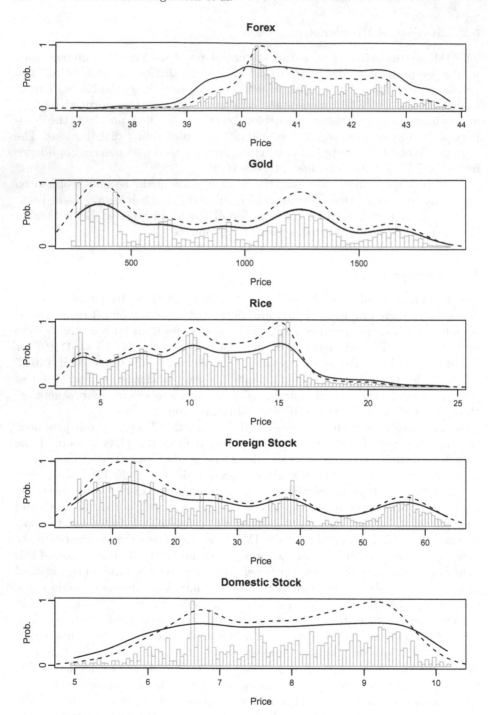

Fig. 2. Histogram (gray), probability (dashed), and plausibility (solid) distributions of sample assets

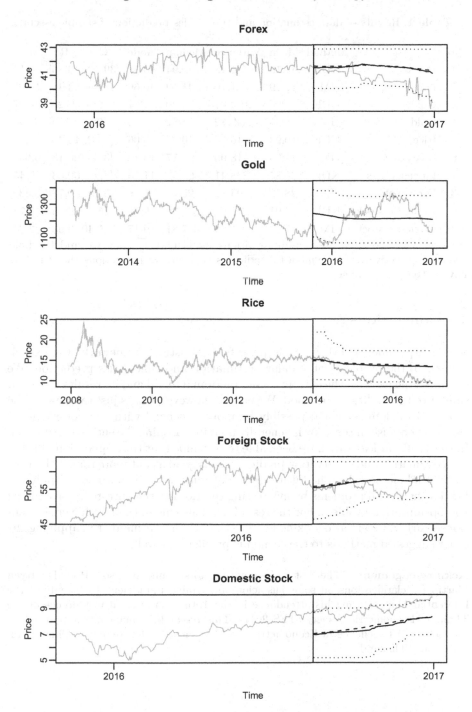

Fig. 3. Actual (gray) vs predicted time-series of sample assets by KDE (dashed), IM (solid) with 95% credible interval (dotted)

Table 1. Results of data estimation and time-series prediction of sample assets

No.	Asset	Method	Mean	Variance	Median	Mode	C.I./C.R	MSE
1	Forex	KDE	41.13	1.18	40.94	40.37	[39.32, 43.30]	1.14
2	Forex	IM	41.19	1.70	41.19	40.66	[38.86, 43.48]	1.06*
3	Gold	KDE	906	214367	920	361	[267, 1729]	9477
4	Gold	IM	955	204212	982	365	[285, 1760]	9508
5	Rice	KDE	10.93	16.76	10.88	15.06	[3.72, 18.46]	7.70
6	Rice	IM	11.25	18.96	11.17	15.10	[3.88, 20.13]	6.63*
7	Foreign stock	KDE	26.85	298.41	22.63	11.85	[4.82, 60.03]	5.45
8	Foreign stock	IM	28.70	294.82	25.45	11.68	[5.11, 60.78]	5.43*
9	Domestic stock	KDE	7.91	1.46	8.00	9.16	[5.79, 9.72]	1.73
10	Domestic stock	IM	7.78	1.74	7.81	9.17	[5.40, 9.93]	2.04

Remark: C.I. and C.R. are the abbreviations for Confidence Interval and Credible Interval respectively. The symbol (*) indicates the cases of predictions that the IMs have better performance.

4   Conclusions

The non-parametric IM which is prior-free statistical framework was experimented in the problems of standard estimation and time-series prediction. We equipped the IM with the entropy-based random interval and its results appeared similar to the traditional method. We believe, however, this is just taking-off. The proposed random set could possibly be more beneficial with the non-singleton assertion because the entropy is generally used in reasoning the entire distribution. In this study, at least, we have been able to conclude that the application of IM in data estimation and observation prediction is as good as but more robust than the traditional method. For the future research topics, we will continue to apply the IM to other social problems by integrating the IM into the auto-regressive model and optimization problem. For the rest of the solutions to the P-value crisis mentioned earlier, we are also looking for the study in similar manner in applying the other suggested methods to the economic problems as well.

Acknowledgements. The first and second authors thank to Assoc. Prof. Dr. Ryan Martin for clarifications on some suspicions pertaining fundamental concepts of the Inferential Model during his attendance in the 10th International Conference of the Thailand Econometric Society (TES 2017). This research is financially supported by the Center of Excellence in Econometrics and the Faculty of Economics, Chiang Mai University, Thailand.

References

1. Nguyen, H.T.: On evidential measures of support for reasoning with integrated uncertainty: a lesson from the ban of P-values in statistical inference. In: Huynh, V.-N., Inuiguchi, M., Le, B., Le, B.N., Denoeux, T. (eds.) IUKM 2016. LNCS (LNAI), vol. 9978, pp. 3–15. Springer, Cham (2016). https://doi.org/10.1007/978-3-319-49046-5_1
2. Zhang, Z., Xu, H., Martin, R., Liu, C.: Inferential models for linear regression. Pak. J. Stat. Oper. Res. **7**(2), 413–432 (2011)
3. Liu, C., Martin, R., Zhang, J.: Situation-specific inference using Dempster-Shafer theory. Preprint (2008)
4. Liu, C., Zhang, J.: Dempster-Shafer inference with weak beliefs. Preprint (2008)
5. Ermini Leaf, D., Hui, J., Liu, C.: Statistical inference with a single observation of $N(\theta, 1)$. Pak. J. Statist **25**, 571–586 (2009)
6. Lawrence, E.C., Vander Wiel, S., Liu, C., Zhang, J.: A new method for multinomial inference using Dempster-Shafer theory. Preprint (2009)
7. Zhang, J., Xie, J., Liu, C.: Probabilistic inference: test and multiple tests (2009)
8. Martin, R., Zhang, J., Liu, C.: Dempster-Shafer theory and statistical inference with weak beliefs. Stat. Sci. **25**(1), 72–87 (2010)
9. Zhang, J., Liu, C.: Dempster-Shafer inference with weak beliefs. Stat. Sin. **21**(2), 475–494 (2011)
10. Martin, R., Liu, C.: Inferential models (2011). http://www.stat.purdue.edu/~chuanhai
11. Martin, R., Liu, C.: Generalized inferential models. Technical report, Purdue University, October 2011
12. Ermini Leaf, D., Liu, C.: Inference about constrained parameters using the elastic belief method. Int. J. Approx. Reason. **53**(5), 709–727 (2012)
13. Martin, R., Liu, C.: Inferential models: a framework for prior-free posterior probabilistic inference. J. Am. Stat. Assoc. **108**(501), 301–313 (2013)
14. Martin, R., Liu, C.: Correction: 'inferential models: a framework for prior-free posterior probabilistic inference'. J. Am. Stat. Assoc. **108**(503), 1138–1139 (2013)
15. Liu, C., Martin, R.: Frameworks for prior-free posterior probabilistic inference. Wiley Interdisc. Rev. Comput. Stat. **7**(1), 77–85 (2015)
16. Liu, C., Xie, J.: Large scale two sample multinomial inferences and its applications in genome-wide association studies. Int. J. Approx. Reason. **55**(1), 330–340 (2014)
17. Liu, C., Xie, J.: Probabilistic inference for multiple testing. Int. J. Approx. Reason. **55**(2), 654–665 (2014)
18. Martin, R., Liu, C.: Conditional inferential models: combining information for prior-free probabilistic inference. J. Royal Stat. Soc. Ser. B (Stat. Methodol.) **77**(1), 195–217 (2015)
19. Martin, R., Liu, C.: Marginal inferential models: prior-free probabilistic inference on interest parameters. J. Am. Stat. Assoc. **110**(512), 1621–1631 (2015)
20. Martin, R.: On an inferential model construction using generalized associations. arXiv e-prints, November 2015
21. Martin, R., Lingham, R.T.: Prior-free probabilistic prediction of future observations. Technometrics **58**(2), 225–235 (2016)
22. Jin, H., Li, S., Jin, Y.: The IM-based method for testing the non-inferiority of odds ratio in matched-pairs design. Stat. Probab. Lett. **109**, 145–151 (2016)

23. Nguyen, S.P., Pham, U.H., Nguyen, T.D., Le, H.T.: A new method for hypothesis testing using inferential models with an application to the changepoint problem. In: Huynh, V.-N., Inuiguchi, M., Le, B., Le, B.N., Denoeux, T. (eds.) IUKM 2016. LNCS (LNAI), vol. 9978, pp. 532–541. Springer, Cham (2016). https://doi.org/10. 1007/978-3-319-49046-5_45
24. Thianpaen, N., Liu, J., Sriboonchitta, S.: Time series forecast using AR-belief approach. Thai J. Math. **14**(3), 527–541 (2016)
25. Martin, R., Xu, H., Zhang, Z., Liu, C.: Valid uncertainty quantification about the model in a linear regression setting. arXiv e-prints, December 2014

Econometric Applications

Impact of Trade Liberalization on Economic Growth in ASEAN: Copula-Based Seemingly Unrelated Regression Model

Arisara Romyen[1,2] (ORCID), Jianxu Liu[2,3](✉), and Songsak Sriboonchitta[2,3]

[1] Faculty of Economics, Prince of Songkla University, Songkhla 90110, Thailand
arisara.r@psu.ac.th
[2] Faculty of Economics, Chiang Mai University, Chiang Mai 50200, Thailand
[3] Faculty of Economics, Center of Excellence in Econometrics,
Chiang Mai University, Chiang Mai 50200, Thailand
liujianxu1984@163.com

Abstract. Trade liberalization especially ASEAN-5 countries has raised in globalization economy. To understand the linkage of trade openness and economic growth, we examine the endogenous growth approach participated with trade liberalization indicators. The proposed method involved the multivariate Copula-based Seemingly Unrelated Regression (SUR) model with an unstructured correlation matrix. Using data on the augmented aggregate production function (human, capital and trade proxy) from 1980 to 2015, the findings convey that the multivariate Copula-based SUR model outperforms the conventional SUR model in terms of AIC and BIC. In addition, the results indicate that the trade-outward orientations in ASEAN affect economic growth differently. The Philippines and Thailand perform a significant positive effects of the terms of trade (TOT) and foreign direct investment (FDI) to growth, while Indonesia has a negative impact. Furthermore, the total factor production (TFP) creates a high contribution towards ASEAN economic growth.

Keywords: Multivariate Copula-based SUR model
Endogenous growth model · Trade-outward orientation

1 Introduction

Trade liberalization has been essentially expanded in the global economy. In particular, several developing countries have embarked on openness orientations. Flows of goods and services have continuously increased across countries. Substantial literature relating to the connection between trade openness and economic growth have been constructed using the vector autoregressive models [15,16,27], error correction models [1,5,22], or seemingly unrelated regression (SUR) models [21,29]. Although, these studies based on the conventional techniques, which may have some limitations on good fit to the models. According to

© Springer International Publishing AG, part of Springer Nature 2018
V.-N. Huynh et al. (Eds.): IUKM 2018, LNAI 10758, pp. 349–360, 2018.
https://doi.org/10.1007/978-3-319-75429-1_29

the concept of univariate model in pioneering work of [13], the joint error terms are assumed to be the normalized distribution. In applications, the joint normal distributions are ordinarily difficult to occur. These traditional methods are often imposed unrealistically. To address this issue, the Copula is then applied in this study in order to obtain flexible density models of the joint distribution with two or higher dimensional distributions. Furthermore, we propose the multivariate Copula-based SUR model with an unstructured covariance matrix that not holds constraints on values, unlike other covariance regularizations. Individual variance and covariance is taken into account uniquely. Each variance and covariance values is precise to what information reflect.

The Association of Southeast Asian Nations (ASEAN) have long implemented outward-oriented trade strategies since the 1980s by reducing tariff and other barriers. During that period, ASEAN has grown continuously and generated an aggregated gross domestic production (GDP) of $2.4 trillion. It also performed to be the 3rd speedy economic growth in Asia Pacific from 2006 to 2015 [4]. Advanced ASEAN nations including Indonesia, Malaysia, the Philippines, Singapore and Thailand become top five counties attracting foreign direct investment (FDI). The international trade has expanded dramatically. Therefore, the impact of trade openness policy on economic growth is inquired.

The objective of this paper is to investigate the relation between trade openness and economic growth. In addition, we seek to compare the results of the conventional SUR model against the Copula-based SUR Model. We apply the multivariate Copula-based SUR model with an unstructured correlation matrix. To consider the best-fitting models, AIC and BIC criteria are imposed to ensure the fitted model. The paper contribution is that ASEAN incorporates the trade openness in the market integration, so the economic growth seems to affect each other. Hence, improving the conventional SUR model provides more appropriate outward policies to ASEAN.

This paper is organized as follows: Sect. 2 briefly reviews the empirical literature and recall SUR model as well as Copula criteria. Section 3 details the methodology including model specification, Copula functions and estimation method. Section 4 reports the empirical results, discussions and the implications and Sect. 5 concludes.

2 Background

In this section, we first survey the literature involved to the association between trade openness and economic growth. Then, we recall the SUR model in Sect. 2.2. Some general definitions on Copula are briefly summarized in Sect. 2.3.

2.1 Trade Liberalization and Economic Growth

Trade liberalization is broadly determined as a driving force towards economic growth. Until recently the linkage between external trade and economic growth has been theoretically controversial. The conventional wisdom addressed a growth enhancing effect of trade [6,12,14,17,31]. Technology and innovation transformation in particularly can be delivered through FDI [2,3,8]. Reference [18] notified that only high-tech manufacturing through FDI can provide positive spillover effect of knowledge and technology. On the contrary, references [20,26] argued that opportunities of drawbacks such as poverty gap, economic volatility would exist. Reference [18] premised that trade with barriers may bring many gains to a host country. Imposing trade protections can provoke the higher rate of outgrowth comparing to country without restriction in long run.

How we can ensure that the measurements of trade liberalization are appropriate? Reference [33] concerned that trade openness measurements remain sensitive as a proxy indicator for growth. The robustness of previous conclusions needed to ensure statistically sensitive specifications of protectionist regimes. Reference [28] proposed the trade openness classifications consisting of tariff rate, non-tariff barriers, black-market exchange rate, nation monopoly on export and a public economic system. Not only the favorite indicators of outward oriented policies of tariffs and non-tariff barriers are concerned, but other mechanisms such as foreign investment policies should be determined. Consequently, the trade liberalization indicators in this study consist the TOT and FDI.

2.2   Seemingly Unrelated Regression Model

Initially introduced by Arnold Zellner in 1962, the SUR model is a generalization of multivariate equations using a vectorized parameter model. This model is frequently applied to a system of uncorrelated equations. We determine the SUR model consisting of M multiple regressions, such that

$$y_{it} = \sum_{j=1}^{k_i} x_{ijt}\beta_{ij} + \varepsilon_{it}, \quad t = 1, 2, \ldots, T; \quad i = 1, 2, \ldots, M; \quad j = 1, 2, \ldots k_i, \quad (1)$$

where y_{it} is the dependent variable in the ith regression, x_{ijt} is the tth observation on the jth independent variables, β_{ij} is the coefficients participated with X_{ijt}, and ε_{it} is random error terms in each equation. The M equations can be expressed in a vector form as:

$$\begin{bmatrix} y_1 \\ y_2 \\ \vdots \\ y_M \end{bmatrix} = \begin{bmatrix} X_1 & 0 & \cdots & 0 \\ 0 & X_2 & & \vdots \\ \vdots & & \ddots & 0 \\ 0 & \cdots & 0 & X_M \end{bmatrix} \begin{bmatrix} \beta_1 \\ \beta_2 \\ \vdots \\ \beta_M \end{bmatrix} + \begin{bmatrix} \varepsilon_1 \\ \varepsilon_2 \\ \vdots \\ \varepsilon_M \end{bmatrix} \quad (2)$$

The classical assumptions of the SUR model are as follows: $E(\varepsilon_i) = 0$ and $E(\varepsilon_i') = \sigma_{ij}I_T; i, j = 1, 2, \ldots, M$. The σ_{ij}^2 is the variance of disturbances between i^{th} and j^{th}. Compactly, we write the covariance matrix for the system in the following:

$$E(\varepsilon) = 0,$$

$$E\left(\varepsilon\varepsilon'\right) = \begin{bmatrix} \sigma_{11}I_T & \sigma_{12}I_T & \cdots & \sigma_{1M}I_T \\ \sigma_{21}I_T & \sigma_{22}I_T & \cdots & \sigma_{2M}I_T \\ \vdots & \vdots & \ddots & 0 \\ \sigma_{M1}I_T & \cdots & \cdots & \sigma_{MM}I_T \end{bmatrix} = \sum \bigotimes I_T = \Omega, \tag{3}$$

where $\sum$ is $(M \times M)$ a positive definite symmetric matrix, $\bigotimes$ is Kronecker product operator, I_T is an identity matrix, and Ω defines $(M \times M)$ matrix. Then, the estimation of this system equation can be expressed by:

$$\beta_{SUR} = \left(X'\Omega^{-1}X\right)^{-1}X'\Omega^{-1}Y. \tag{4}$$

2.3   Copula

According to the fundamental Copula approach, which was initially introduced by Sklar in 1959, the Copula function is applied to deal with each specified marginal to form a multivariate distribution. H denotes d-dimensional joint distributions of the random variable x_d corresponding to marginal distributions F_d. The d-Copula C then exists for all x_d;

$$H(x_1, \ldots, x_d) = C(F_1(x_1), \ldots, F_d(x_d)). \tag{5}$$

Due to the multivariate distribution function F, we can obtain the marginal distribution and joint dependence individually. The inversion method inverts the relation in Eq. 6 and the Copula is written as:

$$H(u_1, \ldots, u_d) = C(F_1^{-1}(u_1), \ldots, F_d^{-1}(u_d)), \tag{6}$$

where u is uniform distribution on $[0, 1]^d$.

For the Elliptical Copula in class of the Gaussian Copula, the advantages of the Gaussian Copula are the flexibility to obtain degrees of positive and negative dependences [25]. By using the probability integral transform, the Gaussian Copula can be constructed from a multivariate normal distribution through R^d. For a given correlation matrix $R \in [-1, 1]^{d \times d}$, the Gaussian Copula with parameter matrix R can be defined as:

$$C_R^{Gauss}(u) = \Phi_R(\Phi^{-1}(u_1), \ldots, \Phi^{-1}(u_d)), \tag{7}$$

where Φ_R is the joint cumulative distribution function (C.D.F.) of a multivariate normal distribution with zero mean vector and covariance matrix R, Φ^{-1} is the inverse C.D.F. of the probability density function of the standard normal distribution. The density can be expressed as:

$$C_R^{Gauss}(u) = \frac{1}{\sqrt{det\ R}} exp\left(-\frac{1}{2} \begin{pmatrix} \Phi^{-1}(u_1) \\ \vdots \\ \Phi^{-1}(u_d) \end{pmatrix}^T \cdot (R^{-1} - I) \cdot \begin{pmatrix} \Phi^{-1}(u_1) \\ \vdots \\ \Phi^{-1}(u_d) \end{pmatrix} \right), \tag{8}$$

where R^{-1} is an unstructured correlation matrix, and $\mathbf{I}$ is the identity matrix.

3   Methodology

In this section, the model specification is introduced to estimate the influence of trade liberalization on growth. Then, we detail the multivariate Copula-based SUR model, which was applied to the production function in Sect. 3.2.

3.1   Model Specification

According to the endogenous growth approach examining long-run economic growth to a host country, the Cobb-Douglas production function is employed. The trade liberalization then is taken into account to be the augmented aggregate production function as follows,

$$Y_{it} = A_{it}{}^{\beta_{i1}} K_{it}{}^{\beta_{i2}} L_{it}{}^{\beta_{i3}} HC_{it}{}^{\beta_{i4}} LIB_{it}{}^{\beta_{i5}} e^{\varepsilon_{it}}, \tag{9}$$

where Y_{it} is output measured GDP per capita of a country i at year t; A_{it} is the multi-factor productivity represented by total factor productivity (TFP) of a country i at year t; K_{it} is physical capital counted by gross fixed capital formation; L_{it} is labor measured by labor force, and HC_{it} is human capital measured by literacy rate in youth. For the trade liberalization (LIB), LIB_{it} is a proxy of trade liberalization, which is represented by the terms of trade (TOT) and the foreign direct investment (FDI). β_{it} is the coefficients of this model, and ε_{it} reflects an error term. The transformation of Eq. 9 formed to be linear in parameters gives

$$lny_{it} = \beta_{i0} + \beta_{i1}lnA_{it} + \beta_{i2}lnK_{it} + \beta_{i3}lnL_{it} + \beta_{i4}lnHC_{it} \\ + \beta_{i5}lnTOT_{it} + \beta_{i6}lnFDI_{it} + \varepsilon_{it}. \tag{10}$$

Based on the SUR model, the system equation for Indonesia, Malaysia, the Philippines, Singapore and Thailand can be expressed, respectively, by Eq. 10 for $i = 1, \ldots, 5$.

3.2   Multivariate Copula-Based SUR Model

The multivariate normal density can be expressed as:

$$f_{(p)}(x) = \frac{1}{|R|^{\frac{1}{2}}} exp\left\{-\frac{1}{2}u'(R^{-1} - I)u\right\} \prod_{i=1}^{P} \frac{1}{\sigma_i} \varphi(u_i), \tag{11}$$

where $f_p(x)$ is a multivariate density, R is a correlation matrix, u_i equals to $\Phi^{-1}(F_i(x_i))$, and $\varphi(u_i)$ is standard normal density.

Suppose $u_i = \Phi^{-1}(F_i(x_i))$. So, F_i can be an arbitrary distribution function. The density of a Gaussian Copula is defined as shown in Eq. 12.

$$c(x) = \frac{1}{|R|^{\frac{1}{2}}} exp\left\{-\frac{1}{2}u'(R^{-1} - I)u\right\}. \tag{12}$$

The margin of variables uniformly bounds into monotonic sequences converge in which the correlation matrix assigns the dependency structure. There are four classes regularly adopted to dispersion structures including to autoregressive of order one (ar1), exchangeable (ex), toeplitz (toep) and unstructured (un) [32]. For the dimension of p = 5, we apply the unstructured (un) form to determine the standardized dispersion matrix. Since each margin of variables is taken into account in the correlation matrix, which can be written in Eq. 13.

In this study, the five-dimensional joint C.D.F. associated with marginal distribution has the Copula C as shown in Eq. 6. The multivariate distribution function is written in Eq. 14. The Gaussian Copula is then employed to generate the joint density as expressed in Eq. 15.

$$\rho = \begin{pmatrix} 1 & \rho_{12} & \rho_{13} & \rho_{14} & \rho_{15} \\ \rho_{21} & 1 & \rho_{23} & \rho_{24} & \rho_{25} \\ \rho_{31} & \rho_{32} & 1 & \rho_{34} & \rho_{35} \\ \rho_{41} & \rho_{42} & \rho_{43} & 1 & \rho_{45} \\ \rho_{51} & \rho_{52} & \rho_{53} & \rho_{54} & 1 \end{pmatrix} \qquad (13)$$

$$F(y_{1t}, \ldots, y_{5t}) = C(F_1(\varepsilon_{1t}; \sigma_1), \ldots, F_5(\varepsilon_{5t}; \sigma_5); \rho). \qquad (14)$$

$$f(y_{1t}, \ldots, y_{5t}) = c(F_1(\varepsilon_{1t}; \sigma_1), \ldots, F_5(\varepsilon_{5t}; \sigma_5); \rho) \prod_{i=1}^{5} f_i(\varepsilon_{it}; \sigma_i). \qquad (15)$$

Therefore, the log-likelihood function of the multivariate Copula-based SUR model can be given as follow:

$$\begin{aligned} logL(A, .., FDI; \beta, \sigma, \rho) &= \sum_{t=1}^{T} log[c(F_1(\varepsilon_{1t}; \sigma_1), \ldots, F_5(\varepsilon_{5t}; \sigma_5); \rho)] \\ &+ \sum_{t=1}^{T} \sum_{i=1}^{5} log(f_i(\varepsilon_{it}; \sigma_i)). \end{aligned} \qquad (16)$$

Then, we utilize the maximum likelihood method to maximize the multivariate Copula-based SUR model to obtain the estimated parameters.

4   Data and Empirical Results

4.1   Data

The multivariate Copula-SUR model involves annual data taken from the leader ASEAN countries including Indonesia, Malaysia, the Philippines, Singapore and Thailand over the period from 1980 to 2015. The dependent variable is GDP per capital (Y). The independent variables make up of the TFP, the physical capital (K) measured by gross fixed capital formulation, the labor (L) represented by labor force and the human capital (HC) proxy by literacy rate, were collected from index organized by the Penn World Tables (PWT) database. Whilst the proxy of trade liberalizations in terms of the TOT and FDI were assessed from the World Bank Development Indicators (WDI) database.

Figure 1 shows the examples of scatter plots of the outputs and some explanatory variables such as the TFP, the TOT and the FDI. For Indonesia and the Philippines, the TFP and FDI correspond to a positive dependence with the GDP per capita, whereas the TOT performs the negative correlation (Figs. a–c, and g–i). For Malaysia, all mentioned variables have a positive trend (Figs. d–f).

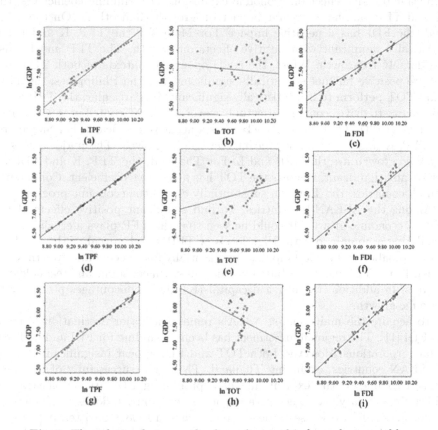

Fig. 1. The relation between the dependent and independent variables

4.2   Empirical Results

To address the influence of trade liberalization on economic growth, we built the SUR models consisting of the conventional SUR model and the multivariate Copula-based SUR model. Table 1 presents the results from the conventional SUR model and the multivariate Copula-based SUR model. The estimations from the two models are quite similar both of values and signs. Furthermore, it can be seen that the multivariate Copula-based SUR model is preferable since the standard errors achieve the smaller numbers. For the model selection, the AIC and BIC of multivariate Copula-based SUR model are -888.08 and -967.25. Besides based on the conventional SUR model, there are -875.68 and -942.58, respectively. Hence, we favor the multivariate Copula-based SUR model due to the minimum AIC and BIC execution compared. The results from Table 1 indicates that each ASEAN country performs differently in effect of trade openness to growth as detailed in the following:

For Indonesia, the endogenous growth model (the TFP, K, L and HC) has a significant positive effect on economic expansion. Determining coefficients, the HC and TFP become a greater impact on growth than others. On the other hand, the FDI has a negative impact. For Malaysia, the TFP, K and L are statistically significant with positive effects on growth. The TFP and L raise considerably to growth. Whist, the trade openness contexts of both TOT and FDI are positive but not statistically significant. For the Philippines, the TFP, L and TOT perform to be statistically significant. In Particular, the TFP is the main source to drive economic growth; however, the TOT contributes few benefit effects on growth. On contrast, the HC has a negative coefficient. For Singapore, the TFP, K and L have a significant increase in growth. The major causes of economic growth are the TFP and L. For Thailand, the TFP, K and FDI are statistically significant, whereas the TOT has a negative coefficient. Consistency with all countries, the TFP can substantially create the economic progress.

Among the ASEAN-5 countries, we find significant positive effects of the TFP on economic growth. It could be seen that the TFP plays a crucial role in stimulating economic growth. High levels of TFP, which reveal to the weighted average of labor and capital inputs, can essentially foster economic growth. Consequently, the TFP can contribute towards long-term economic and technological progress. In addition, the physical capital (K) always encourages positively to economic growth.

To acquire the main points, ASEAN remains a major destination of trade and FDI [11]. The regional investment has been expanding for FDI and multinational corporations. However, both TOT and FDI appear insignificant in most of ASEAN countries, excluding Thailand. They gain successfully of manufacturing FDI due to the explicit aim of a promoting FDI attraction strategy. Whilst, Indonesia yields a negative impact on economy. Indonesia encountered the outflows of FDI because of the economic crisis in 1998. Consistently [7,19,23] postulated that export-led growth is sophisticated regimes. The export-oriented policies provide both advantages and disadvantages. Currently, main manufacturing sectors in ASEAN involve to primary manufacturing and labor intensive tradable such as textiles, rubber and plastics, automobile equipment, electrical machinery, etc. Most of FDI industries operate based on low and middle levels of innovation and R&D. As the results, the technology transfer intensities vary substantially across the ASEAN economy.

Table 2 shows the correlation matrix from the multivariate Copula-based SUR estimation. The correlations between the two countries report both positive and negative relations but these are moderate. Singapore demonstrates a

Table 1. Estimated results of the multivariate Copula-based SUR and GLS-SUR models

		Multivariate Copula-based SUR		SUR (GLS)	
		Coefficient	Std. error	Coefficient	Std. error
Indonesia	Constant	−6.69***	0.481	−6.56***	0.494
	lnTFP	0.45***	0.047	0.46***	0.040
	lnK	0.11***	0.007	0.11***	0.009
	lnL	0.24***	0.044	0.23***	0.048
	lnHC	1.72***	0.123	1.67***	0.141
	lnTOT	0.01	0.003	0.01**	0.005
	lnFDI	−0.01***	0.002	−0.01***	0.002
Malaysia	Constant	0.840	0.743	1.10	0.780
	lnTFP	0.59***	0.068	0.66***	0.069
	lnK	0.06***	0.012	0.06***	0.012
	lnL	0.53***	0.052	0.48***	0.065
	lnHC	−0.04	0.207	−0.23***	0.217
	lnTOT	0.010	0.003	0.01	0.004
	lnFDI	0.010	0.003	0.01	0.003
Philippine	Constant	6.86**	2.384	6.27	2.648
	lnTFP	0.93***	0.034	0.94***	0.055
	lnK	0.02	0.011	0.01	0.017
	lnL	0.21***	0.023	0.20***	0.040
	lnHC	−1.99***	0.512	−1.89***	0.53
	lnTOT	0.01*	0.004	0.01**	0.004
	lnFDI	0.01	0.003	0.01	0.003
Singapore	Constant	−3.15	2.011	−0.90	2.936
	lnTFP	0.74***	0.04	0.72***	0.044
	lnK	0.06***	0.011	0.06***	0.013
	lnL	0.43***	0.05	0.44***	0.052
	lnHC	0.86	0.642	0.41	0.689
	lnTOT	0.01	0.001	0.01	0.002
	lnFDI	0.01	0.006	0.01	0.006
Thailand	Constant	−10.04	9.874	−5.57	10.646
	lnTFP	0.93***	0.042	0.95***	0.059
	lnK	0.07***	0.011	0.07***	0.012
	lnL	−0.02	0.064	−0.05	0.076
	lnHC	1.8	2.215	0.80	2.391
	lnTOT	−0.01	0.001	−0.01*	0.001
	lnFDI	0.02***	0.004	0.02***	0.005
σ	σ_1	0.01***	0.001		
	σ_2	0.02***	0.002		
	σ_3	0.01***	0.001		
	σ_4	0.02***	0.002		
	σ_5	0.02***	0.002		
θ	θ_1	0.73*	0.476		
	θ_2	−0.22	0.438		
	θ_3	0.16	0.487		
	θ_4	−0.88*	0.501		
	θ_5	−0.53	0.511		
	θ_6	1.64***	0.512		
	θ_7	−1.3***	0.426		
	θ_8	−0.12	0.591		
	θ_9	0.08	0.520		
	θ_{10}	−1.41***	0.420		

Source: Calculation, Significance codes: * = 0.1, ** = 0.05, *** = 0.01.

positive correlation to Indonesia and the Philippines but a negative correlation with Malaysia and Thailand.

Singapore has a highly implemented trade-oriented market economy. The major sectors of its economy are manufacturing and financial hubs. In addition, they also has long promoted trade agreements with several trade partnerships within region and other countries. However, Singapore also losses competitiveness due to uncertainty of global economy. Meanwhile, Malaysia and Thailand coordinate strong growth prospects of cross-border economy. They have established the Bilateral Cooperation to enhance comprehensive cooperation. However, Indonesia and the Philippines present a negative correlation with Thailand and Malaysia. Due to the lower cost of the labor-intensive manufacturing sector, those neighbors then experience competition against lower-cost countries. Other factors such stable macroeconomic conditions as inflation, interest rate, political stability are also determined as an exact role of incentives in attracting FDI.

It can be seen that each country has a distinct impact of trade liberalization on economic growth. In particular, the key role of FDI that can enhance economic growth needs more conducive conditions. Host countries should concern its sufficient absorptive capability to capture the innovation and advanced technology from foreign firms, unless FDI may offer an adverse effect on economic progress [9,24,30]. In addition, the growth prospect has been promoted by a speedy acceleration of fixed investment, physical capital and capital accumulation in ASEAN, where the investment rates tend to increase intensely. It is similar to the study of [10] that high physical capital should translate into rapid growth comparing to less capital accumulation. Ultimately, the reinforcement of capital accumulation also sharply translates into improvement of the TFP.

Table 2. Correlation matrix from the Copula-SUR estimation

	Indonesia	Philippine	Malaysia	Singapore	Thailand
Indonesia	1	0.344	−0.106	0.079	−0.402
Philippine	0.344	1	−0.252	0.661	−0.559
Malaysia	−0.106	−0.252	1	−0.058	0.037
Singapore	0.079	0.661	−0.058	1	−0.594
Thailand	−0.402	−0.559	0.037	−0.594	1

Source: Calculation, Significance codes: * = 0.1, ** = 0.05, *** = 0.01.

5   Conclusions and Policy Implementations

In this paper, we examined the endogenous growth model, which augmented in the contexts of trade openness consisting of the TOT and FDI influencing growth. We proposed the multivariate Copula-SUR model along with the unstructured standardized dispersion matrix. This method can be an alternative technique. It releases the assumption of normal distribution for the dependence between the error terms in the system equation, instead of conventional SUR model. Then, the maximum likelihood approach was utilized to estimate the model and the criterion for model selection involved to AIC and BIC. The main findings of this study are as follow: (1) the impacts of trade liberalization to growth are diversified among ASEAN members. The Philippines and Thailand have a significant positive effect in terms of the TOT and FDI, respectively, whereas Indonesia has a negative association between the FDI and growth. (2) the TFP can substantially stimulate growth across the region. (3) the multivariate Copula-based SUR model yields less standard errors and more stable parameter estimation.

This study has some policy implications; it is essential to recognize the diverse determinant for an individual country. From the results, there are both consistency and distinction effects regarding to trade openness policies imposed by policy makers. These outward orientations should be properly established based upon the consequences of internal conditions. The recipient country can seek more precise advantaged consequences regarding to their own endowment. The governments in this region then can accomplish towards competence positive-spillover effects and mitigate the negative spillovers. Furthermore, the prosperity of high capital accumulation also motivates the TFP, and the potential growth has then yielded.

References

1. Ahmed, N.: Export response to trade liberalization in Bangladesh: a cointegration analysis. J. Appl. Econ. **32**(8), 1077–1084 (2010)
2. Alesina, A., Tella, R.D., Culloch, R.M.: Inequality and happiness: are Europeans and Americans different? J. Public Econ. **88**, 2009–2042 (2004)
3. Almeida, R., Fernandes, A.: Openness and technological innovations in developing countries: evidence from firm-level surveys. J. Dev. Stud. **44**(5), 701–727 (2008)
4. Asian Development Bank Institute: ASEAN Economic Integration Through Trade and Foreign Direct Investment: Long-Term Challenges, ADBI Working Paper 545 (2015)
5. Awokuse, T.O.: Causality between exports, imports and economic growth: evidence from transition economies. Econ. Lett. **94**, 389–395 (2007)
6. Baldwin, R.E., Braconier, H., Forslid, R.: Multinationals, endogenous growth, and technological spillovers: theory and evidence. Rev. Int. Econ. **13**(5), 945–963 (2005)
7. Bello, W.: The Anti-Development State: The Political Economy of Permanent Crisis in the Philippines. Zed Books, London (2005)
8. Bond, S.R., Hoeffler, A., Temple. J.: GMM estimation of empirical growth models. CEPR Discussion Paper Series. Center for Economic Policy Research (CEPR), London. No. 3048 (2001)
9. Borensztein, E., de Gregorio, J., Lee, J.W.: How does foreign direct investment affect economic growth? J. Int. Econ. **45**, 115–135 (1998)
10. Collins, S., Bosworth, B.: Economic growth in East Asia: accumulation versus assimilation. Brook. Pap. Econ. Act. **2**, 135–203 (1996)
11. Diaconu, L.: The foreign direct investment in South-East Asia in the context of the 1997 and 2007 crises. Prodedia-Soc. Behav. Sci. **109**, 160–164 (2014)
12. Edwards, S.: Openness, trade liberalization, and growth in developing countries. J. Econ. Lit. **31**(3), 1358–1393 (1993)

13. Engle, R.F.: Autoregressive conditional heteroscedasticity with estimates of the variance of United Kingdom inflation. Econometrica **50**, 987–1007 (1982)
14. Grossman, G.M., Helpman, E.: Innovation and Growth in the Global Economy. MIT Press, Cambridge (1991)
15. Heydariyan, A.H.: To determine the causal relationship between exports and economic growth in the VAR model using seemingly unrelated procedure repeated (ISUR) in Iran. M.Sc. thesis, University of Technology, School of Industrial and Systems (1996)
16. Hozouri, N.: The effect of trade liberalization on economic growth: selected MENA countries. Int. J. Econ. Fin. **9**, 88–95 (2017)
17. Krueger, A.O.: Trade policy and economic development: how we learn. Am. Econ. Rev. **87**, 1–22 (1997)
18. Rodriguez, F., Rodrik, D.: Trade policy and economic growth: a skeptics guide to the cross-national evidence. Working paper 7081, NBER (1999)
19. Lee, H.H., Tan, H.B.: Technology transfer, FDI and economic growth in the ASEAN region. J. Asia Pac. Econ. **11**(4), 394–410 (2006)
20. Lucas, R.E.: On the mechanic of economic development. J. Monet. Econ. **46**, 167–182 (1988)
21. Makki, S., Somwaru, A.: Impact of foreign direct investment and trade on economic growth. Am. J. Agric. Econ. **86**, 795–801 (2004)
22. Mamun, K.A., Nath, H.K.: Export-led growth in Bangladesh: a time series analysis. Appl. Econ. Lett. **12**(6), 361–364 (2005)
23. Merican, Y.: Foreign direct investment and growth in ASEAN-4 nations. Int. J. Bus. Manag. **4**(5), 46–62 (2009)
24. Moura, R., Forte, R.: The effects of foreign direct investment on the host country economic growth-theory and empirical evidence. FEP Working papers 390, pp. 1–32 (2010)
25. Nelson, R.B.: An Introduction to Copulas. Springer, New York (2006). https://doi.org/10.1007/0-387-28678-0
26. Redding, S.: Dynamic comparative advantage and the welfare effects of trade. Oxf. Econ. Pap. **51**, 15–39 (1999)
27. Saaed, A.J., Hussain, M.A.: Imports on economic growth: evidence from Tunisia. J. Emerg. Trends Econ. Manag. **6**, 13–21 (2015)
28. Sachs, J.D., Warner, A.: Economic reform and the process of global integration. Brook. Pap. Econ. Act. **6**, 1–118 (1995)
29. Sumer, K.K.: Testing the validity of economic growth theories with seemingly unrelated regression models: application to Turkey in 1980–2010. Appl. Econ. Int. Dev. **12**, 63–72 (2012)
30. Toulaboe, D., Terry, R., Johansen, T.: Foreign direct investment and economic growth in developing countries. Southwest. Econ. Rev. **42**, 155–169 (2009)
31. Warner, A.: Once more in to the breach: economic growth and integration. Center for Global Development. Working Paper No. 34 (2003)
32. Yan, J.: Enjoy the joy of copulas: with a package copula. J. Stat. Softw. **21**(4), 1–21 (2007)
33. Zahonogo, P.: Trade and economic growth in developing countries: evidence from sub-Saharan Africa. J. Afr. Trade 1–16 (2017, in press)

Investigating Dynamic Correlation in the International Implied Volatility Indexes

Panida Fanpaeng, Woraphon Yamaka, and Roengchai Tansuchat[✉]

Faculty of Economics, Center of Excellence in Econometrics,
Chiang Mai University, Chiang Mai, Thailand
roengchaitan@gmail.com

Abstract. This paper investigates dynamic interaction among international volatility indexes, consisting of VIX, VSTOXX, VDAX, VFTSE, VNVIXN, VHSI and VKOSPI. This paper also extends the multivariate normal distribution and multivariate student-t distribution based dynamic conditional correlation (DCC) model to a multivariate skew distribution. We then apply this extended model to estimate the dynamic volatility and correlation in international volatility indexes. The empirical results of model comparison reveal the multivariate skewed student-t distribution based CGARCH-DCC model to perform the best in our real data analysis. This indicates that the time-varying conditional correlation coefficients as well as volatility are skewed and fat tailed or leptokurtic in characteristic.

Keywords: Implied volatility indexes · Dynamic correlation and volatility
Multivariate skew distribution

1 Introduction

Volatility indexes are the economic indicators of risk assessment in the financial markets and they are designed to measure the market's expectation of future volatility implied by options prices. Moreover, volatility indexes are able to estimate the expectation of the future volatility over the next 30 days. The implied volatility indexes were introduced and have been calculated and published by Chicago Board Options Exchange (CBOE) since 1993. CBOE proposed the volatility index or VIX methodology to minimize risk on the portfolio of investment. In addition, Badshah [3] stated that the volatility indexes and the stock market returns have negative relationship; therefore, VIX is advantageous for investors to manage their risks. There are several related studies and writings emphasizing the volatility indexes of various worldwide financial markets (see e.g., Kaeck and Alexander [11]; Bugge et al. [4]; Psaradellis and Sermpinis [15]; Huskaj and Larsson [10]).

Nowadays, there are many famous volatility indexes, for instances, VIX, VNX, VXD VSTOXX, VFTSE, VCAC 40, VSTOXX and VKOSPI. These indexes are computed and provided on a 60–s basis as an average of implied volatilities in at–the–money options with a residual time–to–maturity equal to 30 days. We can observe that all volatility indexes have similar fluctuation pattern and they changed over time from early 2008 until the early 2009, corresponding to the credit crunch and liquidity crises.

© Springer International Publishing AG, part of Springer Nature 2018
V.-N. Huynh et al. (Eds.): IUKM 2018, LNAI 10758, pp. 361–372, 2018.
https://doi.org/10.1007/978-3-319-75429-1_30

Badshah [3] found that the high volatility indexes can put high pressure on stock market and thereby reducing return of stock.

Many Researchers believe that the different volatility indexes are likely to have correlation. Therefore, it is very important to know the volatility indexes spillover from one market to another or others. In addition, studying volatility indexes spillover phenomenon across all markets will contribute a great benefit for risk management, international portfolios and options traders. Several studies have investigated volatility spillover using historical volatility indexes (e.g., Hamao et al. [9]; Badshah [3]; Gamba-Santamaria et al. [7]). Äijö [1] studied the relationship among various European volatility indexes (VDAX, VSTOXX, and VSMI) and found that these volatility indexes not only are highly correlated but also vary over time. In addition, Badshah [3] found that the volatility indexes (VIX, VXN, VDAX and VSTOXX) are positively correlated.

Furthermore, some studies suggested that other assets and securities in capital market can be the factor affecting the volatility indexes. Khositkulporn [12] revealed that the oil price hike has shocked the global economy. For example, when the oil price increased to above USD 114 per barrel, the global economy faced recession and the equity markets became volatile during global financial crisis in 2008. Moreover, Kumar [13] examined the return and volatility spillover between gold price and Indian stock sector by assuming that the error term followed the student-t distribution. Although this study could not found any significant spillover from gold to stock but it found a negative dynamic correlation between these two variables, especially during the crisis. We expect that we should consider the other factors that may affect volatility indexes. Finally, we add oil and gold price to further investigate the factors affecting volatility indexes.

In this paper, we employ a multivariate generalized autoregressive conditional heteroskedasticity (GARCH) with exogenous variables based dynamic conditional correlation to investigate the correlation among volatility indexes and also find the effect of oil and gold on conditional mean and variance of the volatility indexes. However, this study has a concern that the symmetric assumption of the multivariate normal and student-t distributions might not be adequate in reality. To tackle such unrealistic assumption, we extend symmetric based dynamic conditional correlation (DCC) model to a multivariate skew distribution. The aim here is to allow for possible departure from symmetry to produce more flexible and more realistic families of distributions. In this study, a multivariate skew-normal and skew-student-t distributions, presented in Azzalini [2], are considered to construct a likelihood function of DCC model. Consequently, our model will give more flexibility to embrace the skewed and fat tailed or leptokurtic characteristics of volatility index.

The next section briefly outlines the methodology. Section 3 is the empirical part presenting data description, model selection and the results. The last section provides the conclusion of this work.

2 Econometric Methodology

2.1 Brief Review of Generalized Autoregressive Conditional Heteroskedasticity (GARCH) Families

2.1.1 GARCH with Exogenous Variables

The generalized autoregressive conditional heteroskedasticity (GARCH) process is an econometric model developed in 1982 by Engle to describe an approach to estimating conditional volatility in financial markets. In this study, we aim to investigate the effect of exogenous variables on the volatility index return therefore we employ a general GARCH with exogenous variables which contain exogenous variables in both mean and variance equations. Our model reads

$$y_t = c + \sum_{k=1}^{K} \phi_k x_{kt} + \sigma_t \varepsilon_t, \tag{1}$$

$$\sigma^2 = \varpi + \sum_{k=1}^{K} \varphi_k x_{kt} + \sum_{i=1}^{p} \alpha_i \varepsilon_{t-i} + \sum_{j=1}^{q} \beta_j \sigma_{t-j}^2. \tag{2}$$

where y_t is the return, x_{kt} represents $K \times T$ matrix of exogenous variables and σ^2 is time varying volatility obtained from the GARCH process in Eq. (2). It is quite obvious the structure of GARCH(p, q) consists of two parts. It has a polynomial $\beta(L)$ of order p- the autoregressive term, and a polynomial $\alpha(L)$ of order q - the moving average term. The parameter α_i and β_j are assumed to be less than 1 and their summation must be less than 1. In addition, parameter ϕ_k and φ_k are the coefficients of the exogenous variable k in mean and variance equation, respectively. Note that the mean equation is applied to every GARCH type model with exogenous variables.

2.1.2 The GJR-GARCH

The model was proposed by Glosten, Jagannathan and Runkle [8] to model an asymmetry in the ARCH process. The GJR-GARCH with exogenous variables model is represented by the expression

$$\xi_t^2 = \varpi + \sum_{k=1}^{K} \varphi_k x_{kt} + \sum_{i=1}^{p} \alpha_i \varepsilon_{t-i} + \sum_{j=1}^{q} \beta_j \sigma_{t-j}^2 + \sum_{i=1}^{p} \gamma_i I_{t-i} \varepsilon_{t-i}, \tag{3}$$

where $I_{t-i} = \begin{cases} 1 & if\ \varepsilon_{t-i} < 0 \\ 0 & if\ \varepsilon_{t-i} \geq 0 \end{cases}$.

2.1.3 Exponential GARCH

The exponential GARCH (EGARCH) may generally be specified as

$$\sigma^2 = \varpi + \sum_{k=1}^{K} \varphi_k x_{kt} + \sum_{i=1}^{p} \alpha_i \varepsilon_{t-i} + \sum_{j=1}^{q} \beta_j \ln \sigma_{t-j}^2. \tag{4}$$

This model differs from the variance equation in GARCH structure because of the log of the variance. The following specification also has been used in the financial literature Dhamija and Bhalla [5].

2.1.4 Integrated GARCH

IGARCH model applies both autoregressive and moving average structures to the variance, σ^2. The IGARCH is specified as

$$\sigma^2 = \varpi + \sum_{k=1}^{K} \varphi_k x_{kt} + \sum_{i=1}^{p} \alpha_i \varepsilon_{t-i} + \sum_{j=1}^{q} \beta_j \sigma_{t-j}^2. \tag{5}$$

where the sum of coefficients (α, β) must be less than 1.

2.1.5 Component GARCH

The Component GARCH model (CGARCH) can be written as:

$$\sigma_t^2 = q_t + \sum_{k=1}^{K} \varphi_k x_{kt} + \sum_{i=1}^{p} \alpha_i (\varepsilon_{t-i}^2 - q_{t-i}) + \sum_{j=1}^{p} \beta_j (\sigma_{t-j}^2 - q_{t-j}),$$

$$q_t = \omega + \rho q_{t-1} + \phi (\varepsilon_{t-1}^2 - \sigma_{t-1}^2) \tag{6}$$

where effectively the intercept of the GARCH model is now time-varying following first order autoregressive type dynamics. The sum of coefficients (α, β) must be less than 1 and $\rho < 1$ (effectively the persistence of the transitory and permanent component).

2.2 Dynamic Conditional Correlation (DCC)

The DCC–GARCH model can be best understood by recalling the best fit GARCH type model in Subsects. 2.1.1, 2.1.2, 2.1.3, 2.1.4 and 2.1.5. The difference is that the DCC-GARCH model is for multivariate volatility modeling. The advantage of the DCC model is that we can examine the time-varying correlation between many dimensions of a time series instead of a constant correlation. We again consider a k-dimensional innovation ε_{it} to the asset return series y_{it}, $i = 1, \ldots, N$. Let $\boldsymbol{\eta}_t = (\eta_{1t}, \ldots, \eta_{it})$ be the marginally standardized innovation vector ($\eta_{it} = \varepsilon_{it} / \sqrt{\sigma_{ii,t}}$). The DCC model can be formulated as the following statistical specification:

$$\mathbf{Q}_t = (1 - \theta_1 - \theta_2) + \theta_1 \mathbf{Q}_{t-1} + \theta_2 \boldsymbol{\eta}_{t-1} \boldsymbol{\eta}_{t-1}', \tag{7}$$

$$\mathbf{R}_t = \mathrm{diag}\{\mathbf{Q}\}_t^{-1} \mathbf{Q}_t \mathrm{diag}\{\mathbf{Q}\}_t^{-1} \quad \text{and} \quad \mathbf{H_t} = \mathbf{J}_t \mathbf{R}_t \mathbf{J}_t. \tag{8}$$

Here, $\mathbf{R}_t$ is the correlation matrix, $\mathbf{H_t}$ is the conditional covariance matrix of returns, and this correlation matrix is allowed to vary over time. Moreover, $\mathbf{Q}_t \equiv \{\sigma_{it}\}$ is the conditional covariance matrix of $\boldsymbol{\eta}_t$, θ_i are non-negative real numbers satisfying $0 \leq \theta_1 + \theta_2 < 1$, and $\mathbf{J}_t = diag\{\sqrt{\sigma}_{1t}, \ldots, \sqrt{\sigma}_{it}\}$.

2.3   Estimation

In this study, we are concerned that the large number of parameters in the model could bring a difficult optimization. Thus, the two-stage estimation method is used, following (Engle [6]). This method allows the model to be estimated more easily even when the covariance matrix is very large. The model is estimated in two steps: firstly, various GARCH type models are estimated and then dynamic conditional correlation parameters are estimated in the second step. In other words, the parameters to be estimated in the correlation and GARCH processes are independent Engle [6]. Under reasonable regularity conditions, consistency of the first step will ensure consistency of the second step (see, Newey and McFadden [14]). In the two-step method, we can maximize the DCC likelihood function conditional on the estimated parameters from GARCH-type models in the first step.

$$\hat{\theta} = \arg\max(L_D(\theta|y, \hat{\Theta})), \text{ where } \hat{\Theta} = \arg\max(L_G(\Theta|y)) \tag{9}$$

is the maximization of likelihood function of GARCH type models. In this study, we extend the multivariate normal and student-t distributions based dynamic conditional correlation (DCC) model to skew-normal and skew-student-t distributions. The study herein tries to propose a skew likelihood function in the DCC-GARCH type models with exogenous variables and their performances are compared based on Akaike and Bayesian information criteria. In addition, we are concerned about the consistency of the two-step estimator, hence the likelihood distribution of the GARCH-type model in the first step and DCC model in the second step are assumed to have the same distribution. In this estimation, we consider multivariate normal, skew-normal, student-t, and skew-student-t distributions function, which are justified in Azzalini [2], are employed to construct the likelihood function in DCC part.

In this paragraph, several likelihood functions are presented, giving simple consistent but inefficient estimates of the parameters of the model. Here, the likelihood functions of volatility part (GARCH types), $\hat{L}_V(\Theta)$, and correlation part (DCC), $\hat{L}_C(\theta|\Theta)$, are written as in the followings:

(1) Normal likelihood function

$$\hat{L}_{Vn,i}(\Theta) = \prod_{t=1}^{T} \left(\frac{1}{\sqrt{2\pi(\sigma_{it}^2)}} \exp\left(-\frac{\sigma_{it}\varepsilon_{it}}{2(\sigma_{it}^2)} \right) \right), \tag{10}$$

for return $i, i = 1, \ldots, N$ and

$$\hat{L}_{Cn}(\theta|\hat{\Theta}) = \prod_{t=1}^{T} \frac{\exp(-\frac{1}{2}(\varepsilon_t)^{\mathrm{T}}\mathbf{H}^{-1}(\varepsilon_t))}{\sqrt{|2\pi\mathbf{H}|}}, \tag{11}$$

where ε_t is $N \times T$ matrix of error term.

(2) Skewed normal likelihood

$$\hat{L}_{Vsn,i}(\Theta) = \prod_{t=1}^{T} \left(\frac{2}{\xi_i + (\xi_i)^{-1}} \cdot f_n(z_i/\Upsilon_i) \right), \tag{12}$$

$$z_i = \frac{\varepsilon_{it}}{(\sigma_{it}^2)} \sqrt{((1 - m_1^2)(\xi_i^2 + 1/\xi_i^2) + 2m_1^2 - 1)} + mu_i,$$

$$m_1 = 2/\sqrt{2\pi}, \quad \Upsilon_i = \xi_i^{sign(z_i)}, \quad mu_i = m_1(\xi_i - 1/\xi_i),$$

where ξ_i is skew parameter of return i, $f_n(\cdot)$ is the probability density function of the normal distribution for return i and $sign(\cdot)$ is a function that returns a vector with the signs of the corresponding elements of z_i (for example, the sign of a real number is 1, 0, or -1 if the number is positive, zero, or negative, respectively). Then, the correlation part is

$$\hat{L}_{Csn}(\theta|\hat{\Theta}) = \prod_{t=1}^{T} \left(\frac{2}{\xi + (\xi)^{-1}} \cdot f_n(\mathbf{z}/\Upsilon) \right), \tag{13}$$

$$\mathbf{z} = \frac{\varepsilon_t}{(\mathbf{H})} \sqrt{((1 - m_1^2)(\xi^2 + 1/\xi^2) + 2m_1^2 - 1)} + mu,$$

$$m_1 = 2/\sqrt{2\pi}, \quad \Upsilon = \xi^{sign(\mathbf{z})}, \quad mu = m_1(\xi - 1/\xi),$$

where ξ is skew parameter of DCC likelihood.

(3) Student's t likelihood

$$\hat{L}_{Vt,i}(\Theta) = \prod_{t=1}^{T} \left[\frac{\Gamma\left(\frac{v_i + 1}{2}\right)}{\sqrt{(v_i - 2)\pi}\Gamma\left(\frac{v_i}{2}\right)} \left(1 + \frac{\varepsilon_{it}^2}{(v_i - 2)(\sigma_{it}^2)}\right)^{\frac{-v_i + 1}{2}} \cdot \left(\frac{1}{\sigma_{it}^2}\right) \right], \tag{14}$$

where v_i is degree of freedom of return i and Γ is gamma distribution. Then, the correlation part is

$$\bar{L}_{Ct}(\theta|\Theta) = \prod_{t=1}^{T}\left(\frac{\Gamma\left(\frac{v+1}{2}\right)}{\sqrt{(v-2)\pi}\Gamma\left(\frac{v}{2}\right)}(1+\frac{\varepsilon_t^2}{(v-2)(\mathbf{H}_t)})^{\frac{-v+1}{2}}\cdot\left(\frac{1}{\mathbf{H}_t}\right)\right), \qquad (15)$$

where v is degree of freedom of DCC likelihood.

(4) Skewed student's t likelihood

$$L_{Vsstd,i}(\Theta) = \prod_{t=1}^{T}\left[\frac{2}{(\xi_i+1)/\xi_i}f_{std}(z_i/\xi_i^{sign(z)},v_i)F_i\right], \qquad (16)$$

$$z_i = \frac{\varepsilon_{it}}{\sigma_{it}^2}F_i + \left[\frac{2\sqrt{v_i-2}}{(v_i-1)}\left(beta\left(0.5,\frac{v_i}{2}\right)\right)^{-1}\right]\left(\frac{\xi_i-1}{\xi_i}\right),$$

$$F_i = \sqrt{1-\left[\frac{2\sqrt{v_i-2}}{(v_i-1)}\left(beta\left(0.5,\frac{v_i}{2}\right)\right)^{-1}\right]^2\left(\frac{\xi_i^2+1}{\xi_i^2}\right)+2\left[\frac{2\sqrt{v_i-2}}{(v_i-1)}\left(beta\left(0.5,\frac{v}{2}\right)\right)^{-1}\right]^2}-1,$$

where $f_{std}(\cdot)$ is a density of student-t distribution. $beta(\cdot)$ is beta distribution. Then, the correlation part is

$$L_{Csstd}(\theta|\Theta) = \prod_{i=1}^{T}\left[\frac{2}{(\xi+1)/\xi}f_{std}(z/\xi^{sign(z)},v)F\right] \qquad (17)$$

$$z = \frac{\varepsilon_t}{\mathbf{H}}F + \left[\frac{2\sqrt{v-2}}{(v-1)}\left(beta\left(0.5,\frac{v}{2}\right)\right)^{-1}\right]\left(\frac{\xi-1}{\xi}\right)$$

$$F = \sqrt{1-\left[\frac{2\sqrt{v-2}}{(v-1)}\left(beta\left(0.5,\frac{v}{2}\right)\right)^{-1}\right]^2\left(\frac{\xi^2+1}{\xi^2}\right)+2\left[\frac{2\sqrt{v-2}}{(v-1)}\left(beta\left(0.5,\frac{v}{2}\right)\right)^{-1}\right]^2}-1$$

3 Empirical Study

3.1 Data

This paper focuses on seven implied volatility indexes: the volatility index for S&P 500 stock index in the United States (VIX), the volatility index for Dow Jones EURO STOXX 50 stock index in Europe (VSTOXX), the volatility index for DAX 30 stock index in Germany (VDAX), the volatility index for FTSE stock index in United Kingdom (VFTSE), India's volatility index (INVIXN), the volatility index for HSI stock index in Hong Kong (VHSI), and the volatility index for KOSPI stock index in South Korea (VKOSPI). All of the 7 volatility indexes are calculated as the expectations of the future stock volatility index in the options. This study collected the data from

Bloomberg. We obtained the daily time series closing price data for VIX, VSTOXX, VDAX, VFTSE, VNVIXN, VKOSPI, and VHSI. In addition, the daily gold price and oil price are also collected from Thomson Financial DataStream. These data sets are collected from 1 November 2007 to 11 August 2017 covering 22,149 daily observations. The detailed description of variables and descriptive statistics for daily time series data are presented in Table 1.

Table 1. Data description

	VIX	VSTOXX	VDAX	VFTSE	VNVIXN	VHSI	VKOSPI	Gold	Oil
Mean	−0.0001	−0.0001	0.0000	−0.0002	−0.0005	−0.0006	−0.0004	8.7000	−0.0001
Median	−0.0025	−0.0009	−0.0004	−0.0008	0.0000	−0.0015	0.0000	0.0002	4.1900
Max	0.1761	0.1423	0.1327	0.1868	0.2158	0.1997	0.2116	0.0445	0.0712
Min	−0.1523	−0.1888	−0.1607	−0.1957	−0.2042	−0.0911	−0.1315	−0.0413	−0.0567
Std. dev.	0.0320	0.0278	0.0256	0.0314	0.0250	0.0237	0.0232	0.0052	0.0109
Skew	0.6916	0.4628	0.4801	0.2023	0.4593	1.5349	1.4267	−0.2543	0.1670
Kurtosis	6.6421	6.3000	5.9407	5.9820	13.8863	11.7211	13.3697	9.2134	7.5162
JB-test	1556.42	1204.55	981.335	928.629	12238.9	8765.40	11861.1	3985.34	2102.90
Prob.	0.0000	0.0000	0.0000	0.0000	0.0000	0.0000	0.0000	0.0000	0.0000
ADF	−54.2***	−38.1***	−36.8***	−38.0***	−39.8***	−28.4***	−28.4***	−48.9***	−52.5***
Sum	−0.1751	−0.2358	−0.0919	−0.4653	−1.3130	−1.4756	−1.0116	0.2142	−0.2822

Note: Table reports the descriptive statistics of the volatility indexes, gold and oil prices. The Jarpue-Bera (JB) test and the Augmented Dickey Fuller (ADF) test values are reported. ***, **, * denote rejection of null hypothesis at 1%, 5% and 10% significance respectively.

In this study, all raw data are transformed into log-returns. The table shows that the skewness and kutosis of volatility index returns and oil returns are positively skewed while gold returns show a negative skew. In addition, normality of all returns are rejected by the Jarque–Bera test, prob. = 0.0000. We also investigate the stationarity of our data. We find that all returns data are rejected at 1% significance level hence all nine returns are stationary.

3.2 Model Selection

3.2.1 GARCH Types and Distribution Selection
In this section we investigate and compare the performance of each model specification in the real data analysis. Tables 2 and 3 present the values of AIC and BIC, respectively. We observe that CGARCH with the skewed normal distribution is the best fit model in this study because the lowest values of AIC and BIC are obtained.

Table 2. Akaike Information Criterion (AIC)

	SGARCH	GJR-GARCH	EGARCH	IGARCH	CGARCH
norm	−89798.02	−90174.11	−91010.33	−88457.73	−89548.78
STD	−103990.1	−104290.3	−104776.1	−101971.5	−103965.7
SSTD	−100992.7	−101244	−101672.4	−99037.72	−101006.9
snorm	−378529.6	−360434.5	−337667.4	−346934.6	**−380235.7**

Source: Calculation

Table 3. Bayesian Information Criterion (BIC)

	SGARCH	GJR-GARCH	EGARCH	IGARCH	CGARCH
norm	−89420.48	−89755.92	−90592.13	−88080.19	−89089.92
STD	−103571.9	−103831.4	−104317.2	−101553.3	−103466.2
SSTD	−100533.7	−100744.5	−101172.9	−98578.86	−100466.7
snorm	−378111.4	−359975.7	−337208.6	−346516.4	**−379736.2**

Source: Calculation

3.3 Estimation Results

The estimated optimal parameters from the best model, DCC-CGARCH model with skewed normal student-t distribution, are presented in Table 4. We investigate the relationships between volatility indices. The most important parameters are sum of α_1 and β_1 in the model. They are close to 1 indicating a high degree of volatility persistence of transitory and permanent components. All volatility indexes exhibit high persistency,

Table 4. Estimation DCC-CGARCH with skewed normal distribution results

	VIX	VSTOXX	VDAX	VFTSE	VNVIXN	VHSI	VKOSPI
C	0.00039	-0.00003	-0.00008	-0.00003	0.00009	0.00004	-0.00006
	(0.0005)	(0.0005)	(0.0005)	(0.0006)	(0.0004)	(0.0004)	(0.0004)
ar1	-0.1011[a]	-0.01413	0.01158	-0.0376[a]	-0.1028[a]	-0.0868[a]	-0.0798[a]
	(0.0219)	(0.0220)	(0.0214)	(0.0218)	(0.0330)	(0.0201)	(0.0227)
$\emptyset_1$	0.05857	0.02216	-0.15981[a]	0.22820[a]	-0.06203	0.06329	0.04794
	(0.1143)	(0.1080)	(0.0969)	(0.1163)	(0.0871)	(0.0829)	(0.0794)
$\emptyset_2$	-0.68356[a]	0.08256	0.12489[a]	-0.71572[a]	-0.29274[a]	-0.20490[a]	-0.20135[a]
	(0.0574)	(0.0527)	(0.0476)	(0.0587)	(0.0471)	(0.0386)	(0.0413)
ω	0.000001[a]	0.000001[a]	0.000002[a]	0.000004[a]	0.000006[a]	0.000002[a]	0.000003[a]
	(0.00)	(0.00)	(0.00)	(0.00)	(0.00)	(0.00)	(0.00)
α_1	0.15330[a]	0.07384[a]	0.06335[a]	0.06252	0.10971	0.08262[a]	0.13461[a]
	(0.0185)	(0.0147)	(0.0118)	(0.0162)	(0.0096)	(0.0162)	(0.0055)
β_1	0.70185[a]	0.79304[a]	0.85382[a]	0.62168[a]	0.59677	0.87675[a]	0.76803[a]
	(0.0408)	(0.0521)	(0.0222)	(0.1234)	(0.0738)	(0.0238)	(0.0259)
ρ	0.99914[a]	0.99928[a]	0.99748[a]	0.99523[a]	0.98899[a]	0.99647[a]	0.99464[a]
	(0.0002)	(0.0060)	(0.00)	(0.00)	(0.00)	(0.00)	(0.00)
$\emptyset$	0.00201[a]	0.00017	0.00037[a]	0.01138[a]	0.02070[a]	0.00025	0.00755[a]
	(0.0006)	(0.0001)	(0.00)	(0.0030)	(0.0026)	(0.0006)	(0.0034)
φ_1	0.00000	0.00000	0.00000	0.00000	0.00000	0.00000	0.00000
	(0.0010)	(0.0002)	(0.0003)	(0.0012)	(0.0008)	(0.0006)	(0.0007)
φ_2	0.00000	0.00000	0.00000	0.00000	0.00000	0.00000	0.00000
	(0.0003)	(0.0002)	(0.0002)	(0.0005)	(0.00)	(0.0003)	(0.0003)
ν	1.27469[a]	1.15575[a]	1.15627[a]	1.05636[a]	1.17309[a]	1.42484[a]	1.27925[a]
	(0.0304)	(0.0260)	(0.0262)	(0.0235)	(0.0233)	(0.0343)	(0.0291)
DCC	$\theta_0 = 0.0035^a$, $\theta_1 = 0.9943^a$, $\xi = 9.9986^a$						

Source: Calculation.
[a]denotes rejection of null hypothesis at 1%, 5% and 10% significance.

indicating a strong volatility of VHSI, VDAX, VKOSPI, VSTOXX, VIX and VNVIXN. In addition, the degree of asymmetry in volatility is highest in VSTOXX ($\rho = 0.99928$) and lowest in VKOSPI ($\rho = 0.99464$). Moreover, we also found that oil and gold have effects on return of volatility indexes. For example, the value of coefficient of gold price on VIX is 0.05857, meaning that the change in oil return by 1% will change the return of VIX about 5.857% in the same direction. To estimate DCC, the value of θ_0, θ_1 and ξ are parameters used to find integrated and mean reverting models. It presents dynamic correlation for volatility indexes. In this case it is fitted as the skewed normal distribution.

3.4   Dynamic Correlation from the Best GARCH Fit

3.4.1   Correlation Analysis

Due to the space limit, we only present the dynamic correlation results of some interesting pairs. We compute time series daily correlation between the changes in volatility indexes for instances, VIX, VSTOXX, VDAX, VFTSE, VNVIXN, VHSI and VKOSPI

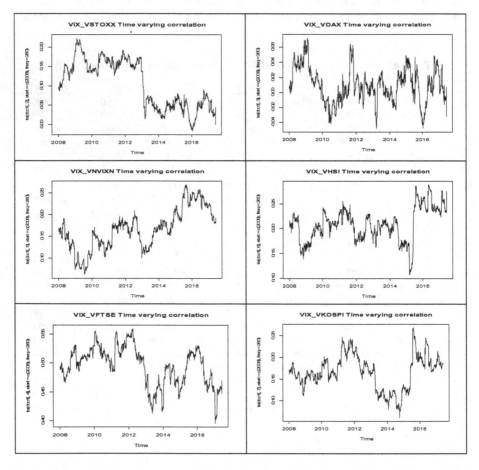

Fig. 1.   Daily correlation between volatility indices from 1 November 2007 to 11 August 2017.

from 1 November 2007 to 11 August 2017 and only the pairs between VIX and other indexes are presented in Fig. 1.

There are a number of significant results and find that most volatility indexes have positive correlation. The correlation for VIX and VSTOXX is strongest in 2009 around 0.20, and then the value becomes close to zero after 2012, indicating a weak positive correlation. Similarly, the VIX and VDAX correlation reached the highest level in 2009 at 0.06 and it turned to be negative correlation around −0.04 in 2010, 2013, and 2016, and this indicates that there are both positive and negative correlations between them over time. Additionally, the correlation between VIX and VFTSE reached the highest level in 2012 around 0.55 indicating a high positive correlation. Meanwhile, the correlation for VIX and VNVIXN, VIX and VHSI, VIX and VKOSPI reached high level after 2014 around 0.25.

4 Conclusion

Based on the results, we suggest the best fit model for estimating the optimal parameters to be DCC-CGARCH based on the skewed student-t distribution. We observe that, U.S. volatility index (VIX) has the positive dynamic correlation with other volatility indexes (VSTOXX, VFTSE, VHSI, VKOSPI, VDAX, VNVIXN) meaning that if VIX increases, it will lead VSTOXX to increase. Additionally, this paper includes exogenous effects on volatility indexes consisting of oil and gold. We found that there is a significant effect of oil and gold on volatility indexes.

References

1. Äijö, J.: Implied volatility term structure linkages between VDAX, VSMI and VSTOXX volatility indices. Glob. Finance J. **18**(3), 290–302 (2008)
2. Azzalini, A.: The Skew-Normal and Related Families, vol. 3. Cambridge University Press, Cambridge (2013)
3. Badshah, I.: Asymmetric return-volatility relation, volatility transmission and implied volatility indexes (2009)
4. Bugge, S.A., Guttormsen, H.J., Molnár, P., Ringdal, M.: Implied volatility index for the Norwegian equity market. Int. Rev. Financ. Anal. **47**, 133–141 (2016)
5. Dhamija, A.K., Bhalla, V.K.: Financial time series forecasting: comparison of neural networks and ARCH models. Int. Res. J. Finance Econ. **49**, 185–202 (2010)
6. Engle, R.: Dynamic conditional correlation: a simple class of multivariate generalized autoregressive conditional heteroskedasticity models. J. Bus. Econ. Stat. **20**(3), 339–350 (2002)
7. Gamba-Santamaria, S., Gomez-Gonzalez, J.E., Hurtado-Guarin, J.L., Melo-Velandia, L.F.: Stock market volatility spillovers: evidence for Latin America. Finance Res. Lett. **20**, 207–216 (2017)
8. Glosten, L.R., Jagannathan, R., Runkle, D.E.: On the relation between the expected value and the volatility of the nominal excess return on stocks. J. Finance **48**(5), 1779–1801 (1993)
9. Hamao, Y., Masulis, R.W., Ng, V.: Correlations in price changes and volatility across international stock markets. Rev. Financ. Stud. **3**(2), 281–307 (1990)

10. Huskaj, B., Larsson, K.: An empirical study of the dynamics of implied volatility indices: international evidence. Quant. Finance Lett. **4**(1), 77–85 (2016)
11. Kaeck, A., Alexander, C.: Continuous-time VIX dynamics: on the role of stochastic volatility of volatility. Int. Rev. Financ. Anal. **28**, 46–56 (2013)
12. Khositkulporn, P.: The factors affecting stock market volatility and contagion: Thailand and South-East Asia evidence. Doctoral dissertation, Victoria University (2013)
13. Kumar, D.: Return and volatility transmission between gold and stock sectors: application of portfolio management and hedging effectiveness. IIMB Manag. Rev. **26**(1), 5–16 (2014)
14. Newey, W., McFadden, D.: Large sample estimation and hypothesis testing. In: Engle, R., McFadden, D. (eds.) Handbook of Econometrics, vol. 4, pp. 2113–2245. Elsevier Science, New York (1994)
15. Psaradellis, I., Sermpinis, G.: Modelling and trading the US implied volatility indices. Evidence from the VIX, VXN and VXD indices. Int. J. Forecast. **32**(4), 1268–1283 (2016)

Markov-Switching ARDL Modeling of Parboiled Rice Import Demand from Thailand

Roengchai Tansuchat[✉] and Woraphon Yamaka

Faculty of Economics, Center of Excellent in Econometric,
Chiang Mai University, Chiang Mai, Thailand
roengchaitan@gmail.com

Abstract. In this paper, we develop a Markov Switching autoregressive distributed lag (MS-ARDL) model in which short- and long-run nonlinearities are introduced. The model is used to investigate the import demand of Nigeria for parboiled rice from Thailand. We demonstrate that the model is estimable by Maximum likelihood estimator and then a reliable long-run inference can be achieved by bound testing regardless of the integration orders of the variables. Furthermore, we first examine the accuracy of the model using a simulation study, and then the salient features of the model are employed to investigate the Thai parboiled rice demand from Nigeria.

Keywords: Markov switching · ARDL · Thai parboiled rice · Nigeria

1 Introduction

Rice is one of the major cereal and food sources of the world. People in Asian and Middle East countries including many countries in Africa consume rice as their main staple. Rice is also a major economic crop that has enormous implications for the world population. With the traditional rice culture and the growing preference for rice consumption in many regions of the world, the demand for rice is likely to increase in the long run. In addition, the market demand for rice is more complex due to the growth of urban and emerging middle class in many countries resulting in higher demand for rice quantity and quality. While the demand for low quality rice is declining.

In the past decade, rice was traded small in value and volume compared with other agricultural commodities such as wheat, corn (maize), and soybeans. The global rice exports totaled US$ 15.166 billion in 2016, up 50.68% from 2007 or with average growth 8.744% per year. However, during the final years of the decade, the value of global rice export dropped by 15.7% between 2014–2015 and 9.073% between 2015–2016 (Global Trade Atlas, 2017). In 2016, prices of most agricultural commodities increased except rice dropped 7.54% [1]. In the world rice market, rice can be classified according to grain size: long, medium and short-grain. Types of rice in most markets are brown rice, white rice and parboiled rice.

Parboiled rice (also called converted rice) is rice that has been partially boiled in the husk. The three basic steps of parboiling are soaking, steaming and drying. These steps also make rice easier to process by hand, boost its nutritional profile and change its

© Springer International Publishing AG, part of Springer Nature 2018
V.-N. Huynh et al. (Eds.): IUKM 2018, LNAI 10758, pp. 373–384, 2018.
https://doi.org/10.1007/978-3-319-75429-1_31

texture. In this market only India and Thailand are major producers, exporters, competitors as well as price leaders. Currently, Thai parboiled rice is virtually reliant on exports. The major export markets are the countries in Africa and the Middle East such as Nigeria, South Africa, Saudi Arabia and United Arab Emirate. For Thai rice, in year 2016 parboiled rice was the second largest in terms of quantity export, at 2.079 million tones, behind white rice, at 2.806 million tons; while being the third largest in terms of value export, at 795 million US$, after white rice and Hom Mali rice, at 1,130 and 992, respectively [2].

Based on international trade theory, there are many factors affecting export demand which are applicable in parboiled rice export modeling, namely income of import country, export price, competitors' export price, inflation rate and exchange rate [3]. In addition, there are dummy variables which represent government intervention policy such as rice export ban, pledging scheme, and tariff or quota policies. Figure 1 displays the quantity and price of Thai parboiled rice during 2006–2016. From 2008 to 2011, Thailand is the major parboiled rice exporting country since India's government decided to ban the export of non-basmati rice since Q3:2007–Q3:2011 because of India's food security problem.

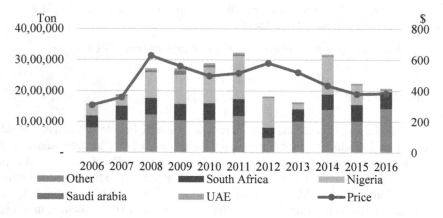

Fig. 1. Quantity and price of Thai parboiled rice during 2006–2016 (Source: Thai Custom (2017))

Later Q3:2011–Q1:2014, Thai government implemented a rice pledging scheme in which Thai government guaranteed unlimited purchase of paddy from farmers at 15,000 baht ($488) per ton. Consequently, the cost of paddy for parboiling has increased dramatically causing Thailand to lose its competitiveness. In addition, parboiled rice millers could not directly procure paddy from farmers because farmers prefer to sell paddy to the government rather than rice millers. From Fig. 1 the quantity of Thai parboiled rice exported to Nigeria has decreased significantly in 2013 and 2016 because of Nigeria's policy to reduce food imports to achieve food security for nations by increasing import duty and import levy in Q3:2013, and imposing trade barriers to prevent rice importers from buying foreign currency from official Nigerian financial institutions, measures to ban rice imports through Nigeria's land border, and no rice quotas announcement in 2016.

In order to model the rice import demand, many different econometrics techniques have been applied to estimate export supply and import demand functions in many different countries. [4] analyzed the rice import demand in Cyprus by using unrestricted error-correction model and found long run relationship among relative prices. The estimated model displayed the price inelastic of demand, but negative income inelastic of demand for imported rice in Cyprus. [5] studied the demand function of the world rice market using annual data from 1994 to 2007 with various models such as OLS, Instrumental Variables, Generalized Method of Moments and Seemingly Unrelated Regression, and found that the economic growth, foreign direct investment and the population of importing countries had a direct impact on national income, which in turn had a direct impact on rice consumption in the same direction. Co and Boosarawongse [3] compared the performance of artificial neural networks (ANNs) with exponential smoothing and ARIMA models in forecasting rice exports from Thailand and found that ANNs outperformed the latter to model the dynamic non-linear trend and seasonality. Toshiyuki [17] studied the import of Jasmine rice of China from Thailand.

In order to model the import demand for Thai parboiled rice, this study develops the Markov Switching Autoregressive Distributed Lag (ARDL) to explain the nonlinear behavior in the rice trade. There are a number of empirical evidences suggesting that the behaviors of economic and agriculture time series may exhibit different patterns over time (see, [6–8]). Thus, the linear model might not be appropriate to explain the nature of data.

To understand the proposed Markov Switching ADRL model, let's describe briefly the concept of ADRL approach which has been used in econometrics for decades and employed as a method of examining cointegrating relationships between variables (see, [9, 10]). The advantage of this approach is that it can deal with short run spurious regression problem without transforming the data in order to achieve stationarity. To resolve this spurious problem, the error correction term, which is used as the estimation of the long-run and the speed of adjustment to equilibrium, is incorporated in the ARDL model and thereby having both long-run and short-run exist in the model.

However, as we already mentioned about nonlinear behavior in the economic and agriculture trade data, instead of using linear ADRL model, it is natural to employ nonlinear model to present these patterns. Therefore, in this study, a Markov switching model which was first considered in [1] is considered and extended to a linear Autoregressive Distributed Lag (ARDL) model. Thus, we shall focus on the Markov Switching ARDL model. The model has an ability to capture the structural change in the import demand for Thai parboiled rice, and distinguish two different regimes, one is parboiled rice market expansion and the other rice market contraction.

The rest of the paper proceeds as follows. Section 2 introduces Markov Switching ADRL model. Section 3 employs a Monte Carlo simulation to investigate the accuracy of the model. Section 4 presents the data and results of our empirical illustration. Lastly, Sect. 5 offers some concluding remarks.

2   Methods and Procedures

2.1   The Markov Switching Model of ADRL

In this section, firstly, we illustrate the structure of ARDL using a simple model and then discuss about the Markov switching ARDL model specifications. Secondly, we explain the import demand model.

2.2   Simple ARDL Model

Before developing the full representation of the Markov Switching ARDL model, we introduce the following linear ARDL model:

$$y_t = \alpha_0 + \sum_{i=1}^{p} \beta_i y_{t-i} + \sum_{j=1}^{k} \sum_{l=1}^{q} \delta_{j,l_j} x_{j,t-l_j} + \varepsilon_t \tag{1}$$

The regressors of the model Eq. (1) may include lagged values of the dependent variable, y_{t-i}, and lagged values of explanatory variables, $x_{j,t-l_j}$. α_0 is an intercept term, β_i and δ_{j,l_j} are the coefficients of y_{t-i} and $x_{j,t-l_j}$, respectively. In other words, the model is related not only contemporaneously, but also across lagged values. ε_t is the error term assumed to have normal distribution with mean zero and variance σ^2.

In general, when non-stationary variables are regressed in a model we may get results that are spurious. Without differencing the data to achieve stationarity, one way to resolve the spurious equation problem in the ARDL is to include the error correction term (EC) in the model. Following Pesaran et al. [10], the error correction term (EC) is the estimation of the long-run or the speed of adjustment to equilibrium. Thus, we can rewrite Eq. (1) in the error correction form as

$$y_t = \alpha_0 + \phi EC_{t-1} + \sum_{i=1}^{p} \beta_i y_{t-i} + \sum_{j=1}^{k} \sum_{l=1}^{q} \delta_{j,l_j} x_{j,t-l_j} + \varepsilon_t, \tag{2}$$

where $\alpha_0, \phi, \beta_i, \delta_{j,l_j}$ are the estimated parameters, then the error correction term is defined by

$$EC_t = y_t - \sum_{j=1}^{k} \theta_j x_{j,t-1} - \psi w_{t-1} \tag{3}$$

where w_t is a $s \times 1$ vector of deterministic variables such as the intercept term, time trends and explanatory variables with the fixed lags. The EC_t shows how much of the disequilibrium is being adjusted in the ADRL Eq. (2). The sign and value of EC_t represents a different adjustment. A positive value indicates a divergence, while a negative value indicates convergence. If the estimate of EC_t equals to 1, then 100% of the adjustment takes place within the period t, if the estimate of EC_t equals to 0.5, then 50% of the adjustment takes place in period t. Lastly, if the estimate of EC_t equals to 0, it indicates that there is no adjustment, and no long-run relationship exists in this model [11]. Then, by substituting Eq. (3) into Eq. (2) and rearranging it, we finally obtain the following conditional ECM:

$$y_t = \alpha_0 + \phi\left(y_t - \sum_{j=1}^{k}\theta_j x_{j,t-1} - \psi w_{t-1}\right)_{t-1} + \sum_{i=1}^{p}\beta_i y_{t-i} + \sum_{j=1}^{k}\sum_{l=1}^{q}\delta_{j,l_j}x_{j,t-l_j} + \varepsilon_t \qquad (4)$$

The difficulty of this model is a choice of lag length of p and q. Thus the various lag length specification is selected by the Akaike information criteria (AIC) and the lowest value is preferred.

2.3 Markov Switching ARDL

The study extends the ADRL of [9, 10] to Markov switching model of [12]. In this study, two regimes are considered namely, rice market expansion and contraction. In this section, we first illustrate the features of Markov Switching using a simple model and then discuss more general model specifications.

Let $s(t)$ denote an unobservable state or regime variable assuming the value one or zero. A simple Gaussian Markov Switching ADRL model involves two ADRL specifications:

$$y_t = \alpha_{0,s(t)} + \phi_{s(t)}(y_t - \sum_{j=1}^{k}\theta_{j,s(t)}x_{j,t-1} - \psi_{s(t)}w_{t-1})_{t-1} + \sum_{i=1}^{p}\beta_{i,s(t)}y_{t-i} + \sum_{j=1}^{k}\sum_{l=1}^{q}\delta_{j,l_j,s(t)}x_{j,t-l_j} + \varepsilon_{t,s(t)}, s(t) = 1,2 \qquad (5)$$

where $\alpha_{0,s(t)}, \phi_{s(t)}, \theta_{j,s(t)}, \psi_{s(t)}, \beta_{i,s(t)}$ and $\delta_{j,l_j,s(t)}$ are state or regime dependent parameters, $\varepsilon_{t,s(t)}$ are independently and identically distributed random variables with mean zero and variance $\sigma^2_{s(t)}$. In this study, the normal distribution assumption is hold thus we have $\varepsilon_{t,s(t)} \sim i.i.d. N(0, \sigma^2_{s(t)})$. This model allows two dynamic structures at different levels, depending on the value of the state or regime variable $s(t)$. This state or regime variable is governed by first order Markov chain; in other words, the switching between the states or regimes at time t is governed by time $t - 1$. In addition, $s(t)$ is also given to follow the transition matrix (**P**):

$$\mathbf{P} = \begin{bmatrix} P(s(t) = 0), (s(t-1) = 0) & P(s(t) = 1), (s(t-1) = 0) \\ P(s(t) = 0), (s(t-1) = 1) & P(s(t) = 1), (s(t-1) = 1) \end{bmatrix},$$
$$= \begin{bmatrix} p_{00} & p_{10} \\ p_{01} & p_{11} \end{bmatrix} \qquad (6)$$

where p_{00}, p_{10}, p_{01} and p_{11} denote the transition probabilities of $s(t) = 1$ given that $s(t) = 0$, and vice versa. Clearly, the transition probabilities satisfy $p_{00} + p_{10} = 1$, and $p_{01} + p_{11} = 1$, The transition matrix governs the random behavior of $s(t)$, and here we can estimate only two parameters p_{00} and p_{11}, to achieve this transition matrix. The Markovian switching mechanism was first considered in [12, 13] which presented thorough analysis of the Markov switching model and its estimation method; see also [14].

In the estimation method of this model, there are various ways to estimate the Markov switching model. In this study, we consider only the maximum likelihood estimator to estimate all of unknown state or regime dependent parameters. Given the Eqs. (5 and 6), the vector of regime dependent parameters is $\theta = (\alpha_{0,s(t)}, \phi_{s(t)}, \theta_{j,s(t)}, \psi_{s(t)}, \beta_{i,s(t)}, \delta_{j,l_j,s(t)}, P_{00}, P_{11})$, we can construct the normal likelihood function as

$$L(\theta|y,x) = \sum_{j=1}^{2} \left(\prod_{i=1}^{n} \left(\frac{1}{\sqrt{2\pi\sigma_{s(t)=j}^2}} \times \exp\left[\frac{\left(\varepsilon_{t,s(t)=j}\right)^2}{2(\sigma_{s(t)=j}^2)} \right] \right) \left(\Pr(s(t)=j|\theta_t) \right) \right) \quad (7)$$

where $\varepsilon_{t,s(t)=j}$ is obtained from Eq. (5) and the $\Pr(s(t)=j|\theta_t)$ is the filtering probabilities which are based on the past and current information. To compute these probabilities, we employ the Hamilton filter of [12]. For $s(t) = 0, 1$, the filtering probabilities of $s(t)$ are

$$\Pr(s(t)=j|\theta_t, Z_t) = \frac{f(y,x|s(t)=j, \theta, Z_{t-1})\Pr(s(t)=j|\theta, Z_{t-1})}{\sum_{j=1}^{2} f(y,x|s(t)=j, \theta, Z_{t-1})\Pr(s(t)=j|\theta, Z_{t-1})}, \quad (8)$$

where $Z_t = \{y_t, \dots, y_{t-p}, x_t, \dots, x_{t-q}\}$ denote the collection of all the observed variables up to time t and $f(y,x|s(t)=j, \theta, Z_{t-1})$ is the normal density function. To predict a probability in next time t, we can predict by

$$\Pr(s(t+1)=j|\theta_t, Z_t) = p0j \Pr(s(t)=0|\theta_t, Z_t) + p2j \Pr(s(t)=1|\theta_t, Z_t), \quad j=0,1 \quad (9)$$

By iterating Eqs. (8 and 9) $t = 1, \dots, T$, we can obtain all filtered probabilities.

2.4 Export Demand Model

From the literature review above we assume Thai's parboiled rice export quantity depends on parboiled rice export prices of exporting countries, parboiled rice export prices of competing countries, income of importing countries (GDP), inflation and exchange rate. Therefore, the ARDL and error correction model of demand for Thai's parboiled rice export in double log form is shown as follows:

$$\ln EX_t = \alpha_0 + \beta_1 \ln pthai_t + \beta_2 pindia_t + \beta_3 \ln GDP_t + \beta_4 \ln EX_t + \\ \theta_1 DIN + \theta_2 DThai + \theta_3 DNI + \gamma_1 DS_1 + \gamma_2 DS_2 + \gamma_3 DS_3 + \varepsilon_t$$

$$\Delta \ln EX_t = \gamma + \delta_i \sum \Delta \ln pthai_{t-i} + \varphi_j \sum \Delta pindia_{t-j} + \phi_k \sum \Delta \ln GDP_{t-k} + \\ \varpi_l \sum \Delta \ln EX_{t-l} + \mu ECM_{t-1} + e_t$$

where, EX_{it} is export quantity from Thailand to country i, $pthai_{it}$ is Thai parboiled rice export price to country i, $pindia_{it}$ is India parboiled rice export price to country i, GDP_{it} is a measure of the domestic income of importing country i, EX_{it} is a measure of the exchange rate between country i and Thailand, Inf_{it} represents inflation of importing country i. In this paper, country i is Nigeria. In addition, in international rice market, there are some exogenous variables that may affect Thai rice export such as the ban on India exports of non-basmati rice (DIN) during Q3:2007–Q3:2011, the rice-pledging scheme of Thailand ($DThai$) during Q3:2011–Q1:2014, and trade barriers of Nigeria (DNI) in 2013 and 2016.

3   Simulation Study

The simulation study is conducted to evaluate the performance and accuracy of our model. We provide the Monte Carlo experiment with 100 and 500 simulated values generated from the following two models:

$$
\begin{aligned}
y_t &= 1 - 0.3(y_t - 0.5 - 0.5x_{t-1} - 1y_{t-1})_{t-1} + 1y_{t-1} + 0.5x_{1,t-1} + \varepsilon_{t,s(t)}, \quad s(t) = 0 \\
y_t &= 2 - 0.3(y_t - 1 - 0.7x_{t-1} - 0.5y_{t-1})_{t-1} + 2y_{t-1} + 0.8x_{1,t-1} + \varepsilon_{t,s(t)}, \quad s(t) = 1
\end{aligned}
\tag{10}
$$

Table 1. Simulation results

Parameter	True	N = 100		N = 500	
		$s(t) = 0$	$s(t) = 1$	$s(t) = 0$	$s(t) = 1$
$\alpha_{0,s(t)}$	[1, 2]	1.0333	1.5734	1.2736	2.3298
		(0.0616)	(0.2195)	(0.1295)	(0.2143)
$\phi_{s(t)}$	[−0.3, −0.3]	−0.3292	−0.2752	−0.3016	−3.0981
		(0.0143)	(0.1036)	(0.0475)	(0.0698)
$\theta_{0,s(t)}$	[0.5, 1]	0.5108	1.5172	0.4763	1.2768
		(0.0096)	(0.1496)	(0.0011)	(0.1005)
$\theta_{1,s(t)}$	[0.5, 0.7]	0.6290	0.6515	0.5640	0.7706
		(0.1570)	(0.0826)	(0.1578)	(0.0824)
$\psi_{s(t)}$	[1, 0.5]	0.8809	0.3275	1.0786	0.4099
		(0.0614)	(0.1912)	(0.0340)	(0.1366)
$\beta_{i,s(t)}$	[1, 2]	1.2380	1.7242	1.0179	2.3475
		(0.1495)	(0.1158)	(0.0658)	(0.1009)
$\delta_{j,l_j,s(t)}$	[0.5, 0.8]	0.1455	0.8678	0.5017	0.3475
		(0.3445)	(0.0832)	(0.0658)	(0.1009)
$\sigma^2_{s(t)}$	[0.8, 1]	0.6663	1.3951	1.0115	1.0287
		(0.1114)	(0.1245)	(0.0112)	(0.0125)
p_{11}	0.95	0.9994 (0.1256)		0.9728 (0.0125)	
p_{22}	0.95	0.9552 (0.0245)		0.9386 (0.0222)	

Source: Calculation
Note: () is standard error

Here, we consider two regimes. We simulate the value $x_{1,t-1}$ from $N(0, 1)$. The relationship between right hand side variables, $x_{1,t-1}$ and y_{t-1}, and left hand side variable y_t using iteration from $t = 1, \ldots, T$ for two regimes which are specified as in Eq. (10) whereas $\varepsilon_{t,s(t)} \sim [0, 1]$. In addition, the simulated filtered probabilities for two-regime model, $s(t)$, is generated from $U[0, 1]$ where the transition probabilities $p_{11} = p_{22} = 0.95$. In this simulation, we consider sample size $T = 100$ and $T = 500$.

In Table 1 we report the parameter estimates based on our simulations. Columns denote the true values, parameter estimates for regime 1 and 2 when $T = 100$, and parameter estimates for regime 1 and 2 when $T = 500$, respectively. From these tables, we see that the parameter estimates are close to the true values and when the sample size increases, the standard error of the parameters are lower, indicating that the parameters obtained by maximum likelihood estimator is reliable.

4   Economic Implication

In this section, we aim to illustrate the performance of the MS-ADRL model in the real data analysis. As we mentioned in the introduction, the study will apply this model to investigate the demand for Thai rice of Nigeria, which is one of the biggest rice importers of Thai parboiled rice.

4.1   Data

The data are quarterly time series consisting of Thai parboiled rice export quantity (Q_{Thai}) and price (P_{Thai}) to Nigeria, and India parboiled rice export price (P_{India}) to Nigeria including GDP, consumer price index (CPI) and exchange rate (EX) of Nigeria for the period from Q1-2000 to Q4-2016 totally 68 observations. The quantity and price were collected from Global Trade Atlas. The macroeconomic variables were obtained from Thomson Reuters DataStream, from Financial Investment Center (FIC), Faculty of Economics, Chiang Mai University. All variables were transformed into natural logarithms before estimation in order to reduce the heteroscedasticity. In addition, the estimated parameters of which model have elasticity meaning.

Table 2 shows ADF test and HEGY test for unit root and seasonal unit root. The statistically insignificant ADF test means that we would not reject the null hypothesis of a unit root or $I(0)$, but reject the null hypothesis under first difference transformation. Due to quarterly data, the HEGY test is applied to test the seasonal unit root. Since $t(\pi_2)$ and every joint F test for π_2, π_3 and π_4 are statistically significant, the series is stationary in both seasonal and nonseasonal frequencies. Table 3 shows the results of Engle and Granger test [15], Phillips and Ouliaris test [16], and bound test of cointegration [10]. Under the null hypothesis of no long-run relationship, the t-statistic, F-statistic and Bound test of Nigeria and South Africa are statistically significant implying that the null hypothesis of no cointegration is rejected.

Table 2. Unit root test and seasonal unit root test

ADF test		Qthai	Pthai	Pindia	EX	GDP	CPI
Nigeria (Level)	None	−0.672	0.302	0.840	1.114	4.607	1.7868
	c	−4.181***	−1.192	−2.214	−0.674	−0.270	1.050
	c &T	−4.535***	−1.456	−6.935***	−2.0501	−2.430	−1.317
HEGY test							
Nigeria	$t(\pi_1)$:	−2.77	−1.38	−2.86	1.25	−1.97	0.52
	$t(\pi_2)$:	−4.98***	−4.35***	−2.66*	−2.61*	−4.53***	−4.29***
	$F(\pi_3, \pi_4)$:	19.22***	20.53***	12.72***	7.23***	24.59***	4.16
	$F(\pi_2, \pi_3, \pi_4)$	22.93***	39.67***	11.51***	16.59***	49.79***	10.03***
	$F(\pi_1, \pi_2, \pi_3, \pi_4)$	18.62***	29.84***	10.92***	18.08***	39.23***	7.59**

Note: c and T mean constant and time trend, respectively. Significance levels ***, ** and * correspond to 1%, 5% and 10% respectively.

Table 3. Cointegration test

	Engle and Granger test		Phillips and Ouliaris test		Bound test
	tau-statistic	z-statistic	tau-statistic	z-statistic	
Nigeria	−6.551105***	−52.66968***	−6.612777***	−53.22766***	6.461815***

Note: Significance levels ***, ** and * correspond to 1%, 5% and 10% respectively. The critical value of Bound test at 1% = 3.41–4.68

The estimated Markov switching ARDL models of Thai parboiled rice export to Nigeria in both short-run and long run are presented in Table 4. Prior we discuss our result, the Markov switching ARDL is compared to conventional ARDL to confirm our 2 regimes model in this study. To compare these two model, we consider Akaiki Information criteria (AIC) and the lowest AIC is preferred. According to Table 4, we can see that the AIC of Markov switching ARDL is lower than the conventional one, indicating the higher performance of our proposed model.

Based on regime separation by residual standard error of long-run model, the higher residual standard error is regime 1 and called upturn export situation, and lower residual standard error is regime 2 and called downturn export situation. In the long-run equilibrium model of regime 1, the significant variables are first lag of the export volume of Thai parboiled rice, first and second lag of Indian parboiled rice export prices, and first lag of exchange rate.

In regime 1, the cross-price elasticity between the export parboiled rice volume from Thailand to Nigeria and Indian export parboiled rice price from India to Nigeria was negative and inelastic. This is because in the first and third periods of regime 1 (2001Q4–2005Q1 and 2013Q4–2015Q1) although India has reduced rice export prices, Thailand also reduced rice export prices as well, so the price gap between two countries decreased. However, Thai parboiled rice quality is better than that of India making Nigeria prefer to import parboiled rice from Thailand rather than India. In the second period of regime 1 (2009Q1–2012Q1), since India abstained from exporting rice, then Thailand became the only major exporter in the world.

In the second regime, the own export price elasticity is significant and less than one thus corresponding to the law of demand and meaning that if export price of Thailand

Table 4. Markov switching ARDL model of Nigeria demand for Thai rice

Long-run			Short-run		
	Regime 1	*Regime 2*		*Regime 1*	*Regime 2*
Intercept	12.3263***	12.4984***	Intercept	−0.0312*	0.012
lnQThai(−1)	0.3799***	0.2181**	ΔlnQThai(−1)	−1.159***	−0.1749
lnPThai(−1)	0.5269	−0.8626	ΔlnPThai(−1)	2.2056***	−0.9228*
lnPIndia(−1)	−0.3424**	−1.0304**	ΔlnPIndia(−1)	0.1649	0.0442
lnPIndia(−2)	−0.3138*	0.0947	ΔlnPIndia(−2)	0.2001	0.0292
lnEX(−1)	−0.8752*	1.3248**	ΔlnEX(−1)	−2.2113*	−0.054
DTH	−0.0953	−1.5625***	ECM(−1)	0.05	−0.0705**
DIN	0.5344	0.2788	ECM(−2)	−0.151***	−0.079***
DNI	0.8034**	−2.3919***			
DS1	−0.9042***	−0.5394*			
DS2	−0.9092***	0.2518			
DS3	−0.109	−0.7331***			
SE(residual)	0.38	0.257		0.048	0.097
	Regime 1	*Regime 2*		*Regime 1*	*Regime 2*
Regime 1	0.8288	0.2055	*Regime 1*	0.6821	0.2436
Regime 2	0.1712	0.7945	*Regime 2*	0.3179	0.7564
AIC-MS-ADRL	154.7276				
AIC-ARDL	168.0578				

Note: Significance levels ***, ** and * correspond to 1%, 5% and 10% respectively.

decreases, the export volume should increase. In addition, the factors affecting Thai rice export are the same as those in regime one except for Indian parboiled rice export prices of the two-past quarter, and second quarter coefficient of dummy variable. In regime 2, the pledging scheme of Thai government dummy is also significant meaning pledging scheme caused Thai rice export volume to decrease significantly. In short run model only the speed of adjustment coefficient of both regime 1 and 2 are significant and between $(-1, 0)$ meaning long run relationship exists between Thailand export volume of parboiled rice and explanatory variables (Fig. 2).

Transition probability matrices, which are also reported in Table 4, provide evidence that the regimes are persistent and the ADRL parameters are affected by the regime. We denote the probabilities $\Pr(s(t) = 0 | s(t) = 0)$ by p_{00} and $\Pr(s(t) = 1 | s(t) = 1)$ by p_{11}. We can notice that both of the regimes are persistent because of the high values obtained for the probabilities $p_{11} = 0.8288$ and $p_{22} = 0.7945$. The duration corresponding to regime 1 and 2 are respectively 5.8411 and 4.8661 quarters $(1/p_{ij})$.

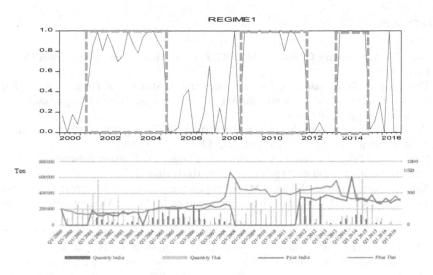

Fig. 2. Regime probabilities (upper panel) and parboiled rice volume (lower panel)

5 Conclusion

The study investigates the demand of Nigeria for Thai parboiled rice. Indeed, the behaviors of economic and agriculture time series may exhibit different patterns over time. Therefore, the Markov Switching ADRL approach is proposed in this study since it has the ability to capture and explain nonstationarity in variable and nonlinearity in model structure and it has recently played a prominent role in econometric research. The desirable feature of this approach is that it can deal with short run spurious regression problem, without the need for data transformation in order to achieve stationarity, using the error correction term. This error correction term which is estimation of the long-run part and the speed of adjustment to equilibrium and thereby having both long-run and short-run exist in the model.

Hence, these key strengths of the Markov Switching ADRL framework have been demonstrated in the case of the long- and short-run behaviors of the demand for Thai parboiled rice from Nigeria. The results suggest that the model can capture the two regimes of Thai parboiled rice demand namely rice market expansion and rice market contraction. We also find that long run relationship exists between Thailand export volume of parboiled rice and explanatory variables. In addition, we find that Thai parboiled price played an import role in Thai parboiled rice demand from Nigeria in both two regimes and also in both long run and short run. This means that the government needs to emphasize the price of rice and boost the demand for Thai rice using export price policy.

Acknowledgement. We are grateful for financial support from Puay Ungpakorn Centre of Excellence in Econometrics, Faculty of Economics, Chiang Mai University.

References

1. IMF.: World Economic Outlook, April 2017: Gaining Momentum? International Monetary Fund (2017)
2. Thai Custom: Trade Statistics, Customs Department, Ministry of Finance (2017). http://www.customs.go.th/statistic_report.php?ini_content=statistics_report&lang=en&left_menu=menu_report_and_news
3. Co, H.C., Boosarawongse, R.: Forecasting Thailand's rice export: statistical technique vs. artificial neural networks. Comput. Ind. Eng. **53**, 610–527 (2007)
4. Pattichis, C.: Price and income elasticities of disaggregated import demand: results from UECMs and an application. Appl. Econ. **31**(9), 1061–1071 (1999)
5. Kang, H., Kennedy, P.L., Hilbun, B.M.: An empirical estimation of the import demand model and welfare effects: the case of rice importing countries. In: Annual Meeting, Atlanta, Georgia (2009)
6. Pastpipatkul, P., Maneejuk, P., Sriboonchitta, S.: Welfare measurement on Thai rice market: a Markov switching Bayesian seemingly unrelated regression. In: Huynh, V.N., Inuiguchi, M., Demoeux, T. (eds.) International Symposium on Integrated Uncertainty in Knowledge Modelling and Decision Making, vol. 9376, pp. 464–477. Springer, Cham (2015). https://doi.org/10.1007/978-3-319-25135-6_42
7. Pastpipatkul, P., Panthamit, N., Yamaka, W., Sriboochitta, S.: A copula-based Markov switching seemingly unrelated regression approach for analysis the demand and supply on sugar market. In: Huynh, V.N., Inuiguchi, M., Le, B., Le, B., Denoeux, T. (eds.) IUKM 2016, vol. 9978, pp. 481–492. Springer, Cham (2016). https://doi.org/10.1007/978-3-319-49046-5_41
8. Tansuchat, R., Maneejuk, P., Wiboonpongse, A., Sriboonchitta, S.: Price transmission mechanism in the Thai rice market. In: Huynh, V.N., Kreinovich, V., Sriboonchitta, S. (eds.) Causal Inference in Econometrics, vol. 622, pp. 451–461. Springer, Cham (2016). https://doi.org/10.1007/978-3-319-27284-9_29
9. Pesaran, M.H., Shin, Y.: An autoregressive distributed lag modelling approach to cointegration analysis. In: Strom, S. (ed.) Econometrics and Economic Theory: The Ragnar Frisch Centennial Symposium. Cambridge University Press, Cambridge (1998)
10. Pesaran, M.H., Shin, Y., Smith, R.J.: Bounds testing approaches to the analysis of level relationships. J. Appl. Econom. **16**, 289–326 (2001)
11. Nkoro, E., Uko, A.K.: Autoregressive Distributed Lag (ARDL) cointegration technique: application and interpretation. J. Stat. Econom. Methods **5**(4), 63–91 (2016)
12. Hamilton, J.D.: A new approach to the economic analysis of nonstationary time series and the business cycle. Econometrica **57**, 357–384 (1989)
13. Goldfeld, S.M., Quandt, R.E.: A Markov model for switching regressions. J. Econom. **1**, 3–16 (1973)
14. Hamilton, J.D.: Time Series Analysis. Princeton University Press, Princeton (1994)
15. Engle, R.F., Granger, C.W.J.: Co-integration and error correction: representation, estimation and testing. Econometrica **55**, 251–276 (1987)
16. Phillips, P.C.B., Ouliaris, S.: Asymptotic properties of residual based tests for cointegration. Econometrica **58**, 165–193 (1990)
17. Toshiyuki, M.: The Chinese market and Thai fragrant jasmine rice: why does China, the world's largest rice producer, import rice from Thailand? (Japanese) (No. 11005) (2011)

The Role of Agricultural Commodity Prices in a Portfolio

Chatchai Khiewngamdee[1,3]($\boxtimes$), Quanrui Song[2], and Somsak Chanaim[2,3]

[1] Department of Agricultural Economy and Development, Faculty of Agriculture,
Chiang Mai University, Chiang Mai, Thailand
getliecon@gmail.com
[2] Faculty of Economics, Chiang Mai University, Chiang Mai 50200, Thailand
[3] Center of Excellence in Econometrics, Chiang Mai University,
Chiang Mai 50200, Thailand

Abstract. This paper aims to investigate whether including agricultural commodities can improve the portfolio performance by comparing the risk and return of multi-asset portfolio with and without an agricultural commodity price. To achieve our goal, we propose fitting a C-Vine copula based AR-GARCH model to interval data which allows us to capture uncertain characteristics that cannot be sometimes fully described with single data series. By using a convex combination method, we can obtain expected marginal distribution and joint density function, respectively. We then evaluate the portfolios' risk and return using the expected shortfall concept. The results present that the average risk and return of non-agricultural portfolio outperforms agricultural portfolios. However, considering the one step ahead forecasting efficient frontier, the portfolio with soybean futures becomes superior to other portfolios.

Keywords: Interval data · Convex combination · GARCH
Expected shortfall

1 Introduction

Since the introduction of the commodity futures and commodity index funds, investments in commodity markets have emerged as an attractive asset class and grown rapidly over the last decades. The reason is that investors have gained the potential diversification benefits from investing in commodities. It is due to the low correlations of commodities with traditional assets, namely stocks and bonds, and hence provide diversification benefits in a mixed-asset portfolio (Paraschiv et al. [1]). Geman [2] and Daskalaki et al. [3] also explain the low correlation between commodities and stocks is because risk factors that drive the commodity prices, such as weather, geopolitical events, and supply conditions, are distinct from those determine the value of stocks and bonds. Moreover, commodities are usually seen as an inflation hedge (Bodie and Rosansky, [4]; Erb and Harvey, [5]; Gorton and Rouwenhorst, [6]). As a result, many investors and hedge funds include commodities into their portfolios with the aim of diversifying and reducing the downside risk.

© Springer International Publishing AG, part of Springer Nature 2018
V.-N. Huynh et al. (Eds.): IUKM 2018, LNAI 10758, pp. 385–396, 2018.
https://doi.org/10.1007/978-3-319-75429-1_32

There are a large number of studies suggesting that including commodities into portfolio can help improve the portfolio's performance. For instances, Conover et al. [7], Jensen et al. [8], and Idzorek [9] found a significant benefit from adding commodities into an equity portfolio. Fortenbery and Hauser [10] also found that the addition of agricultural futures contracts to the portfolio can reduce the portfolio's risk. Furthermore, Anson [11] and Georgiev [12] performed portfolio optimization and found that adding commodities to a portfolio of stocks and bonds increases the Sharpe ratio of optimal portfolios.

Recently, portfolio analysis has concerned the co-movement between assets since it can improve the estimated accuracy for portfolio's risk and return. Many studies, for examples, Cheung and Miu [13], Daskalaki and Skiadopoulos [14], and Silvennoinen and Thorpy [15] show evidence of increased co-movement between the commodity and stock markets because commodities are being hugely financialized. Therefore, copula approach, which provides a measure of market co-movements, has been used in many portfolio optimization problems. For instances, Hürlimann [16], Wei and Zhang [17], He and Gong [18] and Wang et al. [19] employed multivariate copula based GARCH models to evaluate the risk of the portfolio. They found that copula based GARCH model is practicable and more effective in capturing the dependence structure among asset returns and hence accurately measures the risks.

This paper aims to investigate whether including commodities, in particular, agricultural futures can improve portfolio performance. Our portfolio considers S&P 500 future, Japanese yen, gold futures, and Brent crude futures including agricultural futures namely, corn, wheat, and soybean. We apply interval valued data and C-vines copula approach. Additionally, we handle interval valued data using convex combination introduced by Somsak et al. [20]. We first employ GARCH model to filter the asset returns and then estimate joint distribution among asset returns using C-vines copula approach. Finally, we evaluate the risk and return of the portfolios based on the concept of expected shortfall and find the optimal weights of the portfolios.

The reminder of this paper is constructed as follows. Section 2 presents the methodology used in this study. The data used in this study are explained in Sect. 3. Section 4 provides the empirical results and final section gives conclusions.

2   Methodology

Let $X = [\underline{X}, \overline{X}]$ and $Y = [\underline{Y}, \overline{Y}]$ are the interval data. If $\underline{X}, \underline{Y} > 0$ and $\overline{X}, \overline{Y} < \infty$. we can define division by

$$\frac{1}{X} = [\frac{1}{\overline{X}}, \frac{1}{\underline{X}}],$$

$$\frac{X}{Y} = [\frac{\underline{X}}{\overline{Y}}, \frac{\overline{X}}{\underline{Y}}].$$

For logarithm function is

$$\log(X) = [\log(\underline{X}), \log(\overline{X})].$$

For more details about interval arithmetics see Moore et al. ([21], Chaper 2) or Nguyen et al. [22]. We can define interval return r_t by

$$r_t = [\underline{r}_t, \overline{r}_t] = \left[\log\left(\frac{\overline{s}_{t+1}}{\overline{s}_t}\right), \log\left(\frac{\overline{s}_{t+1}}{\underline{s}_t}\right)\right],$$

where $\underline{r}_t$ is a lower return, $\overline{r}_t$ is a higher return, $\underline{s}_t$ and $\overline{s}_t$ are stock price or index of the stock market at lower and higher level respectively at time t.

2.1   Autoregressive - GARCH Model

Many previous studies suggested that volatility of financial return data is not constant over time, but is rather clustered. This issue can be tackled using volatility modeling. Within a class of autoregressive processes with white noises having conditional heteroscedastic variances, this paper considers a GARCH (1,1) model to estimate the dynamic volatility. It is the workhorse model and mostly applied in many financial data. The model is able to reproduce the volatility dynamics of financial data. Thus, in this study, we consider AR(p)-GARCH(1,1) which can be written as

$$r_t = \phi_0 + \sum_{i=1}^{p} \phi_i r_{t-i} + \epsilon_t \tag{1}$$

$$\epsilon_t = \sigma_t \eta_t \tag{2}$$

$$\sigma_t^2 = \omega_0 + \omega_1 \sigma_{t-1}^2 + \omega_2 \epsilon_{t-1}^2 \tag{3}$$

where $\phi_0, \phi_1, \cdots, \phi_p \in \mathbb{R}$ and the root of polynomial $x^p - \phi_1 x^{p-1} - \cdots - \phi_p$ lie inside of the unit circle. η_t is t distribution with degree of freedom ν, σ_t^2 is the conditional variance in GARCH process by Bollerslev [23]. Some conditions on the parameters are given by

$$\omega_0 > 0, \ \omega_1 > 0, \ \omega_2 > 0, \ \omega_1 + \omega_2 \ < 1.$$

2.2   Convex Combination Method

The convex combination method is applied to deal with the interval return data, where the appropriate value over the range of interval can be computed by

$$r_t = \alpha_1 \underline{r}_t + (1 - \alpha_1)\overline{r}_t,$$
$$r_{t-1} = \alpha_2 \underline{r}_{t-1} + (1 - \alpha_2)\overline{r}_{t-1},$$

$$\vdots$$

$$r_{t-p} = \alpha_{p+1}\underline{r}_{t-p} + (1 - \alpha_{p+1})\overline{r}_{t-p},$$

where $\alpha_1, \alpha_2, \cdots, \alpha_{p+1} \in [0, 1]$. In this study, we consider the cases $\alpha_1 = \alpha_2 = \cdots = \alpha_{p+1} = \alpha$ and we consider the discrete value of $\alpha = \{0, 0.01, 0.02, \cdots, 0.00, 1\}$ for each AR(p) GARCH(1,1) model. We can use standard software in finance or econometrics to estimate value of parameter. Then, we select the best model at weight α by using the minimum value of Akaike information criterion (AIC). For example, in the case of AR(2)-GARCH(1,1), we can rewrite the equation into

$$\alpha \underline{r}_t + (1 - \alpha)\bar{r}_t = \phi_0 + \phi_1 \left(\alpha \underline{r}_{t-1} + (1 - \alpha)\bar{r}_{t-1}\right) + \phi_2 \left(\alpha \underline{r}_{t-2} + (1 - \alpha)\bar{r}_{t-2}\right) + \epsilon_t$$

where

$$r_t = \phi_0 + \phi_1 r_{t-1} + \phi_2 r_{t-2} + \varphi_1 \epsilon_{t-1} + \epsilon_t,$$
$$\epsilon_t = \sigma_t \eta_t, \eta_t \sim t(\nu),$$
$$\sigma_t^2 = \omega_0 + \omega_1 \sigma_{t-1}^2 + \omega_2 \epsilon_{t-1}^2, \quad \omega_0, \omega_1, \omega_2 > 0, \quad \omega_1 + \omega_2 < 1.$$

2.3   Copula

In this paper, we use Canonical vines copula (C-vines) to approximate joint density of portfolio return. For 4 dimensional, Canonical vines copula has the form

$$h(r_1, r_2, r_3, r_4) = \left(\prod_{i=1}^{4} f_i(r_i)\right) \times c(F_1(r_1), F_2(r_2), F_3(r_3), F_4(r_4)), \qquad (4)$$

$$c(F_1(r_1), \cdots, F_4(r_4)) = \left(\prod_{j=2}^{4} c(r_1, r_j)\right) \times \left(\prod_{j=3}^{4} c(r_2, r_j | r_1)\right) \times c(r_3, r_4 | r_1, r_2),$$
$$(5)$$

and 5 dimensional, Canonical vines copula has the form

$$h(r_1, r_2, r_3, r_4, r_5) = \left(\prod_{i=1}^{5} f_i(r_i)\right) \times c(F_1(r_1), F_2(r_2), F_3(r_3), F_4(r_4), F_5(r_5)), \quad (6)$$

$$c(F_1(r_1), \cdots, F_5(r_5)) = \left(\prod_{j=2}^{5} c(r_1, r_j)\right) \times \left(\prod_{j=3}^{5} c(r_2, r_j | r_1)\right) \times \left(\prod_{j=4}^{5} c(r_3, r_j | r_1, r_2)\right)$$
$$\times c(r_4, r_5 | r_1, r_2, r_3), \qquad (7)$$

where $f_i(\cdot)$ $F_i(\cdot)$, $i = 1, 2, \cdots, 5$ are probability density function and cumulative distribution function respectively, $c(\cdot, \cdot)$ is a bivariate copula density function and $c(\cdot | \cdot)$ is the conditional bivariate copula density function. For details about copula and C-vines copula see Nelson [24] and Aas et al. [25].

2.4   Optimal Portfolio with Expected Shortfall

As the expected shortfall or conditional value at risk (CVaR) is a coherent risk measure, risk measurement in this way can help us diversify the risk. We start our calculation of CVaR of an equally weighted portfolio and then, the optimal portfolio can be constructed by minimizing CVaR subject to maximum returns. For the procedure of optimization, we refer to the paper by Autchariyapanitkul [26]. The following formula is used:

$$\min CVaR = E[r_p | r \leq r_\theta],$$

subject to

$$E(r_p) = w_1 E(r_1) + w_2 E(r_2) + \cdots + w_n E(r_n),$$

$$\sum_{i=1}^{n} w_i = 1, \ w_i \in [0, 1], \ \forall i = 1, \cdots, n,$$

where r_θ is the lower $\theta - quantile$, and r_i is the return on individual asset at time $t + 1$. We use C-vines copula to estimate dependence structure between return for stock price, index, exchange rate or commodity price and used the structure of C-vines copula to simulate the portfolio. Finally, we find the optimal solutions for the expected returns with minimum CVaR.

3   Data

This study uses the weekly log return of S&P 500 futures index, Japanese yen, gold futures price, Brent crude oil futures price, corn futures price, wheat futures price, and soybean futures price collected from Thomson Reuters Database. Our sample covers the period from June 24th, 1988 to June 9th, 2017 totally 1,512 observations. We divide the data into four portfolios :

Table 1. Summary statistics

	S&P500		JPY		Gold		Brent	
	r_l	$\bar{r}_l$	r_l	$\bar{r}_l$	r_l	$\bar{r}_h$	r_l	$\bar{r}_h$
mean	−0.029	0.0319	−0.0239	0.0237	−0.0253	0.0267	−0.064	0.0655
median	−0.0212	0.0273	−0.0196	0.0204	−0.0196	0.0221	−0.0512	0.0566
variance	0.0283	0.0225	0.0191	0.0154	0.0241	0.0234	0.0535	0.0454
min	−0.3644	−0.0068	−0.2749	−0.0024	−0.2346	−0.0423	−0.5013	−0.0163
max	0.0035	0.2179	0.001	0.1553	0.051	0.2294	0.0474	0.395
skewness	−3.3283	2.2984	−3.5635	2.0689	−2.0783	2.0114	−2.2788	1.6822
kurtosis	24.7688	12.6325	31.774	11.9281	11.747	12.2492	12.3635	8.3188

	Corn		Wheat		Soybean	
	r_l	$\bar{r}_h$	r_l	$\bar{r}_h$	r_l	$\bar{r}_h$
mean	−0.0495	0.0496	−0.0158	0.0161	−0.0572	0.0573
median	−0.0408	0.0405	−0.0148	0.0137	−0.05	0.0485
variance	0.0413	0.0389	0.0367	0.0371	0.0401	0.0421
min	−0.4452	−0.2185	−0.231	−0.1709	−0.4463	−0.012
max	0.0485	0.3054	0.134	0.1787	0.0156	0.4155
skewness	−2.5304	1.2313	−0.2619	0.3034	−1.9387	1.8263
kurtosis	16.3267	7.8436	5.1371	5.0293	11.7873	10.0368

(1) Portfolio 1: S&P 500 futures index, Japanese yen, gold futures price, and Brent crude oil futures price.
(2) Portfolio 2: Portfolio 1 adding corn futures price
(3) Portfolio 3: Portfolio 1 adding wheat futures price
(4) Portfolio 4: Portfolio 1 adding soybean futures price

The summary statistics of the interval asset returns are presented in Table 1.

Note that the model used here is the C-vines copula based AR-GARCH model. This model can be estimated using a step-wise procedure. We first specify the marginal distributions for each return using AR-GARCH with student-t distribution. Next, we look for the best C-vines copula structure and the combination of bivariate copula function for each pair.

Since we consider the interval valued data of our asset returns, we apply convex combination approach to AR-GARCH model in order to obtain the expected value of the interval volatility as well as the marginal distribution. To find the best AR-GARCH specification, the Akaiki Information criterion (AIC) is conducted and the lowest value is preferred.

4   Empirical Results

4.1   Marginal Distribution

In this first step, the marginal parameters are estimated using AR-GARCH with convex combination, then we transform the standardized residuals into uniform

Table 2. Parameter estimate

	S&P500		JPY		Gold		Brent	
	value	SD.	value	SD.	value	SD.	value	SD.
weight	0.42		0.38		0.5		0.42	
constant	0.0055	0.0004	0.0037	0.0003	0.0002	0.0003	0.0064	0.0007
AR(1)	0.2125	0.0267	0.3267	0.0266	0.293	0.0274	0.307	0.0269
AR(2)	−0.1027	0.0268	−0.0958	0.0271	−0.1157	0.0275	−0.0983	0.0277
AR(3)							0.0762	0.0269
ω_0	6.5×10^{-6}	1.76×10^{-6}	2.9×10^{-6}	1.52×10^{-6}	4.77×10^{-6}	2.34×10^{-6}	2.24×10^{-5}	8.73×10^{-6}
ω_1	0.858	0.0211	0.8942	0.0197	0.8828	0.0218	0.877	0.0179
ω_2	0.1213	0.01975	0.0834	0..0148	0.1055	0.0203	0.1088	0.017
df	7.6528	1.4071	10.7753	2.4608	8.2029	0.9954	10.6941	2.2137
LogLike	4,243.10		4,735.88		4,186.64		3,063.63	
AIC	−8,472.21		−9,457.77		−8,359.28		−6,111.26	

	Corn		Wheat		Soybean	
	value	SD.	value	SD.	value	SD.
weight	0.54		0.56		0.54	
constant	−0.0019	0.0006	−0.0018	0.0008	−0.0036	0.0007
AR(1)	0.2589	0.0278	0.023	0.0268	0.2991	0.0269
AR(2)	−0.0499	0.9575			−0.1212	0.0279
AR(3)					0.0288	0.0292
AR(4)					−0.0537	0.0265
ω_0	7.11×10^{-5}	1.76×10^{-5}	3.423×10^{-5}	1.33×10^{-5}	4.8×10^{-5}	1.54×10^{-5}
ω_1	0.7181	0.0388	0.8971	0.0225	0.8421	0.0299
ω_2	0.2151	0.0337	0.0783	0.0171	0.1089	0.0196
df	6.5847	0.9575	7.766	1.4563	11.9192	2.4438
LogLike	3,334.43		2,980.87		3,195.48	
AIC	−6,654.87		−5,949.74		−6,371.60	

distribution [0,1]. These marginals are then used as inputs to the Copula selection routine. The parameter estimates of AR-GARCH model are shown in Table 2. According to the AR-GARCH parameter estimation, the best fit AR-GARCH specification for S&P 500 futures index, Japanese yen, gold futures price, and corn futures price is AR(2)-GARCH(1,1) and for wheat futures price, Brent crude oil futures price, and soybean futures price is AR(1)-GARCH(1,1), AR(3)-GARCH(1,1), and AR(4)-GARCH(1,1), respectively.

4.2 Dependence Modelling Using C-Vines Copula Estimation Results

We use C-vine copula to map dependence structures between some of the returns in four portfolios. These, in turn, can be used for portfolio evaluation and risk modelling. In this second step, we fit a C-vine copula structure to the set of the standardized residuals while taking into account the constraints imposed by the factorial structure we retain a prior. The bivariate copulas are chosen from a set of families, namely Elliptical (Gaussian and Student t) as well as Archimedean (Clayton, Gumbel, Frank, Joe, BB1, BB6, BB7 and BB8) copulas to cover all of possible dependence structures. In this study, the rotated Archimedean copula families are included to cover negative dependence as well.

Table 3 presents the approximate joint density function with C-vine copula for each portfolio. According to the lowest AIC value, the appropriate joint copula density for our portfolios are selected. The results show that, for the portfolio consisting of S&P 500 futures index, Japanese yen, gold futures price, and Brent crude oil futures price (Portfolio 1), the best order joint copula density function is in a form, referred to Eqs. 5, 7, that $c(u_{r_{JYP}}, u_{r_{S\&P500}}, u_{r_{comex}}, u_{r_{Brent}})$. Additionally, for the following portfolios: Portfolio 2, Portfolio 3 and Portfolio 4, the most appropriate joint copula density functions are, respectively, $c(u_{r_{JYP}}, u_{r_{S\&P500}}, u_{r_{corn}}, u_{r_{comex}}, u_{r_{Brent}})$, $c(u_{r_{wheat}}, u_{r_{S\&P500}}, u_{r_{JYP}}, u_{r_{comex}}, u_{r_{Brent}})$, $c(u_{r_{S\&P500}}, u_{r_{soybean}}, u_{r_{JYP}}, u_{r_{comex}}, u_{r_{Brent}})$. Note that, the value of second estimated parameter, which is degree of freedom, for $c(u_{r_{S\&P500}}, u_{r_{soybean}})$ and $c(u_{r_{S\&P500}}, u_{r_{comex}})$ in Portfolio 4 are equal to 30. This indicates that we can use Gaussian copula as alternative to student-t copula. The dependency parameters are shown in par 1 and par 2 column. The result shows that the dependency of all pairs are mostly positive, indicating a positive relationship in these returns.

Then, the co-dependencies calculated by C-vine copula can be used for portfolio Value at Risk quantification. We construct an equally weighted portfolio to explore the use of C-vine copula in modelling CVaR of our portfolios. We select subsample of size 1,512, dated from June 24th, 1988 to February 24th, 2014, as the training set for the parameters estimation for models and the remaining sample of size 500 weekly data, from September 17th, 2007 to June 9th, 2017 is used as the test set or for out-of sample forecasting. We use a 500 weeks moving window approach to forecast the CVaR for this equally weighted portfolio which results in 500 forecasts for each portfolio.

Table 3. C-vine copula

Joint copula density function Portfolio 1	copula family	par 1	sd	par 2	sd	AIC
$c(u_{r_{JYP}}, u_{r_{S\&P500}})$	t	0.2630	0.0314	4.1474	0.6835	−124.86
$c(u_{r_{JYP}}, u_{r_{comex}})$	t	0.4372	0.0236	8.9983	2.7543	−229.03
$c(u_{r_{JYP}}, u_{r_{Brent}})$	BB1	0.1025	0.0494	1.1242	0.0298	−66.05
$c(u_{r_{S\&P500}}, u_{r_{comex}} \| u_{r_{JYP}})$	Gaussian	−0.1339	0.0298			−17.06
$c(u_{r_{S\&P500}}, u_{r_{Brent}} \| u_{r_{JYP}})$	t	0.0605	0.0325	7.4807	2.0089	−14.72
$c(u_{r_{comex}}, u_{r_{Brent}} \| u_{r_{JYP}}, u_{r_{S\&P500}})$	Gaussian	0.1615	0.0280			−28.86
Total AIC						**-480.59**
Portfolio 2						
$c(u_{r_{JYP}}, u_{r_{S\&P500}})$	t	0.2630	0.0314	4.1474	0.6835	−124.86
$c(u_{r_{JYP}}, u_{r_{corn}})$	survival BB7	1.1145	0.0383	0.1838	0.0478	−43.40
$c(u_{r_{JYP}}, u_{r_{comex}})$	t	0.4372	0.0236	8.9983	2.7543	−229.03
$c(u_{r_{JYP}}, u_{r_{Brent}})$	BB1	0.1025	0.0494	1.1242	0.0298	−66.05
$c(u_{r_{S\&P500}}, u_{r_{corn}} \| u_{r_{JYP}})$	Frank	0.6510	0.1831			−10.59
$c(u_{r_{S\&P500}}, u_{r_{comex}} \| u_{r_{JYP}})$	Gaussian	−0.1339	0.0298			−17.06
$c(u_{r_{S\&P500}}, u_{r_{Brent}} \| u_{r_{JYP}})$	t	0.0605	0.0325	7.4807	2.0089	−14.72
$c(u_{r_{corn}}, u_{r_{comex}} \| u_{r_{JYP}}, u_{r_{S\&P500}})$	Frank	0.8857	0.1804			−21.91
$c(u_{r_{corn}}, u_{r_{Brent}} \| u_{r_{JYP}}, u_{r_{S\&P500}})$	survival Joe	1.0908	0.0313			−8.55
$c(u_{r_{comex}}, u_{r_{Brent}} \| u_{r_{JYP}}, u_{r_{S\&P500}}, corn)$	Gaussian	0.1539	0.0278			−26.63
Total AIC						**-562.82**
Portfolio 3						
$c(u_{r_{wheat}}, u_{r_{S\&P500}})$	survival Joe	1.0426	0.0292			−0.52
$c(u_{r_{wheat}}, u_{r_{JYP}})$	Gaussian	0.1503	0.0312			−19.48
$c(u_{r_{wheat}}, u_{r_{comex}})$	t	0.7203	0.0123	7.0123	1.3433	−916.23
$c(u_{r_{wheat}}, u_{r_{Brent}})$	survival Joe	1.1055	0.0321			−12.16
$c(u_{r_{S\&P500}}, u_{r_{JYP}} \| u_{r_{wheat}})$	t	0.2615	0.0311	4.3584	0.7402	−120.55
$c(u_{r_{S\&P500}}, u_{r_{comex}} \| u_{r_{wheat}})$	Frank	0.5428	0.1667			−8.59
$c(u_{r_{S\&P500}}, u_{r_{Brent}} \| u_{r_{wheat}})$	t	0.1179	0.0332	5.3435	1.0896	−36.69
$c(u_{r_{JYP}}, u_{r_{comex}} \| u_{r_{wheat}}, u_{r_{S\&P500}})$	Frank	0.6043	0.1588			−12.48
$c(u_{r_{JYP}}, u_{r_{Brent}} \| u_{r_{wheat}}, u_{r_{S\&P500}})$	Gaussian	0.1943	0.0269			−44.98
$c(u_{r_{comex}}, u_{r_{Brent}} \| u_{r_{wheat}}, u_{r_{S\&P500}}, u_{r_{JYP}})$	BB8	1.2383	0.2065	0.8028	0.2241	−5.71
Total AIC						**-1177.39**
Portfolio 4						
$c(u_{r_{S\&P500}}, u_{r_{soybean}})$	t	0.0989	0.0323	30.0000	NA	61.96
$c(u_{r_{S\&P500}}, u_{r_{JYP}})$	t	0.2630	0.0314	4.1474	0.6835	−21.85
$c(u_{r_{S\&P500}}, u_{r_{comex}})$	t	0.0989	0.0323	30.0000	NA	16.05
$c(u_{r_{S\&P500}}, u_{r_{Brent}})$	Gaussian	0.1242	0.0318			−11.94
$c(u_{r_{soybean}}, u_{r_{JYP}} \| u_{r_{S\&P500}})$	Frank	0.9075	0.1646			179.21
$c(u_{r_{soybean}}, u_{r_{comex}} \| u_{r_{S\&P500}})$	Gaussian	0.9999	0.0001			−538.51
$c(u_{r_{soybean}}, u_{r_{Brent}} \| u_{r_{S\&P500}})$	Frank	0.5738	0.1723			−1.02
$c(u_{r_{JYP}}, u_{r_{comex}} \| u_{r_{S\&P500}}, u_{r_{soybean}})$	Gaussian	−0.3168	0.0516			148.17
$c(u_{r_{JYP}}, u_{r_{Brent}} \| u_{r_{S\&P500}}, u_{r_{soybean}})$	survival Gumbel	1.1140	0.0226			−30.81
$c(u_{r_{comex}}, u_{r_{Brent}} \| u_{r_{S\&P500}}, u_{r_{soybean}}, u_{r_{JYP}})$	survival Joe	1.0926	0.0479			4.82
Total AIC						**-193.91**

Calculation by using package "CDvine" by Brechmann and Schepsmeier [27] in program R.

The main steps of the approach are as follows:

1. After we estimate a C-vine copula, we generate a simulation using the fitted C-vine copula model. We generate 100,000 simulations per return for forecasting a week ahead CVaR. 2. Convert the simulated uniform marginals to standardized residuals. 3. Simulate returns from the simulated standardized residuals from the best fit AR-GARCH with convex combination (see, Sect. 4.1). 4. Generate a

series of simulated weekly portfolio returns to forecast 5% CVaR. 5. Repeat step 1 to 5 for a moving window. The result is explained in the next section.

4.3 Portfolio Allocation

As we mentioned before, we measure the returns and risks of the equally weighted portfolio by employing the expected shortfall (ES) and use the principle of the weekly rolling window forecasting. A plot of the results of this empirical study is shown in Figs. 1 and 2. This presents that the 5% CVaR forecasts and return of four portfolios obtained from our method share similar movements. We also notice that the return of all portfolios was reached the lowest return as well as a highest CVaR around 2008, corresponding to the financial crisis in 2008. Moreover, we also observe that CVaR of Portfolio 2 was suddenly high in 2013, corresponding to European debt crisis. Note that Portfolio 2 consists of S&P 500 futures index, Japanese yen, gold futures price, Brent crude oil futures price, corn futures price. Therefore, we can say that corn futures price cannot be viewed as a hedge asset reducing a risk of portfolio.

Finally, we report the average values of expected return and risk, as shown in Table 4. The results show that Portfolio 1, which does not include any agri-

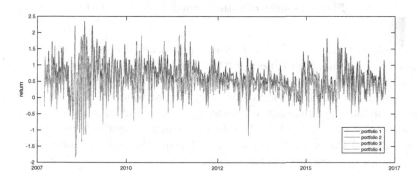

Fig. 1. Rolling window forecast return of portfolio with equal weight

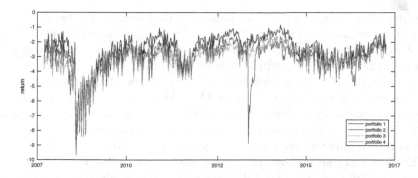

Fig. 2. Rolling window forecast CVaR of portfolio with equal weight

Table 4. Mean of expected return and mean of expected CVaR

	Mean of expected return	Mean of expected CVaR
Portfolio 1	0.5235	−2.5019
Portfolio 2	0.4955	−3.2258
Portfolio 3	0.4308	−3.2982
Portfolio 4	0.4881	−2.8468

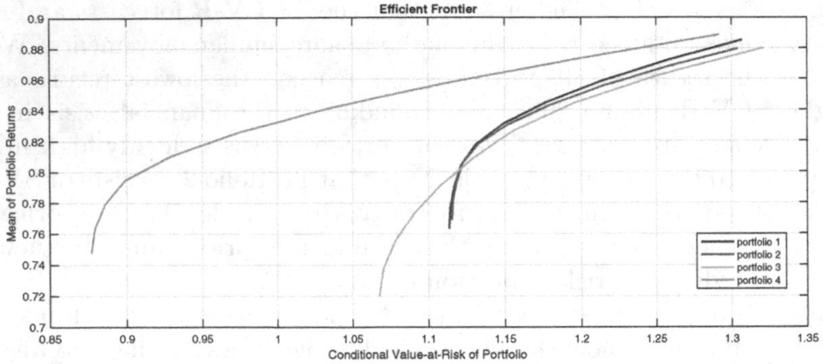

Fig. 3. Frontier of optimal portfolio at time t+1

cultural assets, yields the highest average return and lowest average risk. This means that, for equally weighted portfolio, including agricultural futures does not help improve the portfolio performance.

For optimal weighted portfolio, however, the portfolio which includes soybean futures (Portfolio 4) obviously outperforms other portfolios according to the efficient frontier in Fig. 3, while the efficient frontiers resulted from Portfolio 1, Portfolio 2, and Portfolio 3 are quite indifferent. We estimate one step ahead forecasting for optimal weighted portfolio.

5   Conclusions

The objective of this paper is to examine whether including agricultural commodities can improve portfolio performance. Our original portfolio analysis consists of S&P 500 futures index, Japanese yen, gold futures price, and Brent crude oil futures price. Then, we include corn futures price, wheat futures price, and soybean futures price as an additional asset to our portfolio, respectively. We apply convex combination approach to AR-GARCH model to get the value of the interval asset returns. Then, using C-vine copula approach, we approximate the joint density function among asset returns. Finally, we measure the portfolio risk and return applying the concept of expected shortfall.

The empirical results show that, by employing the rolling window approach, the average return and risk of equally weighted portfolio which excludes an agricultural futures price outperforms any other agricultural portfolios. Furthermore, we estimate the efficient frontier of our portfolios using one step ahead forecasting and find that the portfolio with soybean futures price becomes superior to other portfolios as it yields higher return at the same level of risk.

Acknowledgement. The authors would like to thank Mr. Woraphon Yamaka for a rolling window suggestions. Last but not least, we would like to thank all the referees for giving comments on the manuscript.

References

1. Paraschiv, F., Mudry, P.A., Andries, A.M.: Stress-testing for portfolios of commodity futures. Econ. Model. **50**, 9–18 (2015)
2. Geman, H.: Commodities and Commodity Derivatives: Modeling and Pricing for Agriculturals Metals and Energy. Wiley, Hoboken (2009)
3. Daskalaki, C., Kostakis, A., Skiadopoulos, G.: Are there common factors in individual commodity futures returns? J. Bank. Financ. **40**, 346–363 (2014)
4. Bodie, Z., Rosansky, V.I.: Risk and return in commodity futures. Financ. Anal. J. **36**(3), 27–39 (1980)
5. Erb, C.B., Harvey, C.R.: The strategic and tactical value of commodity futures. Financ. Anal. J. **62**(2), 69–97 (2006)
6. Gorton, G., Rouwenhorst, K.G.: Facts and fantasies about commodity futures. Financ. Anal. J. **62**(2), 47–68 (2006)
7. Conover, C.M., Jensen, G.R., Johnson, R.R., Mercer, J.M.: Is now the time to add commodities to your portfolio? J. Invest. **19**(3), 10–19 (2010)
8. Jensen, G.R., Johnson, R.R., Mercer, J.M.: Efficient use of commodity futures in diversified portfolios. J. Futur. Mark. **20**(5), 489–506 (2000)
9. Idzorek, T. M.: Commodities and strategic asset allocation. In: Intelligent Commodity Investing, pp. 113–177 (2007)
10. Fortenbery, T.R., Hauser, R.J.: Investment potential of agricultural futures contracts. Am. J. Agric. Econ. **72**(3), 721–726 (1990)
11. Anson, M.J.: Maximizing utility with commodity futures diversification. J. Portf. Manag. **25**(4), 86–94 (1999)
12. Georgiev, G.: Benefits of commodity investment. J. Altern. Invest. **4**(1), 40–48 (2001)
13. Cheung, C.S., Miu, P.: Diversification benefits of commodity futures. J. Int. Financ. Mark., Inst. Money **20**(5), 451–474 (2010)
14. Daskalaki, C., Skiadopoulos, G.: Should investors include commodities in their portfolios after all? New evidence. J. Bank. Financ. **35**(10), 2606–2626 (2011)
15. Silvennoinen, A., Thorp, S.: Financialization, crisis and commodity correlation dynamics. J. Int. Financ. Mark., Inst. Money **24**, 42–65 (2013)
16. Hürlimann, W.: Multivariate Frchet copulas and conditional value-at-risk. Int. J. Math. Math. Sci. **2004**(7), 345–364 (2004)
17. Wei, Y.H., Zhang, S.Y.: Multivariate Copula-GARCH Model and its applications in financial risk analysis. App. Stat. Manag. **3**(008) (2007)
18. He, X., Gong, P.: Measuring the coupled risks: a copula-based CVaR model. J. Comput. Appl. Math. **223**(2), 1066–1080 (2009)
19. Wang, Z.R., Chen, X.H., Jin, Y.B., Zhou, Y.J.: Estimating risk of foreign exchange portfolio: using VaR and CVaR based on GARCHEVT-Copula model. Phys. A: Stat. Mech. Appl. **389**(21), 4918–4928 (2010)

20. Chanaim, S., Sriboonchitta, S., Rungruang, C.: A convex combination method for linear regression with interval data. In: Huynh, V.-N., Inuiguchi, M., Le, B., Le, B.N., Denoeux, T. (eds.) IUKM 2016. LNCS (LNAI), vol. 9978, pp. 469–480. Springer, Cham (2016). https://doi.org/10.1007/978-3-319-49046-5_40
21. Moore, R.E., Kearfott, R.B., Cloud, M.J.: Introduction to Interval Analysis, pp. 7–18. SIAM, Philadelphia (2009)
22. Nguyen, H.T., Kreinovich, V., Wu, B., Xiang, G.: Computing statistics under interval and fuzzy uncertainty. Studies in Computational Intelligence, vol. 393. Springer, Heidelberg (2012). https://doi.org/10.1007/978-3-642-24905-1
23. Bollerslev, T.: Generalized autoregressive conditional heteroskedasticity. J. Econom. **31**(3), 307–327 (1986)
24. Nelsen, R.B.: Introduction. An Introduction to Copulas. Springer, New York (1999). https://doi.org/10.1007/978-1-4757-3076-0
25. Aas, K., Czado, C., Frigessi, A., Bakken, H.: Pair-copula constructions of multiple dependence. Insur.: Math. Econ. **44**(2), 182–198 (2009)
26. Autchariyapanitkul, K., Chanaim, S., Sriboonchitta, S.: Portfolio optimization of stock returns in high-dimensions: a copula-based approach. Thai J. Math. 11–23 (2014)
27. Brechmann, E.C., Schepsmeier, U.: Modeling dependence with C-and D-vine copulas: the R-package CDVine. J. Stat. Softw. **52**(3), 1–27 (2013)

Assessing Consumers' Perceptions of Packaging Attributes and Packaging Label of Community's 'Green' Chili (Pepper) Products in Thailand

Jakkreeporn Sannork[1]([⊠]) [iD], Aree Wiboonpongse[2,3,4],
and Tzong-Ru Lee[5]

[1] Agricultural Systems Management Program, Faculty of Agriculture,
Center for Agricultural Resource System Research, Chiang Mai University,
Chiang Mai, Thailand
nn_cmu@hotmail.com
[2] Faculty of Economics, Prince of Songkla University,
Hatyai, Songkhla, Thailand
[3] Faculty of Agriculture, Chiang Mai University, Chiang Mai, Thailand
[4] Faculty of Economics, Centre of Excellence in Econometrics,
Chiang Mai University, Chiang Mai, Thailand
[5] Department of Marketing, College of Management,
National Chung Hsing University, Taichung 402, Taiwan R.O.C.

Abstract. This study aims to assess and identify the key factors that contribute in developing packaging attributes and label information of 'green' chili pepper products of farmers' community enterprises to make them more appealing to consumers in Thailand. The study was conducted in five provinces in Thailand. The data were obtained through interviewing 200 respondents using a standardized questionnaire. Gray Relational Analysis (GRA) and Binary Logit Regression model were applied to analyze the data. The analyses show that consumers pay more attention on information label than packaging attributes of the product when they buy the product. It was also found that consumers give importance to safety features of the product when they select the product. Therefore, the producers of 'green' chili products should include food safety features on packaging label of their product. The study findings might be beneficial to farmers in branding and selling their products with greater success.

Keywords: Consumers' perception · Attributes · Packaging label
Community product ('green' chili)

1 Introduction

Chili pepper is an important element in cuisines of many nations, particularly in South Asia and Southeast Asia including Thailand [4, 20]. Therefore, chili pepper cultivation is a part of Thai agriculture [20]. Many small farmers in different areas of the country have formed groups as a community business to produce and sell this product in the market. However, now-a-days, there is an increasingly and rapidly growing tendency of the

© Springer International Publishing AG, part of Springer Nature 2018
V.-N. Huynh et al. (Eds.): IUKM 2018, LNAI 10758, pp. 397–407, 2018.
https://doi.org/10.1007/978-3-319-75429-1_33

consumers to select and buy safe food or organic products [3, 18, 30]. Accordingly, some farmer groups produce and sell safe chili pepper products. However, in most of the cases, cultivation of safe/organic products cannot be continued as a common practice due to several reasons. Firstly, farming safe/organic products are expensive. For example, the cultivation of chili peppers without using pest and disease control chemicals is extremely difficult as chili peppers are highly prone to pests and diseases [13]. Secondly, safe/organic products cannot be sold in the market at a higher price than the conventionally produced outputs. Moreover, the farmers are facing a number of marketing communications problems including the packaging of the product. Packaging is the first thing visible to consumers when they select any product to buy. Packaging is not only a way of knowing about a product or marketing it attractive but also defines the quality of the product [7]. Therefore, packaging plays a very crucial role in successful marketing of a product [1, 15, 21–25]. In fact, packaging has become a contemporary tool for integrated marketing and the most influential promotional tool at the point of sale [32]. Other studies on the design of consumer product also found that consumers place much emphasis on packaging on the shelf-life [8, 31]. Because it shows the information about freshness of the product. The studies also revealed that consumers place less emphasis on the features such as ease of use, durability of product as well as shape and color of packaging [8, 31]. Another study with this line reported that information on label of the product such as nutritional fact panel or safety logo (for example, organic logo, origin of the product) increases probability of the product being purchased by consumers [2]. Therefore, it is evident that there is a great connection between product packaging and consumer perception and purchasing habit [17]. In sales and marketing of a product, packaging can be viewed as a sale person who influences consumer's purchasing decision through attractive colors and messages [11].

This study aims to assess and identify the key factors that contribute in developing the packaging attributes and label information of 'green' chili pepper products of farmers' community enterprises to make them more appealing to consumers in Thailand. The rest of the paper is organized as follows: Sect. 2 briefly summarizes the methodology. The results and discussions are reported in Sect. 3. Finally, Sect. 4 focuses on the limitations and conclusions.

2 Methodology

2.1 Sample Size, Survey Design and Data Collection

The study was conducted in five provinces in Thailand. The study population consisted of consumers having green consumption behavior such as buyers of toxin-free products and pesticide-free vegetables in the markets. The sample size for the study was 200 respondents. The sample size was determined using the formula for large and unknown population [12]. The formula is as below:

$$n = \frac{P(1 - P)(Z^2)}{e^2} \tag{1}$$

Where,

Desired confidence level = 95%
Desired level of precision = 5%
Estimated proportion of an attribute that is present in the population (p) = 0.15
Sample size = 195.92 (or approximately 200)

Accidental sampling method was applied to select the samples from the population of the study. To ensure the representation of target consumers, the samples were selected from various markets in large provinces in northern, northeastern, and central regions of Thailand including Chaiyaphum Province which is the large producer of 'green' chili pepper and thus should be the target market of 'green' chili pepper products. The data for this study were collected through field survey in the study areas.

A comprehensive and standardized questionnaire was prepared for the field survey. The questionnaire was pre-tested through a preliminary survey. The questionnaire was finalized after making necessary correction and modification based on the results and comments of pre-test. After that it was used in the final survey. The data were obtained through interviewing (i.e. face-to-face interview) the respondents. The data included socio-demographic characteristics of the respondents and their chili pepper consumption behavior, packaging attributes that influence consumers' decision to buy 'green' chili pepper products, and information on packaging labels that assist consumers' decision making to consume 'green' chili pepper products.

2.2 Data Analysis

To fulfill its objectives, the study employed grey relational analysis (GRA) and binary logit regression model in analyzing the data.

Grey Relational Analysis (GRA). GRA was proposed by Professor Deng Julong in 1985 [5, 27]. After that, it has been applied in many management and marketing fields [14, 28, 29]. Furthermore, GRA has been widely used to identify key factors influencing consumers' buying decision for a particular product such as food products, medicines and investment products [6, 9, 10, 16]. Therefore, the present study employed GRA to find out the key factors that contribute in developing the packaging design and label information of community's "green" chili pepper products to influence consumers in buying the product.

The concept of GRA application for this study can be shown in Fig. 1.

To identify the key factors for inclusion on packaging label and the packaging attributes of 'green' chili pepper products, the study preliminary identified 12 factors or criteria based on previous literature. Subsequently, the 12 factors were separated into two groups to investigate. First group contained 6 factors involving packaging label while the second group comprised of the rest 6 factors concerning packaging attributes.

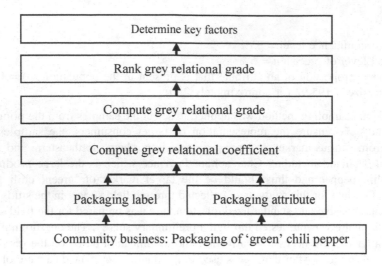

Fig. 1. GRA framework

After that grey relational coefficient and grey relational grade were calculated by applying the following equations [26, 33]:

$$\gamma_{01}(j) = \frac{\min_{i=1}^{n} \min_{j=1}^{m} \Delta_{0i}(j) + \zeta \times \max_{i=1}^{n} \max_{j=1}^{m} \Delta_{0i}(j)}{\Delta_{0i}(j) + \zeta \times \max_{i=1}^{n} \max_{j=1}^{m} \Delta_{0i}(j)} \qquad (2)$$

$$\Gamma_{0i} = \frac{1}{n} \sum_{j=1}^{n} w_j \gamma(x_0(j), x_i(j)) \qquad (3)$$

Where,

$\sum_{j=1}^{n} w_j = 1$ and with equal weights

w_j = normalized weight of criterion j

Γ_{0i} = grey relational degree that indicates the magnitude of correlation (i.e. similarity) measured between the compared sequence (i.e. the ith data sequence) and the reference sequence.

All calculated values are in the range of (0,1). Any j[th] criterion having a calculated value equal to 1 or most close to 1, represents the best choice [26]. Therefore, for this study, any criteria with calculated value equal to 1 or most close to 1 was treated as a key factor that should appear on the packaging label of 'green' chili pepper products.

Binary Logit Regression. In order to identify the target markets for 'green' chili pepper products, the study further explored the following hypotheses:

H1: Consumers in upcountry areas are less attentive to the statement 'safe chili pepper product' than those in Bangkok (Capital city in central region).

H2: Male is different from female in paying attention to the statement 'safe chili pepper product'.

H3: Consumers in different age groups have different level of attention to the statement 'safe chili pepper product'.

H4: Consumers in student occupational groups have greater attention than those in other occupational groups to the statement 'safe chili pepper product'.

To test the hypotheses, the study employed binary logit regression model as stated below [19].

$$P_i = 1/1 + e^{-zi}$$ (4)

Where,

P_i = probability of attention on statement 'safe chili pepper product' of the ith respondent

e^{zi} = stands for the irrational number e raised to the power of Zi

Zi = is a function of N-explanatory variables and expressed as:

$$Zi = \beta_0 + \beta_1 X_1 + \beta_2 X_2 + \ldots + \beta_n X_n + \mu_i$$ (5)

Where,

β_0 = Constant term

$\beta_1, \ldots, \beta_n$ = Regression co-efficient

3 Empirical Results

3.1 Consumers' Socio-Demographic Characteristics and Chili Pepper Consumption Behavior

The findings show that more than two-third (77.5%) of the respondents are females. Majority of the study participants are in the young adult and adult age group. It was found that most of study consumers usually buy small amount of chili pepper products (on average, 250 g at a time) due to their concerns about expiry date and safety of the product. It was also found that 31% of the consumers purchase the product for the purpose of household consumption and 61.5% did so for household consumption and giving out as gift. Majority of the surveyed consumers (nearly 75%) purchase the chili pepper products in fresh food market, and the rest (25%) from convenience store. The findings indicate that almost every Thai household generally like to have chili in cooking their food.

3.2 Key Factors Involving Packaging Label of 'Green' Chili Pepper Products

Table 1 presents the GRA scores of the factors that contribute in developing the packaging label of community's 'green' chili pepper products. The analyses show that the most important key factor for packaging label is the 'information on manufacturing

Table 1. GRA scores for packaging label information of 'green' chili pepper products

Factor considered when buying 'green' chili pepper products	GRA score						
	All locations N = 200	Chiang Mai N = 40	Chaiyaphum N = 40	Nakhonratchasima N = 40	Khonkaen N = 40	Bangkok N = 40
1. You will choose, considering the manufacturing and expiry dates (L04)	0.886	0.895	0.914	0.854	0.821	0.863
2. You will choose, considering the presence of information on ingredients (L05)	0.774	0.825	0.770	0.716	0.715	0.814
3. You will choose the product having certification of standards from Thailand FDA, TISI, etc. (L01)	0.770	0.764	0.739	0.823	0.745	0.740
4. You will choose the product providing creditable information on where it is produced/manufactured (L03)	0.764	0.783	0.737	0.807	0.665	0.761
5. You will choose the product with a statement indicating it is a safe food (L02)	0.761	0.787	0.773	0.837	0.718	0.693
6. You will choose, considering the statement about health benefits and uses of chili pepper (L06)	0.706	0.694	0.733	0.720	0.636	0.705

Source: Calculation

date and expiry date'. The 2nd and 3rd important key factors are 'information on ingredients' and 'evidence of standard certification given by agencies such as Thailand FDA (Food and Drug Administration)'. The factors that ranked 4th and 5th as per their importance are the 'information on geographical area of manufacturer' and 'the statement indicating that the product is 'green' or organic chili pepper'. The findings suggest that the five key factors mentioned above should be displayed on the packaging label of 'green' chili pepper products as they contribute to the make the products more appealing to the consumers.

3.3 Key Factors Involving Packaging Attributes of 'Green' Chili Pepper Products

Table 2 shows the GRA scores of the factors that contribute in developing the packaging attributes of 'green' chili pepper products. It was found that the most important key factor for packaging attributes is 'easy and convenient in opening/closing the container' while the attribute for 'environmentally friendly feature' was found to be 2nd most important key factor for packaging. The results indicate that consumers give high priority to the functional characteristics of packaging rather than the external designs. Therefore, packaging designers of 'green' chili pepper products should put

Table 2. GRA scores for packaging attributes of 'green' chili pepper products

Factor considered when buying 'green' chili pepper products	GRA score					
	All locations N = 200	Chiang Mai N = 40	Chaiyaphum N = 40	Nakhonratchasima N = 40	Khonkaen N = 40	Bangkok N = 40
1. You will choose the product that has container which is easy and convenient to open/close (P06)	0.761	0.804	0.716	0.786	0.712	0.748
2. You will choose the product considering its package that is made from environmentally friendly raw materials such as glass, paper, and other natural raw materials (P05)	0.738	0.767	0.642	0.811	0.640	0.693
3. You will choose the product considering its package that looks simple yet luxurious and stylish (P02)	0.666	0.649	0.609	0.729	0.658	0.630
4. You will choose the product considering its package that looks simple (P04)	0.645	0.611	0.638	0.633	0.643	0.504
5. You will choose the product considering its package that has striking shape and colors (P01)	0.623	0.623	0.582	0.691	0.596	0.624
6. You will choose the product considering its package that has unique or unconventional look and design as distinct from packaging of other products in the same category (P03)	0.610	0.585	0.598	0.689	0.579	0.601

Source: Calculation

prior importance on how the packaging of the product can be more easy and convenient to use as well as environmentally friendly.

3.4   Discussions

Findings of the study indicate that 'green' chili pepper products have market potentials in the country. However, it was observed that consumers pay relatively more attention to information label compared to packaging attributes. This is because certain information is indicative for product quality such as manufacturing and expiry dates, creditable statement about manufacturing place, as well as palatability of the product like ingredients and brand. The findings of this study are similar to other study results which concluded that consumers usually tend to read information label whether the product has any standard certification rather than look at packaging design when they make purchase decision.

Table 3. Definition of variables and descriptive statistics of the samples

Variable definitions	Variable name	Mean	Standard deviation
Independent variables			
Attention on statement 'safe chili pepper product'; scores 4–5 = 1; 0 otherwise	LEBEL	0.73	0.448
Explanatory variables			
Information collection location			
Chaiyaphum province = 1; 0 otherwise	CHAI	0.20	0.401
Chiang Mai province = 1; 0 otherwise	CM	0.20	0.401
Nakhonratchasima province = 1; 0 otherwise	KORAT	0.20	0.401
Khonkaen province = 1; 0 otherwise	KK	0.20	0.401
Bangkok = 1; 0 otherwise	BKK	0.20	0.401
Female = 1; 0 = Male	FEMALE	0.78	0.419
Age	AGE	45.13	14.698
Occupation			
Private business employee, self-employed, government worker = 1; 0 otherwise	OCC_PBG	0.75	0.437
Housewife = 1; 0 otherwise	OCC_HOUSE	0.07	0.256
Retired government worker = 1; 0 otherwise	OCC_RETIRE	0.10	0.301
Student = 1; 0 otherwise	OCC_STUDENT	0.08	0.272

Furthermore, in order to identify the target markets for 'green' chili pepper products, the study explored several hypotheses pointed out in the methodology section. Binary logit regression analyses were conducted to test the hypotheses (Table 3). The results of the analyses (Table 4) lead to the rejection of the last hypothesis while accepting the first three hypotheses. Apparently, female and people in older age group place more importance on food safety than male and young people do. These findings might help the farmers groups to consider using the statement 'safe chili pepper product' to attract more consumers. In essence, it is necessary for them to include the statement 'safe chili pepper product' in the information label of community's 'green' chili pepper products in order to expand their markets further to the neighboring provinces.

Table 4. Estimates binary logit model for attention on label information 'safe chili pepper product'

Variables[a]	Estimated coefficients	Wald statistic	p-values
Constant	−.825	−.925	˙.3550
CM	.783	1.437	.1508
CHAI	1.034*	1.926	.0541
KORAT	1.931***	2.973	.0029
KK	.363	.695	.4867
FEMALE	.746*	1.876	.0607
AGE	.029*	1.703	.0886
OCC_PBG	−.874	−1.239	.2152
OCC_HOUSE	28.045	.000	1.0000
OCC_RETIRE	−1.042	−.953	.3408

Goodness of fit test: McFadden Pseudo R-squared .1434748 Chi squared 33.75496 Prob[ChiSqd > value] = .9859856E-04 Prediction accuracy 91.5%

*** Statistically significant at the 0.01 level and * at the 0.10 level.
[a] Dummy variables BKK and OCC_STUDENT were treated as baseline.

4 Limitations and Conclusions

This study attempts to identify the key factors that contribute in developing information label and packaging attributes of 'green' chili pepper products in Thailand. However, the study included only six factors/criteria for each of the investigations. It creates a limited scope for better learning about consumers' perception towards 'green' chili pepper products because in reality consumers may consider more other factors than the study included in its investigation. Therefore, further study and more discussion on the standardized questionnaire that present study employed in its investigation is needed. It will make a beneficial contribution to the readers. In addition, future in-depth research can be carried out to include more factors in the information label and packaging attributes of the products. Furthermore, the definition of safe chili pepper product can be expanded to distinguish semi-finished product (for example, dried chili) from finished product (i.e. ready-to-eat food) to sharpen the analysis of key factors.

The findings of the study show that consumers pay more attention on information label than packaging attributes of the product. The information about manufacturing and expiry date of the product was found to be the most important factor. It was also found that consumers give importance to safety features of the product when they make decision to buy the product. It indicates that the producer groups of 'green' chili pepper products can create value addition to their products through including food safety features on the information label of the product. This can act as a selling point for their products in emerging economies like Thailand. Furthermore, the farmers should take initiatives to create product identity by using ingredients that make the food tastes different from competing products and putting this information on the packaging label can help branding their products.

Acknowledgement. Financial support from the Thailand Research Fund through the Royal Golden Jubilee Ph.D. Program (Grant No. PHD/0120/2554) to Jakkreeporn Sannork and Prof. Dr. Aree wiboonpongse is acknowledged.

References

1. Ahmed, A., Ahmed, N., Salman, A.: Critical issues in packaged food business. Br. Food J. **107**(10), 760–780 (2005)
2. Azucena, G., Tiziana, D.M.: Consumer preferences for food labeling: what ranks first? Food Control **61**, 39–46 (2016)
3. Bruschi, V., Shershneva, K., Dolgopolova, I., Canavari, M., Teuber, R.: Consumer perception of organic food in emerging markets: evidence from Saint Petersburg. Russ. Agribus. **31**(3), 414–432 (2015)
4. Choi, S.E., Chan, J.: Relationship of 6-n-propylthiouracil taste intensity and chili pepper use with body mass index, energy intake, and fat intake within an ethnically diverse population. J. Acad. Nutr. Diet. **115**(3), 389–396 (2015)
5. Dogan, M.: Measuring bank performance with gray relational analysis: the case of Turkey. Ege Acad. Rev. **13**(2), 215–225 (2013)
6. Duan, L.-Z., Duan, G.-N., Zhai, G.-Q., Zhang, Y., Xuan, C.-Y., Geng, H.: The grey relational analysis of influential factors for Chinese medicine in general hospital. Grey Syst.: Theory Appl. **2**(2), 311–323 (2012)
7. Gomez, M., Martin-Consuegra, D., Molina, A.: The importance of packaging in purchase and usage behaviour. Int. J. Consum. Stud. **39**(3), 203–211 (2015)
8. Heide, M., Olsen, S.O.: Influence of packaging attributes on consumer evaluation of fresh cod. Food Qual. Prefer. **60**, 9–18 (2017)
9. Hsu, T.-H.: Study on supplier product core competencies based on customers' needs: an application of GRA and QFD methods. Int. J. Adv. Comput. Technol. **5**(5), 12–20 (2013)
10. Huang, M., Wang, B.: Evaluating green performance of building products based on gray relational analysis and analytic hierarchy process. Environ. Prog. Sustain. Energ. **33**(4), 1389–1395 (2014)
11. Ilyuk, V., Block, L.: The effects of single-serve packaging on consumption closure and judgments of product efficacy. J. Consum. Res. **42**(6), 858–878 (2016)
12. Israel, G.D.: Determining sample size. In: Cochran, W.G. (ed.) Sampling Techniques, 2nd edn. Wiley, New York (1963). University of Florida. https://edis.ifas.ufl.edu/pd006. Accessed 11 Apr 2014
13. Jalili, M., Jinap, S.: Natural occurrence of aflatoxins and ochratoxin A in commercial dried chili. Food Control **24**, 160–164 (2012)
14. Kuo, Y., Yang, T., Huang, G.-W.: The use of grey relational analysis in solving multiple attribute decision-making problems. Comput. Ind. Eng. **55**, 80–93 (2008)
15. Lamba, B., Raheja, S.: Rural India: innovative marketing strategies. Int. J. Mark. Technol. **4**(5), 110–123 (2014)

16. Lin, C.-T., Lee, T.-R., Kao, C.-K., Wu, J.: Application of grey relational analysis to determine key factors for investment of overseas branch by logistics industry–U.S. market as an example. J. Supply Chain Oper. Manag. **12**(1), 14–29 (2014)
17. Lo, S.C., Tung, J., Huang, K.-P.: Customer perception and preference on product packaging. Int. J. Organ. Innov. **9**(3), 3–15 (2013)
18. Rihn, A.L., Yue, C.: Visual attention's influence on consumers' willingness-to-pay for processed food products. Agribusiness **32**(3), 314–328 (2016)
19. Mohammed, N.U., Bokelmann, W., Entsminger, J.S.: Factors affecting farmers' adaptation strategies to environmental degradation and climate change effects: a farm level study in Bangladesh. Climate **2**, 223–241 (2014)
20. Ooraikul, S., Siriwong, W., Siripattanakul, S., Chotpantarat, S., Robson, M.: Risk assessment of organophosphate pesticides for chili consumption from chili farm area, Ubon Ratchathani Province. Thailand. J. Health Res. **25**(3), 141–146 (2011)
21. Robert, L.U., Ozanne, J.L.: Is your package an effective communicator? A normative framework for increasing the communicative competence of packaging. J. Mark. Commun. **4**(4), 207–220 (1998)
22. Robert, L.U., Klein, N.M., Burke, R.R.: Packaging communication: attentional effects of product imagery. J. Prod. Brand Manag. **10**(7), 403–422 (2001)
23. Robert, L.U., Klein, N.M.: Packaging as brand communication: effects of product pictures on consumer responses to the package and brand. J. Mark. Theory Pract. **10**(4), 58–68 (2002)
24. Silayoi, P., Speece, M.: Packaging and purchase decisions. Br. Food J. **106**(8), 607–628 (2004)
25. Silayoi, P., Speece, M.: The importance of packaging attributes: a conjoint analysis approach. Eur. J. Mark. **41**(11/12), 1495–1517 (2007)
26. Tseng, M.-L., Chiu, A.S.F.: Evaluating firm's green supply chain management in linguistic preferences. J. Clean. Prod. **40**, 22–31 (2013)
27. Wang, C.-H., Lee, T.-R., Lin, W.-S., Sinnarong, N., Dadura, A.M., Li, J.-M.: Improving B2C business for online store – the case study of agriculture products e-market shop in Taiwan. Int. J. Serv. Stan. **7**(2), 95–118 (2011)
28. Wang, W., Cao, Q., Huang, X.: Identifying key influential factors of transforming regional science and education advantages into industrial advantages by grey relational analysis. J. Grey Syst. **25**(1), 63–75 (2013)
29. Wu, C.-H.: On the application of grey relational analysis and RIDIT analysis to Likert scale surveys. Int. Math. Forum **2**(14), 675–687 (2007)
30. Wu, L., Gong, X., Qin, S., Chen, X., Zhu, D., Hu, W., Li, Q.: Consumer preferences for pork attributes related to traceability, information certification, and origin labeling: based on China's Jiangsu Province. Agribusiness **33**, 424–442 (2017)
31. Wyrwa, J., Barska, A.: Packaging as a source of information about food products. Procedia Eng. **182**, 770–779 (2017)
32. Zerzyk, E.: Design and communication of ecological content on sustainable packaging in young consumers' opinions. J. Food Prod. Mark. **22**(6), 707–716 (2016)
33. Zhai, L.Y., Khoo, L.P., Zhong, Z.W.: Design concept evaluation in product development using rough sets and grey relation analysis. Expert Syst. Appl. **36**, 7072–7079 (2009)

Macroeconomic News Announcement and Thailand Stock Market

Saowaluk Duangin$^{(\boxtimes)}$, Woraphon Yamaka, Jirakom Sirisrisakulchai, and Songsak Sriboonchitta

Faculty of Economics, Chiang Mai University, Chiang Mai, Thailand
saowaluk.econ@gmail.com

Abstract. This paper investigates the effect of Thailand surprising macroeconomic news announcement on Thailand stock market. We adopt the MSGARCH-jump with augmented news intensity model, which is an extension to the original MSGARCH-jump for switching across two regimes. This model was applied to high-frequency data set from January 2011 through the end of December 2016. The results show that (1) MSGARCH-jump is better than MSGARCH-jump augmented with news intensity model, (2) 2-h data set is significant to explain the volatility in both models, (3) Thailand macroeconomic news such as foreign trade export, foreign trade import, CPI, GDP, and trade balance have the same effect on Thailand stock market, (4) Thailand stock market was less affected by macroeconomic news announcement.

Keywords: MSGARCH-jump · Thailand stock market
Macroeconomic news · High-frequency data

1 Introduction

According to the efficient market hypothesis (EMH), market price always instantly reacts to available information. When a new information occurs, the stock market will respond to both bad and good news. Traders will try to allocate their portfolio and decide on risk management. Macroeconomic reports, including Gross domestic product (GDP), inflation rate, and unemployment rate, are vital sources of information transcribing the changes in economic fundamental indicators or policies into different metaphors of return and volatility of stock market.

In this research area, [1, 2] used a macroeconomic news announcement to test the concept of EMH and found an evidence to confirm that macroeconomic news, such as GDP, consumer price index (CPI), and trade balance, have a significant impact on the stock market. In addition, the study of [3] also found the significant jumps on news day more than no-news day and market response varied with economic situation. [3–6] found that the release of macroeconomic news has immediate influence on stock market causing a sudden and quick change in stock prices.

In the recent years, the GARCH-type model which focusses on conditional return and conditional variance of stock market returns has been applied and widely used to study the effect of macroeconomic news on stock market. [7] proposed a GARCH with constant jump-diffusion model which has become popular and has been discussed in

© Springer International Publishing AG, part of Springer Nature 2018
V.-N. Huynh et al. (Eds.): IUKM 2018, LNAI 10758, pp. 408–419, 2018.
https://doi.org/10.1007/978-3-319-75429-1_34

many studies. However, this model fails to deal with absence of autocorrelations, leptokurtosis, long memory and volatility persistence, mean-reverted volatility, and leverage effect. To improve the model, [8, 9] allowed the jump component to vary overtime and tried to capture abnormal extreme events through Poisson distributed jumps [10]. The study of [2, 11] analyzed the behavior of price jump by applying GARCH-jump model, which allows the sudden jumps in stock market, to capture the volatility when jump occurs. Another advantage is that, this model allows us to add news in the model to capture the effect of each news. The performance of this model over the conventional GARCH is examined by [12]. They compared these two models and found that GARCH-jump model augmented with news intensity performs slightly better than GARCH model without jumps.

However, the persistence of the conditional variance using GARCH-jump model may be higher in some periods, while in other periods it is less volatile. Following this idea, the GARCH jump model of [8] was extended to have a structural break in both GARCH component and jumps by incorporating a regime-switching approach of [13], and thereby proposing a Markov Switching GARCH jump (MSGARCH-jump) model. In the two regimes case, the model will have an ability to capture a low jump intensity regime and high jump intensity regime by using a Markov process to govern the GARCH and jump components to switch between regimes.

In this study, we are also concerned about the persistence of the conditional variance of the model and thus the MSGARCH-jump model is considered. However, the strong assumption of normal distribution in this model might fail to capture volatility clustering and leptokurtosis of asset returns. Therefore, investigating the choices of the most suitable distribution amongst a variety of distributions should be considered. To the best of our knowledge, it is difficult to find the papers which have compared the conditional density distribution of GARCH-jump, especially under the context of GARCH-jump with regime-switching model. Therefore, the main contribution of this paper is to investigate whether alternative error distributions can outperform MSGARCH-jump.

Therefore, in this study, we investigate four major issues. First, we investigate and estimate magnitude and duration of the effect of Thailand macroeconomic news announcement on Stock Exchange of Thailand (SET) index[1]. Second, we also explore whether positive, negative, or size of surprising news is the main factor affecting SET. Third, we examine the performance of the model across different high-frequency data sets (30-min, 1-h, and 2-h data set) from January 2011 to December 2016. Finally, this paper investigates how the MSGARCH-jump performs when it includes news intensity.

The remainder of this study is organized as follows: Sect. 2 describes the methodology used in this study. Section 3 provides data description. Section 4 describes empirical results. Finally, Sect. 5 contains conclusion.

---

[1] Bloomberg [14] "SET Index is a capitalization-weighted index of stocks traded on the Stock Exchange of Thailand. The index was developed with a base value of 100 as of April 30, 1975."

2 Methods

2.1 Markov Switching GARCH-Jump Augmented with News Intensity

The model structure of Markov regime-switching GARCH (1,1) jump can be set as.

$$R_t = \mu(s_t) + \varepsilon_{1t}(s_t) + \varepsilon_{2t}(s_t), \tag{1}$$

$$\mu(s_t) = \mu_0(s_t) + \sum_{q=1}^{Q} \mu_q(s_t) Size_{q,t-1}, \tag{2}$$

$$\varepsilon_{1t}(s_t) = z_t(s_t)\sigma_t(s_t), \tag{3}$$

$$\sigma_t^2 = \alpha_0(s_t) + \alpha_1(s_t)\varepsilon_{1t-1}^2 + \beta_1\sigma_{t-1}^2(s_t), \tag{4}$$

where $s_t = \{1, 2, \ldots, H\}$ is unobserved state variable with K regimes being governed by first order Markov process that is defined by the transition probabilities matrix P. R_t is the stock market return. $\mu(s_t)$ is the regime dependent conditional expected return of R_t. $Size_{q,t-1}$ is size of surprising news which calculate from standardized news surprise (See Eq. 20). $\mu_0(s_t)$ is regime dependent constant term of mean equation. $\mu_0(s_t)$ is regime dependent coefficient of size of jump. $z_t(s_t)$ is regime dependent standardized residual, $\sigma_t^2(s_t)$ is regime dependent conditional variance.

Consider the error of the model in Eq. 1, this term consists of two error components $\varepsilon_{1t}(s_t)$ and $\varepsilon_{2t}(s_t)$, where $\varepsilon_{1t}(s_t)$ denotes regime dependent normal news while $\varepsilon_{2t}(s_t)$ denotes an unusual news. Thus the error component can be written as

$$\varepsilon_t(s_t) = \varepsilon_{1t}(s_t) + \varepsilon_{2t}(s_t), \tag{5}$$

where $\varepsilon_{1t}(s_t)$ is assumed to have normal, Student's t, skewed normal, and skewed Student's t distribution, and ε_{2t} is defined by

$$\sum_{k=1}^{\eta_t} Y_{t,k}|(s_t) - \mathsf{E}(\sum_{k=1}^{\eta_t} Y_{t,k}|I_{t-1}(s_t)) \tag{6}$$

where $Y_{t,k}$ is the size of k-th jump that occurs from time $t-1$ to t, $1 \leq k \leq \eta_t$. η_t is a Poisson random variable with conditional jump intensity. $I_{t-1}(s_t)$ is the information set at previous time for each regime. In this model, it is supposed that the jump size $Y_{t,k}|(s_t)$ is realization of either normal distribution $(Y_{t,k}|(s_t) \sim N(\theta(s_t), \delta^2(s_t)))$, Student's t distribution $(Y_{t,k}|(s_t) \sim sT(\theta(s_t), \delta^2(s_t), d(s_t)))$, skewed normal distribution $(Y_{t,k}|(s_t) \sim sN(\theta(s_t), \delta^2(s_t), \xi(s_t)))$, and skewed Student's t distribution $(Y_{t,k}|(s_t) \sim ssT(\theta(s_t), \delta^2(s_t), d(s_t), \xi(s_t)))$, and

$$\mathsf{E}(J_t(s_t)|I_{t-1}(s_t)) = \theta(s_t)\lambda_t(s_t), \tag{7}$$

where $J_t(s_t)$ is the expected jump. $\theta(s_t)$ is mean parameter. $\lambda_t(s_t)$ is jump intensive parameter.

Thus, the error of unexpected events and is responsible for jumps in volatility becomes

$$\varepsilon_{2t}(s_t) = \sum_{k=1}^{n_t} Y_{t,k}|(s_t) - \theta(s_t)\lambda_t(s_t), \tag{8}$$

where $\lambda_t(s_t)$ is assumed to vary over time following Autoregressive Moving average (ARMA) process.

$$\lambda_t(s_t) = A(s_t) + B(s_t)(\lambda_{t-1}) + C(s_t)\vartheta_{t-1}(s_t), \tag{9}$$

where $A(s_t)$, $B(s_t)$, and $C(s_t)$ are regime dependent estimated parameters for conditional jump intensity ARMA process. $\vartheta_{t-1}(s_t)$ is the approximate error term, which is the expected number of jumps given all information set at $t-1$, $\vartheta_{t-1}(s_t) \sim \mathrm{E}[\eta_{t-1}|\Theta_{t-1}(s_t)] - \lambda_{t-1}(s_t)$. Note that η_t is a Poisson random number of jumps occurring between t and $t-1$ conditional on s_t. Thus, the conditional density of η_t is

$$P(\eta_t = j|\mathrm{I}_{t-1}(s_t)) = \frac{\exp(-\lambda_t(s_t))\lambda_t^j(s_t)}{j!}, \tag{10}$$

where j is the number of jumps ($j = 0, 1, \ldots$). To keep the value of $\lambda_t(s_t) > 0$ for all t, a sufficient condition is that $A > 0$, $B > 0$, $C > 0$, and $A > B$. This specification of the conditional jump intensity removes the problem of regime path dependence, and allows the jump intensity to be autocorrelated, which can explain for the phenomenon of jump clustering around significant news events.

Note that s_t is assumed to follow a first-order Markov process as in [13]. Thus,

$$p_{ab} = \mathrm{P}((s_{t+1}) = a|(s_t) = b) \text{ and } \sum_{a,b=1}^{H} p_{ab} = 1; \quad a, b = 1, \ldots, H. \tag{11}$$

Then, the transition probabilities in the transition matrix P is,

$$P = \begin{bmatrix} p_{11} & p_{21} & \cdots & p_{h1} \\ p_{12} & p_{22} & \cdots & p_{h2} \\ \vdots & \vdots & \cdots & \vdots \\ p_{1h} & p_{2h} & \cdots & p_{hh} \end{bmatrix}, \tag{12}$$

where p_{ab} is the probability of switching from regime a to b.

Note that, the Markov Switching GARCH-jump model excluding news can be estimated by ignoring $\varepsilon_{2t}(s_t)$ and $Size_{q,t-1}$.

2.2   Estimation

In this study, we consider two-regime Markov Switching GARCH-jump model and we estimate the unknown parameters, $\psi(s_t)$, in the model using a maximum likelihood estimator (MLE). Thus,

$$\underset{\psi(s_t)}{\arg\max} = \widehat{L}(\psi(s_t)). \tag{13}$$

Here, we allow the model to have either normal, Student's t, skewed normal or skewed Student's t distribution. Therefore, likelihood function of our model can be written as in the following:

(1) Normal likelihood

$$\widehat{L}_n(\psi(s_t)) = \left(\sum_{s_t=1}^{2} \prod_{t=1}^{T} \left(\frac{1}{\sqrt{2\pi(j\delta^2(s_t) + \sigma_t^2(s_t))}} \exp\left(-\frac{R_t - \mu(s_t) - j\theta(s_t)}{2(j\delta^2(s_t) + \sigma_t^2(s_t))} \right) \right) \cdot P(\eta_t = j|\psi_{t-1}(s_t)) \right), \tag{14}$$

where $\psi_{t-1}(s_t)$ is all information, including $I_{t-1}(s_t)$.

(2) Skewed normal likelihood

$$\widehat{L}_n(\psi(s_t)) = \left(\sum_{s_t=1}^{2} \prod_{t=1}^{T} \left(\frac{2}{\xi(s_t) + (\xi(s_t))^{-1}} \cdot f_n(z/\Upsilon) \right) \cdot P(\eta_t = j|\psi_{t-1}(s_t)) \right), \tag{15}$$

$$z = \frac{R_t - u(s_t) - j\theta(s_t)}{2(j\delta^2(s_t) + \sigma_t^2(s_t))} \sqrt{((1 - m_1^2)(\xi(s_t)^2 + 1/\xi(s_t)^2) + 2m_1^2 - 1)} + \Upsilon,$$

$$m1 = 2/\sqrt{2\pi}, \text{ and } mu = m1(\xi(s_t) - 1/\xi(s_t)),$$

where $\xi(s_t)$ is regime dependent skew parameter and $f_n(\cdot)$ is density of normal distribution.

(3) Student's t likelihood

$$\widehat{L}_t(\psi(s_t)) = \sum_{s_t=1}^{2} \left(\prod_{t=1}^{n} \left(\frac{\Gamma(\frac{d(s_t)+1}{2})}{\sqrt{(d(s_t)-2)\pi}\Gamma(\frac{d(s_t)}{2})} (1 + \frac{R_t - u(s_t) - j\theta(s_t)}{(d-2)(j\delta^2(s_t) + \sigma_t^2(s_t))})^{\frac{-d+1}{2}} \cdot (\frac{1}{\sigma_{s(t)}^2}) \right) P(\eta_t = j|\psi_{t-1}(s_t)) \right), \tag{16}$$

where $d(s_t)$ is regime dependent degree of freedom, Γ is gamma distribution.

(4) Skewed student's t likelihood

$$\hat{L}_t(\psi(s_t)) = \sum_{s_t=1}^{2} \left(\prod_{t=1}^{n1} \left(\frac{2}{\xi(s_t) + 1/\xi(s_t)} f_t(\xi(s_t)\mu(s_t)) \right) P(\eta_t = j|\psi_{t-1}(s_t)) \right) \quad \text{for } x(s_t) < 0, \text{ and}$$

$$\hat{L}_t(\psi(s_t)) = \sum_{s_t=1}^{2} \left(\prod_{t=1}^{n2} \left(\frac{2}{\xi(s_t) + 1/\xi(s_t)} f_t(\frac{\xi(s_t)}{\mu(s_t)}) \right) (\Pr(\eta_t = j|\psi(s_t))) \right) \quad \text{for } x(s_t) \geq 0,$$

$$(17)$$

where $\xi(s_t)$ is regime dependent skew parameter, $\mu(s_t)$ is the regime dependent expected mean of the model and $f_t(\cdot)$ is density of student's t-distribution, while $n1$ and $n2$ are the number of observations in each case. And the second part of each likelihood can be derived as follows:

$$P(\eta_t(s_t) = j|\psi_{t-1}(s_t)) = P(\eta_t = j|\psi_{t-1}(s_t))P((s_t) = s_t|\psi_{t-1}(s_t)), \quad (18)$$

where $P(\eta_t = j|\psi_{t-1}(s_t))$ can be derived from Eq. (10) and in this estimation, we employ the Hamilton's filter to predict the filtered probabilities, $P((s_t) = s_t |\psi_{t-1}(s_t))$, $s_t = 1, \ldots, H$ regime. Then, we maximize the likelihood function to obtain the estimated parameters of our model.

3 Data and Descriptive Statistics

3.1 The Stock Exchange of Thailand

Because we are interested in the short-live effects of macroeconomic news announcement on SET, we have studied SET index high-frequency data sets: 30-min, 1-h, and 2-h data set give 17762, 8240, and 5091 observations respectively which are obtained from Bloomberg database. The sample period is from the first trading day of 2011 to the last trading day of 2016, between 9.55 am to 4.40 pm in Bangkok time (GMT + 7.00). The data has been removed for the lunch break time.

We define the percentage return of SET ($SET_{i,t}$) on the interval i at trading day t as the logarithm of the ratio of the current price ($SET_{i,t}$) over the price of the previous period ($SET_{i,t-1}$) times 100.

$$R_{i,t} = \ln\left[\frac{SET_{i,t}}{SET_{i,t-1}}\right] x100 \qquad (19)$$

Based on the results presented in Table 1, the p-value of the Jarque-Bera test (JB test) are close to zero. Then, we can summarize that returns of all three SET index sample sets are non-normally distributed.

Table 1. Data description: SET index

	Min	Max	Mean	SD	S	K	JB test
30-min	−0.0543	0.0774	0.0000	0.0029	0.1683	58.31	2.2e−16
1-h	−0.0471	0.0477	0.0000	0.0036	−0.4417	16.96	2.2e−16
2-h	−0.0459	0.0383	0.0000	0.0046	−0.5802	9.58	2.2e−16

3.2   Macroeconomic News Announcement

In this study, we examine the effect of macroeconomic news announcement on return of SET by considering a set of five Thailand macroeconomic news releases - Thailand Customs Department's Foreign Trade Exports (CEx), Thailand Customs Department's Foreign Trade Imports (CIm), Consumer price index (CPI), Gross domestic product (GDP), and Customs Trade balance (CTB) which are announced between SET trading hours (See Table 2).

Table 2. Data description: macroeconomic news announcement

	Unit	Min	Max	Mean	# good news	# bad news	Release cycle
CEx	YoY%	−12.40%	38.30%	2.77%	30	42	Monthly
CIm	YoY%	−26.20%	44.00%	2.20%	32	37	Monthly
CPI	MoM%	−0.59%	1.38%	0.13%	43	15	Monthly
GDP	QoQ%	−10.70%	11.00%	0.74%	13	10	Quarterly
CTB	Million$	−5487	4986	−77.375	37	35	Monthly

We focus only on the surprising news because normal news does not render new information to stock market. Following [15, 16], we translate macroeconomic news announcements into surprising news by using standardized news surprise, $S_{q,t}$, which are the difference between the expected value, $E_{q,t}$, and the actual value, $A_{q,t}$, divided by their standard deviation (σ_q) for each news, q, in period t.

$$S_{q,t} = \frac{A_{q,t} - E_{q,t}}{\sigma_q}, \quad q = 1, \ldots, 5. \tag{20}$$

According to the EMH, only the unexpected part of the announcements should have an impact on stock return. Then, we also classify the news to be good news when $A_{q,t} > E_{q,t}$ and bad news when $A_{q,t} < E_{q,t}$. However, CPI announcement has reverse impact[2]. Table 2 shows the description of macroeconomic surprises for the January 2011 to December 2016. Most of the data are announced monthly as percentage change from previous period and the announcements are made without fixed time schedule. But GDP announcement is made at 9.30 am as a lower frequency - four times per year.

---

[2] CPI announcement will be classified as good news when $A_{q,t} < E_{q,t}$ and bad news when $A_{q,t} > E_{q,t}$.

4 Empirical Results

From the objective of this paper, there are three main steps being studied. First is model selection for subsequent investigation by the fit model. In the second step, we show the performance of these two models. In the third step, we show the impact of macroeconomic news announcement on stock market.

4.1 Model Selection

We determine the best model by comparing the values of the Akaike Information Criterion (AIC) of the four alternative distributions. The model with the lowest value of AIC is the best model specified for the sample data sets. For MSGARCH-Jump model with 30-min and 1-h sample sets, they are fit with normal distribution while 2-h sample set fit with skew normal distribution (See Table 3). The results in Table 3 also indicate that MSGARCH-Jump augmented with news intensity model of all sample sets are fit with skewed Student's t distribution.

Table 3. Akaike Information Criterion (AIC) value of MSGARCH-jump and MSGARCH-jump augmented with news intensity for log return of SET index

	Normal distribution	Student's t distribution	Skew normal distribution	Skewed Student's t distribution
MSGARCH-Jump				
30-min	12373.96*	188588.4	12377.77	184623.2
1-h	10859.61*	93182.28	10865.60	91350.31
2-h	9298.12	58762.44	9191.19*	57961.71
MSGARCH-Jump augmented with news intensity				
30-min	371813.6	213557.6	367970.1	205654.6*
1-h	158210.8	100197.3	153821	96813.71*
2-h	61540.54	61540.51	88959.35	60082.85*

Note: * is the smallest AIC value for each sample set.

4.2 MSGARCH-Jump and MSGARCH-Jump Augmented with News Estimation

In this section, we present the results of MSGARCH-jump and MSGARCH-jump augmented with news intensity estimation. In Table 5, we present the parameter estimates of the two-regime MS-GARCH-jump and two-regime MS-GARCH-jump augmented with news intensity model, respectively, for all frequencies to examine the regime dependent volatility of stock.

First, we consider the volatility persistence of the models which can be obtained by summing ARCH, α_{1,S_t}, and GARCH, β_{1,S_t}, parameters. We distinguish the volatility persistence into two regimes are high and low volatility regime. Table 5 also shows the volatility persistence coefficients, $\alpha_{1,S_t} + \beta_{1,S_t}$, for all sample sets of MSGARCH-jump and MSGARCH-jump augmented with news intensity estimation. We find that the volatility persistence of 30-min sample set in regime 1 and 1-h sample set in regime 2 of MSGARCH-jump are more than 1, indicating that these results are non-stationary. For MSGARCH-jump augmented with news intensity, the volatility persistence of all sample set are less than one, indicating that these results are stationary. From results, we will interpret regime 1 as high volatility regime while regime 2 as low volatility regime. We observe that the parameters in both models are mostly significant except the parameters in the augmented news intensity term in MS-GARCH-jump augmented with news intensity. We found that all parameters of augmented news intensity term in both regimes are insignificant. It implies that there is no effect of news in both low and high volatile periods.

In addition, let's consider the transition matrix of two models in Table 4. We denote the probabilities $\Pr(s_t = 1|s_{t-1} = 1)$ by p_{11} and $\Pr(s_t = 2|s_{t-1} = 2)$ by p_{22}. We can observe the low persistence in both MSGARCH-jump and MSGARCH-jump augmented with news intensity models because of the low value of some transition probabilities. We also notice that MSGARCH-jump augmented with news intensity model shows a lower persistence when compared to pure MSGARCH-jump. Thus, it leads us to confirm that incorporating news intensity might lead to a high volatility in the market returns.

Table 4. Transition matrix

	MSGRACH-jump			MSGRACH-jump augmented with news intensity		
	30 min	1-h	2-h	30 min	1-h	2-h
p_{11}	0.9722*	0.9966*	0.9447*	0.9000*	0.9000*	0.9000*
p_{22}	0.0001*	0.2798*	0.0001*	0.9000*	0.9000*	0.9000*
Duration1	35.9712	294.1177	18.0832	10	10	10
Duration2	1.0001	1.3885	1.0001	10	10	10

Note: * indicate significant value.

4.3   The Effect of Macroeconomic News Announcement on Stock Market

To capture the impact of Thailand macroeconomic news announcement on stock market, we estimate MSGARCH-jump augmented with news intensity model to measure the duration of the impact. Table 5 shows the estimation results that $B(s_t)$ and $C(s_t)$ are similar, indicating that the impacts from the number of negative and positive news on the jump in SET returns are not different. Table 5 also shows the measured size of surprise. The estimated coefficient of standardization variable, μ_q, shows a positive and significant result but close to zero for all news. This suggests that SET index responds indifferently to all macroeconomic news variables in terms of magnitude and sign.

Then, the conditional volatility of SET index was less affected by macroeconomic news, but there existed jump shock with augmented news intensity (See Fig. 1). We expect that incorporating news intensity in the volatility leads to the low persistence in

Table 5. Maximum likelihood estimates of MSGARCH-jump and MSGRACH-jump augmented with news intensity

	MSGARCH-jump						MSGRACH-jump augmented with news intensity					
	Regime 1 ($s_t = 1$)			Regime 2 ($s_t = 2$)			Regime 1 ($s_t = 1$)			Regime 2 ($s_t = 2$)		
	30-min	1-h	2-h	30-min	1-h	2-h	30-min	1-h	2-h	30-min	1-h	2-h
$\mu_0(s_t)$	0.0136*	0.0255*	0.0299*	0.2274*	0.0294	0.4842	0.0032*	0.0080*	0.0147*	0.0026	0.0067	0.0122
$\alpha_0(s_t)$	0.0213	0.0403	0.0390	0.5254	0.5955	0.5961**	0.0010*	0.0018*	0.0031*	0.0009*	0.0015*	0.0026
$\alpha_1(s_t)$	0.0519	1.0000	0.0391	0.6700*	0.7327	0.9020*	0.0414*	0.0460*	0.0505*	0.0345*	0.0383	0.0040
$\beta_1(s_t)$	0.9524*	0.9393*	0.9374*	0.2816*	0.5705*	0.0487*	0.9560*	0.9460*	0.9432*	0.7967*	0.7884*	0.7860*
$\xi(s_t)$		1.0000*	1.0000*			0.1000*	0.9615*	0.9481*	0.9563*	0.8012*	0.7901*	0.7969*
$d(s_t)$	−0.0224	−0.033*	−0.0417	−0.351*	−0.6240	−0.783*	2.7479*	3.1999*	3.1635*	2.2899*	2.6665*	2.6363*
$\theta(s_t)$	0.0962*	0.1157*	0.1243	0.4760*	0.5380	0.1788*	0.1000*	0.1000*	0.1000*	0.0833*	0.0833	0.0833*
$\delta^2(s_t)$	1.0000*	1.0000*	1.0000	1.0000*	1.0000*	1.0000*	0.1000*	0.1000*	0.1000*	0.0833*	0.0833	0.0833*
$\lambda(s_t)$			1.0000*	1.0000*		1.0000*						
$A(s_t)$							0.10000*	0.10000*	0.10000*	0.08333*	0.08333*	0.08333*
$B(s_t)$							0.00001*	0.00001*	0.00001*	0.08333*	0.08333*	0.08333*
$C(s_t)$							0.00001*	0.00001*	0.00001*	0.08333*	0.08333*	0.08333*
$\mu_1(s_t)$							0.00001*	0.00001*	0.00001*	0.00001*	0.00001*	0.00001*
$\mu_2(s_t)$							0.00001*	0.00001*	0.00001*	0.00001*	0.00001*	0.00001*
$\mu_3(s_t)$							0.00001*	0.00001*	0.00001*	0.00001*	0.00001*	0.00001*
$\mu_4(s_t)$							0.00001*	0.00001*	0.00001*	0.00001*	0.00001*	0.00001*
$\mu_5(s_t)$							0.00001*	0.00001*	0.00001*	0.00001*	0.00001*	0.00001*
$\alpha_1(s_t) + \beta_1(s_t)$	1.0043	1.9492	0.9765	0.9516	1.3032	0.9686	0.9974	0.9920	0.9937	0.8312	0.8267	0.7900

Note: * indicate significant value.

Thai stock market. This result indicates that the conditional volatility not only responds asymmetrically to normal innovations, but also to jump shocks with augmented news intensity. From literature review, Thailand macroeconomic news announcement has a significant effect on Thailand stock market has been supported by previous studies. [17], using daily and monthly data and found that SET responded to Thailand's surprising news. [18] suggested that the important domestic macroeconomic announcements such as GDP and inflation had significant impact on SET. We expect that our empirical results fail to capture the main assumption because there are many important macroeconomic news variables while the variables used in this study might not have correlation or not major surprises with Thai stock market.

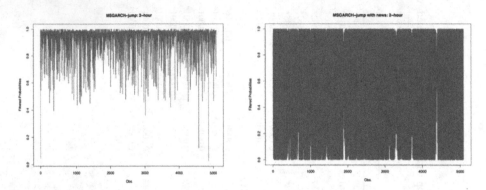

Fig. 1. Filtered probabilities of SET index: 2-h data set. Vertical scales for each panel are different. Left panel is filtered probabilities plots of MSGARCH-jump model. Right panel is filtered probabilities plot of MSGARCH-jump augmented with news mode

5 Conclusions

In this paper, we investigate the MSGARCH-jump model and suggests MSGARCH-jump augmented with news intensity model to estimate the effect of five Thailand macroeconomic news on Thailand stock market using different high-frequency data set from January 2011 to December 2016. The empirical results demonstrate that the Markov Switching GARCH-Jumps augmented with news intensity model has higher volatility but does not perform better than pure Markov Switching GARCH model with jumps. We expect that the adding of macroeconomic news announcements might lead to a higher likelihood and thereby leading to a higher AIC. We also found that Skew Students' t distribution is suitable for MSGARCH-jump augmented with news intensity model. Secondly, only 2-h sample set is significant to explain the volatility in market regimes of both models.

Thirdly, the estimated coefficient of number of news indicates that the effects of negative and positive news on SET are the same. Finally, standardization coefficients have small positive significant value. This suggests Thailand stock market was less affected by the five macroeconomic news announcements concerning foreign trade exports, foreign trade imports, consumer price index, gross domestic product, and customs trade balance.

Acknowledgement. This work was supported by Centre of Excellence in Econometrics, Faculty of Economics, and Chiang Mai University, without which the present study could not have been complete.

References

1. Andersen, T.G., Bollerslev, T., Diebold, F.X., Vega, C.: Real-time price discovery in global stock, bond and foreign exchange markets. J. Int. Econ. **73**(2), 251–277 (2007)
2. Rangel, J.G.: Macroeconomic news, announcements, and stock market jump intensity dy- namics. J. Bank. Finan. **35**(5), 1263–1276 (2011)
3. Huang, X.: Macroeconomic News Announcements, Systemic Risk, Financial Market Volatility and Jumps (2015)
4. Hussain, S.M.: Simultaneous monetary policy announcements and international stock markets response: an intraday analysis. J. Bank. Finan. **35**(3), 752–764 (2011)
5. Mastronardi, R., Patan, M., Tucci, M.P.: Macroeconomic News and Italian Equity Market (2013)
6. Gurgul, H., Wjtowicz, T., Suliga, M.: The reaction of intraday WIG returns to the US macroeconomic news announcements. Metody Ilosciowe w Badaniach Ekonomicznych **14** (1), 150–159 (2013)
7. Merton, R.C.: Option pricing when underlying stock returns are discontinuous. J. Finan. Econ. **3**(1–2), 125–144 (1976)
8. Chan, W.H., Maheu, J.M.: Conditional jump dynamics in stock market returns. J. Bus. Econ. Stat. **20**(3), 377–389 (2002)
9. Duan, J.C., Ritchken, P., Sun, Z.: Approximating GARCH-jump models, jump-diffusion processes, and option pricing. Math. Finan. **16**(1), 21–52 (2006)
10. Maheu, J.M., McCurdy, T.H.: News arrival, jump dynamics, and volatility components for individual stock returns. J. Finan. **59**(2), 755–793 (2004)
11. Lu, X., Kawai, K.I., Maekawa, K.: Estimating bivariate garch-jump model based on high frequency data: the case of revaluation of the chinese Yuan in July 2005. Asia-Pac. J. Oper. Res. **27**(02), 287–300 (2010)
12. Sidorov, S.P., Revutskiy, A., Faizliev, A., Korobov, E., Balash, V.: GARCH model with jumps: testing the impact of news intensity on stock volatility. In: Proceedings of the World Congress on Engineering, vol. 1 (2014)
13. Hamilton, J.D.: A new approach to the economic analysis of nonstationary time series and the business cycle. Econometrica: J. Econom. Soc. **57**(2), 357–384 (1989)
14. Bloomberg L.P: Thailand stock market definiton (2017). Bloomberg terminal. Accessed 8 Aug 2017
15. Balduzzi, P., Elton, E.J., Green, T.C.: Economic news and bond prices: evidence from the US Treasury market. J. Finan. Quant. Anal. **36**(4), 523–543 (2001)
16. Cui, J., Zhao, H.: Intraday jumps in China's Treasury bond market and macro news announcements. Int. Rev. Econ. Finan. **39**, 211–223 (2015)
17. Sidney, A.E.: The impact of information announcements on stock volatility. Assumpt. Univ. J. Manag. **10**(1) (2012)
18. Gok, I.Y., Topuz, S.: The impact of the domestic and foreign macroeconomic news announcements on the Turkish stock market. Finan. Stud. **20**(3), 95–107 (2016)

The Optimizing Algorithm for Economic Cycles in ASEAN Stock Indexes

Satawat Wannapan[✉], Pattaravadee Rakpuang, and Chukiat Chaiboonsri[✉]

Faculty of Economics, Puay Ungphakorn Centre of Excellence in Econometrics,
Chiang Mai University, Chiang Mai, Thailand
lionz1998@gmail.com, chukiat1973@gmail.com

Abstract. This paper is aimed to employ econometric tools for clarifying switching regimes inside time-series trends of ASEAN's stock indexes as well as suggesting the proportional target to invest in an optimal portfolio The tools are the Markov-Switching model (MS-model) and Markovian optimization for portfolio model selecting, respectively. Daily sampled stock variables in five sectors such as banking system, energy, financial agriculture, telecommunication, and real estates were collected during 2010 to 2017. Technically, the condition of stationarity in collected data is verified by using the ADF unit-root test. Empirically, the findings of the switching states estimation show that financial markets in ASEAN countries have been continuously growing since 2010. In other words, there are 1,358 times standing for bull periods and they are more than bear periods which are 322 times during total collected 1,680 days. The second crucial process is the portfolio optimization. The empirical results indicate that the most efficient choice to minimize the risk values in portfolios following optimal solutions is to focus on long-term investments rather than speculative investments. For bull periods, the stock exchange regarding real estate (LogProb) in Singapore potentially provides the best opportunity to invest. Conversely, the banking stock index (Maybank) in Indonesia is the stock that provides the most safety choice and lowest risk value when the ASEAN stock markets are in bear periods.

Keywords: MS-model · Bull markets · Bear markets
Markovian optimization · ASEAN stock indexs

1 Introduction

The investment attraction is inevitable that a lot of new investors in the world financial market are intensively focused. At the present moment, continentally economic integrations such as North American Free Trade Agreement (NAFTA), European Economic Community (EEC), even Association of Southeast Asian

© Springer International Publishing AG, part of Springer Nature 2018
V.-N. Huynh et al. (Eds.): IUKM 2018, LNAI 10758, pp. 420–432, 2018.
https://doi.org/10.1007/978-3-319-75429-1_35

Nation (ASEAN) are being connected by financial transferences; we can call this phenomenon like *Global Money Flows*. Obviously, to increase opportunity of investments, improve bargaining power, even exchange resources for making high returns and reducing risks are the principle that investors always keep in their rules. Interestingly, there is another point that we should not ignore. The question is where is the most considerable investment area in the world that we can recommend investors to pay their attentions.

Considering the new potential emerging market in the recent days, financial investments in ASEAN continent is the talk of the town. Because the conditions of plenty natural resources, populations of skilled labors, even rapid rates of economic expansions, these positively impact ASEAN financial markets to grow up simultaneously. Consequently, this paper is proposed to study on problems of investments due to choose the best choice for creating portfolio models, which are relied on five types of ASEAN predominant stock indexes, including a financial sector; Bank MayBank (Indonesia), an energy sector; PPB GROUP BERHAD (Malaysia), food & beverage index; Golden Agri-Resource (Singapore), telecommunications; SingTel (Singapore), and real estates; Global Logistic Properties (Singapore). Importantly, before launching investments, real business cycles should be theoretically reviewed. Helpfully, the real business cycle theory can explain on the period of economic movements (expansions, recessions and contractions) as well as show the time-varying model. Reasonably, we can apply this theory combining with computationally econometric analyses to investigate the complexity of financial markets. Hence, the result of this research would be an efficient suggestion for deciding to invest in ASEAN stock indexes.

2 The Objective and Scope of Research

This research is aimed to consider five stock indexes in ASEAN including Bank MayBank Index, PPB GROUP BERHAD index, Golden Agri-Resource index, SingTel index and Global Logistic Properties index. All of indexes were observed as daily time-series data during 2010 to 2016. Methodologically, dispositions of the indexes as regimes (bull markets and bear markets) by the MS-VAR model are employed, computationally investigate extreme events by GPD extreme value analysis, and showed a percentage of each index for deciding the portfolio investment by the Value at Risk model (VaR). To clarify all technical processes, details were graphically displayed in Fig. 1.

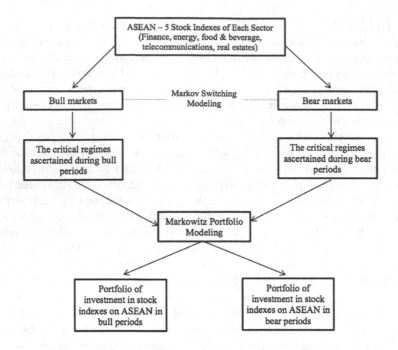

Fig. 1. The conceptual framework of methodological processes of research

3   Literature Review

The theoretical idea to estimate economic systems is Real Business Cycle Theory (RBC) that implies economic cycling movements. This issue is interested by economists to econometrically study. For instances, Plosser (1989) and Vecchi (1999) examined the concept of real business cycles to clarify advantages and disadvantages of regimes in economy. Besides, Dobrescu and Palcu (2012) studied on the core of the classical real business cycle theory and applied it to implement relevant economic policies. Moreover, there are many academic papers which were relied on the RBC model, for example, Gregory (1989) and Jakub (2010).

According to the economic system based on the real business cycling theory, fluctuated regimes can be technically divided into boom period (bull markets) and recession period (bear markets) by the Markov theory. Historically, the study of Engel (1994) and Nielsen and Olesen (2000) were applied the Markov Switching Model (MS-model) to econometrically investigate rates of quarterly currency exchanges and stock indexes, and the result indicated the fluctuated trends of these two variables can be more precisely estimated. Moreover, Otranto (2001) and Cermeo (2002) employed MS-model to demonstrate the panel relation of growth per capita rates in the OECD economy. Consequently, it is undeniable that studying cycling movements in economic systems is being required to combine with econometric tools in the current moment.

In this paper focusing on financial economics, another important concept is the Markowitz portfolio optimization. This idea relied on portfolio planning as well as the formula involved the maximization of estimated returns and minimization of risks. Historically, there were many applied financial researches employing this optimal concept to do risk management, for instance, Mangram (2013), Mirucek et al. (2015), and Lee et al. (2016).

4   Methodology

4.1   Augmented Dickey-Fuller Unit-Root Testing

Fundamentally, the ADF test is given by

$$y_t = \beta' D_t + \phi y_{t-1} + \sum_{j=1}^{p} \psi_j \Delta y_{t-j} + \varepsilon_t \tag{1}$$

From Eq. 1, the ADF test trials the null hypothesis there a time series y_t is $I(1)$ against the alternative that is $I(0)$. That test is based on evaluating the test regression. As the details in Eq. 1, A vector of deterministic terms is D_t. The vector that used to estimate the ARMA structure or the errors are p and Δy_{i-j}. Besides, The value of p can set from the error of a vector ε_t. The deterministic terms is assumed to conduct the vectors y_t. According to the assumption, it is $I(1)$ which implies that $\phi = 1$,

$$ADF_t = t_{\phi-1} = \frac{\hat{\phi} - 1}{SE(\phi)}, \tag{2}$$

$$ADF_n = \frac{T(\hat{\phi} - 1)}{1 - \hat{\psi}_1 - \ldots - \hat{\psi}_p}. \tag{3}$$

From those formulas shown in Eqs. 2 and 3, they are t-statistics and normalized bias statistics, which are based on the least squares estimates of the ADF equation. The ADF regression of the alternative formulation is

$$\Delta y_t = \beta' D_t + \pi y_{t-1} + \sum_{j=1}^{p} \psi_j \Delta y_{t-j} + \varepsilon_t. \tag{4}$$

Considering details in Eq. 4, the parameter of historical information is $\pi = \phi - 1$. Under the null hypothesis, Δy_t, is $I(0)$ if the lag parameter, $\pi = 0$, has no impact to the model and this implies data is stationary.

4.2   The Generalized Concept of the Multivariate Markov Switching Model

This research is aimed to employ the multivariate Markov switching model, which is to clarify that our model has an effect on changes of any factor. For the data set,

$y_t = (y_{1,t}, \ldots, y_{p,t})$, the vector p is the number of vectors of sort $(1, p)$. It explained states on $S_t = [1, \ldots, M]$ as unobserved variables, allowing Markov chain in the first order condition. The first state, $S_t = 1$, stands for the time series data that is implied the lowest (resp. the highest) regimes. The matrix is illustrated as (Hamilton 1994)

$$P = \begin{bmatrix} p11 & \ldots & pM1 \\ p12 & \ldots & pM2 \\ . & \ldots & . \\ . & \ldots & . \\ p1M & \ldots & pMM \end{bmatrix}, \quad \text{with} \sum_{j=1}^{M} p_{kj} = 1, \text{ and } p_{kj} \geq 0, \forall k, j \in \{1, \ldots, M\},$$

where $I_{t-1} = (y_{t-1}, \ldots, y_1)$, the data set is accessible and caused by time variables, $t - 1$. This can be expressed as

$$p_{kj} = P(S_t = j | S_{t-1} = k, \ldots, S = l, I_{t-1}) = P(S_t = j | S_{t-1} = k), \quad (5)$$

where $\forall k, j \in \{1, \ldots, M\}$. The *Shaddow Random Variable*, ζ_t, is signified for a vector $(M, 1)$ which is j is the constituent described as follows

$$\zeta_t^j = I_{s_{t-j}}, \quad \text{and } \zeta_t = (\zeta_t^1, \ldots, \zeta_t^M), \quad (6)$$

when $S_t = j$ then $\zeta_t = e_j'$. e_j is the jth column vectors of the individuality matrix, I_M,

$$\zeta_t = \sum e_j' \times \zeta_t^j. \quad (7)$$

The conditional likelihood connected to the state j is forwarded as

$$P(S_t = j | I_t) = E(\zeta_t^j | I_t), \quad (8)$$

and the vector of filtered possibilities is

$$P(S_t | I_t) = E(\zeta_t | I_t) = \hat{\zeta}_{t/t} = (E(\zeta_t^1) | I_t), \ldots, E(\zeta_t^M)). \quad (9)$$

In this case, we assumed the vectors of data sets as $x_t = (x_{1,t}, \ldots, x_{n,t})$ and $z_t = (z_{1,t}, \ldots, z_{q,t})$ to be the vectors of exogenous regressors subjected to switching regimes. The parameters, $\beta_{S_t} = (\beta_{S_t}^1, \ldots, \beta_{S_t}^p)$ are entered into the matrix of regression-coefficients which are dependent regimes, and $\delta = (\delta^1, \ldots, \delta^p)$ are added into the matrix of regression-coefficients which are independent regimes. These can be expressed as

$$\Gamma_{S_t} = (\beta_{S_t} | \delta)', \quad A_t = (x_t | z_t). \quad (10)$$

The majority of models evaluated with MS-VAR estimating are called a Markov Switching Gaussian Model, which is explained as

$$y_t = A_t \Gamma_{S_t} + u_t, \quad u_t | S_t \, N(0, \sum_{s_t}), \quad (11)$$

with

$$\sum\nolimits_{S_t} = \begin{bmatrix} \sigma_{11}(S_t) & .. & \sigma_{1M}(S_t) \\ . & . & . \\ . & . & . \\ \sigma_{M1}(S_t) & .. & \sigma_{MM}(S_t) \end{bmatrix},$$

such that

$$\sigma_{ii}(S_t) = \sigma_i^2(S_t), \quad \sigma_{ii}(S_t) \geq 0$$
$$\sigma_{ii}(S_t) = \rho_{ij}(S_t \sigma_i(S_t) \sigma_j(S_t)), \quad \forall i \neq j \in \{1, \dots, M\}.$$

From the expressed details, these are adaptably independent and dependent regime regressors.

4.3   Markowitz Portfolio Model Selection

The theory of optimal selection of portfolios was developed by Harry Markowitz in the 1950's, which earned him the 1990 Nobel Prize for Economics. Basically, the idea of his work considers about an investor who has a certain amount of money to be invested in a number of different factors (bonds, equities, and macroeconomic variables) with random returns. For each variable, $i = 1, \dots, n$, estimates of its expected outcome, u, and variances, σ_i^2 are given. In addition, the correlation coefficient p_{ij} between two variables i and j is technically assumed to be known. According to this research, the proportion of the total amount in variable i by x_i, one can compute the expected return and the variance of the resulting portfolio $x_i = (x_i, \dots, x_n)$ as follow Cornuejols and Tütüncü (2006). This can be expressed in Eq. (12),

$$E[x]_{BOOM}, = x_1 u_i + \dots + x_n u_i = u^T x, \tag{12}$$

then

$$Var[x]_{BOOM}, = \sum_{ij} \rho_{ij} \sigma_i \sigma_j x_i x_j = x^T x, \tag{13}$$

and

$$E[x]_{RECESS}, = x_1 u_i + \dots + x_n u_n = u^T x, \tag{14}$$

such that

$$Var[x]_{BOOM}, = \sum_{ij} \rho_{ij} \sigma_i \sigma_j x_i x_j = x^T x, \tag{15}$$

where $\rho_{ij} = 1, Q_{ij} = \rho_{ij} \sigma_i \sigma_j$.

Generally, the portfolio vector x must satisfy $\sum_i = 1$ and there may or may not be added the feasible constraints. A feasible portfolio x is explained as efficient. However, economically, since economic systems have many latent variables as well as equilibrium points only occur theoretically. Consequently, the vector x is allowed to be more than 1, $\sum_i x_j \geq 1$.

5   Empirical Results of Research

5.1   Descriptive Information

In the first section, the basic indicators are demonstrated in Table 1. Technically, the collected data is transformed into log-return formation. The total observations are 1,680 samples, which are daily time-series trends of highly potential stock indexes in ASEAN during 2010 to 2017. There are five types of financial markets such as Maybank (banking), PPBgroup (energy), GoldAgri (agriculture), SingTel (telecommunication), and LogProb (real estate), and they were collected from three countries in ASEAN, including Malaysia, Indonesia and Singapore. Additionally, these five indicators contain the condition of normality checked by Jarque-Bera testing, and all of them are stationary by employing the ADF unit-root test. As a result, collected data is ready to computationally explore by econometric applications.

Table 1. Presentation descriptive data of ASEAN stock indexes during 2010 to 2017 (1,358 days)

	Maybank (Banking)	PPBgroup (Energy)	GoldenAGRI (Agriculture)	SingTel (Telecommunication)	LogProb (Real estate)
Mean (log return rate)	0.00001	−0.00001	−0.00024	0.00014	0.00016
Median (log return rate)	0.000000	0.000000	0.000000	0.000000	0.000000
Maximum (log return rate)	0.29523	0.11368	0.10821	0.06130	0.08305
Minimum (log return rate)	−0.14921	−0.05942	−0.08168	−0.07853	−0.09228
Std. dev	0.02549	0.01205	0.01993	0.01170	0.01590
Skewness	3.11911	0.70660	0.34371	−0.10080	0.11294
Kurtosis	34.15417	11.35068	4.76926	5.92728	6.12893
Jarque-Bera	70,664.86	5,021.175	252.1995	602.6725	688.8862
Probability	0.000000	0.000000	0.000000	0.000000	0.000000
ADF testing (t-statistics)	0.000000 (−44.60448)	0.000000 (−38.19114)	0.000000 (−39.54392)	0.000000 (−41.49348)	0.000000 (−37.85026)
Observations	1,680	1,680	1,680	1,680	1,680

5.2   Expected Durations of Switching Regimes by the Markov Switching Model

In the section of switching regime calculating, the MS model is a powerful econometric tool to clarify the factors relied on Real Business Cycle theory (RBC model). As seen in Fig. 2, daily time-series facts of ASEAN stock exchanges are dramatically fluctuated. This can be implied that collected data contains a booming situation and a recessing period. In this paper, the up and down situations are financially defined as bull and bear markets. This classification is graphically shown in Fig. 3. State 1 which stands for bull markets occurs when the smoothed states probabilities equal one. On the other hand, bear markets would be risen when the smoothed states probabilities of State 2 equal one. Empirically, the estimated results of switching states are 1,358 and 322 times, which refer to bull markets and bear markets, respectively. The facts are presented in Table 2. Consequently, because the collected data contains bull situations more that bear markets. The result can be indicated that financial markets in ASEAN countries have been growing since 2010.

5.3 Markovian Portfolio Selection

The last section is the Markovian optimization for portfolio selecting. This optimal model is employed to clarify a suitable choice for investing in stock markets. At the beginning, the general facts shown in Table 3 represent the average return and risk value in bull and bear situations. Expressly, the stock of the real estate (LogProb) is defined to be the sector that provides the highest return (mean) when bull periods occur, which equals 0.00028. Conversely, the stock of energy companies (PPBgroup) is indicated to be the index which contains the lowest risk value (variance) when bull periods arise during 7 years, which is 0.00008. Considering the market recessions, the stock of banking systems (Maybank) is found to be the sector that involves the highest stock return (mean) when bear situations appear, which is 0.00576. The stock of telecommunications (SingTel), on the other hand, is defined to be the index that contains the lowest risk (variance) in bear periods, which equals 0.000294.

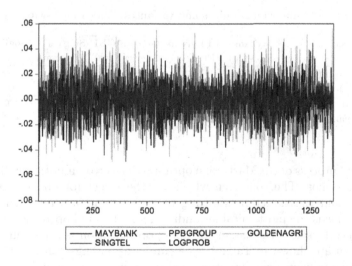

Fig. 2. Descriptive trends of ASEAN stock exchanges

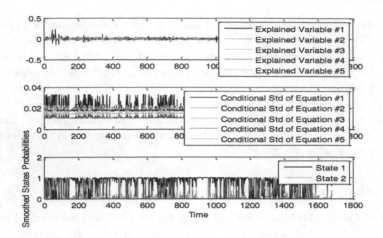

Fig. 3. Presenting the estimated durations (smooth probabilities) of switching regimes

Table 2. Showing estimated switching regimes of ASEAN stock exchanges

	ASEAN stock index (Maybank, PPBgroup, GoldenAGRI, SingTel, LogProb)
Bull periods	1,358
Bear periods	322

Table 3. Presentation average values and variances divided into switching regimes

Stage	Details	Maybank (Banking)	PPBgroup (Energy)	GoldenAGRI (Agriculture)	SingTel (Telecom)	LogProb (Real estate)
Bull periods	Mean (u)	−0.00135	0.00010	−0.00046	0.00013	0.00028
	Variances (σ)	0.00016	0.00008	0.00030	0.00010	0.00018
Bear periods	Mean (u)	0.00576	−0.00044	0.00065	0.00020	−0.00031
	Variances (σ)	0.00267	0.000421	0.0008	0.000294	0.000558

The next process of the Markovian optimization is to computationally explore optimal portfolios. The condition which are the target minimizing risk values (variances) and the target maximizing returns (mean) are included to verify that which one is the most suitable condition and it would optimize the benefits of portfolios. To select the target for doing optimization is relied on the sharp ratios, which are shown in Table 4. The empirical findings can be implied that the risk value management (optimal minimizing variance) should be mentioned as the condition for calculating the optimal portfolio in bull periods. The result shows that the sharp ratio by using the target minimizing variances, which equals

Table 4. Presentation mean-variance portfolio optimization

Stage	Details	Neutral stock markets	Portfolio optimal calculation (Minimizing variances)	Portfolio optimal calculation (Maximizing means)
Bull periods	The average return, u	−0.00026	0.00028	0.00010
	The risky value of whole markets, σ	0.00596	0.00617	0.00370
	The sharp ratio of whole market, u/σ	−0.04374	0.04504	0.02636
Bear periods	The average return, u	0.00117	0.00576	−0.00044
	The risky value of whole markets, σ	0.01428	0.02319	0.00877
	The sharp ratio of whole market, u/σ	0.08195	0.24842	−0.05021

0.04504, is the highest value when we compare with the condition maximizing returns and non-optimal calculation. Similarly, the target minimizing risk values is the most suitable condition for calculating the optimal portfolio in bear periods. The result states that the sharp ratio in which minimizing variances equals 0.24842 and it is the highest ratio when we compare with maximizing returns and non-optimal calculation. Accordingly, it is sensible to conclude that employing the optimal condition for selecting the high-return portfolio is better and more efficient than investing by personal perceptions.

The eventuated process in the optimal portfolio management is to statistically compute the proportions for portfolio investing. Considering bull periods which details are represented in Table 5, the conditional target to minimize risk values for portfolio selecting, relying on the decision making from the sharp ratio in Table 4, is employed to estimate the proportion of portfolio investments. The empirical result presents that the financial market of real estates in Singapore potentially has a good opportunity. The optimal estimated portfolio, in other words, provides 100% for directly investing in real-estate companies, especially the LogProb index. Consequently, this can be implied real properties in Singapore contain the lowest risk when ASEAN's capital markets are continuously expanded.

According to the optimal portfolio management during recession situations, details are presented in Table 6 and the target to minimize risk values for portfolio selecting is chosen by comparing the sharp ratio in Table 4 to calculate the proportion of portfolio investments. The empirical findings show the financial market of banking systems in Indonesia potentially has a good chance to invest. The optimal estimated portfolio, in other words, provides 100% for directly investing in banking companies. Hence, this can be indicated financial investments in banking sectors in Indonesia, for example, Maybank contain the lowest risk when ASEAN's capital markets are continuously declined.

Table 5. Showing Markovic portfolio proportions considering the bull market

Stock indexes	Investment portions in bull markets		
	Normal cycling movement (%)	Minimizing risk optimization (variance) (%)	Maximum return optimization (mean) (%)
Maybank (Banking): Indonesia	20.00	0.00	0.00
PPBgroup (Energy): Malaysia	20.00	0.00	100.00
GoldenAGRI (Food/bevarage): Singapore	20.00	0.00	0.00
SingTel (Telecoms): Singapore	20.00	0.00	0.00
LogProb (Real estate): Singapore	20.00	100.00	0.00

Table 6. Showing Markovic portfolio proportions considering the bear market

Stock indexes	Investment portions in bear markets		
	Normal cycling movement (%)	Minimizing risk optimization (variance) (%)	Maximum return optimization (mean) (%)
Maybank (Banking): Indonesia	20.00	100.00	0.00
PPBgroup (Energy): Malaysia	20.00	0.00	100.00
GoldenAGRI (Food/bevarage): Singapore	20.00	0.00	0.00
SingTel (Telecoms): Singapore	20.00	0.00	0.00
LogProb (Real estate): Singapore	20.00	0.00	0.00

6 Conclusion

Since the ASEAN continent is considerably mentioned as a new potential emerging market and the literature review has confirmed that ASEAN's stock markets have been continuously growing during 2010 to 2017, which are collected as daily time-series indexes. This paper is proposed to apply economic tools, which are the Markov-Switching model (MS-model) and Markovian optimization for portfolio model selecting. The former model is employ to investigate switching regimes inside time-series trends. The latter approach is applied for suggesting the proportional target to invest in an optimal portfolio. Technically, five types of ASEAN stock exchanges such as banking system, energy, financial agriculture, the stock index of communication, and real properties are all stationary and observed in Malaysia, Indonesia. Empirically, the findings of the first section estimated by the MS model are indicated that there are 1,358 days standing for bull periods and 322 days referring to bear periods. This can be implied that financial markets in ASEAN countries have been continuously expanding since 2010.

Moreover, the results of the second approach is statistically calculated by the Markovian optimization with the conditional target with minimizing the risk values (variance minimization). To express optimal portfolio estimation, it is reasonable to confirm that employing the optimal condition for making decision on the non-risky portfolio is better and more efficient than investing by personal perceptions. This is backed up by the calculation of the sharp ratios in Table 4. Additionally, the computational results are stated that the stock exchange regarding real properties (LogProb) in Singapore potentially provides a good opportunity for beneficially investmenting in bull periods. On the other hand, the banking stock index (Maybank) in Indonesia is the one that provides the most safety choice and lowest risk value when the ASEAN stock markets are in bear periods.

In conclusion, the most efficient way to minimize the risk values in portfolios following solutions in this paper is to focus on long-term investments rather than speculations. To conduct more sensibly optimal proportions in portfolio managements requires some advanced econometric tools, including a structural dependent estimation, an asymptotically computational approach, even a non-parametric method. Interestingly, this is not only the benefit for financial sectors, but it can be applied in other academic fields to clarify the optimal proportions and solutions which are elusive.

References

Bellone, B.: Classical Estimation of Multivariate Markov-Switching Models using MSVARlib. Working Paper, Paris (2005)

Cermeno, R.: Growth convergence clubs: evidence from Markov-switching models using panel data. Divisin de Economa, CIDE, Mexico (2002)

Cornuejols, G., Tütüncü, R.: Optimization Methods in Finance. Carnegie Mellon University Press, Pittsburgh (2006)

Dickey, D., Fuller, W.: Distribution of the estimators for autoregressive time series with a unit root. J. Am. Stat. Assoc. **74**, 427–431 (1979)

Dobrescu, M., Palcu, C.E.: New approaches to business cycle theory in current economic science. Theor. Appl. Econ. **7**(572), 147–160 (2012)

Engel, C.: Can the Markov switching model forecast exchange rates? J. Int. Econ. **36**, 151–165 (1994)

Hamilton, J.: Time Series Analysis. Princeton University Press, Princeton (1994)

Jakub, G.: Real business cycle theory methodology and tools. Econ. Sociol. **3**(1), 42–48 (2010)

Lee, H.S., Cheng, F.F., Chong, S.C.: Markowitz portfolio theory and capital asset pricing model for Kuala Lumpur stock exchange: a case revisited. Int. J. Econ. Fin. Issues **6**, 59–65 (2016). https://www.econjournals.com/

Mangram, M.E.: A simplified perspective of the Markowitz portfolio theory. Global J. Bus. Res. **1**(7) (2013)

Gregory, M.N.: Real business cycles: a new Keynesian perspective. J. Econ. Perspect. **3**(3), 79–90 (1989)

Nielsen, S., Olesen, J.O.: Regime-switching stock returns and mean reversion. Working Paper 11. Department of Economics - Copenhagen Business School Solbjerg Plads 3 (2000)

Otranto, E.: The stock and Watson model with Markov switching dynamics: an application to the Italian business cycle. Stat. Appl. **13**(4), 413–428 (2001)

Plosser, C.I.: Understanding real business cycles. J. Econ. Perspect. **3**(3), 51–78 (1989)

Sirucek, M., Kren, L.: Application of Markowitz portfolio theory by building optimal portfolio on the US stock market. AcTa universitatis Agriculturae et Silviculturae Mendelianae Brunensis, **63**(4) (2015)

Vecchi, M.: Real business cycle : a critical review. J. Econ. Stud. **26**(2), 159–172 (1999)

European Real Estate Risk and Spillovers: Regime Switching Approach

Nisara Wongutai, Woraphon Yamaka, and Roengchai Tansuchat[✉]

Faculty of Economics, Center of Excellence in Econometrics,
Chiang Mai University, Chiang Mai, Thailand
roengchaitan@gmail.com

Abstract. In this study, we employ an international version of the Capital Asset Pricing Model (ICAPM) to measure and examine the relationship between each of the 13 European national real estate returns and the European stock market returns over the period from January 1999 to December 2016. Our measure of this relationship is the time-varying European real estate beta, which is an index of European real estate's systematic risk. This time varying beta of the ICAPM for a country's real estate market is computed by the ratio of the covariance between the expected returns of each country and the European stock market portfolio to the variance of the expected returns on the European stock market portfolio. For this purpose, we first find the location of the crisis and period in time using a Markov Switching approach, Then the AR(1)-GJR(1)- Dynamic Conditional Correlation model is estimated to obtain both covariance and variance of each pair return.

Keywords: European real estate · Markov-switching regression
Spillover effect · Beta risk · ICAPM

1 Introduction

Real estate is one of the most important sectors in the analysis of economic and financial crises [15]. Several crises occur because of the bubbles in this sector. The present study deals with real estate in Europe. From Fig. 1, we can observe that the real estate price was in a boom in 2004 and then dramatically declined and reached the lowest around 2010 corresponding to the global financial crisis in America and Europe. According to the data from Royal Institution of Chartered Surveyors (2009), European real estate price went down by 9.1% in early 2008. At the same time, Parmy Olson, a well-known journalist on subprime crisis and working for Forbes, said that public sector was troubled with debt and then a massive debt became the European sovereign debt crisis in 2008. Therefore, we can assume that the movement of the real estate price can be the factor responding to the crisis.

European Central bank states that European real estate markets are confirmed to have a highly significant systemic risk during financial crisis. The main reasons of systemic risk are leveraged financing of construction project and house purchases. Construction sector is considerably important in economy and changes slowly in real estate supply. The best example of systemic financial crisis is the boom-bust cycles in

© Springer International Publishing AG, part of Springer Nature 2018
V.-N. Huynh et al. (Eds.): IUKM 2018, LNAI 10758, pp. 433–444, 2018.
https://doi.org/10.1007/978-3-319-75429-1_36

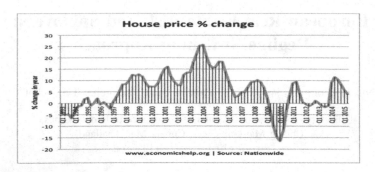

Fig. 1. Housing price percentage change in Europe.

real estate market around 1990s which happened in Greece, Ireland and Spain [11]. Finance Management of Austin Bank of Chicago reported that crises always cause problems to economy and finance because there is a positive relationship between them. Following Blaug [2] studies, there are four periods of property cycle. The first one is boom phase which is the shortest period in the cycle. The price of real estate rapidly increases and the investors would know this phase as a starting for increasing price of real estate. The second one is slump phase where there is an oversupply of property. This situation is a result of the increase in vacancy rate and the decrease in rental return. Next is stabilization phase which is a short period. The economic factors are catching up with each other and stability in market occurs. The last phase is the upturn or recovery phase. The vacancy rate is falling slowly, rental return is rising and property values are starting to increase. Therefore, the nonlinear behavior and the dynamic movement should be considered in this study and thereby employing the Markov switching and dynamic correlation approach.

In this study, we consider the real estate stock index as the representative of the real estate price and it is reasonable to capture the upturn and downturn in the cycle of real estate stock return. The detection of upturn or downturn is beneficial to investors as a signaling period of investment. Normally, investors would buy in the downturn and sold in the upturn. In other cases, if downturn is continually falling, which is called crisis, investors could lose their wealth. Nevertheless, the investors still face with the risk and uncertainty in the market. To quantify this risk, the Capital Asset Pricing Model (CAPM) of Treynor is a good choice [16]. Some examples of the studies that use CAPM include Zabarankin et al. [17]. Another example is Kim and Kim [13]. After that, this model was developed into International Capital Asset Pricing Model (ICAPM) by Fama and French [7] because the conventional CAPM allows no transaction costs and risk averse. Adler and Dumas [1] used ICAPM to measure and examine the relationship between each of the 16 national real estate returns and the global stock market returns. After 1980s, there was ample evidence suggesting that beta risk obtained from CAPM can change over time, see Bollerslev et al. [3], Ferson and Korajczyk [8] and Jagannathan and Wang [12]. According to Bollerslev et al. [3], and Maneejuk et al. [14] time varying beta is appropriate to use expressly with economies that have upturn and downturn where the periods of upturn and downturn can be detected by Markov Switching regression model.

Although previous studies have reported the works on the joint behavior of the respective volatilities between national/regional real estate markets and the global stock market, there has been hardly any works on the joint behavior of the respective volatilities between national real estate markets and the European stock market. To contribute to the existing knowledge on this topic, in this study we systematically analyze two key aspects of European real estate beta for a sample of 13 national real estate markets. In addition, all European markets may have relationship with one another. Consequently, beta with time-varying in one country may have relationship in other countries. In other words, betas of the concerned countries may have correlation with and spillover effects on other countries. Thus, the study outcome is beneficial for investors who are investing in one country of Europe as they would know and beware of the effects and influence from other European countries.

In the next section we outline the methodology. This is followed by presenting the data description in Sect. 3 and empirical results and discussion of the implications in Sect. 4. Section 5 concludes.

2 Methodology

2.1 Identification of Crisis Period Using Markov Switching Regression

As structural change often occurs in the economic and financial market, appropriate econometric model is needed to detect the location of the crisis period in time series. Thus, in this study, we aim to detect the location of the crisis or even the downtrend of the individual market behavior by employing a Markov-switching regression model as proposed in Goldfeld and Quandt [10]. Markov switching Regression is applied to ICAPM to detect upturn and downturn of each national real estate market. Markov process can estimate by using dependent parameters and independent parameters.

$$Y_t = \mu_{s_t} + X_t \phi_{s_t} + \varepsilon_{s_t,t}, \tag{1}$$

where Y_t is a dependent variable or each national real estate market. μ_s is a regime dependent intercept term, ϕ is a regime dependent coefficient term, X_t is an exogenous variable or European market return and $\varepsilon_{s,t}$ is independent identically distributed normal error, $\varepsilon_{s_t,t} \sim i.i.d.\ N\left(0, \sigma_{s_t}^2\right)$. Here, we consider two regimes, consisting of an economic expansion and economic recession, therefore we have two regimes, $s_t = 1, 2$. The latent state variable s_t, that controls the regime shifts, follows a Markov-chain and is assumed to be ergodic irreducible with stationary transition probabilities given by

$$p(s_t = j | s_{t-1} = i) = p_{ij}, \quad \sum_{j=1}^{k} p_{ij} = 1, \text{ for } i, j = 1, \ldots, k. \tag{2}$$

Therefore the first order Markov process of two regimes is written as:

$$p(s_t = 1|s_{t-1} = 1) = p_{11}; \ p(s_t = 2|s_{t-1} = 1) = p_{21}$$
$$p(s_t = 1|s_{t-1} = 2) = p_{12}; \ p(s_t = 2|s_{t-1} = 2) = p_{22}. \tag{3}$$

2.2 Autoregressive Glosten-Jagannathan-Runkle Dynamic Conditional Correlation (AR-GJR-DCC)-GARCH Model

AR-GJR-DCC-GARCH is estimated to capture volatility clustering and the leverage effects (asymmetry) on return volatility arising from both negative and positive shock. Moreover, this model also has an ability to detect possible changes in conditional correlations over time by using DCC part. The DCC model was introduced in Engle [6] to construct the multivariate GARCH. It has the flexibility of univariate GARCH but not the complexity of conventional multivariate GARCH. These models are naturally estimated in two steps namely a univariate GARCH estimation step and the DCC estimation step.

In this study, we also conduct a two-step estimation which was proposed in Engle [5]. In the first step, AR-GJR (1)-GARCH (1,1) of Glosten et al. [9] is estimated for each stock and market returns to obtain an error ε_t^{stock} and ε_t^M. In the second step, we model a bivariate dynamic correlation by $Z_t = diag\{Q_t\}^{-\frac{1}{2}} Q_t diag\{Q_t\}^{-\frac{1}{2}}$ where $diag\{Q_t\}$ is diagonal matrix and Q_t is positive definite to guarantee that Z_t is in the interval $[-1, 1]$.

Then, the DCC equation can be specified as

$$Q_t = (1 - \theta_1 - \theta_2)\bar{Q} + \theta_2 Q_{t-1} + \theta_2 \varepsilon_{t-1}^{stock} \varepsilon_{t-1}^{M'}, \tag{4}$$

$\bar{Q}$ is unconditional correlation matrix of standard error by $\bar{Q} = E(\varepsilon_t^{stock} \varepsilon_t^{M'})$ and also ranges between 1 and −1. The parameter θ_1 and θ_1 must satisfy: $0 \leq (\theta_1 + \theta_2) < 1$.

Finally, the estimated dynamic correlation, $COR_t^{stock,M}$, in the second step and conditional volatility obtained from AR-GJR (1)-GARCH(1,1), h_t^{stock} and h_t^M in the first step are used to compute the time varying beta risk as follows:

$$\beta_t^{stock,M} = \frac{COR_t^{stock,M} \sqrt{h_t^{stock}}}{\sqrt{h_t^M}}. \tag{5}$$

2.3 Forecast Error Variance Decomposition

In this study, we also investigate the beta spillover effect of all stock returns. To do this, we include the all estimated beta risks obtained from Eq. (5) to the Vector Autoregressive (VAR) framework of Diebold and Yilmaz [4]. This VAR model examines the decomposition of forecast error variances through analyzing the total and directional beta spillovers across all stock markets. First of all, let's consider the VAR (p) model structural form:

$$Y_t = V + \sum_{p=1}^{P} \sum_{i=1}^{k} A_i Y_{t-p} + U_t, \qquad (6)$$

where Y_t is the k dimension of endogenous variables; V is $k \times 1$ vector of intercept term; A_i is $kp \times kp$ matrix of autoregressive coefficient, p is number of lag in VAR, and U_t is the error term with normal distribution $(U_t \sim N(0, \Sigma))$. Thus, we can extend (Eq. 7) as follows:

$$\begin{bmatrix} y_{1,t} \\ \vdots \\ y_{k,t} \end{bmatrix} = \begin{cases} v_1 + A_{11}y_{1,t-1} + \ldots + A_{1p}y_{1,t-p} + u_{1,t} \\ \qquad\qquad \vdots \\ v_k + A_{k1}y_{k,t-1} + \ldots + A_{kp}y_{k,t-p} + u_{k,t} \end{cases} . \qquad (7)$$

Then, to calculate the forecast error variance, the mean squared error of the h-step forecast of variable i is

$$\mathbf{MSE}[y_{i,t}(h)] = \sum_{h=0}^{H-1} \sum_{i=1}^{K} (u_i \Theta_h u_i)^2 = (\sum_{h=0}^{H-1} \Theta_h \Theta_h)_{ii} = (\sum_{h=0}^{H-1} \Phi_h \Sigma \Phi_h)_{ii}, \qquad (8)$$

where $\Theta_h = \Phi_h P$, and $\Phi_i = JA^i J'$, P is a lower triangular matrix obtained by a Cholesky decomposition of Σ such that $\Sigma = PP'$ where Σ is the covariance matrix of the errors u_t. J is an identity $k \times kp$ dimensional matrix. The amount of forecast error variance of variable i accounted for by exogenous shocks to variable k is given by

$$\omega_{ik}(h) = \sum_{h=1}^{H-1} (u_i \Theta_h u_i)^2 / \mathbf{MSE}[y_{i,t}(h)]. \qquad (9)$$

3 Data

In this study, we use a weekly data which are the return of each country property index in Europe and European market return index. The data period is from January 1999 to December 2016. Property index in each country in Europe is from S&P index of 13 European countries. They are Austria, Belgium, Denmark, Finland, France, Germany, Ireland, Italy, Norway, Spain, Sweden, Switzerland, and United Kingdom. For the European's index variable, it is the S&P Europe 350. The summary of our data is given in Table 1.

From Table 1, the normality Jarque-Bera test shows that real estate returns of countries in Europe data are non-normally distributed. In addition, Skewness and Kurtosis also show non-normal distribution. Skewness shows a degree of symmetry in the variable distribution. The countries skew to the right are Denmark, Italy, Spain and

Table 1. Data description and stationary test

	Mean	Max	Min	Skew	Kurtosis	JB**	ADF**
Aus	0.0000	0.1168	−0.1848	−1.7682	22.2565	14981.410	−5.3806
Bel	0.0000	0.1011	−0.1602	−1.5761	18.9037	10273.600	−8.6528
Den	−0.0005	0.3238	−0.1408	2.9405	54.6360	105558.80	−29.945
Fin	0.0002	0.0639	−0.1037	−0.7807	6.8773	682.8467	−32.013
Fra	0.0007	0.0671	−0.0854	−0.5016	6.1105	417.4599	−31.967
Ger	0.0000	0.0953	−0.1375	−0.8989	11.6356	3040.8760	−30.284
Ire	0.0000	0.0953	−0.1375	−0.8989	11.6356	3040.8760	−30.284
Ita	−0.0005	0.1771	−0.1187	0.3481	12.2320	3350.3300	−32.644
Nor	−0.0008	0.1514	−0.1722	−0.4212	10.3539	2141.3300	−34.598
Spn	−0.0009	0.2043	−0.1467	0.9936	15.4907	6252.038	−32.234
Swe	0.0010	0.1102	−0.1175	−0.4361	9.0116	1442.155	−33.116
Swit	0.0006	0.0634	−0.0478	0.0159	7.1694	679.4638	−29.530
UK	0.0001	0.0879	−0.1204	−1.0586	12.9764	4065.064	−32.940
Beta	0.0000	0.0836	−0.0676	−0.2926	6.8423	590.3730	−35.057

Source: Calculation

Switzerland whilst the rest skew to the left. The last is stationary test by Augmented Dickey Fuller (ADF)-Statistic. It turns out that every country's real estate return data and European market index are stationary.

4 Empirical Result

4.1 Identification of Crisis Period Using Filtered Probabilities

First of all, filtered probability obtained from the Markov Switching regression is used to detect the crisis regime. In this study, the filtered probabilities of 13 countries are plotted in Fig. 2. Here, we plot only the expansion regime. According to Fig. 2, Austria and Belgium had obviously faced with a recession during 2007–2012, coinciding with the global financial crisis. Denmark had fallen into several turbulent regimes during the period 2001–2002, 2005–2012 and after 2015. Finland and Sweden were Nordic countries facing banking crisis in 2004 and became stable around 2007 but then again have fallen into crisis after 2010. Filtered probability of France went down in 2005 up until now. Germany became economically stable in 2010 to 2015. Spain's filtered probability showed a fluctuating movement during 2007–2015.

In sums, every country in Europe fell into crises during the period 2007–2012 which is corresponding to the US housing crisis which began at the end of 2007 to 2009. After that, the European sovereign debt crisis also occurred from 2010 to the middle of 2012. In some countries, particularly the emerging markets, their wave goes down at the end of 2014 to 2015, which also corresponds to global crisis.

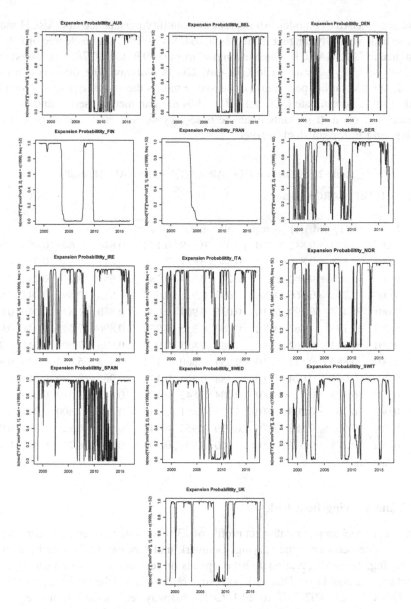

Fig. 2. Expansion period of each market's country.

4.2 European Real Estate Beta Estimates and Dynamic

Table 2 presents the estimates of maximum likelihood for 13 bivariate ARMA(1,1)-GJR-DCC-GARCH models. Here, AR(1) is a parameter of autoregressive, MA(1) is a parameter of moving average, omega is a parameter of variance intercept, alpha is a parameter of ARCH, Beta is a parameter of GARCH and gamma 1 is a parameter of leverage. Summation of ARCH and GARCH parameters is the degree of volatility persistence of GARCH model. All degrees of volatility persistence in GARCH models are close to 1, volatility ranging from 0.8420 to 0.9624. The other is degree of volatility persistence in GJR-GARCH which is ranging from 0.7930 to 0.9778. The lowest of the degree of volatility persistence in GJR-GARCH is Norway. The other countries are above 0.8000. Volatility persistence close to one means the volatility remains for a long period of time. To estimate DCC(1,1), θ_1 and θ_2 are parameters used to find integrated and mean reversion. It presents dynamic time-varying real estate return of European countries and European stock market.

Table 2. Estimation DCC-ARMA(1,1)-GJR-GARCH results

ARMA(1,1)-GJR-GARCH parameter										DCC parameter	
	MU	AR(1)	MA(1)	omega	alpha	Beta	Gamma	skew	shape	θ_1	θ_2
Aus	0.000	0.880	−0.842	0.001	0.051	0.869	0.122	0.844	7.609	0.055	0.934
Bel	0.001	0.888	−0.858	0.001	0.040	0.887	0.119	0.877	8.864	0.030	0.963
Den	0.000	−0.456	0.462	0.001	0.213	0.629	0.284	0.952	3.241	0.288	0.708
Fin	0.000	0.294	−0.366	0.001	0.033	0.921	0.038	0.840	0.595	0.044	0.950
Fran	0.001	−0.417	0.399	0.001	0.036	0.907	0.069	0.880	11.870	0.049	0.094
Ger	0.001	−0.193	0.162	0.001	0.080	0.855	0.056	0.896	7.695	0.041	0.954
Ire	0.001	−0.193	0.162	0.001	0.080	0.855	0.056	0.896	7.695	0.041	0.954
Ita	0.002	0.883	−0.873	0.001	0.034	0.860	0.144	0.918	5.673	0.046	0.941
Nor	0.000	−0.086	0.024	0.001	0.388	0.526	0.145	0.963	3.526	0.310	0.690
Spa	0.002	0.995	−0.978	0.001	0.084	0.842	0.102	0.945	3.380	0.035	0.889
Swed	0.001	0.200	−0.209	0.001	0.027	0.851	0.130	0.858	7.800	0.014	0.981
Swit	0.000	−0.602	0.582	0.001	0.010	0.952	0.040	0.912	6.820	0.052	0.930
UK	0.001	0.288	−0.334	0.001	0.060	0.828	0.109	0.881	8.216	0.012	0.983

Source: Calculation

4.3   Time Varying Beta Risk

Figure 3 is European real estate beta profile of 13 real estate markets. The curves show that real estate betas are time-varying. Denmark and Norway real estate beta are not time-varying. Therefore, real estate beta depends on real estate return. In some periods, their returns are equal to 0. Denmark real estate return has reached 0 during 07/08/2003 to 27/09/2005 and 10/02/2012 to 29/10/2013. Norway real estate return is equal to 0 during 13/07/2004 to 27/09/2005 and 29/11/2005 to 26/09/2006. In addition, most curves are peaking in European real estate beta during crisis period, which was the debt crisis in Europe.

4.4   Effect of Crisis: Expansion/Recession on Time Varying Beta

Table 3 shows conditional risk of European real estate in crisis period. This table is used to study the effect of the crisis with time-varying European real estate beta by using dummy variable that 1 is recession and 0 is expansion. Note that the recession

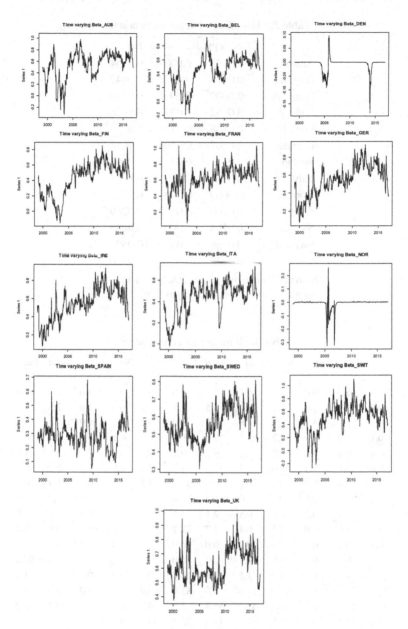

Fig. 3. Time varying beta risk of each market's country.

and expansion periods are detected already in Subsect. 4.1. The results show that coefficient of crisis or recession is positive for six countries. They are Austria, Finland, Ireland, Italy, Spain and United Kingdom. The positive coefficient can give a conclusion that European real estate betas are significantly affected by the crisis period.

Table 3. The effect of crisis on beta risk.

	Intercept	Recession	Beta $_{t-1}$
Austria	0.0133	0.0009	0.9754
Belgium	0.0079	−0.0029	0.9840
Denmark	−0.0002	−0.0001	0.9558
Finland	0.0052	0.0097	0.9728
France	0.0548	−0.0145	0.9183
Germany	0.0207	−0.0103	0.9661
Ireland	0.0105	0.0121	0.9661
Italy	0.0062	0.0075	0.9718
Norway	−0.0009	−0.0008	0.7365
Spain	0.0100	0.0130	0.9289
Sweden	0.0298	−0.0025	0.9485
Switzerland	0.0286	−0.0173	0.9527
UK	0.0320	0.0120	0.9320

Source: Calculation; Note: Dummy variable where 1 = recession, 0 = expansion.

Table 4. Variance Decomposition

Week 10	Aus	Bel	Den	Fin	Fran	Ger	Ire	Ita	Nor	Spa	Swed	Swit	UK
Aus	37.8	13.3	0.0	7.8	5.1	3.3	3.3	4.8	0.0	1.0	0.7	21.6	1.3
Bel	0.0	44.9	0.0	11.1	4.5	5.5	5.5	3.2	0.0	2.6	2.5	17.9	2.3
Den	9.7	7.6	63.3	0.7	1.8	1.0	1.0	0.7	1.0	4.9	2.6	4.5	1.4
Fin	0.1	0.0	0.0	34.7	16.7	9.9	9.9	4.1	0.1	4.6	5.4	10.8	3.7
Fran	0.0	0.0	0.0	0.0	46.3	9.2	9.2	3.7	0.0	7.7	11.1	1.0	11.8
Ger	0.0	0.1	0.0	0.0	0.3	45.0	44.3	2.8	0.0	0.2	3.8	0.2	3.1
Ire	0.0	0.0	0.0	0.0	0.1	0.9	94.3	1.3	0.0	0.1	1.7	0.1	1.4
Ita	0.2	0.0	0.0	0.3	0.0	1.2	1.2	89.2	0.0	1.7	2.3	0.8	3.0
Nor	0.3	0.9	0.2	0.0	0.5	2.1	2.1	0.1	90.2	0.9	1.5	1.0	0.0
Spa	1.5	1.6	0.0	2.5	0.4	2.7	2.7	0.8	0.0	85.0	0.1	2.4	0.3
Swed	0.1	0.2	0.0	0.1	0.2	0.0	0.0	0.0	0.4	0.4	74.9	5.3	18.4
Swit	0.1	0.6	0.0	0.0	0.3	0.0	0.0	0.2	0.0	0.8	0.0	97.8	0.2
UK	1.1	1.3	0.0	1.0	1.1	1.8	1.8	0.3	0.0	0.1	2.0	1.1	88.3
To	13.1	25.6	0.2	23.5	31.0	37.6	81.0	20.0	1.5	25.0	33.7	66.7	46.9
From	62.2	55.1	36.9	65.3	53.7	54.8	5.6	10.7	9.6	15.0	25.1	2.2	11.6
Net	−49.1	−29.5	−36.7	−41.8	−22.7	−17.2	75.4	11.3	−8.1	10.0	8.6	64.5	35.3
Total	50.9	70.5	63.5	58.2	77.3	82.9	175.3	111.2	91.7	110.0	108.6	164.5	135.2

Source: Calculation

While the rest of coefficients show a negative significant effect of recession on European betas.

Last but not least, in this section, we investigate the spillover effect of beta risk in 13 European countries. To achieve this goal, we employed a Forecast Error Variance Decomposition. In this case, according to Table 4, the spillover of one country to other counties is presented. Total is spillovers of one country that is summation of variance in row. Net is the difference between **TO** and **FROM**. Note that a positive value indicates the first market is the spillover leader; whereas a negative value indicates the second market is the spillover follower.

From Table 4, Ireland (81.0), Switzerland (66.7) and United Kingdom (46.9) are the top three that affect the other countries. Austria (−49.1), Finland (−41.8) and Denmark (−36.7) are top three markets that are affected by the other countries.

5 Conclusions

The interdependence and interaction between real estate markets and European stock market have not been fully explored. In this study, we recognize the structural change in the real estate cycle thus the Markov switching regression is used to detect crisis in the cycle. The results of this method show that European countries became worse during US housing crisis and European sovereign debt crisis. Then, we used ARMA (1,1)-GJR-GARCH to model a dynamic volatility and found that all degrees of volatility persistence of both GARCH models and GJR-models are close to 1, indicating a high volatility in a long sample period. Moreover, we also conduct a DCC model to find the dynamic correlation between individual real estate market and European market index. The time varying beta is then computed using the dynamic correlation and volatility from GARCH and DCC process, respectively. The empirical results show that betas of all European countries are time-varying through the time excepting Denmark and Norway of which the betas are mostly close to zero. Time-varying curves also reach the peak during Europe debt crisis period. After that, the time varying betas of all countries are used to find spillover effect using the Forecast Error Variance Decomposition. Estimation of European time-varying real estate beta spillover index shows correlation and connection for each pair of countries. The results show that the top three countries which affect the other countries in Europe are Ireland, Switzerland and United Kingdom. The top three countries that are affected by the other countries in Europe are Sweden, Austria and Finland.

References

1. Adler, M., Dumas, B.: International portfolio choice and corporation finance: a synthesis. J. Fin. **38**(3), 925–984 (1983)
2. Blaug, M.: Henry George: rebel with a cause. Eur. J. Hist. Econ. Thought **7**(2), 270–288 (2000)
3. Bollerslev, T., Engle, R.F., Wooldridge, J.M.: A capital asset pricing model with time-varying covariances. J. Polit. Econ. **96**(1), 116–131 (1988)

4. Diebold, F.X., Yilmaz, K.: Better to give than to receive: predictive directional measurement of volatility spillovers. Int. J. Forecast. **28**(1), 57–66 (2002)
5. Engle, R.: Dynamic conditional correlation: a simple class of multivariate generalized autoregressive conditional heteroskedasticity models. J. Bus. Econ. Stat. **20**(3), 339–350 (2002)
6. Engle, R.F.: Dynamic conditional beta. J. Fin. Econ. **14**(4), 643–667 (2016)
7. Fama, E.F., French, K.R.: Size and book-to-market factors in earnings and returns. J. Fin. **50**, 131–155 (1995)
8. Ferson, W.E., Korajczyk, R.A.: Do arbitrage pricing models explain the predictability of stock returns? J. Bus. 309–349 (1995)
9. Glosten, L.R., Jagannathan, R., Runkle, D.E.: On the relation between the expected value and the volatility of the nominal excess return on stocks. J. Fin. **48**(5), 1779–1801 (1993)
10. Goldfeld, S.M., Quandt, R.E.: A Markov model for switching regressions. J. Econ. **1**(1), 3–15 (1973)
11. Hartmann, P.: Real estate markets and macroprudential policy in Europe. J. Money Credit Banking **47**(S1), 69–80 (2015)
12. Jagannathan, R., Wang, Z.: The conditional CAPM and the cross-section of expected returns. J. Fin. **51**(1), 3–53 (1996)
13. Kim, K.H., Kim, T.: Capital asset pricing model: a time-varying volatility approach. J. Empir. Fin. **37**, 268–281 (2016)
14. Maneejuk, P., Pastpipatkul, P., Sriboonchitta, S.: Analyzing the effect of time-varying factors for Thai rice export. Thai J. Math. 201–213 (2016)
15. Quan, D.C., Titman, S.: Commercial real estate prices and stock market returns: an international analysis. Fin. Anal. J. **53**(3), 21–34 (1997)
16. Treynor, J.L.: Market value, time, and risk (1961)
17. Zabarankin, M., Pavlikov, K., Uryasev, S.: Capital asset pricing model (CAPM) with drawdown measure. Eur. J. Oper. Res. **234**(2), 508–517 (2014)

Volatility Jump Detection in Thailand Stock Market

Saowaluk Duangin[✉], Woraphon Yamaka, Jirakom Sirisrisakulchai,
and Songsak Sriboonchitta

Faculty of Economics, Chiang Mai University, Chiang Mai, Thailand
saowaluk.econ@gmail.com

Abstract. The purposes of this study are threefold. The first is to employ three jump tests (Amed, Amin and BNS jump test) to detect jump in high-frequency return of the Stock Exchange of Thailand (SET) index over the period of five years from 2011 to 2016. The second is the application of the LLP test to detect jump in SET returns in respond to Thai macroeconomic news announcements using various GARCH-type models. The final purpose is to estimate the out-of-sample volatility forecasting and compare the results between GARCH-type models under various distributions using filtered and raw returns. This paper finds that (1) the jumps are significantly detected by Amed, Amin and BNS jump test in frequencies; (2) the number of jump detection in all samples are found between 1–3% of observations and the results also show that 1-h sample set and CGARCH models with Student's t distribution have highest percentage of detected jump around 3%; (3) the simple GARCH-type models estimated using filtered return show more accurate out of sample forecasts of the conditional variance than GARCH estimated from raw return.

Keywords: Jump test · High-frequency data · Stock Exchange of Thailand
Macroeconomic news announcements

1 Introduction

It is widely known that stock market will respond to new information (both bad and good news). Based on the efficient market hypothesis (EMH), market price always instantly reacts to available information. The important macroeconomic news reports such as Gross Domestic Product (GDP), inflation rate, unemployment rate, export, and import value are the sources of information which may lead to changes in return and volatility of stock market. The effect of macroeconomic news announcement on stock market has been shown in many studies such as Rangel [1], and Hussain [2].

Many of the previous studies have been based on the non-parametric jump test. Barndorff-Nielsen and Shephard [3] suggested the price process and proposed bi-power and multi-power variation process to estimate parameter by separating the jump component from the continuous volatility in time series asymptotic distribution data. Anderson et al. [4] proposed the jump test to approximate multiple jumps and distinguish jump from volatility of intraday returns. They also showed the better volatility forecasting when incorporate jump component from return volatility.

© Springer International Publishing AG, part of Springer Nature 2018
V.-N. Huynh et al. (Eds.): IUKM 2018, LNAI 10758, pp. 445–456, 2018.
https://doi.org/10.1007/978-3-319-75429-1_37

Lee and Mykland [5] developed LM jump test which differs from that of Anderson et al. [4] in terms of critical value. They found that the critical values LM jump test were higher than those from Anderson et al. [4] leading to more frequent rejection of the null hypothesis (no jump occurs). However, both tests provided the time and the number of jumps occur within and during the period.

There are some empirical studies on the jump detection and testing with macroeconomic news announcement. Huang [6] further developed Barndorff-Nielsen and Shephard [3] to study the jump in macroeconomic news announcement period by using a realized volatility. He found that there are more jumps on macroeconomic announcement days more than no news days. However, many studies faced with a spurious detection while applying a test to high-frequency financial data. To avoid this identified jump problem, Bajgrowicz et al. [7] provided multiple testing and considered a statistic testing on a certain threshold level to news announcements. They found that their method could eliminate all the spurious jump detections. Laurent et al. [8] proposed a new semi-parametric statistic, LLP test, with standardized conditional volatility based on Andersen et al. [4] and Lee and Mykland [5] to detect jumps in GARCH-type model with non-Gaussian distribution. They suggested that this test will be useful when the return of asset is not liquid enough to be frequently traded, and useful to avoid the spurious detections.

This paper contributes to the existing literature in the econometric way by generating the data from various GARCH specifications. Moreover, in the empirical part of this paper, we also contribute to the literature by fitting the volatility of Stock Exchange of Thailand (SET) using different high-frequency sample sets (2-h, 1-h, 30-min, and 5-min interval) covering five years from January 2011 to December 2016.

The aims of this study are to detecting jump in different high-frequency sample sets of return of SET using Amed test, Amin test, and BNS tests. Secondly, we test and match Thai macroeconomic news announcements with jump in SET returns using LLP test to confirm the significant link between them. Moreover, we investigate the LLP test with four different GARCH-type models along with four different distributions. Thirdly, we compare the volatility forecasting performance among the GARCH-type models and distributions. Lastly, we employ the Model Confidence Set (MCS) approach to find the best volatility forecasting.

This study is organized as follows: Sect. 2 describes the model used in this study. Section 3 empirical analysis. Finally, Sect. 4 contains conclusions.

2 Model and Test

We start by detecting the jump in the stock market return and standardize our test statistic using the conditional volatility based on GARCH volatility and conditional mean estimate. In this study, we consider four GARCH-type models namely the generalized autoregressive conditional heteroskedastic (GARCH), the exponential GARCH (EGARCH), the Glosen-Jagannathan-Runkie GARCH (GJR-GARCH), and the component GARCH (CGARCH) model. In addition, the different four distribution assumptions

- normal (*N*), Student's t (*sT*), skewed normal (*sN*), and skewed Student's t (*ssT*) distributions - are given for the respective GARCH-type models.

2.1 The Data Generating Process

In this study, we employ the method for additive jumps detection in a Data Generating Process (DGP) satisfying GARCH process, the test proposed by Laurent et al. [8]. They extend the procedure of Franses and Ghijsels [9] for additive outlier detection in GARCH models to make it applicable for DGP. The DGP assumes that the observed return series, r_t^*, consist of a GARCH component and an additive jump component,

$$r_t^* = r_t + a_t I_t, \tag{1}$$

where r_t is the stock market return, I_t is binary variable taking one in case of a jump at period t and zero otherwise, and a_t is the jump size of positive and negative news. a_t and I_t are assumed to be independent. The generalized autoregressive conditional hetero-skedasticity (GARCH) of Bollerslev [10] is proposed to estimate conditional volatility in stock markets, where the conditional mean and volatility equation are

$$r_t = \mu + \sigma_t \varepsilon_t, \tag{2}$$

$$\sigma_t^2 = \alpha_0 + \sum_{i=1}^{p} \alpha_i \varepsilon_{t-i}^2 + \sum_{j=1}^{q} \beta_j \sigma_{t-j}^2, \tag{3}$$

where μ is mean parameter, σ^2 is time varying volatility obtained from the GARCH process in Eq. (2). It is obvious that the structure of GARCH (p, q) consists of two parts. It has a polynomial $\beta(L)$ order p - the autoregressive term, and a polynomial $\alpha(L)$ of order q - the moving average term. The parameter μ, α_i, and β_j are the estimated parameters. Here, some assumptions are made that are $\alpha_i > 1$, $\beta_j > 1$, and $\alpha_i + \beta_j < 1$ to guarantee a stationary process in GARCH model. Note that, we can further extend to account of the leverage effect like in the GJR-GARCH model as originally proposed by Glosten et al. [11], the reaction of asymmetrically good and bad news like in EGARCH model of Nelson [12], the long-run volatility dependencies like the CGARCH model of Engle and Lee [13].

2.2 Andersen, Bollerslev, and Dobrev Jump Test Based on Minimum Realized Variance and Median Realized Variance

Andersen, Bollerslev, and Dobrev jump test based on minimum realized variance (Amin test) and median realized variance (Amed test) was proposed by Andersen et al. [14] for detecting jumps by estimating integrated volatility in the presence of jumps based on the nearest neighbor truncation. The null hypothesis of the test is that there is no jump in the realization of the process at a certain time t. The minimum realized variance

(*MinRV*) and median realized variance (*MedRV*) eliminate jumps by taking respectively the minimum and the median over adjacent returns:

$$MinRV_t = 2.75 \frac{M}{M-1} \sum_{j=2}^{M_t} \min(\left|r_{t,j}\right|, \left|r_{t,j-1}\right|)^2, \tag{4}$$

and

$$MedRV_t = 1.42 \frac{M}{M-1} \sum_{j=2}^{M_t} \text{med}(\left|r_{t,j}\right|, \left|r_{t,j-1}\right|, \left|r_{t,j-2}\right|)^2. \tag{5}$$

In this study, we also consider the same test statistic as BNS procedure, and the jump test is given by

$$Amin \, jump \, test = \frac{1 - MinRV_t/RV_t}{\sqrt{1.81 \, \delta \, \max(1, < MinRQ_t/RV_t^2)}} \rightarrow N(0, 1), \tag{6}$$

and

$$Amed \, jump \, test = \frac{1 - MedRV_t/RV_t}{\sqrt{0.96 \, \delta \, \max(1, < MedRQ_t/RV_t^2)}} \rightarrow N(0, 1), \tag{7}$$

where $MinRQ_t = 2.21 \frac{n^2}{n-1} \sum_{j=2}^{M} \min(\left|r_{t,j}\right|, \left|r_{t,j-1}\right|)^4$ is the minimum realized quarticity and $MedRQ_t = 0.92 \frac{n^2}{n-1} \sum_{j=3}^{M} \min(\left|r_{t,j}\right|, \left|r_{t,j-1}\right|, \left|r_{t,j-2}\right|)^4$ is the median realized quarticity that estimates the integrated quarticity, where n is number of observations.

2.3 Barndorff-Nielsen and Shephard Jump Test

Barndorff-Nielsen and Shephard [3] introduced the non-parametric jump test for detecting jump in the high-frequency data. BNS proposed using the realized variance (RV) and bi-power variation (BV), the multi-power variation and integrated variance to indicate remarkable and rare jumps of price process

$$dP_t = \mu_t dt + \sigma_t dW_t + dJ_t, \tag{8}$$

where μ_t is constant term, σ_t is the spot volatility, and W_t is a standard Brownian motion. $J_t = \sum_{j=1}^{N_t} a_{tj}$ is the jump process, where N_t is the number of jumps up to time t and a_{tj} is the size of the jump. The null hypothesis for this test can be stated as there is no jump in period t. The alternative hypothesis is there exists at least one jump in period t. In this study, we will consider the following feasible difference- and ratio- statistics;

$$\text{BNS jump test} = \frac{1 - BV_t/RV_t}{\sqrt{0.61\delta \, \max(1, TQ_t/BV_t^2)}} \rightarrow N(0, 1).$$ (9)

TQ_t is the realized tri-power quarticity (TPQ) which is defined as;

$$TQ_t = M1.74\left(\frac{M}{M-2}\right) \sum_{j=3}^{M} \left|r_{tj-2}\right|^{4-3} \left|r_{tj-1}\right|^{4/3} \left|r_{tj}\right|^{4/3},$$ (10)

RV_t is the realized variance which is defined as;

$$RV_t = \sum_{j=2}^{M_t} r_{tj}^2,$$ (11)

and BV_{t}, is the realized bi-power variation defined as

$$BV_{t,M} = \sum_{j=2}^{M_t} \left|r_{t\,j-1}\right|\left|r_{t,j}\right|,$$ (12)

where r_{tj} is the return (with $j = 2, \dots, M$) in day t. Indeed, the interval $[0, t]$ is split into M equal subintervals.

2.4 The Jump Detection

The objective of this study is to examine whether detected jumps can explain the fluctuation of Thailand is stock market. We also conduct a Laurent et al. [8] (LLP) test for additive jumps detection in Data Generating Process (DGP). Unlike the above three jump tests, this test enables us to identify the number of jumps and the exact time that jumps occur. The null hypothesis is that there is no jump occurring in period $t, H_0{:}a_t I_t = 0$, against the alternative hypothesis that there is at least one jump occurring in period $t, H_a{:}a_t I_t \neq 0$. The test consists of two steps:

(1) Compute standardized returns

$$|\tilde{J}_t| = \frac{|r_t^* - \tilde{\mu}_t|}{\tilde{\sigma}_t},$$ (13)

where $\tilde{\mu}_t$ and $\tilde{\sigma}_t$ are the estimated parameters from GARCH process.

(2) In this test, H_0 is rejected if $|\tilde{J}_t| > g_{t,\lambda}$, where $g_{t,\lambda}$ is the critical value of test:

$$g_{t,\lambda} = \log(-\log(1 - \lambda))(1/\sqrt{2\log T}) + (2\log T)^{1/2},$$ (14)

where λ is the critical level.

3   Empirical Analysis

3.1   Data

We use jump tests on 17762, 8240, and 5091 observations of 2-h, 1-h, and 30-min of Stock Exchange of Thailand (SET) returns. The data obtained from Bloomberg data base start from the first trading day of 2011 to the last trading day of 2016. We define the return of SET as $R_{i,t} = \ln \left[\dfrac{SET_t - SET_{t-1}}{SET_{t-1}} \right]$.

Moreover, we focus on the Thailand macroeconomic news announcements of Foreign Trade Exports (FTEx), Foreign Trade Imports (FIm), Consumer price index (CPI), Gross Domestic Product (GDP), and Customs Trade Balance (CTB), announced during SET trading hours, 9.55 A.M. to 4.40 P.M. in Bangkok time (GMT+7.00). We follow Balduzzi et al. [15] to identify surprising Thailand macroeconomic news using standardized news surprise, $S_{t,k}$, at period t for each news k, $(k = 1, \ldots, 5)$ by differencing between the expected value and the actual value divided by the standard deviation of each news.

3.2   Jump Tests

We apply the Amed, Amin and BNS jump tests on the four data sets. Table 1 shows p-value of three jump tests for all sample sets are less than the critical value. These imply that the jumps are significantly detected by Amed, Amin and BNS jump tests.

Table 1.   Amed, Amin, and BNS statistic testing

Test/sample	2-h	1-h	30-min	15-min
Amed	11.60999	13.87397	10.78637	10.78637
	(0.0000)	(0.0000)	(0.0000)	(0.0000)
Amin	10.69246	10.45082	7.565578	7.565578
	(0.0000)	(0.0000)	(0.0000)	(0.0000)
BNS	14.84717	16.27361	18.47596	18.47596
	(0.0000)	(0.0000)	(0.0000)	(0.0000)

Note: Numbers in parentheses show p-value.

3.3   Jump Detection with Macroeconomic News Announcements Based on Various GARCH-Type Models

In this section, we investigate the relationship between Thailand macroeconomic news announcement and jump by applying a Laurent, Lecourt and Palm (LLP) test on four different GARCH-type models including GARCH (1, 1), GJR-GARCH (1, 1), EGARCH (1, 1), and CGARCH (1, 1). We also consider four different distributions including normal, Student's t, skewed normal, and skewed Student's t for each GARCH-type model, in total 16 GARCH specifications for each data set.

We consider the Akaike Information Criterion (AIC) for model selection; the lowest value of AIC, the best fit specification. The AICs are presented in Table 2. Due to the space limit, we only present the GARCH-type models with Student's t distribution since all GARCH-type models with Student's t distribution perform better than those with other distributions. The result suggests that the EGARCH model is the best fit among the four GARCH models with Student's t distribution as it shows the lowest AIC in all sample sets.

Table 2. Model selection under Student's t distribution assumption

Sample\model	GARCH	GJR-GARCH	EGARCH	CGARCH
2-h	−8.3259	−8.3447	−8.3471*	−8.3261
1-h	−8.8127	−8.8283	−8.8333*	−8.8132
30-min	−9.4375	−9.4513	−9.4611*	−9.4367
15-min	−10.094	−10.102	−10.114*	−10.097

Note: * is the best model among GARCH-type models for each sample set.

We apply LLP test on the EGARCH model as it is the best fit model. The results on the percentage of jump and the jump detection in sample returns are shown in Table 3. The jump-detection percentages of the 1-h sample set are about 3.7–3.8% (the highest among those of other sample sets), while those of the 30-min and 15-min sample sets are around 1%. It appears that the percentages of jump detection decrease with high frequency data. A plot of the jump detection for 1-h frequency is shown in Fig. 1, as an example.

Table 3. Percentage of detected jumps from LLP test

Sample/model	GARCH	GJR-GARCH	EGARCH	CGARCH
2-h	3.0827% (157)	3.0827% (157)	3.0630% (156)	3.1416% (160)
1-h	3.7985% (313)	3.7985% (314)	3.7136% (306)	3.8471% (317)
30-min	1.5481% (275)	1.4636% (260)	1.4186% (252)	1.5593% (278)
15-min	1.0730% (348)	1.0730% (348)	0.9065% (294)	1.1377% (369)

Note: Numbers in parentheses show number of significant detected jumps by LLP test.

3.4 Matching Jump Detection with Macroeconomic News Announcement

Based on the previous finding, we focus some significant detected jumps to match with macroeconomic news announcement of 1-h frequency return (8240 observations). In this section, we consider the period between April 2013 to October 2016 as this interval contains the highest percentage of jump (Gray shaded area in Fig. 1). We can observe that the LLP test can detect both positive and negative jumps along that period.

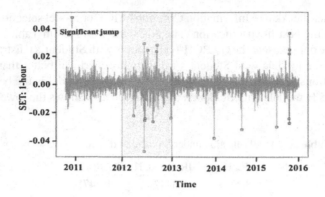

Fig. 1. The filtered returns in percentage of the 1-h sample set over the period to 2016 and detected jumps.

We expect that surprising news announcements will contribute to a high jump in SET return. The results show that most of the significant detected jumps corresponding to the news announcement period will be detected at time t and $t + 1$. In other words, the effect of news announcement can be stored for at least two periods of time. For example, on July 26, 2013, the Thai Customs Department reported that the custom consumer foreign export was 3.38% YoY, while the expected value is 1.60% YoY. This might be bad news to Thai stock market; the standardized news surprise at that time is equal to −93.08%. We match this announcement with detected jump by LLP test, and find that the jump occurs during the announced period at 9.30 P.M. and period after announced at 10.30 P.M.

This result also provides good news surprise jump; for example, on August 1, 2013 at 1.35 P.M., the Ministry of Commerce of Thailand reported that the overall Thailand consumer price index was 0.24% MoM, while the expected value is 0.30%; the standardized news surprise at that period is 0.7%. This might be good news to stock market. We also find that a jump occurs during the announced period at 1.35 P.M. and period after announced at 2.35 P.M.

It is not certain that the above mentioned macroeconomic news announcements caused the jump in stock return and volatility of SET because there were many important economic events around the world in 2013 such as Fed's quantitative easing (QE) policies announcement of US and European debt crisis.

3.5   Volatility Forecasting

In this section, to understand Thailand's stock market behavior we measure volatility forecasting. We examine the forecasting performance of several GARCH-type models estimated from filtered returns against ones estimated from raw returns. According to Laurent et al. [8] the filtered returns can be filtered out from r_t^*, thus

$$\tilde{r}_t = r_t^* - (r_t^* - \tilde{\mu}_t)I_t. \tag{15}$$

In the forecasting procedure, the models are estimated using 2-h, 1-h, 30-min, and 15-min frequency return sample sets to do out of sample volatility forecast. The total number of h-step-ahead volatility forecasts is therefore about 2000 for every model. To compare the performance of competing models, the loss function which is used to measure the prediction error gives no consensus. This study relies on the Mean Absolute Deviation (MAD) approach to measure the forecasting error. However, it is not easy to choose the best model which is always the best under all loss functions or all data samples. Thus, we employ the model confidence set (MCS) approach of Hansen et al. [16] to identify the subset of models that are equivalent in terms of forecasting ability, but outperform all other competing models. This approach consists of a sequence of statistic test which permit us to construct a set of "superior model" where the null hypothesis of equal predictive ability (EPA) is not rejected at specified confidence level.

It is possible to test GARCH-type models in various aspects depending on MAD approach. For each step of the EPA hypothesis, if the null hypothesis is rejected, it will eliminate the worse model specifications. The testing procedure will stop when the null hypothesis cannot be rejected, otherwise, the EPA will continue test after the elimination of worst model. In this MCS test, the range statistics is constructed to test the hypothesis above,

$$T_R = \max_{i,j \subseteq M} \frac{\left| \bar{d}_{ij} \right|}{\sqrt{\text{var}(\bar{d}_{ij})}}, \tag{16}$$

where $\bar{d}_{ij} = \left| \sigma_{t,i}^2 - \hat{\sigma}_{t,j}^2 \right| - \left| \sigma_{t,j}^2 - \hat{\sigma}_{t,j}^2 \right|$, $i,j = 1, \ldots .m$, and the sum is taken over the models in M, σ_t^2 and $\hat{\sigma}_t^2$ are the true and the estimated conditional volatility, respectively, in the out-of-sample samples.

In this test, we set the confidence level of the MCS test to 10% based on bootstrap simulation at 10000 times to compare the performance among the different GARCH-type models and distributions. The results of MCS test are shown in Table 4. Note that, when p-value is greater, more likely to be rejected.

Table 4. MCS test

Sample	Specification p-value	Model based on filtered return		Model based on raw return	
		p-value	MAD	p-value	MAD
2-h	GJRGARCH-sT	1.0000	1.3800e−04	0.4264	1.6707e−04
1-h	GJRGARCH-sN	1.0000	1.0007e−04	<0.1	1.4153e−04
30-min	GJRGARCH-sT	1.0000	6.9848e−03	<0.1	0.0122000
15-min	CGARCH-sT	1.0000	5.2353e−05	<0.1	0.0001071

Note: p-value less than 0.1 indicates that the null hypothesis cannot be rejected.

In to Table 4, we report the best forecasting models based on MCS-test. The p-values of all GARCH-type models using raw return are less than 0.10 imply that we can reject the null hypothesis of EPA. Moreover, the volatility forecasting models using raw return are removed in the MCS inspection process. As expected, the GARCH-type models estimated from raw return are frequently eliminated. Our results are similar to the finding from previous study, e.g., Dumitru and Urga [17] found that the simple GARCH-type models estimated using raw return deliver less accurate out-of-sample forecasts of the conditional variance than GARCH-type models estimated using filtered data. We expect that the logarithmic specification of the conditional volatility is too sensitive to the previous volatility changes.

However, it is quite difficult to compare results and choose the best fit model specification by using p-value because the p-values of all GARCH specification models are equal to 1.000. So, we consider the loss function (or MAD) and find that GJR-GARCH model with Student's t distribution is the best model for the 2-h frequency and 30-min frequency data sample sets, whilst GJR-GARCH model with skewed normal distribution is the best fit model for the 1-h frequency data sample set, and CGARCH model with Student's t distribution is the best fit for the 15 min.

In this section, the best fit models for forecasting volatility in different high frequency data sets are used to plot the out-of-sample forecasts. Figure 2 shows most of the

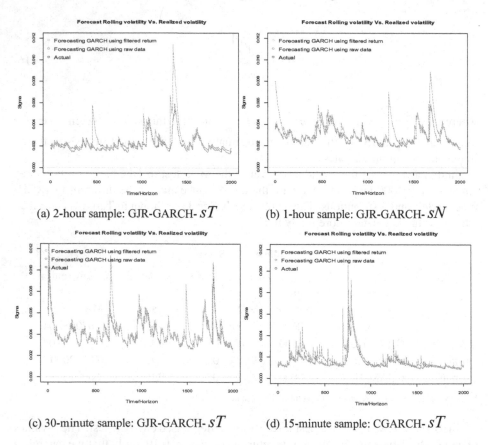

(a) 2-hour sample: GJR-GARCH- sT (b) 1-hour sample: GJR-GARCH- sN

(c) 30-minute sample: GJR-GARCH- sT (d) 15-minute sample: CGARCH- sT

Fig. 2. Forecasted volatility and realized volatility of the out of samples. (Color figure online)

forecasts seem efficient since the forecast value (red dashed line) is close to the actual value (blue dash line). We expect that when the jumps are filtered out from the raw return, the forecasting volatility model will perform better. Moreover, we also compare the forecasting volatility between the best fit GARCH model using filtered data (red dashed line) and raw return (green dashed line). The results show that GARCH model estimated from filtered returns performs better in all frequencies data. We expect that the GARCH model estimated from raw return is only a deterministic function of the past return. However, in the real world, many factors affect the volatility and jump. Therefore, the GARCH models estimated from raw return are not appropriate when using high-frequency data. As a consequence, our study finds a convincing evidence that the results of all GARCH specifications estimated from filtered return present a significantly better than GARCH specifications estimated from raw return. These show a good finding in high-frequency data framework.

4 Conclusions

In this paper, we test jump in high-frequency return of Stock Exchange of Thailand index over five years from January 2011 to December 2016 in different sample frequencies (2-h, 1-h, 30-min, and 5-min interval). We start by investigating the existence of jump in our high-frequency returns using three jump tests, namely Amed, Amin, and BNS jump test. We observe a significant jump in three jump tests for all frequency returns. We then test and match the macroeconomic news announcement with the significant jump SET returns. We employ the LLP test to detect jump in the filtered returns which are generated from GARCH, GJRGARCH, EGARCH, and CGARCH along with normal, Student's t, skewed normal, and skewed student's t distribution. The results show that the jumps in all frequencies are detected between 1-3% of observations. The result also shows that 1-h interval sample set and CGARCH models with Student's distribution have the highest percentage of jump detection around 3%.

To compare the volatility forecasting performance among the GARCH-type models and distributions, we use the MCS test based on bootstrap simulation and real data analysis. The empirical results demonstrate that the simple GARCH-type models estimated using filtered returns deliver more accurate out of sample forecasts of the conditional variance than GARCH estimated from raw return.

Acknowledgement. We are grateful for financial support from Centre of Excellence in Econometrics, Faculty of Economics, Chiang Mai University and Chiang Mai University.

References

1. Rangel, J.G.: Macroeconomic news, announcements, and stock market jump intensity dynamics. J. Bank. Financ. **35**(5), 1263–1276 (2011)
2. Hussain, S.M.: Simultaneous monetary policy announcements and international stock markets response: an intraday analysis. J. Bank. Financ. **35**(3), 752–764 (2011)

3. Barndorff-Nielsen, O.E., Shephard, N.: Econometrics of testing for jumps in financial economics using bipower variation. J. Financ. Econom. **4**(1), 1–30 (2006)
4. Andersen, T.G., Bollerslev, T., Diebold, F.X., Vega, C.: Real-time price discovery in global stock, bond and foreign exchange markets. J. Int. Econ. **73**(2), 251–277 (2007)
5. Lee, S.S., Mykland, P.A.: Jumps in financial markets: a new nonparametric test and jump dynamics. Rev. Financ. Stud. **21**, 2535–2563 (2008)
6. Huang, X.: Macroeconomic news announcements, systemic risk, financial market volatility and jumps (2015)
7. Bajgrowicz, P., Scaillet, O., Treccani, A.: Jumps in high-frequency data: spurious detections, dynamics, and news. Manag. Sci. **62**(8), 2198–2217 (2015)
8. Laurent, S., Lecourt, C., Palm, F.C.: Testing for jumps in conditionally Gaussian ARMA–GARCH models, a robust approach. Comput. Stat. Data Anal. **100**, 383–400 (2016)
9. Franses, P., Ghijsels, H.: Additive outliers, GARCH and forecasting volatility. Int. J. Forecast. **15**, 1–9 (1999)
10. Bollerslev, T.: Generalized autoregressive conditional heteroskedasticity. J. Econom. **31**, 307–327 (1986)
11. Glosten, L.R., Jagannathan, R., Runkle, D.E.: On the relation between the expected value and the volatility of the nominal excess return on stocks. J. Financ. **48**(5), 1779–1801 (1993)
12. Nelson, D.B.: Conditional heteroskedasticity in asset returns: a new approach. Econom.: J. Econom. Soc. **59**, 347–370 (1991)
13. Engle, R.F., Lee, G.: A long-run and short-run component model of stock return volatility. In: Cointegration, Causality, and Forecasting: A Festschrift in Honour of Clive W.J. Granger, pp. 475–497 (1999)
14. Andersen, T.G., Dobrev, D., Schaumburg, E.: Jump-robust volatility estimation using nearest neighbor truncation. J. Econom. **169**(1), 75–93 (2012)
15. Balduzzi, P., Elton, E.J., Green, T.C.: Economic news and bond prices: evidence from the US Treasury market. J. Financ. Quant. Anal. **36**(4), 523–543 (2001)
16. Hansen, P., Lunde, A., Nason, J.: Model confidence sets. Econometrica **79**, 453–497 (2011)
17. Dumitru, A.M., Urga, G.: Identifying jumps in financial assets: a comparison between nonparametric jump tests. J. Bus. Econ. Stat. **30**(2), 242–255 (2012)

Thai Export Efficiency in AFTA: Copula-Based Gravity Stochastic Frontier Model with Autocorrelated Inefficiency

Petchaluck Boonyakunakorn[1(✉)], Pathairat Pastpipatkul[1], and Songsak Sriboonchitta[2]

[1] Faculty of Economics, Chiang Mai University, Chiang Mai, Thailand
petchaluckecon@gmail.com
[2] Center of Excellence in Econometrics, Chiang Mai University, Chiang Mai, Thailand

Abstract. This paper studies the effects of the ASEAN Trade Area (AFTA) on Thai export efficiency using copula-based gravity stochastic frontier gravity model with autocorrelated inefficiency. One of the main assumptions of standard stochastic frontier model is that one-sided technical inefficiency error u and a symmetric error v are independent, instead researches have proposed to relax this assumption using copula to joint distribution of (u, v). Nevertheless, most of previous studies treat technical efficiency as time invariant which is a strong assumption. Therefore, we will include the lag effect of inefficiency to allow inefficiency in one period to be influenced by its previous level. In this paper, we consider the multivariate normal copula and t copula based in term of technical efficiency, and then select the best model based on AIC and BIC criteria. This is the first application of the copula-based model of error terms and temporal dependence of inefficiency with gravity trade model. It is found that Thai export efficiency are in range from 0.58 and 0.67 implying that the Thailand is not taking full benefits of AFTA.

Keywords: AFTA · Stochastic frontier gravity model · Export efficiency

1 Introduction

The ASEAN Free Trade Area (AFTA) aims to transform ASEAN into a region with free movement of goods, services, investment, capital and skilled labor. The AFTA agreement was signed in 1998 by ASEAN's first six members: Brunei, Indonesia, Malaysia, Philippines, Singapore and Thailand. Vietnam, Laos, Myanmar and Cambodia joined later on. AFTA now consists of the ten countries. The main tool for achieving AFTA is the Common Effective Preferential Tariff (CEPT) scheme. The ASEAN memberships will give each other uniform preferential treatment in intra-ASEAN trade. The tariff scheme reduced tariff rates to between 0% and 5% by 1 January 2002 for Brunei, Indonesia, Malaysia, the Philippines, Singapore and Thailand (ASEAN 6), by 2006 for Vietnam, by 2008 for Laos and Myanmar and by 2010 for Cambodia [5]. Eventually all tariffs will be eliminated.

The elimination of tariff of a wide range of products leads to price reduction throughout ASEAN. Trade in intra-ASEAN should increase as a result. The effect of AFTA has on the intra-regional goods trade which can be simply measured by relative

© Springer International Publishing AG, part of Springer Nature 2018
V.-N. Huynh et al. (Eds.): IUKM 2018, LNAI 10758, pp. 457–466, 2018.
https://doi.org/10.1007/978-3-319-75429-1_38

share (RS) and trade intensity index (TII). RS of ASEAN is the share of ASEAN countries' trade in their overall trade. It indicates the degree of dependency on among ASEAN trade. Meanwhile, TII of ASEAN measures the intensification of ASEAN members' relationship indicating that the exchange of goods between ASEAN countries is more intense than what the intra-regional trade share numbers indicate.

Figure 1 shows overall trends of intra-ASEAN trade. RS had an increasing trend, even though RS had a slowing trend since 2010. Meanwhile, the TII after 1998 began to slightly increase according to the event of forming the AFTA in 1988, and then intra-ASEAN trade had an increasing trend since 2002 due to the result of tariff reduction of ASEAN 6 controlling tariff rates to be between 0% and 5% by 1 January 2002. It still maintained a high trade level until 2008. After that intra-ASEAN trade significantly declined because of the global crisis. Overall both RS and TII indicated relatively strong trade relationships among members. These measurements indicate that AFTA brings benefits to increase their export to intra-ASEAN countries.

Fig. 1. Intra-ASEAN trade index between 1991 and 2015

According to RS and TII indicators, Thai export efficiency should be improved. Previous researches have investigated the impact of trade liberalization under AFTA with respect to intra-AFTA trade. Okabe and Urata [12] found that AFTA has been successful in trade creation effects due to the tariff elimination of products. Koh [8] also examined whether AFTA is trade creating or trade diverting by using Brunei's trade data, and found that AFTA brings trade creation effects. Therefore, this paper investigates the effects of AFTA on Thai export efficiency.

To study the Thai export efficiency in AFTA, this paper applies the stochastic frontier gravity model as it represents the upper bound of the trade data. It is unlike Ordinary Least Square (OLS) that estimate the mean of the trade data indicating to potential form of the mean. Meanwhile, the stochastic frontier model (SFM) refers to a maximum possible trade that can happen between any two countries when most trade restrictions are relaxed [4]. The error of a SFM with gravity model consists of two parts; a non-negative error u representing the trade efficiency and a symmetric error v representing to country-specific effects. The standard assumption of its error term (u, v) is assumed to be independent. This is a strong assumption in SFM with gravity model, therefore this paper will relax model to allow the dependent between two errors by using copulas.

Many researches have modeled the SFM with dependent of error term. Burns [3] and Smith [14] modeled the dependent between u and v using copulas. The advantage of copula approach is to provide more flexibility in term of a wide range of joint distributions to deal with various marginal distributions and the copula functions. Consequently, SFM with the copula approach can deliver a variety of explanations for observed output. Smith [14] was first to use the copula approach to model noise-inefficiency correlation in a cross-section and a panel data SFM, and use numerical optimization techniques to evaluate the intractable integrals involved in the maximum likelihood estimation of the model parameters.

In this SFM with gravity model, we employed with panel data since it provides more efficiency and provides an opportunity to study the behavior of trade inefficiency over time. Nevertheless, the standard of SFM considers the efficiency as a constant which refers to stochastic independence over time. This is also a strong assumption therefore we will relax this assumption by including the lag effect of inefficiency in SFM with gravity model. This is more reasonable since the inefficiency in one period can be correlated to past levels of inefficiency. This should become more realistic than using the constant technical efficiency. The paper is organized into the following sections. Section 2 describes the research methodologies; gravity model, stochastic frontier with gravity model using copula where their noise-inefficiency can be dependence, the correlated error-inefficiency modeling using normal and multivariate t copula that allow the lag effect of noise on inefficiency. Section 3 estimates the model consisting of model specification and data. Section 4 contains empirical results. The conclusions is in the Sect. 5.

2   Research Methodology

2.1   The Gravity Model

The gravity model from Newton's law of universal gravitation has been widely applied in international trade study. Its main concept of the gravity model is that every point mass attracts to other point mass with the gravitational power. It is directly proportional to the product of their masses, while it is inversely proportional to the square of the distance between them. Then, the gravity model has been applied to study other fields. Tinbergen [15] was the first person who applied extensively in studying the bilateral international trade flows. The bilateral trade is attracted by the economic size of the two countries (normally measured by Gross Domestic Product (GDP)) and inversely to the distance between two countries $\left(D_{ij}\right)$. It can be expressed as:

$$TRADE_{ij} = G\frac{GDP_iGDP_j}{D_{ij}}, \tag{1}$$

To provide a meaning of elasticity in microeconomics, logarithms is applied to the model. Then, the linear gravity model equation can be expressed as:

$$\ln TRADE_{ij} = \beta_0 + \beta_1 \ln GDP_i + \beta_2 \ln GDP_j - \beta_3 \ln D_{ij} + \varepsilon_{ij}, \tag{2}$$

where $TRADE_{ij}$ is the value of the bilateral trade between country i and j, GDP_i and GDP_j belong to country i and j with respect to national incomes. D_{ij} is a measure of the distance between the two countries, $\ln G$ is a constant of proportionality.

2.2 The Stochastic Frontier Model (SFM)

This study, we employed a SFM with gravity model to investigate export efficiency. The main idea comes from the frontier of production function which used for measuring production efficiency. The production function is a relationship between inputs and outputs. If firm has a technical inefficiency, the output will fall from its potential frontier. The production function frontier is to look for the maximum output based on the given inputs. Aigner et al. [2] and Meeusen and Broeck [10] showed that the error term of production model can be separated as two parts; the first is the non-negative error term that showing the production-efficiency, meanwhile the second is the other error term that reflects the measurement error. The stochastic frontier model (SFM) is given by

$$\log Y = x'\beta - U + V, \tag{3}$$

where the output $Y = y \in \mathbb{R}_+$, x is a vector $k \times 1$ of inputs, the error component $V = v \in \mathbb{R}$ has $G(v) = \Pr(V \le v)$ that is assumed continuous, independent of x, but dependent possibly on unknown parameters that are collected into a vector δ_v. Likewise, β is a vector of unknown parameters. In this model, the error component $U = u \in \mathbb{R}_+$ where $\mathbb{R}_+ = \mathbb{R}_+ \cup \{0\}$ is a random variable with cdf $F(u) = \Pr(U \le u)$ that is assumed continuous and independent with x, but dependent possibly on unknown parameters that are collected into a vector δ_u.

2.3 The SFM with Gravity Model

The stochastic frontier, which is applied with the international trade area, measured the potential bilateral trade. Aigner et al. [2] was the first person who established the stochastic frontier estimation. In order to measure the trade inefficiency, the error of gravity model ε_{ij} from (3) can be decomposed into v_{ij} which is a symmetric random variable, and u_{ij} which is one-sided random variable with $u \ge 0$ representing to trade inefficiency. The equation can be expressed as:

$$\ln TRADE_{ij} = \beta_0 + \beta_1 \ln GDP_i + \beta_2 \ln GDP_j - \beta_3 \ln D_{ij} - u_{ij} + v_{ij}, \tag{4}$$

The two error terms u_{ij}, v_{ij} are normally assumed to be independent. u_{ij} represents the amount by which the trade between two countries fails to reach the optimum (the frontier). Therefore, a country with low efficiencies represents that its optimum trade is higher than actual trade.

Over the last decade, many researches have been applied stochastic frontier with gravity model including Drysdale et al. [4], Kalirajan and Singh [7], Koh [8], and Ravishankar and Stack [13]. All of their works follow the standard assumption of SFM with gravity model implying that the two error terms are independent. This strong assumption may lead to inconsistent result. Wiboonpongse et al. [16] found that the SFM with independence assumption provides the overestimated the technical efficiency of production. They also suggested that the dependence between the two error terms should be considered. Therefore, this paper we will relax assumption of SFM with independence to allow the dependent between two errors by using copulas.

2.4 Copula Functions

With copula functions a variety of multivariate distributions can be revealed from the marginal probability distribution of a set of random variables along with different dependence among the random variables.

Let $C_\alpha(u_1, \ldots, u_m)$, m-variate copula function, be a multivariate distribution function with a support of $[0, 1]^m$ and the marginals are uniform $U(0, 1)$, u_i is uniform distribution between $0, 1$, $i = 1, 2, \ldots, m$, and $\alpha \in \Omega$ the parameter vector, which provides the dependence structure among $u_i's$.

Sklar [11] stated that a unique copula function can capture the dependent structure among the random variables and can also be used as a function of this copula function and marginal distribution functions of these random variables.

Let $F(x_1, \ldots, x_m; \gamma)$ be the joint distribution function of the random variables x. As stated by Sklar's theorem, there should be a unique copula function, $C_\alpha(u_1, \ldots, u_m)$ so that

$$F(x_1, \ldots, x_m; \gamma) = C_\alpha(F_1(x_1; \theta_1), \ldots, F_m(x_m; \theta_m)), \tag{5}$$

where $F_i(x_i; \theta_i)$ is the distribution function of x_i, $\theta = (\theta_1, \ldots, \theta_m)'$ and $\gamma = (\theta, \alpha)$.

2.5 A Panel Data Copula-Based Stochastic Frontier Model

At this stage, marginal probability densities are merged through an independently given copula function depicting dependence structure, we obtain the joint probability distribution of noise and inefficiency.

A typical panel data stochastic gravity frontier model is expressed by

$$y_{ij,t} = f_\beta(x_{ij,t}) \exp(v_{ij,t} - u_{ij,t}), \quad \varepsilon_{ij,t} = v_{ij,t} - u_{ij,t}, \tag{6}$$

where $i = 1, \dots, n$, $j = 1, \dots, n$, $t = 1, .., T$, $u_{ij,t} \geq 0$, $y_{ij,t}$ is the value of trade between the country i and country j at time t, $f_\beta(.)$ is the deterministic trade frontier, and $x_{ij,t}$ is the determinants. The deterministic frontier subjects to the error efficiency.

Let $F_{\pi_{ij,t}}(v_{ij,t})$, $G_{\eta_{ij,t}}(u_{ij,t})$ and $f_{\pi_{ij,t}}(v_{ij,t})$, $g_{\eta_{ij,t}}(u_{it})$ be the distribution function and probability density function of the noise and the inefficiency related with the trade between country i and j at time t, $u_{ij} = (u_{ij,1}, \dots, u_{ij,T})'$, $v_{ij} = (v_{ij,1}, \dots, v_{ij,T})'$ and $\eta_{ij} = (\eta_{ij,1}, \dots, \eta_{ij,T})'$, $\pi_{ij} = (\pi_{ij,1}, \dots, \pi_{ij,T})'$. Supposing that a $2T$ – variate copula presents the dependency of u_i and v_i, therefore the joint probability density function of (u_{ij}, v_{ij}) can be expressed as

$$f_{\gamma_{ij}}(u_{ij}, v_{ij}) = \left(\prod_{t=1}^{T} g_{\eta_{ij,t}}(u_{ij,t}) f_{\pi_{ij,t}}(v_{ij,t}) \right) c_{\alpha_{ij}}(G(u_{ij,1}), \dots, G(u_{ij,T}), F(v_{ij,1}), \dots, F(v_{ij,T})), \qquad (7)$$

where $\gamma_{ij} = (\pi_{ij}, \eta_{ij}, \alpha_{ij})$, and α_{ij} are the vector of copula parameters.

2.6   Correlated Error-Inefficiency Modeling Using Normal and T Copula

To model the error-inefficiency dependence and temporal dependence among the inefficiency, it is necessary to assume that there is no temporal dependence among errors for each country along with three assumptions. The first is that stochastic dependences from the noise and the inefficiency associated with any two trading-countries are the same across all time points. The second is that time dependence of inefficiency for the ij th trading countries at any two consecutive periods is of order one and same for all countries. The third is that it does not have time dependence of orders one or more among the error.

The distribution of u_{ij} and v_{ij} is also under the assumptions indicating that time dependence of a firm's inefficiency can be captured by normal copula or t copula, $v_{ij,t} \sim N(0, \sigma_v^2)$, and $u_{ij,t} \sim N^+(0, \sigma_u^2)$. Consequently, the joint density function u_{ij} and v_{ij} under normal copula can be expressed as

$$f_\gamma(u_{ij}, v_{ij}) = \left(\prod_{t=1}^{T} \frac{1}{\sigma_v} \phi\left(\frac{v_{ij,t}}{\sigma_v} \right) \frac{1}{\sigma_u} \phi\left(\frac{v_{ij,t}}{\sigma_u} \right) \right) \frac{1}{|R|^{\frac{1}{2}}} \exp\left[-\frac{1}{2} \xi'(R^{-1} - I)\xi \right], \qquad (8)$$

where $\xi = \left(\Phi^{-1}\left(2\Phi\left(\frac{u_{ij,1}}{\sigma_u} \right) - 1 \right), \dots, \Phi^{-1}\left(2\Phi\left(\frac{u_{ij,T}}{\sigma_u} \right) - 1 \right) \right), R = \begin{pmatrix} R_{11} & R_{12} \\ R_{21} & R_{22} \end{pmatrix}$,

$\Phi^{-1}\left(\Phi\left(\frac{v_{ij,1}}{\sigma_v} \right) \right), \dots, \Phi^{-1}\left(\Phi\left(\frac{v_{ij,T}}{\sigma_v} \right) \right)', R_{11} = \begin{pmatrix} 1 & \psi & 0 & \dots & 0 \\ \psi & 1 & \psi & \dots & 0 \\ \dots & \dots & \dots & \dots & \dots \\ 0 & 0 & 0 & \dots & 1 \end{pmatrix}$,

$$R_{12} = R_{21} = \begin{pmatrix} \rho & \cdots & \cdots & \cdots \\ \cdots & \rho & \cdots & \cdots \\ \cdots & \cdots & \cdots & \cdots \\ \cdots & \cdots & \cdots & \rho \end{pmatrix}, \quad R_{22} = \begin{pmatrix} 1 & 0 & \cdots & 0 \\ 0 & 1 & \cdots & 0 \\ \cdots & \cdots & \cdots & \cdots \\ 0 & 0 & 0 & 1 \end{pmatrix}$$

and ρ is a copula parameter associated with simultaneous dependence between noise and inefficiency independent, whereas ψ is a copula parameter associated with lagged dependence among inefficiency. Replacing $v_{ij,t} = \varepsilon_{ij,t} + u_{ij,t}$ into above and integrating with respect to u_{ij}, the density function of ε_i is expressed as follows;

$$h(\varepsilon_i) = \int_0^\infty \cdots \int_0^\infty \left(\prod_t f(\varepsilon_{ij,t} + u_{ij,t}) \right) C_{\alpha_{ij}} \left(\prod_t g(u_{ij,t}) \right) du_{ij,t}, \tag{9}$$

where $C_{\alpha_{ij}} = c_{\alpha_i}(G(u_{ij,1}), \ldots, G(u_{ij,T}), F(\varepsilon_{ij,1} + u_{ij,1}), \ldots, F(\varepsilon_{ij,T} + u_{ij,T}))$.

The multivariate t copula can capture the tail dependence to modeling the dependence structure of the Thai export value.

Let ρ be a symmetric, positive definite with diag $\rho = 1$ and $T_{\rho,v}$ the standardized Student's distribution with v degrees of freedom and correlation matrix ρ, the joint density function u_{ij} and v_{ij} under T copula can be written as

$$f_\gamma(u_{ij}, v_{ij}) = |\rho|^{-\frac{1}{2}} \frac{\frac{1}{2}\Gamma\left(\frac{v_{ij,t} + T}{2}\right) \left[\Gamma\left(\frac{v_{ij,t}}{2}\right)\right]^T \left(1 + \frac{1}{v_{ij,t}} \varsigma^T \rho^{-1} \varsigma\right)^{\frac{-v_{ij,t} + T}{2}}}{\left[\Gamma\left(\frac{v_{ij,t} + 1}{2}\right)\right]^T \Gamma\left(\frac{v_{ij,t}}{2}\right) \prod_{t=1}^T \left(1 + \frac{\varsigma_T^2}{v_{ij,t}}\right)^{-\frac{v_{ij,t} + 1}{2}}} \tag{10}$$

where $\varsigma_n = t_v^{-1}(u_n)$, t_v^{-1} is the inverse of the univariate Student's distribution.

3 Estimating the Model

3.1 Model Specification

The stochastic frontier gravity model in Eq. 4, which imposes the variables proposed in this study, can be rewritten as

$$\ln Export_{ij,t} = \alpha_{ij} + \beta_1 \ln GDP_{i,t} + \beta_2 \ln GDP_{j,t} + \beta_3 \ln POP_{i,t} + \beta_4 \ln POP_{j,t} + \beta_5 \ln D_{ij,t} + \beta_6 AFTA_{j,t} - u_{ij} + v_{ij}, \tag{11}$$

for $i = 1, \ldots, N$ where i refers to Thailand, $j = 1, \ldots, N$ where j refers to partner countries, $ij = 1, \ldots, N$ where ij refers to between two countries, $t = 1, \ldots, T$, where $EXPORT_{ij,t}$ corresponds to the value of Thai exports goods and services to ASEAN countries partners at year t. The explanatory variables are defined as follows: $GDP_{i,t}$ and $GDP_{j,t}$ denote the nominal income of Thailand and partner countries respectively. It represents the economic size of country, $POP_{i,t}$ and $POP_{j,t}$ correspond to population in Thailand and partner countries at time t respectively, $D_{ij,t}$ is measured for the distance between Thailand and partner countries. $AFTA_{j,t}$ denotes trade agreement of Thailand with country j in year t (dummy variable).

3.2 Data

This study applied panel data consisting of ASEAN countries. The export values are from Thailand to ASEAN countries from 1991 to 2015. The data panel is strongly balanced and has no missing data. The data set yields together 250 complete observations. Thai exports data are from Foreign Trade Statistics of Thailand, whereas GDP and population data are collected from the World Bank database. The distance data are obtained from the data base for the CEPII Geodist dyadic dataset [9]. ASEAN trade agreement data are taken from asean.org.

4 Empirical Results

This study we propose multivariate Normal copula and t copula based model in term of technical efficiency. To choose the most appropriate model for interpretation, we select model based on Akaike's information criterion (AIC) and Bayesian information criterion (BIC). And the result shows that multivariate t Copula provide minimum both AIC and BIC. Consequently, we will apply multivariate t Copula based model in term of technical efficiency. The estimated results are shown in Table 1. The result shows that Thai GDP, Partner GDPs positively and significantly affects Thai exports at the 1% level of significance. The effects of Thai and partner GDP are 0.623 and 0.487 respectively. This implies that a 1% increase in Thai and partner GDP are associated with a rise in Thai export by 0.699% and 0.441% respectively. It was found that the bilateral trade rises when the GDPs of both countries increase, which is consistent with the study of Anderson [1].

4.1 Thai Export Efficiency

Thai export efficiency scores are generated from this frontier specification of the SFM with gravity model. If Thailand achieve an efficient level of exports to each country in ASEAN members, Thai exports will operate on their maximum export potential with that country. Otherwise, there is a deviation of actual observed data and their maximum trade potential indicating to inefficient levels of exports. The partner's countries in ASEAN members that Thai exports to consists of 9 countries which are Brunei Darussalam (BN), Cambodia (KH), Indonesia (ID), Laos (LA), Malaysia (MY), Myanmar (MM), the Philippines (PH), Singapore (SG), and Vietnam (VN). The Technical Efficiency (TE) of Thai export flows to ASEAN countries showed that the mean TEs for all sample range from 0.58 to 0.67. The high efficiency score of Thai exports indicates that Thai export to a partner country is close to its maximum export potential. Among The ASEAN trading partner countries, Singapore is the highest export efficiency with the highest mean TE of 0.67. This is followed by Indonesia and Malaysia with 0.66 and 0.64 respectively. Cambodia and Philippines are the two lowest technical efficiency with 0.60 and 0.58 respectively.

Table 1. Gravity parameter estimates using multivariate Gaussian and t copula based SFMs panel data

Variables	Normal Copula		Multivariate t Copula	
	Coefficients	S.E.	Coefficients	S.E.
Constant	0.735***	0.096	0.659***	0.061
ln GDP_i	0.699***	0.154	0.623***	0.147
ln GDP_j	0.441***	0.130	0.487***	0.128
ln POP_i	7.929**	3.867	7.353**	3.483
ln POP_j	0.092	0.176	0.096	0.145
ln D_{ij}	0.005	0.007	0.004	0.007
AFTA	−0.195***	0.047	−0.120***	0.035
σ_u	0.462***	0.050	0.438***	0.014
σ_v	0.702***	0.091	0.694***	0.054
AIC	2337.923		1071.38	
BIC	2373.107		1115.318	

Note: ***, ** and * represent significance levels at 1, 5 and 10% respectively.

Overall, the TE measure of Thai export to ASEAN members are quite low with range from 0.58 to 0.67. There is a large deviations of actual observed Thai export to AEAN members from the potential export flows. It implies that Thai export to intra-ASEAN countries still inefficiency indicating that Thailand is still not taking full advantage of AFTA. Meanwhile, AFTA should improve Thai export performance to ASEAN members. Therefore, Thailand should explore the potential of ASEAN members and further expand export to intra-ASEAN countries in order to take the maximum advantage of AFTA. Especially Thailand should focus on improving export to countries with a very low TE such as Cambodia and Philippines.

5   Conclusion

This paper applies stochastic frontier gravity model to investigate the effect of AFTA to Thai exports. However, the standard assumption of two error term of stochastic frontier model is assumed to be independent. This is a strong assumption. Therefore, we extend stochastic frontier gravity model to allow the correlation between error components using copula. Furthermore, we also allow the autocorrelation in the inefficiencies error. This paper we apply multivariate normal copula and t copula based model in term of technical efficiency, and select the best model based on AIC and BIC. The result shows that Multivariate t Copula based model is more appropriate than normal copula. The result indicates that Thai exports to ASEAN countries are inefficiency. AFTA causes negative effects to Thai exports. The mean TE of Thai export flows to ASEAN countries range from 0.58–0.67. This implies that the Thailand is not taking full advantage of the benefits of AFTA.Therefore, Thailand should explore the potential of ASEAN members and further expand export to intra-ASEAN countries in order to take the maximum advantage of AFTA.

References

1. Anderson, J.E.: Trade, Size, and Frictions: The Gravity Model. Mimeo Boston College (2014). https://www2.bc.edu/~anderson/GravityNotes.pdf
2. Aigner, D., Lovell, C.K., Schmidt, P.: Formulation and estimation of stochastic frontier production function models. J. Econom. 6(1), 21–37 (1977)
3. Burns, R.C.J.: The simulated maximum likelihood estimation of stochastic frontier models with correlated error components. Unpublished dissertation, Department of Econometrics and Business Statistics, The University of Sydney, Australia (2004)
4. Drysdale, P.D., Huang, Y., Kalirajan, K.: China's trade efficiency: measurement and determinants. In: Drysdale, P., Zhang, Y., Song, L. (eds.) APEC and Liberalisation of the Chinese economy. Asia Pacific Press, Canberra (2000)
5. Fujita, M., Kuroiwa, I., Kumagai, S. (eds.): The Economics of East Asian Integration: A Comprehensive Introduction to Regional Issues. Edward Elgar Publishing, Cheltenham (2011)
6. Jondrow, J., Lovell, C.K., Materov, I.S., Schmidt, P.: On the estimation of technical inefficiency in the stochastic frontier production function model. J. Econom. 19(2–3), 233–238 (1982)
7. Kalirajan, K., Singh, K.: A comparative analysis of China's and India's recent export performances. Asian Econ. Pap. 7(1), 1–28 (2008)
8. Koh, W.: Brunei Darussalam's trade potential and ASEAN economic integration: a gravity model approach. Southeast Asian J. Econ. 1, 67–89 (2013)
9. Mayer, T., Zignago, S.: Notes on CEPII's distances measures: the GeoDist database (2011)
10. Meeusen, W., Van den Broeck, J.: Efficiency estimation from Cobb-Douglas production functions with composed error. Int. Econ. Rev. 18, 435–444 (1977)
11. Sklar, M.: Fonctions de repartition an dimensions et leurs marges. Publ. Inst. Stat. Univ. Paris 8, 229–231 (1959)
12. Okabe, M., Urata, S.: The impact of AFTA on intra-AFTA trade. J. Asian Econ. 35, 12–31 (2014)
13. Ravishankar, G., Stack, M.M.: The gravity model and trade efficiency: a stochastic frontier analysis of eastern European countries' potential trade. World Econ. 37(5), 690–704 (2014)
14. Smith, M.D.: Stochastic frontier models with dependent error components. Econom. J. 11(1), 172–192 (2008)
15. Tinbergen, J.: Shaping the World Economy; Suggestions for an International Economic Policy: Suggestings for an International Economic Policy. Twentieth Century Fund, New York (1962)
16. Wiboonpongse, A., Liu, J., Sriboonchitta, S., Denoeux, T.: Modeling dependence between error components of the stochastic frontier model using copula: application to intercrop coffee production in Northern Thailand. Int. J. Approx. Reason. 65, 34–44 (2015)

Efficiency Analysis of Natural Rubber Production in ASEAN: The Comparison of Panel DEA and Bootstrapping Panel DEA Analysis Based Decision on Copula Approach

Kewalin Somboon[✉], Chukiat Chaiboonsri, and Songsak Sriboonchitta

Faculty of Economics, Puay Ungphakorn Centre of Excellence in Econometrics,
Chiang Mai University, Chiang Mai, Thailand
kewalin.sb@gmail.com

Abstract. The purpose of this paper is to investigate the technical efficiency obtained from both panel data envelopment analysis (Panel DEA) and bootstrapping panel DEA methods under the assumption of error in the production process by using Copula models to compare the results between initial Panel DEA and bootstrapping Panel DEA methods. This paper provides the best tools to analyze technical efficiency and also represent the best natural rubber production efficient country by using the data of eight ASEAN countries (i.e., Brunei Darussalam, Cambodia, Indonesia, Malaysia, Myanmar, Philippines, Thailand, and Vietnam) over the period of 1961–2014 collected by the Food and Agriculture Organization of the United Nations. This study proposes policy suggestions according to the empirical results.

Keywords: Panel DEA · Bootstrapping panel DEA · Technical efficiency
Copula · Natural rubber · ASEAN

1 Introduction

In recent years, the ASEAN community has become the largest exporter of natural rubber in the world while Thailand, Malaysia, and Indonesia account for 67% of global natural rubber production and 79% of its net export (ITRC 2014). The requirements for uniformity and quality of natural rubber produced have increased significantly by rubber industries around the world. The economic benefit in a unit of rubber production has been reflected by the technical efficiency in rubber production. There have been many studies since on technical efficiency via panel data envelopment analysis (Panel DEA) based on Farrell's (1957) ideas (see, e.g., Charnes et al. (1978) and also the bootstrapping Panel DEA approach (see, e.g., Simar and Wilson 1998). This paper provides the answer for the argument on which is the best technical efficiency estimator between the Panel data envelopment analysis (Panel DEA) approach and the bootstrapping Panel DEA approach. Hence, the aims of this paper are threefold using the essential data of eight ASEAN countries (i.e., Brunei Darussalam, Cambodia, Indonesia, Malaysia, Myanmar, Philippines,

V.-N. Huynh et al. (Eds.): IUKM 2018, LNAI 10758, pp. 467–476, 2018.
https://doi.org/10.1007/978-3-319-75429-1_39

Thailand, and Vietnam) over the period of 1961–2014 (FAO 2014). The first objective is to use the original Panel DEA method to estimate the technical efficiency of natural rubber production in ASEAN; second, use the bootstrapping Panel DEA method to estimate the bias-correction. Finally, the pros and cons between both initial Panel DEA and bootstrapping Panel DEA method will be analyzed to base a decision on the model of Copula.

The paper is structured into sections. Section 2 represents reviews of literature. Section 3 briefly reviews the methodology, including Panel data envelopment analysis (Panel DEA), bootstrapping Panel DEA, and Copula models. Section 4 is the data description. Section 5 empirical results discuss of the pros and cons between initial Panel DEA and bootstrapping Panel DEA base decision on Copula methodology.

Finally, the conclusions and suggestions are presented at Sect. 6.

2   Literature Review

The estimation of production efficiency scores in nonparametrics was introduced by Farrell (1957). Simar and Wilson (1998) also presented that the DEA is a nonparametric estimator based on a finite sample of observations measuring the dependent variable of decision-making units (DMU). They also provided a methodology of bootstrapping in nonparametric frontier models showing the sensitivity analysis of DEA efficiency scores, and claimed that the efficiency scores are similar to an estimated production frontier. Moreover, a simple way to analyze the sensitivity of efficiency scores through bootstraping that simulates data-generating process (DGP) was introduced by Efron (1979). Simar and Wilson (2000) suggested the bootstrap distributions based on DEA/FDH estimators were plausible to estimate sampling distributions; the correction for the bias of efficiency estimators was also allowed. The asymptotic distribution of DEA and the use of smooth and subsampling bootstrap was mentioned by Kneip et al. (2008). Jaffry et al. (2013) found bias-correction in the DEA efficiency scores by using the bootstrap approach from Simar and Wilson (1998). Additionally, Chaitip et al. (2014) provided the idea of the statistical properties in technical efficiency that influenced factor efficiency in the sugarcane industry by using data envelopment analysis in Panel data setting (Panel DEA) and the bootstrapping Panel DEA approach. As Song et al. (2013) expressed, "using the bootstrap method to get a sampling distribution can simulate the distribution of the original sample estimator and correct biased estimates of the efficiency value". In addition, a brief review of copula models, Sklar's theorem (1959), named after Abe Sklar provides the theoretical foundation for the application of copulas. Sklar states that every multivariate cumulative distribution function allows the flexibly decomposed joint distribution function into its individual and dependent marginal that is a copula function. Embrechts et al. (1999) mentioned the dependence could not capture by linear correlation if the distributions depart from the elliptical copula, meaning that the independence among variables cannot be explained by the zero correlations.

3   Research Methodology

3.1   Data Envelopment Analysis (DEA), Panel DEA and the Malmquist Index

Charnes et al. (1978) introduced data envelopment analysis (DEA) as a nonparametric estimator to obtain technical efficiency. The DEA approach was used in this paper to estimate an output based technical efficiency using linear programing and a Malmquist DEA approach to estimate indices of total factor productivity (TFP). The Malmquist TFP index measures productivity change that can be modified into a technical efficiency. The bias and the confidence interval of output based technical efficiency were estimated by using the package DEAP version 2.1 (Coelli 1996).

The formulation of output divided by input is the main idea of the data envelopment analysis approach.

$$TE_i = 0\,Q_i/0P_i \tag{1}$$

From Eq. (1), the technical efficiency of the firm (i = 1, 2...n) is denoted by TE_i and the technical efficiency ratio is denoted by $0Q_i/0P_i$. Additionally, the value of TE_i ratio locates between zero and one. If $TE_i = 0$, it means that there is no technical efficiency in firm i. On the contrary, if $TE_i = 1$, it means that firm i has a maximum technical efficiency.

The panel DEA is represented as follows:

$$TE_{it} = 0Q_{it}/0P_{it} \tag{2}$$

From Eq. (2), the technical efficiency of the firm (i = 1, 2...n) and times (t = 1, 2...n) is denoted by TE_{it}, and the technical efficiency ratio is denoted by $0Q_{it}/0P_{it}$. Further, the value of the TE_{it} ratio locates between zero and one. If $TE_{it} = 0$, it means that there is no technical efficiency in firm i during time t, while $TE_{it} = 1$ meaning that firm i at time t has a maximum technical efficiency (Coelli 1996).

Coelli (1996) also showed the output-based Malmquist productivity change index indicated by Fare et al. (1994) as follows:

$$m_o\left(y_{t+1}, x_{t+1}, y_t, x_t\right) = \left[\frac{d_o^t\left(x_{t+1}, y_{t+1}\right)}{d_o^t\left(x_t, y_t\right)} \times \frac{d_o^{t+1}\left(x_{t+1}, y_{t+1}\right)}{d_o^{t+1}\left(x_t, y_t\right)}\right]^{\frac{1}{2}} \tag{3}$$

The relation between $\left(x_{t+1}, y_{t+1}\right)$ and $\left(x_t, y_t\right)$ in Eq. (3) represents the productivity. Under time t to t + 1, a positive Malmquist TFP will be indicated if its value is greater than one. Since one index uses time t and the other uses t + 1 technology, the index geometric is defined as a two output based Malmquist TFP indices.

3.2   Bootstrapping DEA and Panel DEA Approach

Since frontier deviations efficiency is investigated as inefficiency, ignoring some estimator statistical properties, e.g., statistical noise in the estimation, can cause erroneous results and biased DEA methods. In relation to the argument mentioned by Simar and Wilson (1998), the bootstrap approach is the most practicable approach to construct the statistical property in DEA estimators. In this paper, the analysis of natural rubber producers was also applied by the bootstrap method. The steps below indicate the corrected efficiency values.

$$Bias(\hat{\theta}_k) = E(\hat{\theta}_k) - \hat{\theta}_k \tag{4}$$

$$Bias(\hat{\theta}_k) = B^{-1} \sum_{b=1}^{B} \hat{\theta}_{kb}^* - \hat{\theta}_k \tag{5}$$

The bias-corrected efficiency value is:

$$\tilde{\theta}_k = \hat{\theta}_k - Bias(\hat{\theta}_k) = 2\hat{\theta}_k - B^{-1} \sum_{b=1}^{B} \left(\widehat{\theta^*}_{kb}\right) \tag{6}$$

The α confidence is calculated as follows:

$$P_r\left(-\hat{b}_\alpha \le \widehat{\theta^*}_{kb} - \hat{\theta}_k \le -\hat{a}_\alpha\right) = 1 - \alpha$$

$$P_r\left(-\hat{b}_\alpha \le \hat{\theta}_k - \theta_k \le -\hat{a}_\alpha\right) = 1 - \alpha$$

$$\hat{\theta}_k + \hat{a}_\alpha \le \theta_k \le \hat{\theta}_k + \hat{b}_\alpha \tag{7}$$

In addition, Hoang Linh (2012) mentioned a panel bootstrapping approach to indicate statistical exactness of panel data envelopment (Panel DEA).

$$\bar{y}_{bit}^* = \sum_{it}^{n} \frac{y_{bit}^*}{n} \tag{8}$$

where TE_{it} generated the bootstrapping approach by $\bar{y}_{bit}^*$ until achieving the best properties of the bootstrapping method (Efron 1979).

3.3   Copula Model

In statistics and probability theory, a copula is a multivariate probability distribution function where marginal probability distribution of each variable is uniform. According to Sklar (1959), a unique copula exists when a continuous joint distribution $H(x_1, \ldots, x_n)$ and the marginal are given.

$$H(x_1, \ldots, x_n) = C(F_1(x_1), \ldots, F_n(x_n)) \tag{9}$$

On the contrary, Eq. (5) explains an n-dimensional distribution function for every distribution type $F_1, \dots, F_n$ and every copula C. Moreover, a dependence structure and a marginal distribution represent the probability density function of multivariate probability distribution as follows:

$$
\begin{aligned}
f(x_1, \dots, x_n) &= \frac{\partial F(x_1, \dots, x_n)}{\partial x_1, \dots, \partial x_n} \\
&= \frac{\partial C(u_1, \dots, u_n)}{\partial u_1, \dots, \partial u_n} \times \Pi \frac{\partial F(x_i)}{\partial x_i} \\
&= c(u_1, \dots, u_n) \times \Pi f_i(x_i)
\end{aligned}
\tag{10}
$$

The copula density function is defined by c. Meaning that, the marginal distribution no longer needs to be similar to each other. This paper focuses on both Elliptical copulas and Archimedean copulas, which are families of copula models.

3.3.1 Gaussian Copulas (Normal Copula)

The normal copula is an elliptical copula defined by:

$$
C_{Ga}(u, v; \rho) = \int_{-\infty}^{\Phi^{-1}(u)} \int_{-\infty}^{\Phi^{-1}(v)} \frac{1}{2\pi\sqrt{1-\rho^2}} exp\left(-\frac{x_1^2 - 2\rho x_1 x_2 + x_2^2}{2(1-\rho^2)}\right) dx_1 dx_2
\tag{11}
$$

where u, v are cumulative distribution functions, uniform distribution is between zero and one, Pearson's linear correlation is represented by (ρ), and the inverse cumulative distribution function of a standard normal distribution is (Φ^{-1}).

The notions of upper-tail and lower-tail dependence are utilized to describe the copula behavior for large and small α.

3.3.2 Clayton Copula

The Clayton copula is an asymmetric Archimedean copula defined as:

$$
C_{Cl}(u, v; \theta) = \left(u^\theta + v^\theta - 1\right)^{-\frac{1}{\theta}}
\tag{12}
$$

The Clayton copula can capture lower tail dependence ($\lambda_{low} = 2^{\frac{-1}{\theta}}$).

The two random variables are independent indicated by the parameter $\theta \to 0$, and are perfectly correlated if $\theta = +\infty$. Mainly, Kendall's tau equals $\frac{\theta}{(\theta + 2)}$. However, the types of copula model have been more than mentioned above (a rigorous treatment of copulas can be found in Nelsen 2006).

4   Data Description

This paper collects the data of natural rubber production in eight ASEAN countries (i.e., Brunei Darussalam, Cambodia, Indonesia, Malaysia, Myanmar, Philippines, Thailand, and Vietnam) over the period of 1961–2014 from the Food and Agriculture Organization of the United Nations (FAO 2014) and uses the package DEAP version 2.1 (Coelli 1996) to estimate the Panel DEA technical efficiency (TE). The measure for output is the natural rubber production (tonnes). The inputs include three categories: area

Table 1. Summary statistics for natural rubber production in ASEAN +8

	Variable	Mean	SD	Min.	Max.
	Input and output vectors				
Brunei	Production (tonnes)	579.26	1914.21	70.00	39980.00
	Area harvested (ha)	5333.57	3950.68	1522.00	45000.00
	Rainfall (mm)	3197.07	4.13	1758.66	3197.07
	Temperature (°c)	27.54	1.79	27.20	27.54
Cambodia	Production (tonnes)	3342.11	11058.52	70.00	48781.00
	Area harvested (ha)	7620.39	11309.14	1522.00	50000.00
	Rainfall (mm)	3090.52	3.80	1758.66	3197.07
	Temperature (°c)	27.51	0.09	27.20	27.54
Indonesia	Production (tonnes)	7035.61	16603.59	70.00	52982.00
	Area harvested (ha)	10779.46	16666.33	1522.00	55000.00
	Rainfall (mm)	2983.97	5.16	1758.66	3197.07
	Temperature (°c)	27.49	0.12	27.20	27.54
Malaysia	Production (tonnes)	8592.54	17336.59	140.00	52982.00
	Area harvested (ha)	11968.35	17481.42	1522.00	55000.00
	Rainfall (mm)	2877.42	6.04	1758.66	3197.07
	Temperature (°c)	27.46	0.14	27.20	27.54
Myanmar	Production (tonnes)	9774.44	17312.40	140.00	52982.00
	Area harvested (ha)	13061.94	17576.70	1522.00	55000.00
	Rainfall (mm)	2770.87	6.63	1758.66	3197.07
	Temperature (°c)	27.44	0.16	27.20	27.54
Philippines	Production (tonnes)	10670.00	17146.05	150.00	52982.00
	Area harvested (ha)	14049.76	17380.73	2500.00	55000.00
	Rainfall (mm)	2664.33	7.01	1758.66	3197.07
	Temperature (°c)	27.41	0.17	27.20	27.54
Thailand	Production (tonnes)	11352.41	16915.25	150.00	52982.00
	Area harvested (ha)	14747.41	17122.13	2500.00	55000.00
	Rainfall (mm)	2557.78	7.22	1758.66	3197.07
	Temperature (°c)	27.39	0.17	27.20	27.54
Viet Nam	Production (tonnes)	13144.39	16924.53	150.00	52982.00
	Area harvested (ha)	16643.70	17161.12	2500.00	55000.00
	Rainfall (mm)	2451.23	7.26	1758.66	3197.07
	Temperature (°c)	27.36	0.17	27.20	27.54

Source: The data calculated from FAO (FAO 2014)

harvested (ha), rainfall (mm), and temperature (°c). Summary statistics for these variables are listed in Table 1. After that, we calculated the corrected efficiency value by using the bootstrapping Panel DEA method. Summary average values for both Panel DEA and Panel DEA bootstrapping methods are listed in Table 2.

Table 2. Average TE of Panel DEA and Bootstrapping Panel DEA method

DMU	Original Panel DEA	Bias	Bias-corrected
Malaysia	0.949	−0.026	0.975
Indonesia	0.917	−0.038	0.955
Cambodia	0.920	−0.007	0.927
Brunei	0.934	0.037	0.897
Philippines	0.932	0.055	0.877
Myanmar	0.933	0.084	0.849
Thailand	0.944	0.104	0.840
Vietnam	0.936	0.144	0.792

Source: The data calculated from FAO 2014

5 Empirical Results

5.1 Technical Efficiency

As can be seen from the results, almost all of the Bias-corrected data which represents corrected efficiency values are lower than the original Panel DEA efficiency values, meaning that the results of ASEAN natural rubber production efficiency as estimated by

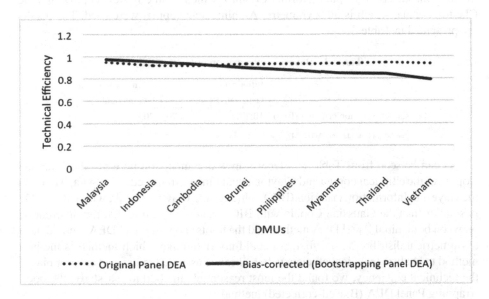

Fig. 1. Average TE of Panel DEA and Bootstrapping Panel DEA method (Source: The data calculated from FAO (2014))

original Panel DEA is overestimated. In order to reveal the relations of initial Panel DEA values and bootstrapping Panel DEA values, the two variations will be illustrated by Fig. 1.

According to the results, the observed production frontier points might be inefficiency scores; also, Hoang Linh (2012) mentions the unknown of true production frontier. These are reasons why the deterministic Panel DEA models have downward biases in efficiency scores. After using the bootstrapping Panel DEA method as mentioned in Simar and Wilson (2000), the estimation of the bias-corrected TE scores is lower than the Panel DEA TE scores, significantly. The ASEAN community is the largest exporter of natural rubber in the world; thus, the results from Panel DEA TE show that the greatest efficiency belongs to Malaysia (0.949), Thailand (0.944), Vietnam (0.936), Brunei Darussalam (0.934), Myanmar (0.933), Philippines (0.932), Cambodia (0.920), and Indonesia (0.917), respectively.

On the contrary, the results from both methods are significantly different since the bootstrapping Panel DEA method represents that the most efficiency still belongs to Malaysia (0.975), but the second order belongs to Indonesia (0.955), Cambodia (0.927), Brunei Darussalam (0.897), Philippines (0.877), Myanmar (0.849), Thailand (0.840), and Vietnam (0.792) respectively. As can be seen from the results, the values from both initial Panel DEA method and bootstrapping Panel DEA method are totally different.

Therefore, this paper investigates the results by using models in Copula to analyze this circumstance as follows.

6 Copulas Method

To compare the results from the initial Panel DEA method and bootstrapping Panel DEA method, the Gaussian Copula (Normal Copula), which is an elliptical copula, and the Clayton copula, which is an asymmetric Archimedean copula, were used. The results are presented in Table 3.

Table 3. Bayesian information criterion (BIC)

	Elliptical copula	Archimedean copula
	Gaussian copula	Clayton copula
Bayesian Information criterion (BIC)	10.4775	3.2037

Source: The data calculated from FAO (2014)

As can be seen from Table 3, after we drew a comparison between the Gaussian Copula in the elliptical copula and Clayton copula in an Archimedean copula, we found the Bayesian Information criterion (BIC) values of the Clayton copula with BIC 3.2037 is smaller than the Gaussian Copula with BIC 10.4775, meaning that the distribution between both initial Panel DEA method and the bootstrapping Panel DEA method have asymmetrical distribution. At first, we could not determine which method is the best method to analyze technical efficiency, but since we assumed to have an error or bias in the technical efficiency, we found the most reasonable method to our study: the bootstrapping Panel DEA (Biased-corrected) method.

For this reason, the answer for our study is that using the bootstrapping Panel DEA method is the best solution to demonstrate technical efficiency. Then we know the best natural rubber production efficiency belongs to Malaysia (0.975), Indonesia (0.955), Cambodia (0.927), Brunei (0.897), Philippines (0.877), Myanmar (0.849), Thailand (0.840), and Vietnam (0.792) respectively.

7 Conclusion

This paper firstly utilized an initial Panel DEA model by using the package DEAP version 2.1 (Coelli 1996) to calculate the technical efficiency values of eight ASEAN countries (i.e., Brunei Darussalam, Cambodia, Indonesia, Malaysia, Myanmar, Philippines, Thailand, and Vietnam) over the period of 1961–2014. This paper used the bootstrap Panel DEA method to correct the technical efficiency values by simulating the same data as the initial Panel DEA method. Then, we found the bias-corrected data using the technical efficiency values. This means that our assumption is true. Moreover, numerous research papers about the study of technical efficiency by using Panel DEA method and Bootstrapping Panel DEA method were analyzed. Meanwhile, with respect to the results in Table 2, we found a huge difference between both the initial Panel DEA method and bootstrapping Panel DEA method. Additionally, the main idea of this paper is to investigate the technical efficiency from the original Panel DEA method and bootstrapping Panel DEA method and to compare the results between both methods under the assumption of the error in the production process by using Copula models, using the Gaussian Copula, an elliptical copula, and the Clayton copula, which is an asymmetric Archimedean copula (see results in Table 3). Then, we concluded that the bootstrapping Panel DEA approach or bias-corrected approach is better than the original Panel DEA method. Following that, we determined the best tools to analyze the technical efficiency belonging to the bootstrapping Panel DEA. Finally, this paper presents the best natural rubber production efficiency which belongs to Malaysia (0.975), Indonesia (0.955), Cambodia (0.927), Brunei (0.897), Philippines (0.877), Myanmar (0.849), Thailand (0.840), and Vietnam (0.792), respectively. Hence, these are reasons why we need to measure the technical efficiency by using the Bootstrapping Panel DEA method rather than the original Panel DEA method.

In future studies, for applying technical efficiency by using the initial Panel DEA or Bootstrapping Panel DEA method under the assumption of biases in the production process, we highly recommend researchers must focus on the correction efficiency by using Bias-corrected data from the bootstrapping Panel DEA approach rather than the initial DEA approach as can be seen from our empirical results.

References

Chaitip, P., Chaiboonsri, C., Inluang, F.: The production of Thailand's sugarcane: using panel data envelopment analysis (Panel DEA) based decision on bootstrapping method. Proc. Econ. Fin. **14**(2014), 120–127 (2014)

Charnes, A., Cooper, W.W., Rhodes, E.: Measuring the efficiency of decision-making units. Eur. J. Oper. Res. **2**, 429–444 (1978)

Coelli, T., Battese, G.E.: Identification of factors which influence the technical inefficiency of Indian farmers. Aust. J. Agric. Econ. **40**, 103–128 (1996)

Efron, B.: Bootstrap methods: another look at jackknife. Ann. Stat. **7**, 1–26 (1979)

Embrechts, P., McNeil, A., Saumann, D: Correlation: pitfalls and alternatives. ETH Zentrum (1999)

FAO: Data of Natural Rubber (2014). http://www.fao.org/faostat/en/#data/QC

Farrell, M.J.: The measurement of productive efficiency. J. R. Stat. Soc. Ser. A **120**(3), 253–290 (1957)

International Tripartite Rubber Council (ITRC) Ministerial Committee Meeting 20 November 2014.   http://globalrubbermarkets.com/21090/irco-media-release-international-tripartite-rubbercouncil-itrc-ministerial-committee-meeting-2014.html

Jaffry, S., Ghulam, Y., Cox, J.: Trends in efficiency in response to regulatory reforms: the case of Indian and Pakistani commercial banks. Eur. J. Oper. Res. **226**, 122–131 (2013)

Kneip, A., Simar, L., Wilson, P.W.: Asymptotic and consistent bootstraps for DEA estimators in nonparametric frontier models. Econ. Theory **24**(6), 1663–1697 (2008)

Nelsen, R.B.: An Introduction to Copulas, 2nd edn. Springer, New York (2006). https://doi.org/10.1007/0-387-28678-0

Sklar, A.: Fonctions de r´epartition `a n dimensions et leurs marges. Publications de l'Institut de Statistique de l'Universit´e de Paris, vol. 8, pp. 229–231 (1959)

Simar, L., Wilson, P.W.: Sensitivity of efficiency scores: how to bootstrap in nonparametric frontier models. Manag. Sci. **44**, 49–61 (1998)

Simar, L., Wilson, P.W.: Statistical inference in nonparametric frontier models: the state of the art. J. Prod. Anal. **13**(1), 49–78 (2000)

Hoang Linh, V.: Efficiency of rice farming households in Vietnam. Int. J. Dev. Issues **11**(1), 60–73 (2012)

Author Index

Printed in the United States
By Bookmasters